A Complete Guide to Building and Plant Maintenance

Second Edition

A Complete Guide to

BUILDING and PLANT MAINTENANCE

Second Edition

by / **THOMAS F. SACK**

Director of Real Estate and Engineering Services

Prentice-Hall, Inc.

Englewood Cliffs, N. J.

PRENTICE-HALL INTERNATIONAL, INC., *London*
PRENTICE-HALL OF AUSTRALIA, PTY. LTD., *Sydney*
PRENTICE-HALL OF CANADA, LTD., *Toronto*
PRENTICE-HALL OF INDIA PRIVATE LTD., *New Delhi*
PRENTICE-HALL OF JAPAN, INC., *Tokyo*

LIBRARY OF CONGRESS
CATALOG CARD NUMBER: 71–126828

30   29   28   27   26   25   24   23

PRINTED IN THE UNITED STATES OF AMERICA
ISBN-0-13-160101-6
B&P

PREFACE TO THE
SECOND EDITION

The new edition of *A Complete Guide to Building and Plant Maintenance* upgrades certain topics and expands the book into additional areas which will prove useful to the maintenance manager, building superintendent, and plant manager. Persons involved in maintenance management are continually trying to improve methods and techniques to reduce plant costs. The new edition provides greater depth into the basics of plant operations and is an excellent training tool for managers, supervisory personnel, and maintenance personnel in acquiring the actual ins and outs of maintenance along with basic subjects and interests.

The second edition has been expanded to encompass several new subjects intended to make the *Guide* an even more valuable and useful reference for those in the building and plant maintenance field. This edition includes comprehensive information on the additional areas of asphalt maintenance; care of terrazzo; estimating landscaping jobs; door closer care; curbing transformer failures; plumbing repairs for water faucets, valves, pipes, tanks; fire extinguishing equipment; approved customs on display of the flag. A revised section on pest control has been incorporated into this edition and contains analysis and definitions of pests as well as suggested control procedures. It is the opinion of the author that a complete up-to-date unit on pest control for plants and buildings should be included to keep maintenance staff and personnel abreast of control and elimination according to the best acceptable methods and standards.

Even though tables are available to the maintenance manager and employee on a random basis, the tables assembled in this edition are those most likely to be referred to by maintenance staff to provide immediate, ready information for estimating, preparing surveys, and in day-to-day building and plant operations. An entirely new segment, Part IV, has been added—entitled *Useful Tables and Rules.* Here is a composite, handy reference containing useful and often vital information. Part IV presents tables and rules on:

- Decimals, Equivalents, Fractions, Per Cents.
- Measures, Weights, Areas, and Volumes—includes linear, square, cubic, and dry measure; measure of angles or arcs.
- Metric Systems, Weights and Measures—contains tables on metric measures of length, surface, capacity, volume, weight, and also contains metric conversion factors on length, area, volume, weight.
- Properties of Materials—includes constant properties of metals, strength and properties of iron, steel, aluminum; weights and specific gravities of metals and masonry.
- Refrigeration and Air Conditioning—with useful information on pipe sizes in ammonia and freon systems.
- Resistance and Inductance.

5

The *Guide* is organized into four major parts:
- Part I–Physical Maintenance.
- Part II–Mechanical Maintenance.
- Part III–Miscellaneous.
- Part IV–Useful Tables and Rules.

Part I provides you with phases of physical maintenance essential in building and plant care. Topics range from:

- Cleaning procedures and practices—how to determine cleaning costs.
- Care of building interior—floors, elevators, escalators, carpets, rugs.
- Lighting maintenance—encompasses relamping and cleaning, spot and group replacement, forms and records for lighting maintenance.
- Care of outside building—asphalt maintenance, both preventive and remedial; fence maintenance.

Other phases of physical maintenance include pest control. This edition has expanded definitions of a variety of pests and recommended measures for their control, types and usage of pesticides, as well as control of rodent and bird pests. Also found in Part I is a glossary of paint terms, paint qualities, surfaces, and formulation.

Mechanical maintenance, as developed in *Part II*, tells how to perform boiler maintenance which may prevent accidents and costly downtime. The "ins" and "outs" of door closer maintenance are discussed, along with methods for curbing transformer failures. The reader will also find useful information on air compressors, air conditioning, automatic controls, electric motors, and plumbing repairs.

Part III, entitled Miscellaneous, takes into account such important segments of building and plant maintenance as fire extinguishing equipment—how to order, install, and keep ready for use. Fire protection is delved into with a discussion of types of sprinklers, smoke detectors, and fire prevention procedures. Also included are safety, security surveys and systems, and training program considerations.

Part IV, Useful Tables and Rules, is designed as a composite reference covering such areas as decimals, fractions, measures and weights, metric system weights and measures, properties of materials, and refrigeration.

THOMAS F. SACK

CONTENTS

7

8 / Contents

SECTION   PAGE

12 / Contents

14 / **Contents**

SECTION   PAGE

16 / **Contents**

18 / **Contents**

Part II / **Mechanical Maintenance**

24 / **Contents**

Part III / **Miscellaneous**

51 Emergencies and Disasters (*Continued*)

Part IV / **Useful Tables and Rules**

A Complete Guide to Building and Plant Maintenance

Second Edition

Part I / **PHYSICAL MAINTENANCE**

1 / CLEANING PROCEDURES AND PRACTICES

(*Source: Buildings* magazine, November 1958, March 1960)

Cleaning costs now represent approximately 33 per cent of the total tab for operating the average building. This fact is leading management men to invest in more administrative study, planning and development time to upgrade their cleaning and maintenance programs and make more efficient use of their cleaning manpower.

With payroll costs representing 90 to 95 per cent of an average building's cleaning costs, man-hour productivity becomes the vital key to stepped-up efficiency and cleaning cost reduction. The *time* it takes to perform the cleaning job is the element which makes or breaks a cleaning program.

How to Determine Cleaning Costs

Comparison of payroll cleaning costs with other buildings can be misleading, as labor rates vary from city to city, and cleaning costs per square foot depend upon quality standards, method of calculating square foot areas, the type of occupancy, the proportion of problem areas, open areas, air conditioning, and so on.

To determine your total annual cleaning cost in dollars, include the following factors:

1. Wages, including overtime.
2. Fringe costs, including payroll taxes, uniforms, vacation pay, pensions, sick leaves, insurance and holiday pay.

The sum of the above items will give your total annual payroll cleaning costs. Next, determine the number of square feet in your building that require cleaning services of any kind, and divide the total annual cleaning cost by the number of square feet.

Now, determine your hours per square foot per year by calculating your average hourly wage. Consider all payroll fringe costs, as well as base rates. Divide your annual payroll cleaning cost per square foot by this average hourly wage. The result—your hours-per-square-foot-per year. This final unit of measurement reflects your true payroll costs. It is the unit by which the results of time and method studies of cleaning operations can be most accurately measured.

This unit of measurement could be called your performance rating figure. Figure 1-1 gives the various classifications of performance ratings. Time study experts suggest that a production rate of .30 man-hours per square foot of rentable area is the acceptable standard for a clean building. If your rating falls below this figure, review the cleaning program—study the job being done, decide how it should be done, and set up the cleaning operation on a regular, controlled basis.

Area vs. Group

There are two common methods of organizing the cleaning personnel:

1. The assignment of specific working areas to individual cleaners.
2. Group, squad or platoon cleaning.

Under the first method, one employee performs all routine cleaning operations within a designated area and is held responsible for the manner in which the area is maintained. Under a group cleaning organization, a gang is assigned certain areas, in which they work together. Each member of the gang is assigned certain operations and the entire group is under the direction of a working supervisor. In some instances, both methods are used simultaneously in the same building. Regular office spaces may be assigned to individuals on an area basis; while the washing of corridors, floors, windows, etc. is done on a group basis.

Figure 1-1

PERFORMANCE RATING FOR OFFICE BUILDINGS

If your figure is:	Your program is:
20-25 per cent	Excellent
25-30 per cent	Very good
30-35 per cent	Good
35-40 per cent	Average
40-45 per cent	Poor
Over 45 per cent	Very poor

The advantages of assigning cleaning work to individuals on an area basis have been enumerated as follows:

1. Reduces monotony, said to be the greatest foe of accomplishment. The variety of duties incident to complete responsibility for a designated area tends to make the work more interesting, removes tiring repetition and increases the amount of work per employee.

2. Work can be arranged to meet conditions of traffic.

3. Diversification of responsibility builds employee interest and provides a "prestige incentive".

4. The diversified work plan improves morale by alternating light and heavy work.

5. Responsibility for poor work, broken articles, possibly petty thievery can best be handled and pinpointed when work is on an assigned area basis.

6. There is less chance to forget important operations. One man, familiar with all operations on "his" floor is less likely to miss something than are a number of men with jobs that parallel and cross each other.

7. Labor turnover is decreased. Training new men is costly, and the diversity of work with its lessened monotony increases tenure. It is the satisfied and encouraged man, not the mechanized man, who does the most work, and the best work and stays longer.

Two methods have been developed for planning individual work assignments; the man-hour method and the Gilbert Formula. Both methods involve the preparation of charts which are used in setting up equitable work loads and are described.

The Man-hour Method

The man-hour method has been widely adapted in the setting up of work schedules and job specifications. A study is made of areas to be cleaned, and work schedules are developed which include each item or operation, the specifications for cleaning and the man-hours required. This guide serves as a logical starting place for a study of cleaning operations and will be helpful in controlling costs. With this method the relative cost of each job is determined. When reductions are necessary, the costs are weighed against the results attained by performance. It also indicates highest cost jobs where studies should be made for labor reduction, and has proved valuable in estimating the staffing and allocation of man-hours for a prospective building operation.

The man-hour production rate for all categories of operation can be determined by filling in actual man-hours for cleaning and maintenance operations (Figure 1-2) and using the formula illustrated in Figure 1-3.

Gilbert Formula

The distinguishing feature of the Gilbert Formula (Figure 1-4) is the adoption of "work units" instead of a designated square foot area as a basis for assigning quotas to cleaning personnel. The immediate objective is a more equitable division of work, with benefits in improved results and increased efficiency. Many buildings have used this formula as a base to develop their own system.

The formula's basic steps are:

1. A complete inventory of each regularly cleaned suite in the building must be made. It must list each piece of furniture and equipment and figure the amount of carpet and other floor coverings.

2. The work load for each room is then calculated by multiplying each item by the allotted number of work units. After these calculations have been made, addition gives the total number of work units for each room.

3. Then find the work load being carried by each cleaner. For instance, if the employee cleans three rooms which, by step two, have been found to contain 300, 430 and 320 work units respectively, then that employee is carrying a work load of 1,050 work units.

4. By comparing "work units," carried by each employee, it is possible to discover any unbalanced distribution of the work load. Actual working experience shows that a cleaner can handle from 240 to 250 work units per hour of work.

5. A reassignment of schedules can be made to equitably distribute the work load on a 250-work unit per hour basis, or 300-work units per hour in air conditioned buildings.

Figure 1-2

MAN-HOUR STUDY CHART

Cleaning	Hours Per Year
Offices and Public Areas (including Supervision)	....
General Maintenance which includes:	
Carpenters	....
Electricians	....
Plumbers	....
Steam Fitters	....
Elevator Mechanics	....
Marble Men	....
Plasterers	....
Tile Setter	....
Painters	....
Wall Washers	....
Utility Men	....
Boiler Operators	....
Air Conditioning Men	....
Freight Elevator Operators	....
Elevators	
Passenger—Operators and Starters	....
If Operatorless—Starters Only	....
Watchmen	....
Window Washers	....
Total (Hours per Year)	....
Building Production Rate (Man-hours per Sq. Ft.)	

$$\frac{\text{Man-hours}}{\text{Rentable Area}} = \qquad \text{...............}$$

$$\frac{\text{Man-hours}}{\text{Gross Area}} = \text{............}$$

Figure 1-3

MAN-HOUR DISTRIBUTION IN BUILDING OPERATION

Number of Employees	Operation	Man-hours per day with Operators	% of Total Hours	Man-hours per day Without Operators	% of Total Hours
59	Cleaning	472	62.1	472	72
6	Watchmen	48	6.3	48	7.3
15	Elevators	120	15.8	16	2.4
11	Maintenance	88	11.6	88	13.4
3	Painters	24	3.1	24	3.7
1	Window cleaners	8	1.1	8	1.2
95		760	100.0	656	100.0

Total Operation with Operators .... .62 man-hours per sq. ft. Rentable Area

Total Operation Without Operators .... .54 man-hours per sq. ft. Rentable Area

Cleaning Operation .... .387 man-hours per sq. ft. Rentable Area

Based on Rentable Area of 317,000 square feet.

FOR DETERMINING THE CLEANING ASSIGNMENT IN ANY GIVEN BUILDING

The application of this formula will reflect by means of work units an equitable distribution of area to be cleaned

STEP NO. 1—You will list on Columnar Form (see sample) room no. of each suite—together with square feet of net Rentable Area. Next secure the no. of sq. ft. of Carpet—Rugs—Linoleum—Concrete or other floor covering in each office and list these amounts in proper column. Next count the no. of Desks—Chairs—Tables—Files—Lamps—Couches—Book Cases—Clothes Trees—Wash Basins—Safes—Cabinets—Shelves in each office and list amounts in proper columns.

STEP NO. 2—Divide total No. of Sq. Ft. of net rentable area in building by your established base No. of Sq. Ft. per Hour allotted to each cleaner. This gives you the No. of Hours work necessary per night—divide that figure by the No. of hours each cleaner works per night. This will give you the No. of cleaners necessary under the Sq. Ft. Allotment System.

STEP NO. 3—Total each column of above form for entire building. Then divide each result by the No. of cleaners shown in Step No. 2. This gives you the Avg. No. of Sq. Ft. of floor covering and the Avg. No. of furniture and equipment units each cleaner should have under the Sq. Ft. Allotment System.

STEP NO. 4—Allot 1½ work units for each ten (10) sq. ft. of Carpet & Rugs shown—Allot one (1) work unit for each ten (10) sq. ft. of Linoleum or other Floor Covering, including Concrete. Allot two (2) work units for each desk—Allot one (1) work unit for each piece of Furniture and Equipment shown. (Note) One work unit to be allotted for each bay of sectional Book Cases and one work unit for each bay of shelving. This will give you the total No. of work units in each of the above columns and also a basis for completing the totals in the last five columns of the form. Now transfer detailed information for each office to individual cleaners schedule card (sample below) and the totals will show the No. of work units each cleaner has under your Sq. Ft. Allotment System. This will show you the inequality in your present system.

STEP NO. 5—Results as shown by Steps 1 to 4 will give you the total no. of work units in the bldg. for each column. You can now determine from these results the base no. of work units each cleaner should be allotted and then rearrange the schedules using WORK UNITS as a base instead of sq. ft.

RENTAL AREA		SQ. FT. FLOOR COVERING				FURNITURE AND OTHER EQUIPMENT													TOTAL FURN. UNITS	FURN. WORK UNITS	FLOOR WORK UNITS	SPACE ADJUST UNITS	TOTAL WORK UNITS
ROOM NO.	NET SQ. FT.	CARPET	RUGS	LINOL.	CONCR.	DESK	CHAIR	TABLE	FILE	LAMP	COUCH	BOOK CASE	CLOTHES TREE	WASH BASIN	SAFE	CABINET	SHELF	MISC					
142X-50	1370	1200	--	--	170	7	34	11	7	1	--	2	--	--	1	--	10	--	73	80	197	--	277
1436	507	--	364	--	143	4	13	5	5	2	--	--	1	1	--	--	6	--	37	41	69	--	.110

INDIVIDUAL CLEANERS SCHEDULE

CLEANER - MARY HUSOCK

Room No.	AREA	CARPET	RUGS	L-C	DESK	CHAIR	TABLE	FILE	LAMP	COUCH	BOOK CASE	CLOTH TREE	WASH BASIN	SAFE	CAB	SHELF	MISC	ADJ. UNITS	TOTAL FURN. UNITS	TOTAL WORK UNITS
SOUTH END 7th FL.	7000	210	718	7000	73	208	27	11	--	--	1	20	--	--	--	--	--	--	338	1219
TOT. WORK UNITS		108	108	700	146	208	27	11	--	--	1	20	--	--	--	--	--	--	--	1219
AVE. WORK UNITS		210	190	392	104	130	38	55	15	2	12	13	6	3	4	24	--	--	--	1200

NOTE—Experience has shown that a woman cleaner should do approximately 200 work units per hour or 1200 work units per six (6) hour night. However, each building can adjust the hourly base up or down where unusual conditions seem to warrant such a change. The hourly allotment given is based on a cleaner doing no scrubbing—no wall washing—no cleaning of venetian blinds—wood and glass partitions—electric light fixtures and no polishing of doors—jam or trim. If cleaners have toilet rooms or other special space to clean proper work units should be assigned in column marked MISC. If space is of very unusual character and exceptionally hard to clean—additional work units may be allotted in space adjustment column.

Figure 1-4 / Gilbert formula.

Group or Squad Cleaning

The group or squad system follows the trend toward specialization. Rather than have one person clean several rooms in a given area and perform all the tasks involved, squads are formed and broken into teams that specialize in one phase of cleaning. Several squads may be employed in one building and each squad is usually comprised of two-man teams. In a typical office building the work of each squad fits into one of five general classifications: (1) vacuum carpets, (2) damp mop floors, (3) empty wastebaskets, (4) dust furniture, and (5) special duties.

A daily work ticket is made up for each team telling exactly what is to be done, where it is to be done, and the time allowed. Each squad, composed of ten to twenty-five men, has a senior cleaner in charge. It is his responsibility to inspect all work.

Users of the squad system report the following advantages:

1. Morale is higher.

2. Closer supervision is possible.

3. More work is done with fewer men and the general quality of the work improves.

4. Less equipment is necessary.

5. There is an equitable distribution of work to each worker.

6. Appraisals of employee proficiency have greater validity.

Supervision and Inspection

From a study of any of the cleaning methods, it is readily seen that the final measure of success is based on the amount and type of supervision and inspection. A good system of inspection or follow up is the only way management can properly control the cleaning operations and insure full value for money spent. The progressive cleaning program incorporates the best tools, training, incentives and organization for the purpose of achieving and maintaining high individual productivity and quality levels. Therefore, the worker must not only be properly educated in efficient basic cleaning techniques, but he must be given proper incentive. Call it morale, job interest, or teamwork—it's the human side of the cleaning program and should be underlined as one of the basic and more important elements.

Rating forms and check lists are frequently used as a basis for checking up on the thoroughness of the cleaning operation. Such a list prevents oversight, as the supervisor is required to check off each item. There may be room on the form for comment on any points that occur to him, either in connection with the work, in changing the system, or in the quality of work compared to preconceived standards of cleanliness.

Formal Training

Cost is one of the very vital characteristics in cleaning—results being, of course, the ultimate. Therefore, the chief possibility of offsetting rising costs lies in training or teaching employees to perform their work better and faster.

There are many different methods of training. A definite course of instructions with the use of classroom technique, for concentrated instruction and for the sake of valuable discussion in which all may join, is particularly effective. Group meetings at the outset should be held at frequent intervals until the best possible results are obtained. After that, refresher meetings should be held as the need indicates. Audio-visual aids—films, slides, charts and practical demonstrations—should be used as much as possible in demonstrating new methods, new tools, and the technique of planning work.

Service manuals have been widely used by management as a guide to cleaning operation and are invaluable in speeding the training process, and cover the following subjects:

1. Compounds and materials used for different kinds of cleaning.
2. The frequency at which each operation is performed.
3. Cleaning organization and personnel.
4. Job specifications of various types of personnel.
5. Detailed instructions on how to perform each task.
6. Cleaning equipment for different purposes and how to care for it.

Scientific Cleaning

Today, outstanding cleaning programs have managed to double and even triple productivity. Costs have been cut from 30 to 50 per cent while quality has reached new heights. What are some of the characteristics of the scientific, closely controlled cleaning program?

Here are some of the fundamentals of these new cleaning programs, which are based squarely on modern management methods, and industrial cost control techniques:

1. Objective measurements of the total job.
2. Proven work standards (which comprise quality standards or degrees of cleanliness; standard cleaning methods; and standard time requirements for each job).
3. Continuous research and testing of new products and equipment.
4. Continuous personnel training, incorporating methods, engineering principles and morale factors.

The work measurement method as advanced by a national cleaning specialist, utilizes the proven practice of time and motion study. This method consists of

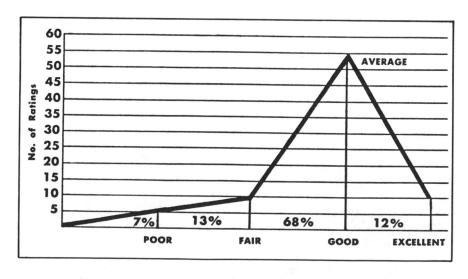

WORKLOAD DETERMINATION FORM

AREA *PROR. MGMT. DEPT. 1570 SQ. FT.*   QUALITY STANDARD *88*

ITEM TO BE CLEANED	CLEANING METHOD (DESCRIBE)	STANDARD TIME PER ITEM	AMOUNT OR NUMBER OF THE ITEM	TOTAL TIME (MINUTES)	CLEANING FREQUENCY	YEARLY TIME (HOURS)
WASTE BASKETS	EMPTY	17 SECONDS	15	4.3	DAILY	18.6
ASH TRAYS	EMPTY & DAMP WIPE	10 SECONDS	11	1.8	DAILY	7.8
STD. DESK GROUP— DESK, CHAIR, TELEPHONE, MAIL BASKET, FILE	DUST-WHISK UPHOLSTERED CHAIRS	42 SECONDS	12	8.4	DAILY	36.4
ASPHALT TILE FLOOR	DUST MOP	8 M/1000	1570 SQ. FT.	12	DAILY	52
ASPHALT TILE FLOOR	COMPLETE BUFFING	20 M/1000	1570 SQ. FT.	30	WEEKLY	25
PROJECT WORK	WAXING, FURNITURE WASHING, SPOT WASHING, ETC.	7 M/1000	1570 SQ. FT.	11	DAILY	47.6
			TOTAL YEARLY TIME (Area Work Load):			

Figure 1-5 / Workload determination form.

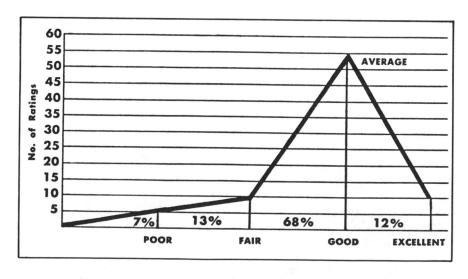

Figure 1-6 / Quality rating of cleaning performance.

standards for production for each of the major cleaning jobs; and standard frequencies for the performance of each job. These practices when properly applied to the building workload, establish the staffing required for cleaning.

The workload is determined by: (1) a complete inventory of work to be accomplished in each work area, and (2) time studies of each item of work.

Time studies are made by: (1) giving trained workers the necessary tools and materials, (2) having them perform the cleaning operation many times at a normal pace, and (3) clocking each complete operation, including make ready, transportation, and put away times. On-job conditions such as soil load and traffic interference are taken into account.

The engineered approach to control of the cleaning program is based upon the principles of applying the best practices and methods available to the job operation. Establishing work standards—the setting up of measures for determining progress and results—is the basis for a cleaning system advanced by a firm of professional cleaning consultants who call it the "Work Standards System."

Figure 1-5 is typical of a form used by these analysts in setting up the workload for each integral area of a building.

The following factors are determined before arriving at the workload:

1. How clean the building is to be kept.
2. How much total space and equipment there is to clean.
3. What methods will be employed to clean it.
4. How much time each particular cleaning job takes.

Under this system, a building is objectively rated on its cleanliness and results are measured by means of a "Productivity Index". By multiplying the sanitation level of the area times the hourly square foot coverage of the cleaning worker, both quality and quantity of actual cleaning output can be computed.

The Quality of Cleaning

Quality of cleaning takes equal rank with cost considerations. A measuring stick for quality is something which the cleaning industry has lacked and engineers have developed. The new basis for measuring quality in industry is known as "statistical quality control". This plan reveals the actual performance as it normally occurs over a period of time. Statistical quality control is a sampling process used to determine the average performance and the occurrence at both extremes of the range of performance. Building managers and maintenance supervisors can apply this principle by sampling and rating performances in their cleaning operations.

For example, you are interested in determining the quality of cleaning on horizontal surfaces. Assume that four different grades should be used to measure quality such as: (1) poor, (2) fair, (3) good, and (4) excellent. Over a specified period of time, you will rate the quality of cleaning on horizontal surfaces. The results might appear this way:

Grades	Number of Ratings	Per Cent of Total
Poor	6	7
Fair	10	13
Good	55	68
Excellent	10	12
Totals	81	100

Figure 1-6 shows how these results appear in graph form.

Frequency Data

OFFICES

Dusting

Desk tops
Ash trays
Horizontal surfaces
Sills
Floor sweep
Treated sweep

1 × day

Partitions
Vertical surfaces
Radiators

1 × week

Damp mop	1 × week
Carpet sweep rugs	3 × week
Vacuum rugs	2 × week

Buff 2 × month

Damp Wipe (Spot Wash)

Walls
Doors
Water fountains
Partitions
Partition glass
Water bottles

1 × day

Sills
Radiators
Desk tops

1 × week

Door saddles 2 × month

Periodic Cleaning

Wash partition glass	1 × 2 weeks
Wash partitions	1 × month
Wash windows	1 × 2 months
Scrub or strip floors	2 × year
Mop floors (wet)	1 × 6 weeks
Refinish floors (1 coat)	1 × 3 months

Wash walls	1 × year
Wash furniture	1 × year
Polish furniture	3 × year
Dust blinds	1 × 3 months
Wash blinds	1 × year
Dust light fixtures	1 × 2 months
Wash light fixtures	1 × year
Clean anemostats (wash)	3 × year
Wash desks (entire)	2 × year
Wash desk floor pads	1 × month

CORRIDORS

Policing

Sand urns	
Floor	4 × day
Floor sweep	1 × day
Treated sweep	
Damp mop tile	1 × week
Terrazzo	1 × day

Damp Wipe (Spot Wash)

Ledges and sills	2 × week
Walls	
Doors	
Door saddles	1 × week
Radiators	
Water fountains	1 × day
Dust light fixtures	1 × 2 months
Wash light fixtures	1 × year
Dust walls	1 × 3 months
Scrub or strip floors (machine)	1 × 3 months
Mop floors (wet)	2 × week
Refinish (1 coat)	1 × 6 weeks
Metal polishing	1 × month
Wash partition glass	1 × month

STAIRWELLS

Damp Wipe (Spot Wash)

Floors	
Sand urns	
Walls	1 × day
Doors	
Handrails	
Treated sweep	
Vacuum clean carpet	
Ledges and sills	1 × week

Periodic Cleaning

Dust blinds	4 × year
Wash blinds	1 × year
Metal polish	1 × month
Dust light fixtures	1 × 3 months
Wash light fixtures	1 × 4 months
Dust walls	1 × 3 months
Mop and rinse steps	1 × day
Hand scrub steps	1 × week

LAVATORIES

Daily

Check soap	
Towels	
Tissues	3 × day
Damp wipe surfaces (porcelain-chrome-stainless steel)	
Empty waste receptacle	1 × day
Spot mop floor	1 × day

Periodic Cleaning

Commodes	
Urinals	2 × day
Wash basins	

Damp Wipe (Spot Wash)

Partitions	
Receptacles	2 × day
Dispensers	
Ledges and sills	2 × week
Radiators	1 × week
Walls	
Doors	1 × day
Wash walls	1 × 4 months
Wash windows	1 × month
Dust blinds	1 × 2 months
Wash light fixtures	1 × year
Wash and rinse floor	1 × day
Machine scrub floor	1 × week

EXECUTIVE OFFICES

Dusting

Sweep
Dust mop

Treated sweep
Vacuum
Ledges and sills
Wall fixtures
Office furniture
Radiators   1 × day
Shelves
Water fountains or bottles
Doors
Walls
Partitions
Partition glass

Damp mop tile   1 × week

Buff   2 × week

Scrub and strip floor (machine)   1 × 2 months

Refinish (1 coat)   1 × month

Dust walls   1 × week

Periodic Cleaning

Dust blinds
Vacuum upholstered furniture   1 × week
Vacuum drapes

Wash furniture   1 × year
Polish furniture   4 × year
Wash blinds   1 × 6 months
Wash light fixtures   1 × year
Wash windows   1 × 2 months
Wash walls   1 × year
Wash partitions   2 × year
Wash partition glass   1 × 2 weeks
Shampoo carpets   2 × year

PUBLIC LOBBIES

Sweep
Dust mop
Treated sweep   1 × day
Damp mop and rinse
Buff

Machine scrub terrazzo   1 × week
Tile   1 × month
Refinish tile (1 coat)   1 × 2 weeks

Dusting

Radiators
Ledges and sills
Doors   1 × day
Signs
Sand urns

Damp Wipe (Spot Wash)

Miscellaneous furniture
Doors
Walls
Water fountains   1 × day
Ledges and sills
Radiators

Periodic Cleaning

Wash walls	1 × 6 months
Dust light fixtures	1 × month
Wash light fixtures	1 × 3 months
Vacuum ceilings	1 × year
Wash ceiling	1 × 2 years
Paint ceiling	1 × 4 years
Wash furniture	1 × 6 months
Wash windows	1 × month
Dust blinds	3 × year
Wash blinds	2 × year
Door glass	1 × day
Polish metal	2 × week

Elevator Lobbies

Sweep
Mop and rinse   1 × day
Treat sweep

Dusting

Doors	1 × day
Walls	1 × week
Signs	1 × day
Wall fixtures	1 × week
Ledges and sills	1 × day

Periodic Cleaning

Wipe sand urns	1 × day
Wipe elevator doors	1 × week
Polish metal	1 × week
Wash walls	1 × year
Wash light fixtures	2 × year
Door glass	1 × day
Wash windows	1 × 2 months
Dust blinds	1 × week
Wash blinds	2 × year
Spot wash walls	1 × week
Spot wash doors	1 × 2 weeks

2 / ALUMINUM CLEANING

(*Source: Plant* magazine)

Architectural aluminum has few maintenance requirements and it adds immeasurably to a plant's appearance and structural longevity. Most aluminum surfaces are protected by a thin, tough film or oxide coating (it forms on exposure to the atmosphere) which guards against rust and corrosion that would quickly destroy other metals.

However, aluminum can get dirty just like any other material exposed to industrial atmosphere and hard usage. Aluminum on the exteriors of industrial structures is subjected to the worst conditions in this respect; smoke and fumes tend to discolor the surface and airborne dust combined with moisture lead to accumulations of dirt.

Ordinarily, this in no way harms the metal but, in most cases, some measure of upkeep is desirable to maintain the most prominent applications—such as entrances, sash, trim, grillework, handrails, similar items in the main reception and office areas—sparkling bright and new looking. While it is not economically feasible to clean the entire exterior of an aluminum clad plant, it is worthwhile to keep the office portion in good order for the purpose of good public relations.

Methods of Cleaning

Periodic cleaning and restoration of aluminum surfaces, even after years of neglect, can be simplified by utilization of proper methods and materials. Methods of cleaning vary with the type and severity of soiling, frequency of cleaning, and other factors. In general, three categories of cleaning techniques have been found acceptable and are described in the following paragraphs. For each specific plant, these methods should be tried in the order described on small but different types of typically soiled areas until satisfactory results are obtained. Once the choice of method has been established, the remainder of the surface can be cleaned accordingly.

1. Mild. Where surfaces are not too dirty or have been cleaned periodically, try the following procedures:

a. Wash with clean water and dry thoroughly. Or,

b. Wash with mild soap and warm water, rinse, and dry. Or,

c. Apply a non-etching chemical cleaner, diluted three parts of water to one part of cleaner, in accordance with the manufacturer's instructions.

2. Medium Severe. If a thick coating of dirt has accumulated, it may be found easier to first loosen the heavy crust with a solvent cleaner, and then try one of the following procedures:

a. Apply a wax base polish cleaner with a clean, soft rag or pad. Make certain the manufacturer's directions are followed closely. Or,

b. Apply a nonwax base polish cleaner. Follow same precautions as above. Or,

c. Apply an abrasive wax with same precautions as above, Or,

d. Apply a mild abrasive cleaner (scouring powder) with a clean, damp cloth. Be sure to rinse well and dry.

3. Severe. If the foregoing methods are not successful, try the following procedure.

Apply liquid or one of the above cleaners with a stainless steel wool pad (size 00 or finer). Mild steel wool can also be used as an applicator, but it is extremely important that no wool particles remain to rust-stain the aluminum surface.

Care should be exercised in using steel wool or abrasive waxes and cleaners as they might damage the appearance of the surface if the metal is scoured too long or too hard. Steel wool or abrasive cleaners should be rubbed in the direction of the metal's finish. A bristle brush, rather than a cloth or pad, is recommended for use with all cleaners on patterned surfaces.

Another precaution is that all cleaner deposits, with the exception of those from wax base preparations, should be removed with clean water followed by a thorough drying of the surface. Wax base cleaners leave a thin protective coating which need not be removed unless the surface is to be lacquered.

Heavy-Duty Cleaning

Should the foregoing methods fail to produce satisfactory results, and the surface still looks streaked and spotted, it will be necessary to resort to even more severe cleaning procedures. All that may be required on parts that have been anodized is a heavy-duty cleaner put on with a damp clean cloth or stainless steel wool pad, rinsed with water, and dried thoroughly. Determination of whether or not a part has been anodized can be made by a check on the original building specifications; or by a "scratch" or "rub" test. If the surface scratches easily or if a rubber eraser leaves a bright spot, an anodic coating has not been applied.

Extremely dirty surfaces not having an anodic coating should be cleaned as follows:

Apply a suitable etching chemical cleaner (phosporic acid type), diluted three parts of water to one part of cleaner, with a clean sponge, rag or brush. Work a small area, just large enough to be covered and keep wet three to five minutes at a time. Be sure to start at bottom of panel and work up, so splashes or rundowns do not get onto uncleaned surfaces. Rinse surface with water before cleaner dries. Make certain that all cracks, grooves, etc. are rinsed. Dry surface thoroughly. Drying may be speeded by swabbing with a 50-50 solution of denatured or wood alcohol and water. When handling chemical cleaners, particular attention should be given to protection of the workmen's eyes and skin.

Preserving the Finish

After cleaning, a coating of liquid wax or clear lacquer can be applied if desired to preserve the restored finish and facilitate future maintenance.

Normally, waxed aluminum can be kept clean by wiping with a damp rag from time to time, or by rinsing with clean water. To apply a liquid wax, first clean the surface with a solvent cleaner, then spread the wax and polish with a soft cloth or pad. As mentioned previously, wax based cleaners also leave a satisfactory protective wax film.

However, after repeated applications of wax, the built-up coating may become soft or tend to yellow. When this occurs, the accumulation must be removed with a solvent or mild abrasive cleaner, and the surface re-waxed.

A good clear methacrylate lacquer coating will protect aluminum surfaces for several years without yellowing or cracking. For best results, the surface should be carefully prepared and a lacquer applied as follows:

1. If old lacquer has worn off in spots, the entire surface should be stripped with a lacquer remover, following the manufacturer's directions.

2. Clean the stripped surface thoroughly with a solvent cleaner.

3. For better lacquer adhesion on bare aluminum, it is advisable to provide a good "tooth" on the surafce by use of an etching type chemical cleaner. Follow the foregoing directions (under "Heavy-Duty Cleaning") on applying this type of cleaner.

4. Apply a thorough, wet coat of lacquer. Spray application is preferable, but brush coating is satisfactory if the lacquer is thinned with a slow evaporating solvent. The coating should be applied as uniformly as possible and after the first coat has dried, a second coat should be applied in the same manner.

Cleaning Schedule

Aluminum surfaces require cleaning more often in some areas than in others; a proper schedule may vary from two weeks to six months. If a plant is located in an area with a concentration of soot, factory fumes or salt sea mist, cleaning must be at more frequent intervals, to insure a bright, attractive appearance. As with any building product, the more aluminum is cleaned, the easier each successive job will be.

3 / ASPHALT MAINTENANCE

(Reprinted with permission from The Asphalt Institute from Manual Series No. 16, *Asphalt in Paving Maintenance*.)

Pavement maintenance is a major activity of every highway, street department, and building owner. Usually, money for maintenance is limited and the maintenance man is called upon to "make one dollar do the work of two." This is not easy.

Large differences in soil types, climate, terrain, traffic, and other factors make for greatly varying problems, even within small areas. Some regions are rugged and mountainous while others are fairly smooth and level; some have heavy rainfall, others are semiarid; some highways and streets must accommodate vehicles carrying coal, ore, logs, or other heavy loads, while others are subjected to only lightweight traffic.

Yet, despite these differences, there are maintenance methods that can be used equally well in all regions. Presenting some of these methods is the purpose of this manual.

Definitions. Some of the terms used are defined here in order that their meanings will be clear.

(1) *Asphalt Concrete*—High-quality, thoroughly controlled hot mixture of asphalt cement and well-graded, high-quality aggregate, thoroughly compacted into a uniform, dense mass typified by Asphalt Institute Type IV mixes. (See *Specifications and Construction Methods for Asphalt Concrete and Other Plant-Mix Types,* Specification Series No. 1 (SS-1), The Asphalt Institute.)

(2) *Asphalt Emulsion Slurry Seal*—A mixture of slow-setting emulsified asphalt, fine aggregate, and mineral filler, with water added to produce slurry consistency.

(3) *Asphalt Fog Seal*—A light application of slow-setting asphalt emulsion diluted with water. It is used to renew old asphalt surfaces and to seal small cracks and surface voids. The emulsion is diluted with an equal amount of water and sprayed at the rate of 0.1 to 0.2 gallon (of diluted material) per square yard, depending on the texture and dryness of the old pavement.

(4) *Asphalt Leveling Course*—A course (asphalt aggregate mixture) of variable thickness used to eliminate irregularities in the contour of an existing surface prior to superimposed treatment or construction.

(5) *Asphalt Overlay*—One or more courses of asphalt construction on an existing pavement. The overlay generally includes a leveling course, to correct the contour of the old pavement, followed by uniform course or courses to provide needed thickness. (Overlays usually are considered construction, not maintenance.)

(6) *Asphalt Pavements*—Pavements consisting of a surface course of mineral aggregate coated and cemented together with asphalt cement on supporting courses such as asphalt bases; crushed stone, slag, or gravel; or on portland cement concrete, brick, or block pavement.

(7) *Asphalt Pavement Structure*—(sometimes called Flexible Pavement Structure)—Courses of asphalt-aggregate mixtures, plus any nonrigid courses between the asphalt construction and the foundation or subgrade. [See also (12), (13), and (15).]

(8) *Asphalt Prime Coat*—An application of low-viscosity liquid asphalt to an absorbent surface. It is used to prepare an untreated base for an asphalt surface. The prime penetrates into the base and plugs the voids, hardens the top, and helps bind it to the overlying asphalt course. It also reduces the necessity of maintaining an untreated base course prior to placing the asphalt pavement.

(9) *Asphalt Seal Coat*—A thin asphalt surface treatment used to waterproof and improve the texture of an asphalt wearing surface. Depending on the purpose, seal coats may or may not be covered with aggregate. The main types of seal coats are aggregate seals, fog seals, emulsion slurry seals, and sand seals.

(10) *Asphalt Surface Treatments*—Applications of asphaltic materials to any type of road or pavement surface, with or without a cover of mineral aggregate, which produce an increase in thickness of less than 1 inch.

(11) *Asphalt Tack Coat*—A very light application of liquid asphalt applied to an existing asphalt or portland cement concrete surface. Asphalt emulsion diluted with water is the preferred type. It is used to ensure a bond between the surface being paved and the overlying course.

(12) *Deep-Lift Asphalt Pavement*—An asphalt pavement structure [see (7)] in which the asphalt base course is placed in one or more lifts of 4 or more inches compacted thickness.

(13) *Deep-Strength Asphalt Pavement*—DEEP-STRENGTH® is a term registered by The Asphalt Institute with the U. S. Patent Office. The term DEEP-STRENGTH (also called "mark") certifies that the pavement is constructed of asphalt with an asphalt surface on an asphalt base and in accordance with design concepts established by the Institute. [See latest edition of *Thickness Design* manual (MS-1).]

(14) *Deflection*—The amount of downward vertical movement of a surface due to the application of a load to the surface.

(15) *Full-Depth Asphalt Pavement*—An asphalt pavement structure [see (7)] in which asphalt mixtures are employed for *all* courses above the subgrade or improved subgrade. A Full-Depth asphalt pavement is laid directly on the prepared subgrade. (The mathematical symbol T_A denotes Full-Depth.)

(16) *Mixed-in-Place (Road-Mix)*—An asphalt course produced by mixing mineral aggregate and liquid asphalt at the road site by means of travel plants, motor graders, drags, or special road-mixing equipment.

(17) *Pavement*—see Pavement Structure. (As used in this section, the word "pavement" means "pavement structure.")

(18) *Pavement Structure*—All courses of selected material placed on the foundation or subgrade soil, other than any layers or courses constructed in grading operations.

(19) *Plant-Mix*—A mixture, produced in an asphalt mixing plant, which consists of mineral aggregate uniformly coated with asphalt cement or liquid asphalt.

(20) *Plant-Mixed Surface Treatments*—A layer, less than 1 inch thick, of aggregate that is coated with asphalt in a plant. Plant-mixed surface treatments are used extensively for providing skid-resistant surfaces.

(21) *Undersealing Asphalt*—A high-softening-point asphalt used to fill cavities beneath portland cement concrete slabs and occasionally to correct the vertical alignment by raising individual slabs.

Maintenance defined. Pavement maintenance is not easy to define. Highway departments agree in general as to what it is but there are some minor differences, chiefly in scope. Some call pavement improvement "maintenance." Others include only the work which keeps the pavement in its as-constructed condition.

There also is some disagreement as to whether repairs made necessary by unusual events such as earthquakes, landslides, forest fires, windstorms, or severe traffic accidents should properly be classified as maintenance. Taking all of these into consideration, the definition which seems most nearly to fit is:

> Pavement maintenance is the routine work performed to keep a pavement, under normal conditions of traffic and normal forces of nature, as nearly as possible in its as-constructed condition.

Why maintenance is necessary. All pavements require maintenance, the chief reason being that stresses producing minor defects are constantly working in all pavements. Such stresses may be caused by change in temperature or moisture content, by traffic, or by small movements in the underlying or adjacent earth. Cracks, holes, depressions, and other types of distress are the visible evidence of pavement wear. They are simply the end results of the process of wear which begins when construction ends. In urban areas, ditches dug through the pavement for water lines and other utilities are a major cause of pavement maintenance.

Preventive maintenance. "A stitch in time . . ." The early detection and repair of minor defects is, without doubt, the most important work done by the maintenance crew. Cracks and other surface breaks, which in their first stages are almost unnoticeable, may develop into serious defects if not soon repaired. This may occur in a very few days on an underdesigned pavement under heavy traffic. For this reason, frequent close inspections of the pavement should be made by qualified men. Indeed, this measure is necessary toward the best use of maintenance money.

An inspection made from a moving vehicle, even one which creeps, is usually not close enough to detect areas where distress may begin. Often the cracks or other surface defects are so small that only a person on foot can spot them. There are other small signs, such as mud or water on the pavement or shoulder, which to an experienced observer may signal future trouble. It is best, then, to walk the pavement for close inspection; or, when there are not enough men available for this purpose, to spot-check selected stretches of roadway.

Upon detection of the warning signs, a detailed investigation, including trenching across the failed area if necessary, should be made to determine the kind of repair called for. If the pavement seems to be moving under traffic, deflection measurements should be carried out to determine the extent of the affected area (use of a *Benkelman Beam* is recommended).

All persons making pavement inspections on foot should take proper safety precautions. They should wear easily seen clothes. They should be protected by adequate warning signs and devices, or followed by a car or truck displaying warning devices. Safety flags, vests, and caps of bright color are very effective.

Drainage maintenance. A form of preventive maintenance is seasonal inspection and cleaning of drainage systems. If drains are kept working properly, some of the major causes of pavement damage are eliminated. Each inspection should include all surface drainage structures, ditches, and channels to insure that they are working as designed. If any part of the system is clogged, it should be cleaned out immediately.

At least twice a year subsurface drains should be examined to make sure they are working as intended. The abnormal appearance of water on the pavement surface may indicate that subsurface drains are improperly located, incorrectly designed, or clogged.

All drain outlets should be well marked on the ground and on maintenance maps. If this is done, they will not be overlooked on inspection trips.

Detailed information about pavement drainage is contained in *Drainage of Asphalt Pavement Structures*, Manual Series No. 15 (MS-15), The Asphalt Institute. Most of the information in this manual applies equally to portland cement concrete pavements.

Make repairs promptly. Repairs should be made as quickly as possible after the need for them is discovered. This is particularly important when the defect makes driving hazardous.

Often, weather conditions make temporary repairs necessary to prevent further damage until more permanent repairs can be made. As examples, crack filling is most likely to be successful during periods of cool, dry weather; chuck (pot) hole patches adhere best when the pavement is warm and dry; and seal coats, or other surface treatments, require warm and dry weather for best results. Selecting the best time to make repairs, therefore, involves the careful balancing of several things and requires both experience and judgment.

Prevention of defect recurrence. In all cases of pavement distress, it is best to determine first the cause or causes of the difficulty. Then repairs can be made which will not only correct the damage but will also prevent or retard its happening again. Time and money spent for such repairs are well spent because the same repairs will not have to be made over and over.

Street maintenance. Streets can develop all types of defects. However, some of these defects are much more of a problem in streets than in any other class of pavement structure. Shoving and corrugating of asphalt pavement surfaces, for example, show up more often in urban areas. Limited speeds on steep grades and frequent traffic lights and stop signs at intersections multiply the need for braking and the result is shoving or corrugating of low-stability pavement surfaces. A heater-planer has been used successfully in repairing these defects.

The importance of skilled maintenance personnel. Maintenance work requires proper supervision, skilled workmen, and good workmanship. Unless all three are employed, it is likely that some repair work will be poorly done and may have to be repeated. Since most pavement repairs involve the use of asphalt, a thorough knowledge of this material is essential for maintenance men. This is especially true for supervisors and inspectors. Successful pavement maintenance requires a knowledge of which asphalts are available and how to use them. Although the basic skills needed for pavement maintenance can be acquired only through experience gained in the actual work, a close study of the literature published by The Asphalt Institute will be found most useful. The following publications are recommended for those engaged in maintenance work:

1. *Asphalt as a Material,* Information Series No. 93 (IS-93).
2. *Specifications for Asphalt Cements and Liquid Asphalts,* Specification Series No. 2 (SS-2).
3. *The Asphalt Handbook,* Manual Series No. 4 (MS-4).
4. *Asphalt Pavements for Airports,* Manual Series No. 11 (MS-11).

5. *Asphalt Surface Treatments and Asphalt Penetration Macadam,* Manual Series No. 13 (MS-13).

6. *Asphalt Mixed-in-Place (Road-Mix) Manual,* Manual Series No. 14 (MS-14).

7. *Drainage of Asphalt Pavement Structures,* Manual Series No. 15 (MS-15).

The importance of weather. Preferably, patching or resurfacing work should be done during warm (50°F and above) and dry weather. When hot or warm mixtures are placed on cold pavements, they may cool so fast that adequate compaction is difficult. This cooling effect is emphasized if the mixture is placed in thin layers. Moreover, asphalt and asphalt mixtures usually do not bond well to damp surfaces.

This does not mean that repairs cannot be made during cold or damp weather. Rather, they require much greater care when made during such periods. They also have much less chance of being satisfactory. It is better, however, when the safety and comfort of the traveling public are concerned, to make the repairs even though they may be only temporary. Also, a delay in repairs may allow small surface breaks to progress into major failures.

Mixtures containing liquid asphalts are slow in curing out when the humidity is high. This is because the air, which already contains a large amount of water vapor, does not readily allow solvent evaporation. Low temperatures also slow up solvent evaporation.

Seal coats and other surface treatments can be affected by moisture during the first few hours after their placement. Rain and/or fast traffic during this critical period will often result in the loss of most of the cover aggregate.

A phone call to the weather bureau may help in scheduling maintenance work during uncertain weather.

Safety. An important contribution to high-quality maintenance is an active safety program. For the maintenance man, safety measures reduce fear of injury, allowing him greater freedom of mind in performing his task. This results in his doing a better job.

For the safety of the workers, the motorist must be warned about what is going on ahead and what he must do as he passes through the work area. Signs and warning devices should be placed far enough ahead for him to grasp their meaning. Yet they should not be so far ahead that they lose their meaning. A sign indicating the end of the work area also is desirable. The use of flagmen near the work is necessary when sight distance is restricted or dangerous driving conditions exist.

The kind of safety equipment to be used by the maintenance men depends upon the type of work they are doing. Examples: If they are subsealing a portland cement concrete slab they should wear clothing and safety gear that leave no skin exposed, obviating injury in the event of hot asphalt blowing back from the hole. If they are merely sweeping a dirty pavement with a power broom, a dust mask and goggles may be all the extra equipment necessary. As appropriate, members of the maintenance crew should be furnished with hard hats, goggles, asbestos gloves, and any other safety apparel that will reduce the possibility of accidents.

Maintenance of Asphalt Pavements

Types of asphalt pavement. Asphalt pavement maintenance, as discussed in this section, applies to all asphalt pavement structures from Full-Depth asphalt to surface treatments. It applies to the traveled way and shoulders of roads, streets, runways and taxiways; to parking lots and aprons; and to other areas, such as driveways. Asphalt overlays on portland cement concrete, brick, or other materials are also included.

This chapter covers the most common types of defects and failures in asphalt pavements, their usual causes, and suggested methods of repair. This does not imply, however, that the subject is completely covered. There may be unusual defects that do not fall into any of the following categories. There also are many good methods of repair that are not described here.

Moisture and granular bases. At the present time, many asphalt pavements consist of an asphalt surface over a granular base. The base materials range from gravel and pit-run products to crushed and processed rock. These bases serve well as long as they are properly drained. But if they become saturated with water, they lose strength rapidly under the weight and action of traffic.

Saturation of granular bases is the cause of many maintenance problems. Among them are asphalt-surfaced pavements on granular bases that become soft and crack in the familiar alligator or chicken-wire pattern. These are problems that won't go away by filling cracks or placing skin patches. The cause of the distress should be eliminated.

Many high-type pavements with granular bases are designed with drainage systems to prevent saturation by ground or surface water. But there are many thousands of miles of sand-clay-gravel roads, now surfaced with asphalt, that become saturated and give trouble. Usually these roads have a high percentage of plastic fine material in them as binder, needed to hold the materials in place when the surface was open. As sand-clay-gravel roads, they became saturated when it rained but dried out rapidly because moisture was free to evaporate. With the addition of the impervious asphalt pavement, this evaporation through the surface is blocked. The result is that water migrating into the base materials from the shoulders and from the subgrade below cannot escape, and the sand-clay-gravel loses strength as it becomes soaked. Cracking, heaving, and other forms of distress take place. Also, in its weakened condition the base, unable to support the traffic, deflects more than normal and cracking is intensified.

Therefore, when investigating surface failures which appear to be related to excessive deflection, the base should be checked for plastic fines or trapped water. If so, repair may call for digging out the broken area to sound material, improving drainage, and patching with asphalt patching mixture.

Benkelman beam. The extent of areas of excessive deflection can be determined quite easily with a device called the Benkelman beam. This device, pictured in Figure 3-1, has a narrow beam that is slipped between the dual tires of the rear axle of a loaded truck. A foot on the end of the beam rests on the pavement between the tires. The truck moves ahead at creep speed and the total pavement rebound deflection is read by means of a dial gauge. (Rebound de-

Photos courtesy of U.S. Bureau of Public Roads

Figure 3-1 / **Benkelman beam.**

flection is the amount of vertical rebound of a surface that occurs when a load is removed from the surface.)

Rebound deflection readings should be taken at locations sufficient to outline the whole area of excessive deflection before repairs are made. Areas of excessive deflection may be estimated by comparing deflection in the distressed area with the average deflection in areas that are performing well.

Patching mixtures. Many patches bleed, become unstable, and are subject to pushing after placement. The cause usually is an excess of asphalt in the patching mixture. Patch instability can also be caused by not allowing the patch (when made with a *stockpiled* patching mixture) to cure before subjecting it to traffic.

For the best patching mixture, a laboratory investigation should be made of the materials proposed for use.

High-quality, hot-mixed patching mixtures, although costing more than other patching materials, result in longer-lasting patches. The major cost of patching lies in placing the patch, not in the cost of the material. Therefore, the use of hot-mix materials for patches outlasting many times those made with other materials is a readily apparent economy.

It is usually possible to get a hot asphalt mixture for patching, even in out-of-the-way areas. One method employs a mix-heater to heat stockpiled pre-mix prior to making the patch. There are several types of these heaters. One type can be suspended from the tailgate of the truck carrying the pre-mix. Another is trailer mounted.

Also available is a small portable mix-plant designed for small jobs and maintenance operations. It is equipped with a small dryer and pug mill. Asphalt is stored in a tank on the mix-plant trailer. Aggregate usually is carried in a truck towing the plant. Output at jobsite is 5-10 tons per hour of hot-mix material.

Prime and tack coats. If the base of a deep patch is made with untreated material, it should be primed with 0.20 to 0.30 gallon per square yard of liquid asphalt. If spray equipment is not available, hand methods can be used to apply the prime. But care must be taken not to apply an excess of asphalt. The amount of asphalt material used to prime the base should be only enough to knit together the top particles.

The prepared edges of the surface surrounding the area being patched should be tack coated to ensure a bond between them and the patch material.

If the prime and tack coat are of asphalt emulsion, enough time should be allowed for the emulsion to "break" and most of the water to dry out before the patch-mix is placed. Similarly, a rapid curing or medium-curing asphalt should be given time to penetrate and cure before the patch-mix is placed.

For a surface patch, a light tack coat is necessary. A slip plane may develop from either the absence of a tack coat or too heavy a tack coat. Application methods are similar to those used for a prime coat except that the quantities used are much smaller.

Placing patching mixtures. After the area to be patched has been properly prepared, including trimming of edges and applying the correct prime coat or tack coat, there remains only the placing and compacting of the mix.

Segregation should be prevented. Patching mixture should never be dumped from the truck into the patch area. It should be shoveled directly from the truck or from a board on to which it has been dumped. The shovelsful of mix should be placed against the edges first rather than piled into the center and raked to the edges.

It should never be necessary to pull material from the center of the patch to the edge in making the joint. If more material is needed at the edge, it should be deposited there and the excess raked away. The quantity of material placed in the patch area should be sufficient to ensure that, after compaction, the patch surface will not be below that of the adjacent pavement. However, if too much material is placed in the patch area a hump will result. A string line and/or a straightedge, used properly, can be a great help in producing a smooth riding surface. (See Figure 3-2, Placing patching mixture.)

Compacting patching mixtures. In compacting the patch, the first pass and return of the roller, vibratory compactor, or maintenance truck wheels (if these are used) should overlap not more than 6 inches onto the patch material at one edge. This should then be repeated on the opposite side to compact the material into the edge joints. Compaction should then proceed from the low side to the high side, with each pass and return lapping an additional few inches onto

Figure 3-2 / **Placing patching mixture.**

the patch. When proper equipment and procedures are used, the surface of the patch should be at the same grade as the surrounding pavement. If hand tamping or other light compaction methods are used, however, the surface of the completed patch should be slightly higher than the pavement. Traffic will compress the patch further. (Figure 3-3, Compacting patching mixture.)

Cracking

General information. Cracking takes many forms. Simple crack filling may be the right treatment in some cases. In others, complete removal of the affected area and the installation of drainage may be necessary before effective repairs can be carried out. To make proper repairs, then, the necessary first step is to determine the cause of cracking.

The repair techniques for the correction of various forms of cracking discussed in this section are not necessarily the only correct ways to do the job. But they are proven ways that should result in neat, long-lasting repairs.

Figure 3-3 / **Compacting patching mixture.**

Alligator cracks. These are interconnected cracks forming a series of small blocks resembling an alligator's skin or chicken wire, Figure 3-4.

(1) *Cause*—In most cases, alligator cracking is caused by excessive deflection of the surface over unstable subgrade or lower courses of the pavement. The unstable support usually is the result of saturated granular bases or subgrade. The affected areas in most cases are not large. Sometimes, however, they will cover entire sections of a pavement. When this happens, it probably is due to repeated loads that exceed the load-carrying capacity of the pavement.

(2) *Repair*—Since alligator cracking usually is the result of saturated bases or subgrades, correction should include removing the wet material and installing needed drainage. Asphalt plant-mixed material can then be used for the full depth for a strong patch. (This may be the least expensive repair because of the single operation with one material.) If the asphalt plant-mixed material is not available, new granular base material in layers not exceeding 6 inches each are compacted in. The granular base should then be primed and patched.

Figure 3-4 / **Alligator cracks.**

When necessary, temporary repairs can be made by applying skin patches or aggregate seal coats to the affected areas. In any event, repairs should be made promptly to avoid further damage to the pavement.

In the case of cracking from overloading, a properly designed overlay will correct the condition. Refer to *A Short, Practical Guide to the Design of Asphalt Overlays,* Information Series No. 139 (IS-139), The Asphalt Institute.

Deep Patch (Permanent Repair)

(a) Remove the surface and base as deep as necessary to reach firm support, extending at least a foot into good pavement outside the cracked area, Figure 3-5. This may mean that some of the subgrade will also have to be removed. Make the cut square or rectangular with faces straight and vertical. One pair of faces should be at right angles to the direction of traffic. A pavement saw makes a fast and neat cut.

(b) If water is a cause of the failure, install drainage.

(c) Apply a tack coat to the vertical faces, Figure 3-6.

(d) For best results, backfill the hole with a dense-graded hot asphalt plant-mix. (Figure 3-7.) Spread carefully to prevent segregation of the mixture, Figure 3-8.

Figure 3-5 / **Removing surface and base.**

Figure 3-6 / **Applying tack coat to vertical surfaces.**

Figure 3-7 / **Backfilling hole with plant-mix.**

Figure 3-8 / **Spreading the mix.**

If the asphalt mixture is not available, make the backfill with a good granular base material. Part of the surface and upper base material removed from the hole, broken into small pieces and mixed thoroughly, can be placed in the bottom of the hole.

(e) Compact in layers if the hole is more than 6 inches deep. Compact each layer thoroughly, Figure 3-9. Compaction should be done with equipment most suited for the size of the job. A vibratory plate compactor is excellent for small patches. A roller may be more practical for large areas.

(f) Full-Depth asphalt mix placed directly on the subgrade needs no prime.

(g) If granular base is used it should be primed. The repair is then completed by placing hot plant-mixed asphalt surfacing material, and compacting to the same grade as the surrounding pavement. If hot-mixed surfacing is not available, plant-mixed material using liquid asphalt can be used.

(h) Use a straightedge or a string line to check the riding quality and the alignment of the patch, Figure 3-10.

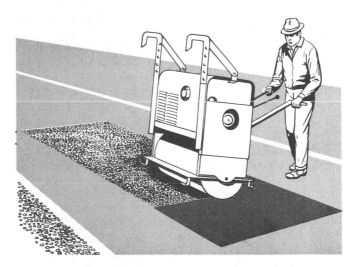

Figure 3-9   /   **Compacting the mix.**

Figure 3-10   /   **Straightedging the patch.**

Skin Patch (Temporary Repair) for Areas with Cracks Wider than ⅛ Inch

(a) Cut a shallow trench around the area to be patched to provide a vertical face around the edge, Figure 3-11.

(b) Clean the cracked area with brooms and, if necessary, compressed air.

(c) Broom plant-mixed, fine-graded asphalt material into cracks, Figure 3-12.

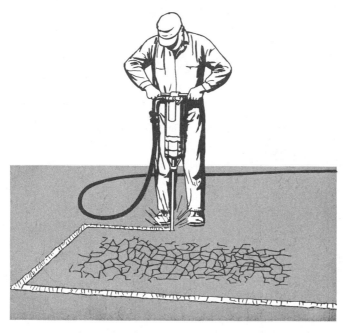

Figure 3-11   /   **Cutting vertical face around cracked area.**

Figure 3-12   /   **Brooming plant-mix into alligator cracks.**

(d) Compact with a vibratory plate compactor or roller, Figure 3-13, or roll with rear wheel of a loaded truck.

(e) Apply a tack coat, Figure 3-14.

(f) Place a skin patch with hot plant-mixed asphalt material, Figure 3-15. If this material is not available, use plant-mix with liquid asphalt. Feather the

Figure 3-13 / **Compacting with vibratory plate compactor.**

Figure 3-14 / **Applying tack coat.**

edges carefully, removing coarse particles with lute and rake before compaction.

(g) Compact the patch with a vibratory plate compactor or roller, Figure 3-16. If neither is available, rolling may be done with the wheels of the truck that carries the mix.

Figure 3-15 / **Placing skin patch of hot plant-mix.**

Figure 3-16 / **Compacting with vibratory plate compactor.**

Figure 3-17 / **Spraying asphalt on alligator cracks.**

*Aggregate Seal Coat Patch (Temporary Repair) for Areas with Cracks
Narrower Than ⅛ Inch*

(a) Clean the cracked area with brooms and, if necessary, compressed air.

(b) Spray the necessary amount of liquid asphalt (either emulsion rapid curing, or medium curing) onto the cleaned area, Figure 3-17. Usually, 0.15 or 0.25 gallon per square yard is enough for the seal coat, but, if an excessive amount is lost in the cracks, slightly more asphalt should be applied.

(c) Apply the cover aggregate *immediately* after spraying the asphalt, Figure 3-18. A good aggregate size for this type of patch is ¼ inch to No. 10 screenings.

Figure 3-18 / **Applying cover aggregate.**

(d) Roll the seal coat with rubber-tired equipment, Figure 3-19. If a roller is not available, the wheels of the truck carrying the cover aggregate can be used.

(e) If it is necessary to build up the patched area to the level of the surrounding pavement, a second seal coat can be applied.

(f) Allow to cure thoroughly before opening to traffic.

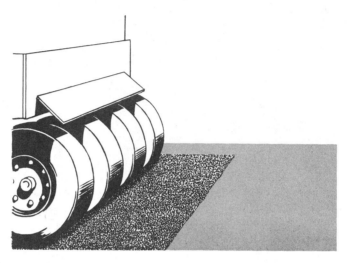

Figure 3-19 / **Rolling seal coat with rubber-tired equipment.**

Slurry Seal Patch (Temporary Repair) for Areas Cracked from Overloading

(a) Clean the cracked area with brooms and, if necessary, compressed air.

(b) Apply an asphalt emulsion slurry seal according to Specification ST-3, *Asphalt Surface Treatments and Asphalt Penetration Macadam,* Manual Series No. 13 (MS-13), The Asphalt Institute.

Edge cracks. These are longitudinal cracks a foot or so from the edge of the pavement, with or without transverse cracks branching towards the shoulder, Figure 3-20.

(1) *Cause*—Usually, edge cracks are due to lack of lateral (shoulder) support. They may also be caused by settlement or yielding of the material underlying the cracked area; which in turn may be the result of poor drainage, frost heave, or shrinkage from drying out of the surrounding earth. In the last case trees, bushes, or other heavy vegetation close to the pavement edge may be a cause.

(2) *Repair*—For temporary repair, fill as for reflection cracks. For more permanent repair, fill cracks with asphalt emulsion slurry or liquid asphalt mixed with sand. If the edge of the pavement has settled, bring up to grade with hot plant-mix patching material.

(a) Improve drainage. Install underdrains, if necessary.

(b) Clean pavement and cracks with broom and compressed air.

(c) Fill cracks with emulsion slurry or liquid asphalt (SS-1, SS-1h, or SM-K) mixed with sand. Wipe with a rubber-edged squeegee.

Photo courtesy of Ohio Highway Department

Figure 3-20 / **Edge crack.**

(d) Apply a tack coat, Figure 3-21.
(e) Bring settled edge up to grade by spreading hot asphalt plant-mixed material, Figure 3-22. Check the smoothness with a straightedge or a string line. Compact with a vibrating plate compactor or a roller, Figure 3-23. Be sure that the edges of the patch are straight and neat.
(f) Remove trees, shrubs, and other vegetation except grass from close to the pavement edge.

Edge joint cracks. An edge joint crack is really a seam. It is the separation of the joint between the pavement and the shoulder, Figure 3-24. It is treated as a crack, however.

(1) *Cause*—A common cause of "cracking" in a pavement-shoulder joint is alternate wetting and drying beneath the shoulder surface. This may result from poor drainage due to a shoulder higher than the main pavement, from a ridge of grass or joint-filling material, or from depressions in the pavement edge, all of which trap water and allow it to stand along and seep through the joint. Other causes are shoulder settlement, mix shrinkage, and trucks straddling the joint.

(2) *Repair*—If water is the cause, the first step is to improve the drainage by getting rid of the condition that traps water. Then repair the crack, see Reflection Cracks.

Figure 3-21 / **Applying tack coat.**

Figure 3-22 / **Spreading hot plant-mixed material on settled edge.**

Figure 3-23 / **Compacting with roller.**

Figure 3-24 / **Edge joint crack.**

Lane joint cracks. Lane joint cracks are longitudinal separations along the seam between two paving lanes, Figure 3-25.

(1) *Cause*—This type of crack usually is caused by a weak seam between adjoining spreads in the courses of the pavement.

(2) *Repair*—See Reflection Cracks.

Figure 3-25 / **Lane joint crack.**

Reflection cracks. These are cracks in asphalt overlays which reflect the crack pattern in the pavement structure underneath, Figure 3-26. The pattern may be longitudinal, transverse, diagonal, or block. They occur most frequently in asphalt overlays on portland cement concrete and on cement-treated bases. They may also occur in asphalt overlays on asphalt pavements whenever cracks in the old pavement have not been repaired properly.

(1) *Cause*—Reflection cracks are caused by vertical or horizontal movements in the pavement beneath the overlay, brought on by expansion and contraction with temperature or moisture changes. They may be caused also by traffic or earth movements and by loss of moisture in subgrades with high clay contents.

Figure 3-26 / **Reflection crack.**

(2) *Repair*—Small cracks (less than ⅛ inch in width) are too small to seal effectively. Large cracks (⅛ inch and over in width) are to be filled with asphalt emulsion slurry or a light grade of liquid asphalt mixed with fine sand. Also, special asphalt compounds or heavier-bodied asphalt material may be used to fill large cracks.

(a) Clean out crack with stiff-bristled broom and compressed air, Figure 3-27.

(b) Large crack. Using a hand squeegee and a broom, fill (do not over-fill) with emulsion slurry or liquid asphalt (SS-1; SS-1h, or SM-K) mixed with sand. When cured, seal with liquid asphalt using a pouring pot and a hand squeegee, Figure 3-28.

Figure 3-27 / Cleaning out crack with broom and air.

Figure 3-28 / Sealing with pouring pot and hand squeegee.

(c) Sprinkle surface of crack filler with dry sand to prevent pickup by traffic, Figure 3-29.

Shrinkage cracks. Shrinkage cracks are interconnected cracks forming a series of large blocks, usually with sharp corners or angles, Figure 3-30.

Figure 3-29　/　**Sprinkling surface with dry sand.**

Figure 3-30　/　**Shrinkage cracks.**

(1) *Cause*—Often it is difficult to determine whether shrinkage cracks are caused by volume change in the asphalt mix or in the base or subgrade. Frequently, they are caused by volume change of fine aggregate asphalt mixes that have a high content of low-penetration asphalt. Lack of traffic hastens shrinkage cracking in these pavements.

(2) *Repair*—Fill cracks with asphalt emulsion slurry followed by a surface treatment or a slurry seal over the entire surface. (Refer to The Asphalt Institute's Specification ST-1 or Specification ST-3, *Asphalt Surface Treatments and Asphalt Penetration Macadam*, Manual Series No. 13.)

  (a) Remove all loose matter from the cracks and pavement surface with brooms and compressed air, Figure 3-31.

Figure 3-31 / **Cleaning shrinkage cracks with compressed air.**

(b) Wet the surface of the pavement and all crack faces with water.

(c) When all surfaces are uniformly damp, with no free water, apply a tack coat of asphalt emulsion diluted with equal parts of water, Figure 3-32.

(d) Prepare the asphalt emulsion slurry mixture.

(e) Pour slurry mixture into cracks and level with a hand squeegee, Figure 3-33. (If cracks are numerous, slurry-seal the whole surface.)

(f) When the slurry is cured until firm, surface-treat or slurry-seal the whole surface with equipment designed for the operation, Figure 3-34.

(g) Allow to cure until firm enough to prevent pickup by traffic.

Figure 3-32 / **Applying tack coat.**

Figure 3-33 / **Filling shrinkage cracks with slurry seal.**

Figure 3-34 / **Slurry-sealing the surface.**

Slippage cracks. These are sometimes crescent-shaped cracks that point in the direction of the thrust of wheels on the pavement surface, Figure 3-35. This does not mean that they invariably point in the direction that traffic is going. For example, if brakes are applied on a vehicle going down a hill the thrust of the wheels is reversed. Slippage occurring in this circumstance will result in cracks pointing uphill.

Figure 3-35 / **Slippage cracks.**

(1) *Cause*—Slippage cracks are caused by the lack of a good bond between the surface layer and the course beneath. The lack of bond may be due to dust, oil, rubber, dirt, water, or other nonadhesive material between the two courses. Usually, such a lack of bond exists when no tack coat has been used. Slippage cracks may also be due to a mixture having a high sand content, and they can occur whether the sand is sharp or rounded. Sometimes slippage may develop under traffic because improper compaction during construction caused the bond layers to be broken.

(2) *Repair*—The only proper way to repair a slippage crack is to remove the surface layer from around the crack to the point where good bond between the layers is found. Then patch the area with plant-mixed asphalt material.

    (a) Remove the slipping area at least 1 foot into the surrounding well-bonded pavement. Make the cut faces straight and vertical. A power pavement saw makes a fast and neat cut, Figure 3-36.

    (b) Clean the surface of the exposed underlying layer with brooms and compressed air, Figure 3-37.

    (c) Apply a light tack coat, Figure 3-38.

Figure 3-36 / **Cutting with power saw.**

Figure 3-37 / **Cleaning surface of exposed layer.**

Figure 3-38 / **Applying tack coat.**

(d) Place enough hot plant-mixed asphalt material in the cutout area to make the surface the same grade as the surrounding pavement when it is compacted, Figure 3-39.

Figure 3-39 / **Placing plant-mix in cut.**

(e) Level the mixture carefully to prevent segregation, Figure 3-40.

(f) Check the riding quality of the patch with a straightedge or a string line, Figure 3-41.

(g) Compact thoroughly with a vibrating plate compactor or a steel-wheeled roller, Figure 3-42.

Widening cracks. Widening cracks are longitudinal reflection cracks that show up in the asphalt overlay above the joint between the old and new sections of a pavement widening, Figure 3-43.

(1) *Cause*—See Reflection Cracks.

(2) *Repair*—See Reflection Cracks.

Figure 3-40 / **Leveling patch mixture.**

Figure 3-41 / **Checking with straightedge.**

Figure 3-42 / **Compacting with roller.**

Figure 3-43 / Widening crack.

4 / CARE OF CARPETS AND RUGS

(*Sources:* American Carpet Institute, New York; General Services Administration, Washington, D. C.)

Regular care is the secret of good carpet and rug maintenance. Clean carpets and rugs daily with a vacuum or carpet sweeper, to keep surface soil from sinking into the pile, where it is difficult to remove and accelerates wear. Light daily cleaning consists of three strokes over a given area. A thorough cleaning consists of seven strokes back and forth over a given area.

Carpet sweepers take up lint, crumbs and other foreign matter from the surface. They do not get at soil that has found its way down between the tufts. Brooms sweep dust from the carpet, tossing it into the air and causing it to fall back on the carpet or other furnishings. A light brooming is helpful in brushing up a matted area. Never use a metal broom or a very stiff brush which might damage looped pile or break carpet tufts. The practice of running a vacuum cleaner over the back of a rug is recommended to be done at least once a year.

Cleaning Carpets and Rugs

There are three types of cleaning which the building superintendent can have performed on carpets or rugs by his own personnel:

1. Powdered cleaners.
2. Sponging with a cleaning fluid.
3. Cleaning with a solution of detergent or shampoo.

Powdered cleaners. Several absorbent powdered cleaners are available for carpet or rug cleaning. These cleaners consist of solvent saturated or detergent saturated sawdust or other inert powdered material. The compound is brushed into the pile in a circular motion, and should be allowed to remain in the rug for several hours until dry. The length of time depends upon the quantity applied and the atmospheric conditions. The carpet or rug should then be vacuumed thoroughly, picking up all of the powder. This method will give the carpet a certain degree of surface brightening. Compounds that contain fine powder may be more difficult to remove from the carpet than a coarser type.

Sponging with a cleaning fluid. Rugs and carpets made of wool or man-made fibers should be sponged with a cleaning fluid, but it is not recommended for cotton rugs. The cleaning fluid dissolves the oil film which makes dirt particles adhere to carpet fibers and thus effectively brightens their colors. This cleaning material will not injure the carpet yarns or the dyestuff, and it evaporates quickly and completely, leaving no residue. Apply the fluid with a sponge, using a gentle wiping motion so that only the top portion of the carpet pile becomes wet. Excess fluid may leave rings or soak into the backing and soften the back sizing. Solvent type cleaners will not remove stains or soil due strictly to water soluble materials, such as sugars or starches. Several dry cleaning solvents are available which have a carbon tetrachloride base. While these solvents are friendly to your carpets, they may be flammable and toxic, or both. Therefore, follow the recommendations and precautions of the manufacturer, and take the following precautions:

1. Avoid inhaling fumes which are toxic.

2. Be sure to have windows open so that liquid evaporates quickly and lessens chance of inhaling fumes.

3. When working with a cleaner which is flammable, avoid having lighted cigarettes or operating electrical appliances near the cleaning operation.

Cleaning with a solution of detergent or shampoo. Wet cleaning is suitable for all types of carpets. Certain precautions must be taken, the most important of which is to avoid the use of soap, ammonia, washing soda or any of the common household cleaning agents intended for use on hard surfaces, such as woodwork, linoleum or tile. Soaps should be avoided, regardless of how pure and mild they are claimed to be, since all soaps are alkaline by nature. Alkaline materials cause dyestuffs to run and may damage carpet fibers. There are many neutral detergents used for dishwashing or home laundering which may be used, along with special shampoos. Two heaping tablespoons of detergent to a gallon of one of these materials in a pail of water will produce a cleaning solution which can be used to clean carpets with a minimum of hazard. Special shampoos generally cover 80 to 100 square feet per gallon of mixed solution, and are derived from sulfated alcohol or sulfonated hydrocarbons.

When cleaning rugs with any solution, guard against over-wetting. Equipment which can be used, is a rug shampoo machine, or a floor polisher with a 15 or 16-inch brush converted by means of a solution tank connected to the handle which has a connecting tube and center feed ring or perforated brush. For best results, use a brush which feeds the solution through its back. Pick up the lather with a vacuum promptly, before it disintegrates and takes the dirt back into the carpeting. If working alone, do not shampoo more than 20 square feet at a time. Use a dry automatic floor brush or a hand pile brush to raise the pile, and allow to dry eight to ten hours before vacuuming.

Professional Cleaning Services

For thorough cleaning that removes all of the soil imbedded in carpets or rugs, reliable professional cleaning should be employed. Professional cleaners' services generally include: (1) plant cleaning, (2) on the floor cleaning, (3) dry cleaning, (4) spot removal, (5) re-dyeing, and (6) repairs.

Plant cleaning. A rug or carpet can always be cleaned most thoroughly and most efficiently in a professional cleaning plant. The plant is equipped to use such processes as dry dusting, shampooing, thorough rinsing and complete drying. Special attention will be given to adhesive seaming or inserts, in order to avoid shrinking or buckling, and in most cases firmness can be restored during shampooing.

On the floor cleaning. When cleaning carpets on the floor, the professional cleaner usually proceeds as follows:

1. Removal of all furniture from the room or area.
2. Preliminary brushing to open the carpet pile and permit removal of imbedded soil by vacuuming.
3. Thorough vacuuming with commercial cleaner.
4. Application of a non-alkaline detergent solution by a rotary scrubbing machine, using a minimum amount of solution.
5. Brushing of the pile to restore it to its original condition.
6. Drying of the carpet by forced drying through the use of heating devices, electric fans, and blowers.

Dry cleaning. Dry cleaning is generally satisfactory, however there are not many dry cleaning plants equipped to handle room-size floor coverings. Dry cleaning is not recommended for rubber-backed floor coverings because the solvent will degrade the rubber. If the carpeting is dry cleaned, the shrinkage is practically nil.

The professional cleaner has the experience for removing such special stains as paints, varnishes, and indelible inks. He also has re-dyeing facilities and a repair department for replacing worn or torn bindings, replacing missing tufts; and repair of holes and tears.

Insect Control

Good housekeeping is the most effective method for protecting carpeting from insect damage. Daily vacuuming, air, and light are beneficial. Periodic cleaning in out of the way areas discourages pest damage. Cleaning the back of the carpet at least once a year is recommended.

Insects which might damage carpets include the larvae (nymphs) of the clothes moth and carpet beetle. They are found in warm, humid, dark areas—generally around baseboards, in cracks of old floors and under radiators. They feed on animal fibers such as wool, fur, hair, and woolen lint, but do not damage cotton and most manmade fibers. Carpets made of blends of wool and manmade fibers are not insect proof because of the wool content.

Insect sprays can offer effective temporary prevention of pest damage. They should be used sparingly due to the general content of an oil or water base. Oil may cling to the carpet and the water may cause dampness. The directions should be read carefully to protect against any effects of a toxin which may be present. When insect damage appears to be a serious problem, professional treatment should be secured. Cleaners have insect repellant which can be applied to carpets or rugs. If insect damage has already occurred, he may be able to make necessary repairs.

Factors to Be Considered in Cleaning

Shrinkage. Shrinkage is natural in wet cleaning, but can be controlled by the use of frames and other equipment designed to retain the original size of the floor covering during or following the drying period.

Shading. Shading refers to areas in carpeting that seem to vary in tone after they are on the floor for a period of time. Walking over carpet pile sometimes pushes the tufts in different directions, and the light reflects from the tufts with various degrees of intensity. There is less shading if a vacuum cleaner is driven in the direction of the pile. Light brooming of cotton rugs raises the pile and therefore corrects shading. Shampooing allows the fibers to relax while wet, and when they are all brushed in one direction and allowed to dry, the shadowing will generally disappear.

Sunlight. Give a carpet normal protection from direct sunlight which is given to any colored fabric.

Pile crushing and corner curling. When carpet pile is crushed because of furniture pressure or walking, it can be raised by means of a hot iron over a damp cloth, or with a steam iron. (Avoid high heat for nylon, saran, and acetate carpet.) Brush up the spot briskly after treatment.

Fluffing. Fluffing is sometimes referred to as shedding. It consists of loose bits of material left in the carpet during manufacture. As the vacuum cleaner or carpet sweeper is used, fluffing disappears. It does not affect the quality or life of the floor covering.

Sprouting. Occasionally a tuft will rise above the pile surface of a carpet.

Snip these high rising tufts off evenly with the pile surface. Do not pull them out.

Stair carpet suggestions. The edges of a stair step take the heaviest amount of wear. To protect the carpeting, lay with an extra foot of length folded under against one or two risers at the top of the stairs. A shift downward can be made when required, to give the edges a rest. As these shifts are made, the excess carpet at the foot of the stairs will fold under against the lowest riser. Another recommendation is an underlay of carpet cushioning with particular attention to the stair step edges.

Humidity factor. During increased periods of humidity, carpets will increase in their size. Excessive humidity or damp weather may cause temporary buckling in the carpet. This will probably disappear with a dryer atmosphere, and if buckling persists, consult the carpet dealer or manufacturer. A carpet layer can sometimes solve the problem by restretching an installation. In dry weather carpet attached to the floors should be layed under tension. If chair pads are used on ground floors of concrete over dirt, they should be taken up periodically to prevent moisture buildup under them, which can cause mildew and rotting. This condition is evident with the use of plastic chair pads which create a vapor barrier.

Static electricity. The friction caused by rubbing almost any two different materials together will cause static electricity. The friction caused by walking across a carpet or rug sometimes generates a charge which is felt as a shock when a person touches a metallic object. Such electricity is not dangerous. The amount will depend upon the amount of friction and atmospheric conditions. Static can be reduced by adding moisture to the atmosphere through humidification; or application of antistatic agents. There are some chemical agents which immediately reduce any static charge generated. Some antistatic agents cause dirt to adhere to the carpet treated with them. This type of treatment should be considered after consultation with a professional cleaner.

Carpet labels. Remove carpet labels as they may be useful if replacement tufts or repair of tape bindings are necessary. Having the pattern number and grade name of the floor covering will be valuable in corresponding with the manufacturer.

Spot Removal

When spots or stains occur:

1. Act quickly to remove spots or stains before they dry or set themselves.
2. Have necessary cleaning equipment at hand.
3. Try to identify what caused the spot or stain, and remove it, using the correct material and procedure.

Often a spilled liquid will dry on a carpet and will not appear to leave a stain. Later, after cleaning the carpet, a brown stain may appear that cannot be removed. Most fruit juices and beverages contain a sugar called "reducing sugar" that is colorless when deposited on a fabric. Later after cleaning or exposure to direct sunlight, the stain will appear as a tan or brown discoloration.

Two cleaning materials that can be safely used for spot cleaning are: (1) One teaspoonful of a neutral detergent to a quart of warm water. To this add one tea-spoonful of white vinegar which is a weak acid and will serve to neutralize any alkaline materials. (2) A dry cleaning fluid (solvent) is useful for some spots. (See precautions described under "Sponging With a Cleaning Fluid".)

GENERAL CLEANING PROCEDURE

Step 1. Remove excess materials. In the case of liquids, absorb with a clean white cloth or other absorbent material. If semisolid, scrape with a knife, spoon, or sponge.

Step 2. Apply solution of detergent, vinegar, and water. Use a clean cloth and wipe gently from the edge of the soiled area toward the center. At intervals, blot with a dry clean cloth to absorb excess solution.

Step 3. Dry the carpet.

Step 4. Apply a dry cleaning fluid again, wiping gently and working from the edges of a spot or stain toward the center.

Step 5. Dry the carpet and brush the pile gently to restore the original texture.

In using the general cleaning procedure, use judgment as to whether both types of cleaning solution are required. If one removes most of the stain, it would be wise to repeat application.

Be sure to dry a wet carpet quickly and completely. This will avoid mildew, or other damage. The cleaned area should be raised from the floor and dried with a fan or other forced circulation. A solvent cleaner will dry quickly; allow for ventilation until all of the fumes have disappeared.

REMOVAL OF SPOTS AND STAINS

Type of Stain	Procedure
Butter, grease, oil, hand cream, ball point pen ink, ammonia	Remove excess materials; apply a dry cleaning fluid; dry the carpet; repeat application of solvent if necessary; dry the carpet and gently brush pile.
Coffee, tea, milk, gravy, chocolate, blood, egg, ice cream, sauces, salad dressings, vomit	Remove excess material, absorbing liquids and scraping semisolids; apply solution of detergent, vinegar, and water; dry the carpet; apply dry cleaning solvent; dry the carpet and brush the pile gently.
Candy, soft drinks, alcoholic beverages, fruit stains, washable ink, urine, excrement	Blot up liquids or scrape off semisolids; apply solution of detergent, vinegar, and water; dry the carpet; apply the solution again, if necessary; dry carpet and brush pile gently.
Gum, paint, tar, heavy grease lipstick, crayon	Remove excess materials; apply a dry cleaning fluid; apply solution of detergent, vinegar, water; apply dry cleaning fluid again; dry carpet and brush pile gently.

Nail polish. Nail polish remover is a satisfactory spot remover for nail polish spilled on a carpet. However, both nail polish and nail polish remover will seriously damage any carpet containing acetate fibers. Such cases generally require professional service. Some dyestuffs are not completely fast to nail polish remover, so test it first by applying it sparingly on some inconspicuous area of the carpet, prior to attempting to remove the nail polish.

Cigarette burns. Cigarette damage cannot be completely remedied except by reweaving. An area of carpet charred in a superficial manner can be improved by this procedure:

1. Carefully clip off blackened ends of tufts using small sharp scissors.
2. Apply solution of detergent, vinegar, and water as described in "General Cleaning Procedure".

Acid substances. Strong acids are not in common use. However, carbolic acid disinfectants and some tile cleaning compounds do fit into this category. If any of these should be spilled on a carpet, the area should be flushed with water, then sponged up. This should be repeated several times until the acid has been diluted and washed away. Then sponge the area with a dilute alkaline solution prepared by adding one tablespoonful of baking soda to a quart of warm water, and rinse again.

Rust. If the stain is fresh, use the detergent, vinegar, and water solution. Generally rust stains require professional service, because the cleaning agents required are hazardous when not properly used.

Permanent ink. Permanent inks are strong fast dyes and cannot be removed by any spotting procedures. Prompt action in blotting up ink stains and washing with water may prove successful. When ink is spilled on a carpet, try to determine if the ink is a washable type. If so, see the procedure described previously.

Job Description

The job of rug cleaning is performed by using a central vacuum cleaning system, a portable type vacuum cleaner, or a domestic type vacuum cleaner.

Usually vacuum cleaning rugs is a part of the room cleaning assignment, except where the presence of a large number of rugs makes specific job assignments more appropriate.

Rugs normally are vacuum cleaned after desks and other articles in the room have been dusted, so that any materials such as paper clips, shavings, and tiny other items that may have fallen on the rug during the dusting operation will be picked up.

Equipment and Material

A. *Equipment:*

   1. *Central Vacuum Cleaning System:*

      a. Elbow joint and valve
      b. Floor handle

c. Floor tool
d. Hose (two or more 25 ft. lengths)
e. Rug tool

2. *Domestic Vacuum Cleaner:*

   Note: Operator should be instructed how to adjust vacuum for rugs with short or long nap in accordance with manufacturer's recommendations.

3. *Portable Vacuum Cleaner:*

   a. Elbow joint
   b. Floor handle
   c. Floor tool
   d. Hose (one or more 25 ft. lengths)
   e. Rug tool
   f. Vacuum cleaner (portable)

4. *Miscellaneous:*

   a. Corn broom
   b. Wedge

B. *Material:*

Wiping cloths (waffleweave)

Job Methods

Step 1

Operation Steps	Explanation
1. Obtain necessary equipment and materials.	a. Obtain (as required) necessary equipment.
	b. Check to see if vacuum dust bag is clean and properly installed.
	c. Inspect vacuum hose for breaks before connecting to machine or outlet.
2. Proceed to assigned work area.	All the equipment should be taken to the starting point of the assigned work area, to be used as needed.

Step 2

1. Prepare vacuum cleaner.	a. Attach hose to valve or elbow joint.
	b. Place tool on handle. Use proper tool for job (carpet or bare floor).
	c. When using either a central vacuum

Operation Steps	*Explanation*
	system or portable vacuum, use a wedge or doorstop to hold the door open, and pull hose into the room.
	d. Inspect plug before inserting the wall receptacle.
	e. Use *only one* 25 ft. hose for ¾ hp machine, and *never* more than two 25 ft. lengths for 1½ hp machine.
2. Prepare for vacuum cleaning.	a. Prepare room by moving as little furniture as necessary.
	b. Start machine by turning on switch and/or valve (in elbow joint).

NOTE

When using central vacuum system the shutoff valve should be closed except when equipment is actually in use.

Step 3

1. Vacuum clean bare floors.	a. Grasp handle with one hand and place tool parallel to floor.
	b. Use rug tool for bare floors around rugs.
	c. For large areas of bare floor, use bare floor tool.
2. Vacuum clean rugs.	a. Vacuum clean rugs by working tool back and forth with the nap of the rug.
	b. The last time over the rug operate the carpet tool in the direction that will lay the nap of the rug most evenly.
	c. Turn off switch and remove plug from wall or hose from outlet. Rewind cable and hose loosely.
	d. Remove machine or hose from room.

PRECAUTIONS

1. Avoid marring furniture or baseboards when operating with floor or rug tool.
2. Avoid marring furniture or baseboards with vacuum hose.
3. Avoid letting doors strike machine when moving machine into

room. Use doorstop or wedge to hold door open.

4. Place vacuum machine switch in OFF position before inserting or removing the plug from wall receptacle.

Operation Steps	*Explanation*
	Step 4
Replace furniture or equipment.	Replace, in their proper locations, any furniture or equipment which becomes disarranged during the vacuum cleaning operation.
	Step 5
Smooth nap on rugs.	Where required use corn broom in short even strokes to smooth nap of rug after vacuum cleaning is completed and furniture replaced in its original position.
	Step 6
Dispose of sweepings.	After each night's use, empty dirt receptacle of vacuum cleaner into paper parcel and place in trash.
	Step 7
Return equipment and materials to storage.	a. Disassemble equipment.
	b. Wrap vacuum hose loosely around machine.
	c. Store cleaning tools (handle with attached carpet tool and elbow joint) with the machine in gear or storage room.

Care of Equipment

A. *Dust Bags:*

In addition to removing sweepings from bags daily, clean and dust bags once a week. Where two or more machines are available use one machine to clean the bag of the other. Where another machine is not available clean the bag with a stiff brush.

B. *Electric Cable:*

1. Make sure cable connection is plugged all the way into electric outlet to avoid voltage drops.

2. Do not splice or replace electric cable. (This is a shop job.)

3. Do not pull on electric cord to remove plug from outlet.

C. *Hose:*

   1. Always operate machine with a cleaning tool on the end of a vacuum hose.

   2. Avoid sharp bends in hose when cleaning. Do not permit platform trucks or other vehicles to run over hose. Report any defects to your supervisor.

   3. Always remove vacuum hose from machine outlet at end of work day.

   4. Coil hose loosely on machine, or in case of central system, hang on hangers provided in gear or storage room.

D. *Machine:*

   1. Upon any signs of smoke or unusual odor stop machine immediately and notify your supervisor.

   2. To avoid damaging wheels, do not bump machine up and down stairs when carrying from one floor to another. Supervisors should make sufficient help available to move machines from floor to floor in buildings where there are no ramps or elevators.

   3. Wipe machine with a damp cloth when day's work is completed.

E. *Wiping Cloths:*

   Wash cloths as required and hang them to dry in proper place.

PRECAUTIONS

   1. *Never* oil or grease machines. (This should be done only by qualified personnel.)

   2. Never use "dry" vacuum unit to pick up liquids.

   3. *Do not tamper* with shutoff valve or other mechanism.

   4. Report needed repairs to your supervisor.

Performance Inspection Guide

A. *Rug Cleaning:*

   Rugs should be thoroughly clean and free from dust, dirt and other debris. Nap on rugs should be laid in one direction.

B. *Bare Floor Cleaning:*

   Bare floors around rugs should have been cleaned with rug tool. Large bare floor areas should have been cleaned with bare floor tool. No dirt should be left in corners, under furniture, or behind radiators or doors.

C. *Furniture and Equipment Replaced:*

   1. All furniture and equipment moved during the cleaning operation should be returned to its original position.

   2. All equipment used during the vacuum cleaning operation should be cleaned and properly placed in its assigned location.

   3. Portable and tank type machines should be wiped clean, dust bags emptied, and vacuum hose coiled loosely on machine or hung on racks provided for that purpose.

Equipment and Material Allocation and Replacement

Items	Allotment per Cleaner	Replacement or Reissue
Broom, corn	1	1 per year
Cloth (waffleweave)	1	4 per year
Vacuum cleaner with attachments (domestic) (or)	1	1 per 10 years*
Vacuum cleaner with attachments (portable)	1	1 per 10 years*
Wedge	To be obtained locally	As required

* Does not apply to replacement of vacuum cleaner attachments.

The allotment and replacement or reissue quantities are based on one rug cleaner working the entire eight hours of each working day on the job.

• Where other duties are assigned to such an employee to secure 100 per cent utilization of his time, consideration should be given to reducing the allotment and replacement quantities in relative proportion to the time spent per hour on other work.

5 / CORRIDOR AND LOBBY POLICING

Source: General Services Administration, Washington, D. C.

Job Description

The job of corridor and lobby policing includes collecting and removing from the assigned corridor and lobby area all loose paper, trash, rubbish, empty bottles and other discarded material; removing wads of gum, spots of tar and other sticky substances from the floors of such area; keeping the jardinieres and/or sand urns in a neat and presentable condition; tidying up as necessary any drinking fountains and glass surfaces that may be located within such area; mopping up any wet areas that may occur during bad weather, or caused by liquids being spilled; and proper care and maintenance of the equipment and material issued to the employee for use in doing the job. Employees assigned to this job are also instructed to notice when lights are out and other defective building items, and to report same to their supervisor.

Equipment and Material

A. *Equipment:*

  1. *For normal policing:*
     a. Dustpan
     b. Dustpan broom

    c. Putty knife
    d. Wiping cloths
    e. Perforated scoop

  2. *For mopping wet areas:*
    a. Bucket with mop wringer
    b. Cotton mop

B. *Material:*

  For equipment care:
  Detergent

Job Methods

Operation Steps	*Explanation*

A. *Normal Policing:*

Step 1

1. Obtain equipment needed.	Equipment needed for normal policing listed under "Equipment and Material".
2. Proceed to work area.	Proceed with the equipment to the assigned area where work is to start.

Step 2

1. Police corridor and lobby floors.	Start at the beginning of the work area and use the broom and dust pan to remove all loose paper, trash and other foreign material from the assigned corridor and lobby areas.
2. Remove gum, tar or other substances from the floor.	Use the putty knife to remove any wads of gum, spots of tar or other foreign substance from the corridor and lobby floors during the policing operation.
3. Police jardinieres and/or sand urns.	While policing the area, collect all cigarette butts, burnt matches, and other discarded material from around the jardiniere or sand urn units located within the corridors and lobbies. Use the perforated scoop to remove the rubbish from the urns, and to sift or stir fresh sand to the top of the unit.

Operation Steps	*Explanation*
4. Police drinking fountains and glass surfaces.	If drinking fountains and glass surfaces such as picture, door, display and directory board glass are located within the assigned area, check these items as they are reached during the normal policing operation and use a damp wiping cloth to remove any unsightly dirt smudges and in general tidy these articles as may be required.

B. *Mopping Wet Areas:*

Step 3

1. Obtain equipment needed.	Secure the mop, mop wringer and bucket from the equipment storage location.
2. Mop up wet spots.	Use the mop to dry thoroughly all wet spots in the entrances, lobbies, and corridors of the assigned area. Squeeze the excess water from the mop frequently.

NOTE

Step 3 is to be performed during periods of rain or snow, and at such other times as may be needed to keep the floors of the assigned area in a dry and presentable condition.

C. *Returning Equipment to Storage Locations:*

Step 4

1. Return all equipment to storage location.	Gather all pieces of equipment and proceed to the location where the items are stored when not in use.
2. Broom, dustpan, putty knife and scoop.	Clean all articles and return them to their proper storage places.
3. Mop and wiping cloths.	Rinse the mop and wiping cloths in clean water, wring or squeeze to remove excess water and put in proper place for drying and storage.
4. Bucket and mop wringer.	Empty the bucket, rinse, wipe dry

Explanation

and return the bucket and wringer to
their proper storage places.

Care of Equipment

A. *Dustpan:*

1. Clean the dustpan of loose dirt and dust before storing.

2. Hang the dustpan by the handle so as not to allow any damaging weight
to be put on it during storage.

3. In dumping the dustpan, avoid banging it with force against the waste
container as such action results in damage to the sides and edge of the pan.

B. *Dustpan Broom:*

1. Comb the broom bristles periodically.

2. Wash a dirty or oily broom in a warm water and detergent solution.
(Use one ounce of detergent per gallon of water.) Rinse in clean water and hang
by the handle to dry. Allow the broom to dry thoroughly before using.

3. Hang the broom in a dry, clean place when stored, so that the weight of
the broom is not upon the bristles.

C. *Putty Knife:*

1. Use the putty knife only for removal of gum and other foreign substances
during policing. *Do not* use it as a screw driver or hammer.

2. Wipe the knife clean and dry and store it in a dry, clean place.

D. *Perforated Scoop:*

1. Wipe the scoop clean and store it in a dry, clean place, either on a shelf
or hanging on a nail.

2. Store it so that no damaging weight is placed on the scoop.

E. *Wiping Cloths:*

1. Rinse all cloths used; or when dirty, wash them in a warm water and
detergent solution (one ounce of detergent per gallon of water), rinse in clean
water and wring or squeeze to remove excess water.

2. Spread or hang cloths to dry in their proper storage place.

3. Avoid leaving wet, dirty cloths piled together in the storage location.

F. *Mop:*

1. After each use and before storage, rinse the mop in clean water and
wring or squeeze to remove excess water.

2. Wash dirty mops in a warm water and detergent solution (one ounce of
detergent per gallon of water); rinse in clean water and wring or squeeze to
remove excess water.

3. Fluff the mop strands by shaking the mophead.

4. Store the mop with the mophead up to allow better drying.

G. *Bucket and Mop Wringer:*

1. Rinse and dry the bucket and wringer thoroughly.

2. Wipe the outside of the bucket free of dirt.

3. Avoid striking the bucket and wringer against objects as such action tends to cause leaks in the bucket and shorten the life of the equipment.

4. Store in a dry, clean place so as to have no damaging weight bearing on the bucket or wringer.

Performance Inspection Guide

A. *Normal Policing:*

1. *Corridor and Lobby Floors*

   a. Corridor and lobby floors of the assigned area should be free of all loose paper, bottles, trash and other foreign materials.

   b. Wads of gum, spots of tar and other foreign substances that tend to stick to the floors should have been removed.

2. *Jardinieres and/or Sand Urns*

   a. Cigarette butts, burnt matches and other discarded material should have been removed from around jardiniere and/or sand urn locations.

   b. Jardinieres should be free of refuse and be presentable in appearance.

   c. Sand urns should be free of refuse and the top sand should present a clean and neat appearance.

3. *Drinking Fountains and Glass Surfaces*

   a. Drinking fountain surfaces should be free of discarded rubbish, ink stains, finger marks and other dirt smudges and should present a neat appearance.

   b. Glass surfaces located in doors, pictures, displays, directory boards and other such items within the assigned area should be tidy and free of finger marks, smudges and other dirt marks.

B. *Mopping Wet Areas:*

1. Entrances, corridor and lobby floors should be free of wet spots and footprints.

2. Wall surfaces and baseboards should be free of water marks resulting from the mopping operation.

C. *Equipment and Material:*

1. All pieces of equipment should be clean and properly placed in their assigned storage locations.

2. Any unused detergent should be returned to its assigned storage location.

Frequency and Production

A. *Frequency and Production Schedule:*

	Main Corridor and Lobby	Secondary Corridor and Lobby
Frequency	4 Times Daily	Daily
Production Rate per Man-day	300,000 sq. ft. of Corridor and Lobby Area	

B. *Qualifying Factors:*

1. *Frequency*

   a. The frequency of corridor and lobby policing in a building is dependent upon the amount of traffic, weather conditions and type of occupancy. Allowances should be made in the frequency of policing at the various locations so as to accomplish the most efficient and economical performance of this job.

   b. The frequencies given in the above schedule are those for a normal standard of corridor and lobby policing under average conditions.

2. *Production Rate per Man-day*

   a. The production rate given in the above schedule is based on the actual corridor and lobby area rather than the gross area.

   b. This rate as established may be subject to modification due to the varying types, sizes and locations of government operated buildings. It does, however, provide a standard from which such deviation (if any) should be figured.

Staffing

The man-day requirements of labor for corridor and lobby policing can be determined by use of the following formula:

$$\text{Man-days (of labor)} = \frac{\text{Quantity}}{\text{Production}} \times \frac{1}{\text{Frequency}}$$

where quantity is the actual area in square feet of corridors and lobbies to be policed; production is the square feet of corridors and lobbies policed per man-day; and frequency is the number of working days between operations.

If the frequency and/or production rate values as established in the Frequency and Production Schedule have been modified to meet the existing conditions at a particular location, then those modified values should be used in the above formula for the computation of the labor requirement.

SAMPLE CALCULATION

Assume a building with 16,000 square feet of main corridor and lobby area, 140,000 square feet of secondary corridor and lobby area, and with such existing conditions that the established values from the Frequency and Production Schedule may be used; then compute the man-days of labor as follows:

$$\text{Man-days} = \underbrace{\frac{\text{Quantity}}{\text{Production}} \times \frac{1}{\text{Frequency}}}_{\begin{array}{c}\textit{Main Corridor}\\\textit{and Lobby Area}\end{array}} + \underbrace{\frac{\text{Quantity}}{\text{Production}} \times \frac{1}{\text{Frequency}}}_{\begin{array}{c}\textit{Secondary Corridor}\\\textit{and Lobby Area}\end{array}}$$

$$= \frac{16,000}{300,000} \times \frac{1}{\frac{1}{4}} + \frac{140,000}{300,000} \times \frac{1}{1}$$

$$= 0.213 \qquad\qquad + 0.467$$

$$= 0.68 \text{ man-days of labor required (jobs)}$$

To this total requirement must be added a factor to compensate for leave in order to arrive at the total number of positions needed to accomplish the work.

Conversion factor including compensation for leave = 1.15
Conversion factor × jobs = total positions required
1.15 × 0.68 = 0.78 positions required

Leave: Average time lost on job through sickness, death in family, and personal absenteeism.

• Where the normal corridor and lobby policing duty does not require the full time of the employee, other duties such as utility work, sidewalk cleaning, and similar work should be assigned to the employee to secure 100 per cent utilization of his time.

Equipment and Material Allocation and Replacement

Items	Allotment per Cleaner	Replacement or Reissue
Bucket (26 qt.)	1	1 per 2 years
Detergent	1 pound	6 pounds per year
Dustpan	1	1 per 2 years
Dustpan broom (hearth size corn broom)	1	1 per year
Perforated scoop	1	1 per 2 years
Mop (1¼ lb.)	1	2 per year
Mop handle	1	1 per year
Mop wringer (squeeze type, size 2)	1	1 per 2 years
Putty knife (1¼″)	1	1 per 2 years
Wiping cloths (waffleweave)	2	24 per year

The allotment per cleaner and replacement or reissue quantities in this section are based on one cleaner having continuous corridor and lobby policing duty.
• Where the situation exists that other related jobs are assigned to the cleaner having this duty to secure 100 per cent utilization of his time, consideration should be given to a reduction in the replacement quantities in accordance with the actual amount of corridor and lobby policing duty assigned.

(*Sources:* Public Buildings Service, General Services Administration, Washington, D. C.; *Buildings* magazine, November 1958, November 1960)

Elevator inspection and maintenance is a highly technical field that requires the service of a trained technician; usually this service is performed under an outside service contract. But by becoming familiar with some of the highlights of inspection and routine maintenance, you can be sure that your equipment is getting the careful attention it deserves.

Maintaining the elevators in a building is another executive responsibility of the building superintendent or plant manager. He must delegate this work to the most responsible employee. Inspections should not be limited to state, city or insurance companies. Inspection service by reputable elevator companies is available in many localities. Where it is not available, the building superintendent or plant manager oversees the inspections. Information as to what points should be inspected is available from the manufacturer, along with recommendations on how to perform preventive maintenance.

Most accidents happen at landings. Make a careful check of the condition of the floors and landing sills. The sills should be of nonslip material, and well illuminated. Particular attention should be given to main floor landings where the difference between outdoor and indoor lighting is severe. Hoistways should be checked frequently, cleaned and greased semimonthly. Vision panels (if present) should be cleaned daily. Counterweights should operate smoothly and sheaves should be properly aligned.

Inspection of the car should include a checkup for structural defects such as loose bolts and other fastenings, excessive play in guide shoes and worn flooring. Test each emergency by opening it. If panels are held in place by thumb screws, they should be removable without the use of pliers; if they are held in place by locks, the key should be in the elevator.

Test the car switch. It should return to neutral position when released, and lock there. If the car is cable operated, examine the cable lock. Check all switch contacts in the car. The glass cover over the emergency relief switch should be in place. To check the clearance of the governor flyballs, if used, lift them above the position at which the governor grips the rope and revolves them at this level. This not only proves them clear of obstruction, but also clears the spindle of congealed oil and dirt. Reliable service personnel should check all safety devices weekly.

All oil buffers lose some oil during normal usage. Oil levels should be checked at each inspection. Alignment and tightness of the bolts in the anchorage should be inspected. Spring buffers should be checked for alignment and for proper seating in cups and mountings. Springs should be examined for flaws and to see that they have not taken a permanent set.

Hoisting machines should be checked for misalignment and cracks. Oil level, motor brushes, commutator, couplings, bolts, set screws, and keys should be inspected. The gear housing should be checked for oil level, and at the same time look for worn teeth on gears. Oil should be removed from the brake drum.

102 / **PHYSICAL MAINTENANCE**

Check pins and bearing surfaces for lubrication and wear. The brake shoes should be relined before rivet heads come in contact with the drum, just as you would reline brakes on an automobile.

Checklist for Elevator Maintenance

Warning: Always open the main line electric power switch to the elevator before cleaning, lubricating or making any adjustments to the moving parts. Make no adjustments or repairs without considering first—the effect on the equipment. Regular care and lubrication with the proper materials are essential to preserve the life of the elevator. The intervals for lubrication specified in the following instructions are based on average operating conditions. For extraordinary conditions consult your municipal elevator inspector, insurance company inspector or the elevator manufacturer. *Always* renew all worn parts.

1. WORM GEARING   *Check monthly*

Use worm gear lubricant. Fill to top of standpipe gauge with machine at rest. With machine running, examine gear through hole in top of case to determine if carrying sufficient lubricant. (Do not tilt gauge to determine oil level.)

2. THRUST BEARINGS   *Check monthly*

Self-lubricated from gear case.

3. DRIP PAN   *Examine frequently*

The drip pan under worm shaft packing gland should be watched for excessive leakage. (Do not pour dripped oil back into gear case.)

4. SHEAVE SHAFT BEARINGS   *Check every two weeks*

Sleeve type (with grease cup—use cup grease).
Sleeve type (with oil reservoir—use bearing oil).

Check monthly

Roller or ball type (use ball and roller bearing grease).
Lubricate at least every six months. (Do not fill over one-half full.)

SHEAVE SHAFT THRUSTS

Self-lubricated from gear case.

5. A/C OR D/C MOTOR   *Check every two weeks*

Use bearing oil. *Bearings:* With machine at rest, pour in oil to gauge height, and see that oil rings are free to rotate with shaft. (Do not overflow reservoir.)

6. BRAKE   *Check monthly*

Magnet cores (D/C only): Use powdered graphite. Wipe clean (use powdered graphite on the core surface and in brass shell) at least every six months. *Pins and linkage:* Use bearing oil, lubricate sparingly.

Check every two months

A/C brake magnets: Use A/C brake magnet oil. Fill magnet case to within one-half inch of top of standpipe gauge. *Brake motor bearings (motor operated brakes)*: Consult manufacturer for proper cleaning and lubrication. *Caution:* Avoid getting oil on brake linings.

7. MOTOR GENERATOR AND EXCITER SETS — *Check every two weeks*

Sleeve type bearings: Use bearing oil. With machine at rest, pour in oil to gauge height, see that oil rings are free to rotate with shaft. (Do not overflow reservoir.)

Ball bearings: Use ball and roller bearing grease. Lubricate at least every six months. (Do not fill over one-half full.)

8. DEFLECTOR SHEAVE — *Check every two weeks*

Shaft lubrication: Use cup grease. Auxiliary sheaves (overhead, compensating, 2:1 etc.). Sleeve bearings (grease cup, use cup grease). Sleeve bearings (oil cup, use bearing oil). Sleeve bearings (chain oiler, use traction machine oil). Ball bearings, use ball and roller bearing grease. Lubricate at least every six months. (Do not fill over one-half full.) Box bearings, use cup grease.

9. ROPES — *Check every two months*

Use wire rope lubricant. Lubricate only when ropes appear dry or show signs of corroding. (Do not oversaturate ropes.)

10. CAR AND COUNTERWEIGHT SAFETY DEVICES — *Check monthly*

Pins, actuating screws, tail rope drum bearings, use bearing oil. Lubricate annually. *Caution:* Do not tamper with safety nor make any adjustments. At least once a year arrange with manufacturer or authorized outside service contractor for a safety operating test.

11. GOVERNOR — *Check monthly*

Gear and grease cup, use cup grease. (Do not use any grease on gear sector of rope jaws.) Oil ports, use bearing oil.

12. GUIDE RAILS — *Check monthly*

Use type "C" lubricator oils. *Important:* Consult manufacturer for proper grade of oil to use on guide rails. *Gibs:* No rail lubrication required.

13. ROLLER GUIDES — *Check every two months*

Roller guides, no rail lubrication required. *Wheel bearings:* Lubricate annually, use ball and roller bearing grease. Stands (for cast shoes and gibs), use cup grease. Be certain that spindles are free and properly lubricated to permit easy action of the guide shoes.

14. CONTROLLER AND RELAY PANEL — *Check every two weeks*

Switch hinge pins, use bearing oil, lubricate sparingly (a drop of oil is sufficient). Time and overload relays and dashpots, use dashpot oil. Refill if necessary to proper gauge level; check oil level every three months. (*Use no other oil.*)

15. SELECTOR AND FLOOR CONTROLLERS — *Check monthly*

Screws and gearings, use cup grease. Pressure type fittings, use ball and roller bearing grease. Oil cups and oil ports, use bearing oil. Square guides, clean and apply coating of Slip-It. *Driving tapes:* Only when dry, use soft brush and apply bearing oil.

16. BUFFERS   *Check every three months*

Use buffer oil. *Important:* Use no other oil (be certain that buffers are filled to proper gauge level at all times).

17. PIT TENSION FRAMES   *Check every two weeks*

Sheave bearing grease cup, use cup grease. Sheave bearing oil port or cup, use bearing oil. Sheave bearing pressure fitting, use ball and roller bearing grease. Weight guides, clean and apply coating of Slip-It.

18. DOOR AND GATE OPERATORS   *Check monthly*

Type "A" gear unit, use worm gear lubricant. Type "O", "S" and "K" gear units, use traction machine oil. *Important:* Use only oil specified and fill to gauge height. Motor bearings, use bearing oil. Link and lever pins, use bearing oil.

19. CYLINDERS   *Check every two months*

Checking cylinders, use door closer oil. Door sill trips, use cup grease.

Note: Clean all of the equipment regularly. Accumulated dust, dirt and oil impairs its efficiency and operation.

6A / ELEVATOR CLEANING

Job Description

Elevator cleaning includes all cleaning work inside the elevator cab, such as vacuum cleaning; mopping or scrubbing; floor waxing and polishing; removing gum or other foreign substances; dusting doors, fans, lights, walls and grilles; cleaning and polishing metal and wood surfaces; cleaning threshold plate; and proper care and maintenance of the cleaning equipment used. Employees assigned to this job should be sure all elevator lights, fans and switches are turned off when the job is completed. They should be instructed to notice defects in flooring, loose railings, and similar service defects, and report them to their supervisors.

Equipment and Materials

A. *Equipment:*

   For elevator cleaning.

   a. Equipment carrier
   b. Vacuum cleaner and attachments
   c. Putty knife
   d. Dust cloths (polyethylene glycol treated cheesecloth)

    e. Wiping cloths
    f. Buckets
    g. Squeeze type wringer
    h. Cotton mops
    i. Steel wool
    j. Wax applicator
    k. Wood wedge

B. *Materials:*

For cleaning elevator.

    a. Lemon oil polish
    b. Metal polish
    c. Detergent, soapless, synthetic
    d. Water emulsion, synthetic, or floor sealer (as directed)

Job Methods

Operation Steps   *Explanation*

Step 1

1. Obtain equipment and materials.

Obtain equipment and materials needed for elevator cleaning. (Listed under "Equipment and Materials".)

2. Proceed to assigned work area.

Push equipment carrier with equipment and materials to the first elevator scheduled to be cleaned. Take the elevator out of service before starting to clean.

Step 2

1. Clean elevator walls, doors, fans and lights.

Use vacuum cleaner with wall brush attachment to clean ceiling, walls, doors, fans and lights. Use treated dust cloth to supplement vacuuming as required.

2. Clean grillwork.

Use vacuum cleaner with radiator brush attachment to thoroughly clean grillwork.

Step 3

1. Vacuum clean elevator floor.

Use vacuum cleaner with bare floor cleaning attachment to thoroughly sweep floor area and threshold plate.

Operation Steps	*Explanation*
2. Remove gum and foreign substances from floor.	Use putty knife to remove any gum or other foreign substances from floor.

PRECAUTIONS

1. Make sure all loose dirt is removed from corners of car during vacuum cleaning operation.

2. Do not pull against electric cord of vacuum cleaner when it is plugged into receptacle. Do not allow machine to run over cord.

Step 4

1. Wall spotting.	Use lemon oil polish, applied with a cloth to remove hand marks and stains from wood surfaces. Rub cleaned area briskly to match luster of the rest of wall surface.
2. Clean and polish metal surfaces.	Use clean cloths and metal polish, as directed, to clean and polish metal handrails, metal doors and other metal surfaces.
3. Clean and polish wood surfaces.	Use clean cloths and lemon oil polish as directed to thoroughly clean and polish wood surfaces.

PRECAUTIONS

Metal polish is not to be used on painted, lacquered, or other specially treated metal surfaces. Do not apply metal polish so heavily that it will clog switch key openings. For such materials as aluminum, stainless steel or white metal trim, special polishes are available and should be used.

Step 5

Mop elevator floor.	a. Mix one ounce of detergent per gallon of warm water in mop bucket.
	b. Use mop and detergent solution to thoroughly clean floor area. Dip mop into solution and wring out excess so mop does not drip. Push mop around edges of floor first, using care to remove all dirt from corners; then mop the rest of floor

Operation Steps   *Explanation*

area. Avoid splashing walls. (If floors are extremely dirty, use scrubbing and polishing machine to scrub floor.)

c. Wring mop and go back over mopped area to pick up excess cleaning solution.

d. Use a second mop and clean water to rinse and mop floor. Frequently rinse and wring mop during this operation.

PRECAUTIONS

1. Do not use so much water that it will tend to splash or to run beneath floor covering.
2. Change the rinse water frequently.
3. Be careful not to splash the doors and walls or scar them with the equipment.

Step 6

Wax elevator floors.

Pour small quantity of wax into pan and use wax applicator to spread wax evenly over floor.

PRECAUTIONS

1. Use only water emulsion wax on rubber and asphalt tile floor coverings.
2. Be sure the floor is thoroughly dry before applying the wax.

Step 7

Polish elevator floor.

Use scrubbing and polishing machine to polish elevator floor. Start at rear of elevator and polish toward the door. Use cloth to polish corners by hand.

PRECAUTION

Do not polish the floor until after wax is thoroughly dry.

Step 8

Clean threshold plate.

Clean grooves of threshold with vacuum floor brush or cloth covered wood wedge. Use steel wool to clean

Operation Steps   *Explanation*

nonskid surface. Clean with metal polish or other approved cleaner as directed.

NOTE

Return each elevator to service when cleaning is completed.

Step 9

1. Return equipment and unused materials to storage location.

Upon completion of assignment, place all equipment and materials on equipment carrier and proceed to proper storage location.

2. Clean and store equipment.

Clean all equipment and store on equipment carrier and proceed to proper storage location, for following cleaning.

3. Unused material.

Return all unused material to proper storage place.

PRECAUTION

Make sure that lids to polish, detergent, wax, and similar containers are on securely.

Care of Equipment

A. *Bucket and Mop Wringer:*
   1. Thoroughly rinse and dry bucket and wringer after use.
   2. Remove mop strands and other foreign matter remaining in wringer.
   3. Wipe dirt from outside bucket.
   4. Avoid striking bucket and wringer against objects, for this tends to cause leaks in the bucket and to shorten the life of the equipment.
   5. Store in dry place in such a way that there is no weight on bucket or wringer.

B. *Dustcloths (polyethylene glycol treated):*
   1. Thoroughly shake or vacuum dustcloths as required to remove loose dust.
   2. Wash dirty cloths in warm detergent solution (mix one ounce of detergent per gallon of warm water). Rinse in clean water and wring or squeeze to remove excess water.
   3. To treat new or washed cheesecloth spray or sprinkle approximately one fluid ounce of treatment (such as polyethylene glycol solution) to three parts water on each square yard of cheesecloth. Cloth should be treated the day before it is to be used.

C. *Equipment Carrier:*

1. Do not allow carrier to strike walls, furniture, or other equipment.

2. Wipe carrier clean of dirt and dust. Keep clean and orderly at all times.

3. Promptly report to the supervisor, any defects or equipment failure.

D. *Mops:*

1. After use and before storing, rinse mops in clean water and wring or squeeze to remove excess water.

2. Wash dirty mops in warm water. Rinse in clean water and wring or squeeze to remove excess water.

3. Fluff mop strands by shaking.

4. Store mops, with mop heads up, to dry.

E. *Polishing Machine:*

1. After each use, wipe machine and cord free of dirt and dust with a damp cloth, using detergent solution and water to remove spots. Wipe machine and cord dry, and rewind cord loosely on hooks provided.

2. Remove brush from the machine. Wash dirty brushes in detergent solution, rinse thoroughly, and shake free of excess water.

3. Wash and rinse wheels of the machine, and wipe with a dry cloth.

4. Use metal polish as directed to clean and polish the floor machine. Store machine and brush, with brush removed, in assigned storage space.

5. Report needed repairs to supervisor.

F. *Putty Knife:*

1. Use putty knife only to remove gum or other foreign substances during sweeping.

2. Store knife in a clean, dry place.

G. *Vacuum Cleaner and Attachments:*

1. Empty and wipe dirt receptacle clean after each use.

2. After use wipe machine and cord free of dirt and dust with a damp cloth, wipe dry, and wind cord loosely on hooks or other means provided.

3. Do not allow machine or attachments to strike walls or other objects.

4. After use, clean attachments and return them with machine to a clean dry storage area where they will not be damaged.

5. Report needed servicing or repairs to supervisor.

H. *Wax Applicator:*

1. Remove and thoroughly wash applicator pad in detergent. (Mix one ounce of detergent to one gallon of warm water.)

2. Rinse applicator pad in clean water.

3. Shake excess water from applicator pad and replace on block to prevent shrinkage.

4. Store applicator where there is free circulation of air and where there will be no weight against it.

I. *Wiping Cloths:*

1. Rinse all cloths used. When cloths are dirty, wash in warm detergent solution (mix one ounce of detergent to one gallon of warm water), rinse and wring out.

2. Spread or hang cloths to dry in their proper storage places.

J. *Wood Wedge or Scraper:*

1. Clean and dry wedge and return it to proper storage location.

Performance Inspection Guide

A. *Normal Policing:*

1. *Vacuum Cleaning and Dusting:*

Interior surfaces of elevator should be free of loose dirt and dust streaks.

2. *Cleaning, Polishing and Wall Spotting:*

a. Handrails, controls and other surfaces should be clean and polished.
b. Walls should be free of finger marks and other smudges.

3. *Mopping, Waxing and Polishing:*

Elevator walls should be free of splash marks. Floor, including corners and threshold plate, should be clean, and floor should be waxed and polished.

B. *Equipment and Materials:*

1. All pieces of equipment should be clean, and properly placed in their assigned storage locations.
2. All polish, wax, soap and other similar leftover materials should be returned to their assigned storage location.

Frequency and Production

A. *Frequency and Production Schedule:*

	Passenger Elevators	Freight Elevators
Frequency	Daily*	Weekly*
Production Rate per Man-day	25	25

* Or as directed by supervisor.

B. *Qualifying Factors:*

1. *Frequency*

a. The frequency of elevator cleaning is dependent upon the type of occupancy in the building, the amount of traffic, size of elevator car, and to some extent, the design of the car.
b. The frequencies given in the following schedules are for a normal standard of cleaning under average conditions.

2. *Production Rate per Man-day*

The production rate used in the schedule is the rate established for cleaning elevators of average size located in one building or in buildings reasonably close to each other. Due to the number of buildings which have only one or two elevators, and because of the varied types of elevators, the production rate as established may be subject to modification at various locations. It does, however, provide a standard from which deviation should be calculated.

Staffing

The man-day requirements of labor for elevator cleaning can be determined by use of the following formula:

$$\text{Man-days (of labor)} = \frac{\text{Quantity}}{\text{Production}} \times \frac{1}{\text{Frequency}}$$

where quantity is the actual number of elevators to be cleaned, production is the number of elevators cleaned per man-day and frequency is the number of workdays between operations.

If the frequency and/or production rate values, shown in the frequency and production schedule, have been modified to meet existing conditions at a particular location, then those modified values should be used in the above formula for the computation of the labor requirement.

SAMPLE CALCULATION

Consider a building that has 16 passenger elevators and three freight elevators of such design and subject to such traffic conditions that the average cleaning values from the frequency and production schedule may be used; and compute the man-days of labor as follows:

$$
\begin{aligned}
\text{Man-days} &= \overbrace{\frac{\text{Quantity}}{\text{Production}} \times \frac{1}{\text{Frequency}}}^{\text{Passenger Elevators}} + \overbrace{\frac{\text{Quantity}}{\text{Production}} \times \frac{1}{\text{Frequency}}}^{\text{Freight Elevators}} \\
&= \frac{16}{25} \times \frac{1}{1} + \frac{3}{25} \times \frac{1}{5} \\
&= .64 \\
&= .664 \text{ man-days of labor required (jobs)}
\end{aligned}
$$

To this figure must be added a factor to compensate for leave in order to arrive at the total number of positions needed to accomplish the work.

Conversion factor including compensation for leave = 1.15
Conversion factor × jobs = total positions required
1.15 × .664 = .764 positions required

• Where the normal elevator cleaning duty does not require the full time of an employee, other duties should be assigned to secure 100 per cent utilization of his time.

Equipment and Material Allocation and Replacement

Items	Allotment per Cleaner	Replacement or Reissue
Buckets	2	1 per 2 years
Cheesecloth	2 yds.	12 yds. per year
Detergent (soapless synthetic)	1 pound	50 lbs. per year
Equipment carrier	1	1 per 10 years
Lemon oil polish	1 pint	4 pts. per year
Metal polish (liquid)	1 pint	4 pts. per year
Mops (1¼ lb.)	2	4 per year
Mop handles	2	2 per year
Mop wringer (squeeze type)	1	1 per 4 years
Polyethylene glycol	1 pint solution	2 pts. glycol per year
Putty knife (1¼")	1	1 per 2 years
Scrubbing & polishing machine	1	1 per 10 years**
Steel wool No. 00	1 pound	6 pounds
Vacuum cleaner and attachments	1	1 per 10 years **
Wax applicator (complete)	1	1 per 2 years
Wax applicator refill pad	As required	3 per year
Wax:	As required	1 drum per year
Liquid spirit (55 gal. drum)		
or		
Water emulsion (55 gal. drum)		
Wiping cloths	2	12 per year
Wood wedge	To be obtained*	As required

* Obtain locally.
** Does not apply to replacement of polishing brushes or vacuum cleaner attachments.

7 / ESCALATOR CLEANING

Job Description

Escalator cleaning includes the cleaning of all surfaces of the escalator and its surroundings that can be reached from the normal walking surface, such as vacuum cleaning landings and treads; removing gum or other foreign substances; dusting fire apparatus doors and ledges; cleaning radiators and grills; cleaning handrails, glass surfaces, and other metal and wood surfaces; wall spotting; cleaning treads and risers; and properly maintaining the equipment used for cleaning the escalator. Mopping and/or scrubbing of the escalator landings should be performed by the regularly assigned corridor mopping and/or scrubbing crews. These crews should be cautioned to use only a damp mop on the access panels to prevent water running through and onto the machinery below.

Employees assigned to this job should be instructed to notice lights out, and similar service defects, and to report them to their supervisor.

Equipment and Materials

A. *Equipment:*

    1. *For cleaning escalator.*

        a. Bucket
        b. Chamois
        c. Dustcloths
           (treated cheesecloth)
        d. Vacuum cleaner and attachments
        e. Wood wedge
        f. Wiping cloths
        g. Wood scraper

B. *Materials:*

    1. *For cleaning escalator.*

        a. Detergent
        b. Lemon oil
        c. White soap (cake)
        d. Metal polish

Job Methods

Operation Steps	*Explanation*
Step 1	
1. Obtain equipment and materials.	Obtain equipment and materials as needed for escalator cleaning (listed under "Equipment and Materials").
2. Proceed to assigned work area.	Take equipment and materials to first escalator scheduled to be cleaned.
Step 2	
1. Vacuum landings and treads.	Start at top landing and use vacuum cleaner with hose and bare floor attachment to vacuum clean all landings and treads.
2. Remove gum, and foreign substances.	Use wood scraper to remove any gum or other foreign substances from landings and treads during vacuum cleaning operation.

Operation Steps	*Explanation*
3. Vacuum clean radiators and grills.	Use vacuum cleaner with radiator brush attachment to remove dirt and dust particles from radiators and grills in escalator areaway.

PRECAUTION

Make sure all loose dirt and dust is picked up with the vacuum equipment, and that no dirt or dust is knocked or brushed into the escalator mechanism.

Step 3

Damp wipe treads and risers.	Start at top landing and clean grooves in treads using a wood wedge, covered with a damp cloth, shaped to fit snugly between the grooves of the treads.

PRECAUTIONS

1. Do not wash treads and risers, but wipe them with a damp cloth.
2. Use separate cloths for treads and risers so that any oil or grease will not be transferred from one to the other.

Step 4

1. Dust with treated dustcloths.	Start at bottom landing and dust fire apparatus, doors, ledges, and grills in the escalator areaway. Use treated dustcloths to dust all surfaces, except walls, that can be reached by standing on landings or treads.

PRECAUTION

Avoid using treated dustcloths on painted wall surfaces.

Step 5

1. Clean glass surfaces.	Use a clean, damp chamois to clean all glass surfaces.
2. Clean handrails.	Use a damp cloth to clean handrails.
3. Clean and polish metal surfaces.	Use clean cloths, and metal polish as directed, to clean and polish metal work except as noted below.
4. Clean and polish wood surfaces.	Use clean cloths and lemon oil polish to clean and polish wood surfaces as directed.

Metal polish is not to be used on painted, lacquered or other metal surfaces that have been specially treated.

Step 6

1. Normal wall spotting.	Use a clean, damp cloth to spot clean the enclosure wall up as high as can be reached while standing on the landing or tread.
2. Remove stubborn spots.	Use white soap on damp cloth to remove any stubborn spots that cannot be removed by normal wall spotting. Carbon tetrachloride is helpful in removing spots from metal surfaces, but should be used only when and as directed.

PRECAUTIONS

1. This step is for spot removal only, and not for washing the entire wall.
2. After rinsing cloths in bucket of clean water, wring them thoroughly to avoid getting an excessive amount of water on the wall. Rinse clothes often.
3. Be sure to wipe any soap particles from wall with a clean damp cloth.

Step 7

1. Return all equipment and unused materials to storage location.	Gather all pieces of equipment and unused materials and proceed to proper storage location.
2. Clean and store equipment.	Clean all particles and return them to proper storage position.
3. Unused materials.	Return all unused materials to proper storage place.

PRECAUTION

Make sure that lids to polish and other containers are on securely.

Care of Equipment

A. *Bucket:*
   1. Rinse and dry bucket thoroughly after each use.
   2. Wipe outside of bucket free of dirt.

3. Avoid striking bucket against objects as this tends to cause leaks and shortens the life of the bucket.

4. Store in dry place, and in such a way that there is no weight on the bucket.

B. *Chamois:*

1. After each use rinse chamois in clean water and squeeze out excess water.

2. Stretch chamois to original shape before placing to dry.

3. Avoid getting any oil, polish or cleaning materials on chamois.

C. *Dustcloths (Polyethylene glycol treated):*

1. Thoroughly shake or vacuum dustcloths to remove loose dust.

2. Wash dirty cloths in warm detergent solution. (Mix one ounce of detergent per gallon of warm water.) Rinse in clean water and squeeze to remove excess water.

3. To treat new or washed cheesecloth, spray or sprinkle each square yard of cheesecloth with approximately one fluid ounce of polyethylene glycol solution (mix three parts water to one part polyethylene glycol). The cloth should be treated the day before it is to be used.

D. *Vacuum Cleaner and Attachments:*

1. Empty dirty receptacle and wipe clean after each use.

2. After use wipe machine and cord free of dirt and dust with a damp cloth. Wipe machine and cord dry, and wind cord loosely on hooks or other means provided.

3. Do not allow machines or attachments to strike walls or other objects.

4. After use, clean attachments and return them with machine to a clean, dry storage area where they will not be damaged.

5. Report needed servicing or repairs to supervisor.

E. *Wood Wedge or Scraper:*

1. Use wood scraper only for removal of gum or other foreign substances.

2. Wipe scraper clean and store it in a clean dry place.

Performance Inspection Guide

A. *Normal Policing:*

1. *Landings and Treads.*

The landings and treads should be free of loose dirt, dust streaks, and gum or other foreign substances.

2. *Cleaning, Dusting, Polishing and Wall Spotting.*

a. The walls and all objects in the area should be free of finger marks and other smudges.

b. Handrails and glass should be clean. Wood and metal surfaces should be cleaned and polished.

B. *Equipment and Materials.*

    1. All pieces of equipment should be clean and properly placed in their assigned locations.

    2. All polish, soap and similar leftover materials should be returned to their assigned storage locations.

Frequency and Production

A. *Frequency and Production Schedule:*

Passenger Escalator

Frequency	Daily*
Production Rate per Man-day	20 Flights (floor to floor)

    * Or as directed by supervisor.

B. *Qualifying Factors:*

    1. *Frequency*

        a. The frequency given in the schedule is for a normal standard of cleaning under average conditions.

        b. Only about 40 per cent of the escalator treads are exposed when the escalator is stopped. Obviously only the exposed section can be cleaned. To assure that each section is cleaned in its turn, the escalator should be divided into three sections, and the beginning of each section numbered consecutively by means of small numbers on the risers. The escalator should be stopped at successive sections for cleaning on successive nights.

    2. *Production Rate per Man-day*

        The production rate used in the schedule is the rate established for cleaning escalators under average conditions. Due to varying designs and conditions, the production rate as established may be subject to modification at various locations. It does, however, provide a standard from which such deviations should be calculated.

Staffing

    The man-day requirements of labor for escalator cleaning can be determined by:

$$\frac{\text{Number of Flights (floor to floor)}}{\text{Production per Man-day}}$$

    If the frequency and/or production rate values as given in the schedule have been modified to meet existing conditions at a particular location, then those modified values should be used for the computation of labor requirement.

SAMPLE CALCULATION

Consider a building that has 16 flights of escalators (floor to floor) and of such design and subject to such traffic conditions that the average cleaning values from the frequency and production schedule may be used, and compute the man-days of labor as follows:

$$\text{Man-days} = \frac{16}{20} = .80 \text{ Man-days of labor required (jobs)}$$

To this figure must be added a factor to compensate for leave in order to arrive at the total number of positions needed to accomplish the work.

Conversion factor including compensation for leave—1.15
Conversion factor × jobs = total positions required.
1.15×80 = positions required—.92

- Where the normal escalator cleaning duty does not require the full time of an employee, other duties should be assigned to secure 100 per cent utilization of his time.

Equipment and Materials Allocation and Replacement

Items	Allotment per Cleaner	Replacement or Reissue
Bucket (12 qt.)	1	1 per 2 years
Chamois	1	1 per year
Cheesecloth	1 yard	12 yards per year
Detergent (soapless synthetic)	1 pound	25 pounds per year
Lemon oil polish	1 pint	4 pints per year
Polyethylene glycol solution	1 pint	2 pints solution per year
Vacuum cleaner and attachments	1	1 per 10 years**
White soap (6 oz. cake)	1 cake	4 cakes per year
Wiping cloths	2	12 per year
Wood scraper	1	2 per year*
Wood wedge	1	2 per year*

* Obtained locally.
** Does not apply to replacement of vacuum cleaner attachments.

8 / EQUIPMENT CARE

(*Sources:* General Services Administration, Washington, D. C.)

This chapter covers instructions for the care, allocation and replacement of general equipment and materials commonly used in the building cleaning operations.

Applicator, Wax

1. Remove applicator pad from block and thoroughly wash.

2. Rinse thoroughly, shake excess liquid from applicator pad and replace on block to prevent shrinkage.

3. Store applicator where there is free circulation of air and where there will be no weight against it.

Bag, Canvas Shoulder Trash

1. Thoroughly empty bag or receptacle of trash and dirt after each day's use.
2. Avoid snagging or tearing the bag.
3. Store in a clean, dry location.

Bag, Wastepaper and Trash Collection

1. Do not empty contents of ash trays into this bag.
2. Avoid snagging the bag on sharp objects or getting it caught in the door.
3. Store neatly.
4. Launder as required.

Basket, Oak Splint

1. Clean the trash basket of trash and dirt after each day's use.
2. Store in such a manner it will not be damaged.

Belt, Safety

Note: The following is an excerpt from standards as they pertain to the care of safety belts:

1. Each safety belt shall be identified and a record of the date it was put in service shall be kept in the office of the building (or facility) superintendent and checked at frequent intervals to insure that proper inspection is made of belts which have been in service for a considerable period. A record shall also be kept of any repairs or replacements made to the belt.

2. Belts shall be maintained in good condition at all times.

3. All safety belts shall be examined by the supervisor at the beginning of each work shift. The person using this equipment shall also examine the belt at the beginning of each work shift and at other times during the day to make certain that no defects have developed.

4. All belts shall be stored, transported and handled so as to prevent corrosion or injury. Belts which have been damaged by mildew, by the action of an acid, by contact with sharp tools or equipment, or by any corrosive or deteriorating agent, shall not be used.

5. Impact carrying parts of safety belts shall be repaired *only by the manufacturer or his designated representative,* unless the regional safety engineer especially authorizes another individual to do so.

6. Persons responsible for supervising the use and care of window cleaning or window maintenance equipment shall make certain that all unsafe pieces of equipment are disposed of at once, and not left lying about where they can be picked up and used.

7. When a belt has been exposed to rain or snow, wipe it off with a clean rag and allow it to dry at room temperature.

8. Never expose belts to heat in excess of ordinary room temperature.

9. Belts shall never be dropped or thrown from one elevation to another, or otherwise mishandled.

10. For standard anchors and terminals a 9/16-inch plug gauge can be used to check wear on the terminals. If the belt terminal slot is so wide at any point that it will accommodate a 9/16-inch plug gauge, the terminal shall be replaced.

Boots, Rubber

1. Rinse off dirt and other matter and wipe dry after each day's use.

2. Store in a clean well ventilated place in such a manner that the boots will not be damaged.

Broom, Corn

1. Soak new broom overnight before using it for first time.

2. Never stand a broom on the straws when storing.

3. Store broom where there is free circulation of air.

4. Wash broom as required, using a warm synthetic detergent solution. Rinse in clear water and hang to dry.

5. Never use the broom for scrubbing or when it is wet.

6. Rotate the broom when using it so that it will wear evenly.

Broom, Dustpan

1. When broom becomes soiled, wash it in warm detergent solution, rinse in clear water and shake free of excess water.

2. Store broom so that its weight is not on the bristles.

Brush, Counter Dust

1. When brush becomes soiled, wash it in warm detergent solution.

2. Rinse in clear water and shake out excess water.

3. Hang up where air circulation is good so brush will dry.

Brush, Deck Scrubbing

1. Rinse brush in clean water, after each day's use and shake dry of excess water.

2. Place for drying and storage in a position so that the bristles do not bear the weight of the brush.

3. Avoid knocking the brush against objects as such action tends to split the brush block.

Brush, Floor Sweep

1. Clean brush daily by combing bristles.

2. Wash oily or very dirty brush as necessary in a warm detergent solution. Rinse in clear water, shake out excess water and hang to dry.

3. Store in a clean dry place so that the bristles do not bear the weight of the brush.

4. Once each week reverse the brush by placing the handle in the other hole of the brush block to increase brush life.

Brush, Radiator Dust
1. Clean brush daily by combing.
2. Store brush on a flat surface so that tufts will be flat and straight.
3. Avoid wetting the brush.

Brush, Sidewalk
1. Wash an oily or dirty brush in warm detergent solution and rinse in clean water. Shake free of excess water and hang to dry with bristles in a downward position and free of any weight.
2. Store in a clean dry place so that the bristles do not bear the weight of the brush.
3. Reverse the position of the handle in the brush block weekly to provide longer brush life.

Brush, Toilet (or toilet mop)
1. Wash daily in a solution of detergent or multipurpose cleaner.
2. Rinse thoroughly and shake out excess water.
3. Hang up to dry where air circulation is good.

Bucket
1. Rinse and dry the bucket inside and outside after each day's use.
2. In storing, place the bucket so that it will not be damaged.

Can, Ash (for collecting non-combustible material)
1. Empty container after each use.
2. Wipe interior and exterior with a damp wiping cloth to remove dirt, dust, etc.
3. Dry surfaces with a dry wiping cloth.
4. Replace container in assigned position so that it will not be damaged and will be ready for next day's use.

Chamois
1. Rinse chamois in clean water and squeeze out excess water after each day's use.
2. Stretch chamois to original shape before placing to dry.
3. Avoid contaminating chamois with oils, polish, etc.

Cloth, Drop
1. Brush or shake free of dust and dirt after each day's use.
2. If cloth becomes wet during use, hang overnight to dry.
3. Avoid snagging or tearing the drop cloth.
4. Fold and store in assigned storage location.

Cloth, Wiping
1. Rinse the wiping cloths in clean water after each day's use.

Dustcloth, Treated
1. Keep the cloth free from grime or oily material.
2. Wash the cloth weekly or as required in a warm detergent solution. Rinse in clear water and wring dry.

3. Treat with prescribed dustcloth treating compound as directed by supervisor.

4. Hang the cloth to dry where there is a good circulation of air.

Dustpan

1. Clean dustpan of dirt and dust before storing.

2. Hang dustpan by the handle so that it will not be bent or otherwise damaged.

Gloves, Rubber

1. After each day's use wash the gloves in warm multipurpose cleaner or detergent solution, rinse in clean water and shake free of excess water.

2. Store gloves in a clean dry place.

3. Always maintain gloves in first-class condition. Discard gloves when holes develop.

4. Be careful not to snag gloves on nails or other sharp-edged objects.

Goggles, Dust

1. Wipe goggles clean after each day's use.

2. Care should be taken not to drop or damage goggles.

3. Goggles should be maintained in first-class condition. When the lenses become chipped or otherwise damaged, the goggles should be replaced.

Hose, Water

1. Drain the hose of any remaining water and wipe the outside free of dirt and excess moisture.

2. Coil and/or properly store the hose in its assigned location.

Knife, Putty

1. Wipe the knife free of dirt and moisture after each day's use.

2. Store in a clean dry place.

3. Use only for the purposes specified. Do not use as a screwdriver or hammer.

Ladders (including platform type)

Ladders should be cleaned and inspected frequently and those which have developed defects should be withdrawn from service for proper repair, or disposal. In either instance, such ladders should be tagged or marked "DANGEROUS! DO NOT USE."

During routine inspections ladders should be carefully checked for the following defects:

1. All Ladders:

    a. Loose steps or rungs (consider step or rung loose if it can be moved at all with the hand).

    b. Loose nails, screws or bolts.

    c. Loose or missing shoes or antislip bases.

    d. Cracked, split or broken uprights, braces, steps or rungs.

    e. Slivers on uprights, rungs or steps.

Litterstick

1. Wipe the stick free of moisture and dirt after each day's use.

2. Place a cork or some similar object over the pointed end before storing.

Mop, Cotton Wet

    1. Rinse a new mop before using.

    2. After each day's use and before storage rinse the mop in clean water and squeeze or wring to remove excess water.

    3. Wash dirty mops in a warm detergent solution, rinse in clean water and wring or squeeze to remove excess water.

    4. Fluff the mop strands by shaking.

    5. Store with the mophead up in a well-ventilated place to allow for better drying.

Mower, Hand Lawn

    1. Wipe all parts free of dirt, moisture and grass cuttings after each day's use.

    2. Store the mower in its assigned location and in such a position as to prevent its being damaged.

    3. The mower should be kept properly lubricated and blades sharpened. Report all defects and/or necessary repairs to the supervisor.

Mower, Power Lawn

    1. Brush off accumulated grass and wipe the entire machine, except the engine, with a dry cloth.

    2. Remove any dirt or grass that has accumulated on the flywheel housing or between the cylinder fins of the engine with a brush.

    3. The mower should be kept properly lubricated and blades sharpened. Report all defects and/or necessary repairs to the supervisor.

Overshoes, Rubber

    1. Rinse or wipe off dirt and other matter and dry thoroughly.

    2. Store in such a manner that they will not be damaged.

Polishing Machine, Floor

    1. Remove brush and wash daily or as required.

    2. Wipe the machine and cord free of dirt and dust with a damp cloth. A cloth dampened with detergent solution may be used to remove spots not otherwise removed. Wipe the machine and cord dry and rewind the cord loosely on the hooks provided.

    3. Wash and rinse the machine wheels and wipe dry.

    4. Use metal polish as directed to clean and polish the machine surfaces.

    5. Store the machine and brush (with brush removed) in assigned storage space.

    6. Report needed repairs to the supervisor.

Raincoat

    1. Wipe the coat free of dirt and excess moisture after each day's use.

    2. Store the coat by hanging on a hanger in a clean well ventilated place.

    3. Avoid snagging on nails and other sharp objects.

Rake, Lawn

    1. Wipe the rake free of moisture, dirt and other debris after each day's use.

2. Store by hanging in a safe and secure manner and so it will not be damaged.

Rake, Heavy Iron (for paper baling)
1. Wipe rake free of dirt and dust and dry thoroughly.
2. Remove any paper that might still be clinging to its spikes.
3. Store rake so the spikes will not be damaged or bent.
4. Avoid hitting the rake against metal or concrete during use.

Respirator, Dust
1. Wipe respirator clean daily.
2. Respirator should be maintained in first-class condition. If it becomes damaged in any way it should be replaced.

Scaffolding
1. Damp wipe scaffolding to remove solution spillage or other liquid.
2. Store the equipment in its assigned storage location. Avoid creating safety hazards in placing the scaffolding for storage.
3. Report defects or needed repairs to supervisor.

Scoop (perforated or sieve)
1. Wipe clean and store in a dry clean place.
2. Store so that no damaging weight is placed on the scoop or sieve.

Scrubber-Vacuum Machine
1. Drain and flush both the cleaning solution and the dirty water tanks.
2. Remove vacuum screen and clean.
3. Clean brushes after each day's use. On dual brush machine, reverse the brushes when replacing them. Install new brushes if required.
4. Clean guide wheel and under carriage.
5. Clean and replace brush shields.
6. If electric scrubber is used, clean and dry wipe electric cable.
7. Wipe dirt and other matter from all machine surfaces except gasoline engine.
8. Store machine in assigned storage position.
9. Report needed repairs to the supervisor. Do not attempt to repair the machine or accessories. (This is a shop job.)

Shovel
1. Wipe dirt and dust from shovel and dry thoroughly before storing.
2. Store shovel so it will not be damaged.
3. Avoid banging shovel against metal or concrete surfaces.

Sign, Warning or Caution
1. Wipe sign free of dirt and other matter.
2. Store in assigned storage position so it will not be damaged.

Snow Plow Attachments
1. Wipe the equipment free of dirt and moisture after each use.
2. Upon storing, coat the bare metal portions of the equipment with a rust

preventive compound as directed by the supervisor and otherwise maintain in a well lubricated and clean condition.

3. Report necessary repairs and defects to the supervisor.

Sponge

1. Rinse the sponge in clean water after each day's use.

2. When dirty, wash in warm detergent or multipurpose cleaner solution, rinse thoroughly and squeeze out excess water.

3. Store in a clean dry location.

Squeegee

1. Rinse the squeegee blade in clean water after each day's use.

2. Wipe dry and store in a clean dry location.

Sweep Mop, Treated

1. Treated sweep mops should be shaken or vacuum cleaned both during use and at close of operation each day. They should be kept free of grime or oily material.

2. Every two weeks or as required, remove the mophead, shake out as much dust as possible and wash mop in a warm detergent solution.

3. Treat the mop with prescribed mop treating material as directed by supervisor.

4. Replace the mophead on the frame and hang mop so that a good air circulation will be obtained to assist in drying the mop.

Sweeper, Power Sidewalk

1. Empty dirt pan.

2. Check vacuum intake housing and vacuum bag.

    a. Keep intake housing clean.

    b. Empty vacuum bag frequently by removing inner bag and shaking it to remove all the accumulated dust.

3. Clean sweeper after each day's use.

    a. Wipe entire sweeper except gasoline engine with dry cloth.

    b. Use a brush as required to reach areas that cannot be reached with cloth.

4. If sweeper is electric powered, return unit to battery charger, otherwise store sweeper in assigned location.

Tamper (for use in wastepaper baling)

1. Wipe tamper free of dirt and dust and dry.

2. Store the tamper so it cannot be damaged or damage other objects by falling.

Tank, Mop

1. Rinse mop unit compartments and wringer thoroughly and wipe dry after each day's use.

2. Remove any remaining mop strands and other matter from the mop unit and wringer.

3. Always leave mop wringer in released position.

4. Store in a clean dry location and otherwise maintain in a clean and orderly condition.

5. Report any defects or necessary repairs to the supervisor.

Tool, Ice Scraping

1. Wipe the tool free of moisture and dirt after each day's use.

2. Store in a safe manner in its assigned location.

Tools, Miscellaneous Shrubbery

1. Wipe tools free of dirt and moisture after each day's use.

2. Maintain tools in a well sharpened and lubricated condition as necessary.

3. Hang tools for storage in such a way that they will not be damaged or endanger personnel.

Truck, Four-wheeled Push

1. Clean the truck of dirt and trash after each day's use.

2. The wheels of the truck should be kept properly lubricated.

3. Report all necessary repairs or defects to the supervisor.

Truck, Stevedore

1. Clean thoroughly each week with a damp wiping cloth and **dry** thoroughly.

2. Avoid striking wall surfaces, radiators, etc. with truck.

3. Store truck out of the way of traffic to prevent anyone falling over it.

4. Arrangements should be made to have the wheels oiled regularly. **Report** any defects or repairs to the supervisor.

Vacuum Cleaner

1. Always operate vacuum equipment with a cleaning tool on the end of hose.

2. Avoid sharp bends in hose during operation. Do not permit platform trucks to run over hose. Report any defects to supervisor.

3. Always remove hose from machine outlet at end of work day.

4. Coil hose loosely on machine, or in case of central system, hang on hangers provided in gear rooms.

5. Remove sweepings from dust bag daily. In addition to removing sweepings daily, clean the inside of the bag once a week. Where two or more machines are available, use one machine to clean the bag of the other. Where another machine is not available, clean the bag with a stiff brush.

6. Wipe the machine with a damp cloth when day's work is completed.

7. Make sure the electric cable connector is plugged all the way into the electric outlet during operation of the machine.

8. Avoid pulling on the electric cable in removing plug from outlet.

9. Do not splice or replace electric cable. (This is shop job.)

10. Upon any signs of smoke or unusual odor stop machine immediately and notify supervisor.

11. To avoid damaging wheels, do not bump machine up and down stairs when carrying from one floor to another. Supervisors should make sufficient help

available to move machines from floor to floor in buildings where there are no ramps or elevators.

Wall Washing Machine

1. Empty the machine, rinse with clear water and wipe dry.
2. Rinse and drain machine hose and trowels, wipe dry and coil hose loosely on trowels.
3. Hang trowels on hooks on side of machine or other assigned location.
4. Store machine in clean dry location where no damage will be done to it.
5. Wash, rinse and place to dry all toweling used in the day's work.

Wedge and/or Scraper, Wood

1. Wipe wedge and/or scraper free of dirt and other matter after each day's use.
2. Store in assigned storage place so it will not be damaged or lost.

Wringer, Mop

1. Remove any mop strands and matter clinging to wringer.
2. Rinse and dry wringer thoroughly.
3. Store so that wringer will not be damaged.

9 / FENCE MAINTENANCE

A fence is generally erected to control access to a restricted area. The maintenance problem of this facility is to make certain that there are no holes in the fence, and that no washouts occur in the ground immediately adjacent to it. Regular inspection should be made, and repairs made when necessary. Also, insofar as appearance and protection against corrosion are concerned, a fence must be kept in a well painted condition unless it is constructed of noncorrosive material. In the interest of plant security, inspection ease, and appearance, the ground for a distance of one to two feet on both sides of a fence traversing an unlandscaped area should be periodically sprayed with a weed killer to prevent the growth of vegetation on or in the close proximity of the fence.

Selection of a paint or coating will depend on the material to be treated and the conditions of atmosphere and use to which it will be subjected. A reliable paint manufacturer or dealer is the best source of information in selecting the right paint for the job. He should be able to take into account the local atmospheric and meteorologic conditions that will affect the surface because of its location. Other conditions that warrant consideration are extreme heat or cold, bright sunlight, chemical fumes, dust abrasion, and salt water spray.

On a wire fence, a long braided fabric roller, dipped in a pan, is often used for painting and enables the painter to cover 100 per cent of the fence surface facing him, and also a considerable amount of the far side of the fence. Electrostatic spray painting is another method.

10 / FLOORS AND FLOOR CARE

(*Sources:* Public Buildings Service, General Services Administration, Washington, D. C.; *Buildings* magazine)

10A / FLOOR FINISHES

Floor finishes include those surface products that are applied regularly, such as wax and emulsified resins, that can be applied quickly and easily, that are prompt drying, and that can be readily removed when necessary. There are two types of wax—the solvent type, including both liquid and paste, and the water emulsion type. Water emulsion waxes are suitable for use on any type of floor except unsealed wood. Solvent waxes, although sometimes considered to be more durable, can be used only on wood, linoleum and cork.

Although not the only measure of quality, the proportion of carnauba wax contained in floor wax will largely determine how well the product beautifies and protects. Solvent waxes seldom exceed 12 per cent in solids. Most water waxes vary in solid content from 12 per cent to 18 per cent.

Many of the improved resin finishes now contain polyethylene or a similar synthetic. When originally developed, many resin floor finishes were easily scratched, and would not respond to buffing. They were also difficult to remove from the floor. Most resin finishes no longer present this problem.

Waxing Precautions

To obtain the best finish for your floor and maintain high quality levels with maximum economy, observe the following precautions when waxing:

1. Never wax an improperly cleansed floor. It only causes unsightly appearance, prevents wax from taking bond to floor. Soil and other impurities mixed with wax can upset the formula and cause a flaky or milky appearance to a new wax coat.

2. Never apply wax to a wet or damp floor. This can prevent bonding of the wax and cause a milky and cloudy appearance.

3. Never apply a heavy coat of wax.

4. Do not rewax every time floors become dull. They can usually be brought to a bright appearance by dry buffing. Dirt cannot be ground into or penetrate a good hard wax film. Therefore, for the most part, dirt appearing on a floor is surface soil and can ordinarily be removed by damp mopping—or if dry, by steel wooling or buffing up with a standard buffing brush.

5. Never permit container of stored wax to stand open. Evaporation of vehicles will increase the solid content of the material, causing a heavier film to be applied. This will cause a faulty film on the floor and, in addition, less coverage per gallon.

6. Never apply wax that is under 70° F., nor apply it to a floor under this temperature. Cold floors or cold wax cause a heavier viscosity, which in turn makes a heavier film. The thinner the wax film, the better the results.

7. Never permit water waxes to freeze. Water waxes are neither an emulsion, nor a suspension, but rest on a point between these two chemical categories. The shock from freezing can unbalance the formulas causing the wax to be valueless.

8. Never use a dirty mop applicator, or receptacle to hold water waxes.

9. Never pour used wax into the original container with unused wax. Contamination can damage water waxes. Pour leftover used wax into a clean gallon or larger jug. This wax should be poured into a funnel that contains steel wool to act as a strainer. Be sure to keep all lids tight on storage containers to prevent evaporation.

Tile Floors and Their Care

Planned maintenance and a good knowledge of floor care is an important phase in a building maintenance program. Floor maintenance programs include a controlled cleaning and finishing schedule set to meet specific requirements under varying conditions. A well planned day-to-day maintenance schedule should take into consideration the type of flooring, amount and nature of traffic, atmospheric soil conditions, appearance and cleanliness level desired.

In order to perform proper maintenance on various types of floors, a knowledge of composition, limitations and adaptability of the floors should be known.

Asphalt tile. Asphalt tile is a mixture of asbestos fibers, pigments, and inert fillers bound together with an asphalt or resin binder. Colors are divided into four groupings, A, B, C, and D, with A as the darkest and D the lightest. A and B generally contain asphalt binders, C and D, resin binders. Asphalt tile is also furnished in a grade designated as greaseproof, obtainable at a slightly higher cost. It is normally the least expensive of the resilient floors, however some thin grades of vinyl-asbestos may be cheaper. With one or two exceptions where special adhesives are used, asphalt tile is the only resilient floor which can be installed on any smooth, solid subfloor, including below grade concrete. It is generally considered moisture proof and decay proof. It is the least resilient of the soft floors, and is easily indented by heavy or pointed objects. Natural enemies of asphalt tile are oils, greases, and such solvents as gasoline, naphtha, turpentine, carbon tetrachloride, and kerosene.

Cracks, checking and breaks in asphalt tile can be blamed directly on the use of alkaline or caustic cleaners which dry out the tile and makes it hard and brittle, eventually causing disintegration. The alkali in the cleaners used on this floor have emulsified the resins and plasticizers of the asphalt tile, causing this brittle, hard, shrinking surface. A drying out effect is obvious at the edges of the tile, indicating that water and alkaline salts have seeped into joints and cracks. In damp mopping that may follow, the wetting and re-wetting of the alkaline salts can cause an almost continuous swelling action. Pitting on asphalt tile is a sign that solvents have come in contact with it. Oil, greases, or solvents will soften and may eventually dissolve the tile. Strong caustic cleaners will also cause pitting.

Do not use varnishes, lacquers, shellacs, or other plastic finishes. These materials usually contain solvents that will permanently injure the floor. Never use oily sweeping compounds or dressings. Asphalt tile should not be installed in areas that are subject to extreme temperature changes.

Any good quality mild soap, neutral cleaner, or synthetic detergent may be used. The alkaline content of such cleaners should have a pH of more than ten. Solvent type cleaners, except alcohol, are detrimental to asphalt tile. After scrubbing with a good neutral cleaner, a thorough rinsing should follow, then a good water emulsion wax or emulsified resin type of floor treatment applied.

Vinyl tile. Vinyl tiles are commonly classified in four main types:

Vinyl asbestos, which is also referred to as semi-rigid or semi-flexible.

Homogeneous, which is also referred to as flexible tile, and better known as vinyl plastic.

Calendered vinyl mix or backed vinyl.

Rotogravure printed or vinyl coated gravure.

Vinyl asbestos, composed through full thickness, of vinyl resins, plasticizers, pigments, fillers and asbestos fibers formed under pressure while hot, is one of the most popular of all the resilient floors. It is more expensive than asphalt tile, which it closely resembles in appearance, but has the following advantages over asphalt:

1. It is practically immune to mineral solvent, oils, and greases, as well as to alkalies and acids.

2. Its indentation resistance is generally 25 lbs. per square inch.

3. Also decay and mildew-proof, it is easier to maintain than asphalt tile since generally, any kind of wax or cleaner can be used.

4. It can be installed on below grade concrete by use of special adhesive.

Vinyl plastic, composed through full thickness, of vinyl resin, plasticizers, pigments and fillers formed under pressure while hot, is considered the best of the vinyls, and in most ratings one of the best of all resilient floors. It is also known as homogeneous vinyl tile. This tile is the most expensive. It possesses all the qualities of vinyl asbestos in being resistant to solvents, oil and other immunities, and is also more resilient, therefore is more comfortable and quieter on which to walk or work. It resembles rubber tile in appearance, and is flexible, colorful and glossy. It is more expensive than rubber tile, and has a load limit of approximately 200 pounds per square inch, which makes it eight times more resistant to indentation than asphalt or vinyl asbestos tile. It may be cleaned and maintained with any of the usual cleaners and waxes, and by use of a special adhesive, may be installed on below grade concrete.

Backed vinyl has a wearing layer composed of vinyl resins, plasticizers, pigments, and fillers, overlaid on a backing of various regular alkali resistant materials.

The four main types of vinyl are generally immune to grease, oil, and ordinary solvents, and resist ordinary acids and alkalies. Either water emulsion wax or spirit wax may be used. Water emulsion wax is more generally prescribed. All the vinyls can be maintained in much the same manner as asphalt and rubber tile, except that they will take solvent type cleaners, waxes, and sweeping compounds. Excess water or flooding will, however, create the danger of seepage between tiles, causing the adhesive to loosen. Some vinyl, when first installed, has a tendency to hold dirt, but ceases after a few months of regular maintenance.

Due to the smooth, dense surface, some vinyls will not take wax properly until after a period of use.

Rubber tile. A mixture of rubber, natural, synthetic, and/or reclaimed, with inert fillers and color pigments, rubber flooring varies widely in form and properties. It is available in tiles and sheets with many degrees of hardness and flexibility. Regular soaps are not recommended for cleaning rubber tile, since fats and oils are detrimental to rubber tile. Synthetic detergents derived from sulfanated hydrocarbons are approved for cleaning rubber tile. Mild alkaline cleaners (not TSP) may be used occasionally, but care should be taken to avoid the use of excess water. Do not wash too soon after installation, or use any substances containing gasoline, kerosene, naphtha, benzine, turpentine, mineral solvents, or harsh alkalies. Avoid the use of sweeping compounds containing oil, abrasives, or chemicals. Never use varnish or lacquer. When necessary to remove stubborn accumulations, mild abrasive cleaners may be used, or steel wool with a synthetic detergent. To remove hardened old wax coatings, steel wool pads and a good wax stripper are effective. Only water wax emulsions should be used on rubber tile. Solvent type, including paste wax, will cause the tile to soften and bleed. Do not install rubber tile in areas of bright sunlight, as the pigment will fade.

Linoleum or linoleum tile. Composed of oxidized linseed oil, fossil, other resins, or other oleo-resinous binder mixed with ground cork, wood flour, mineral fillers and pigments are pressed on burlap or saturated felt backing. It is supplied in rolls six feet wide and up to 90 feet in length, and is available in gauges from .0625″ to .25″ and in which latter class, the popular plain or battleship falls.

Linoleum is made in several classifications, including inlaid, marbleized, jaspe, spatter, etc. These classifications relate more to pattern than grade. Linoleum has a load limit of approximately 75 pounds per square inch and is rated about twice as resilient as asphalt and vinyl asbestos tile. The heavier grades are best adapted for institutional or commercial uses and linoleum should always be installed on suspended floors, never in direct contact with the ground, nor over damp concrete. Linoleum is available in tile form, in which case it is identical in quality with conventional linoleum. There is a special linoleum type tile of greater hardness and durability, having a load limit of approximately 200 pounds per square inch, and a resiliency of about the same as vinyl asbestos. It has a thickness of ⅛ inch and resembles asphalt tile on the floor.

All forms of alkali are injurious to linoleum and it may be stained by grease and oil. Highly alkaline materials will deteriorate the linoleum causing brittleness, and in some cases, result in color fading, whitening or disintegration. The use of excess water will undermine the adhesive and rot the backing. When clear water mopping is necessary, use a mop that is damp, not one that is soaking wet, and pick up rinse water immediately. Abrasives are not recommended, except on isolated stains. Wax accumulations may. be removed by scouring with a non-alkaline cleaner. Be careful to wet only a limited section at a time. A nonalkaline wax stripper (neutral cleaner), used in conjunction with No. 1 steel wool or abradant pad is a good method. Either solvent type (including paste wax) water wax or emulsified resin finishes may be used.

Cork tile. Ground cork bark is molded and compressed to make cork tile.

The natural resins serve to bind the mass together when heat cured under hydraulic pressure. Different shades of color are obtained by varying the baking temperatures. Cork is available in tile or sheet form. It may or may not be waxed or otherwise factory finished.

Sweeping with a soft bristle brush and occasionally buffing with a floor machine should keep cork tile at its best. When discolored, dry clean with No. 0 steel wool or abradant pad. When water is used, it should be used sparingly. Avoid alkaline cleaners of all kinds. Use a neutral synthetic detergent for best results. Never clean cork tile with naphtha, gasoline or similar solvents, oily mops or sprays. When badly soiled, and abrasive cleaning with a cleaning solution does not restore the floor, sand lightly with No. 2 steel wool pad, screen type scouring pad or coarse nylon abradant pad. Then go over floor with vacuum. After treatment with a sealer or solvent type wax, maintain with a water type wax.

Other Types of Floors

Marble floors. Marble can be maintained with comparatively little effort. Routine maintenance includes sweeping or vacuuming or damp mopping which will remove minor soils. When cleaning marble use a neutral cleaner, a small amount should be used in the water when daily damp mopping. An acid cleaner will leave the surface of the marble dull and porous. Strong alkalies tend to penetrate the less highly polished marbles, such as the honed or sand finish types, and upon drying, expand and break the tiny cell walls of the marble, resulting in surface disintegration. Abrasive cleaners should never be used in removing stains or soils unless under recommendation of a floor treating expert. When abrasive cleaners are used, it is almost impossible under regular cleaning methods to remove the gritty residue which may scratch the surface under traffic. Soap type cleaners should never be used, as they are commonly responsible for the dull lifeless appearance of marble floors.

A pre-wetting of the marble will help prevent salts from being absorbed into the pores, and will also facilitate the rinsing procedure. The cleaning solution should be picked up and the floor rinsed thoroughly. Wiping the floor dry will prevent streaking. Seals and waxes are usually not applied to marble floors but in certain cases may be desirable to make a poor floor look better and easier to keep clean. When marble becomes porous or worn, it may be resurfaced by grinding and polishing. Do not try to clean with dirty water or dirty utensils. Oily sweeping compounds may discolor light-colored marbles, as may the continued use of ammonia.

Care of terrazzo.* It is doubtful that there is a flooring material in use today that requires less care than terrazzo. Yet, a number of people have difficulty in maintaining it. Like other materials, there are inherent properties of terrazzo that should be understood . . . once understood, maintenance problems are eliminated and the full beauty as well as economy is realized.

To best understand terrazzo is to first break it down into components . . .

* Reprinted with permission from The National Terrazzo & Mosaic Association, Incorporated.

marble and portland cement. They are mixed together in a ratio of two parts marble and one part cement. During its installation, additional marble is sprinkled on the surface so that a minimum of 70% marble shows on the finished floor.

Protection: When dealing with terrazzo, the use of a pure surface coating (i.e., as most waxes are) is unnecessary and ordinarily *not* recommended. The terrazzo surface is at least 70% marble and marble has a very low porosity. It absorbs very, very little of anything and most staining materials have too thick a consistency to become absorbed. That portion of the terrazzo that needs protection is, therefore, the portland cement; which is porous and will absorb stains. Pure surface waxes will protect . . . but are easily walked off, and will tend to make any smooth surface slippery (particularly with the "spike or stiletto" heel still with us). Terrazzo doesn't need protection from wear, it needs protection from absorption and this is achieved through the use of a penetrating sealer which is absorbed into the portland cement, sealing off its pores.

Internal Protection

Proper protection for terrazzo is then accomplished internally rather than "on the surface" and being internal, the process of waxing and rewaxing is eliminated and only periodic resealings are necessary. It is wise to note here that purely surface protection holds dirt and adds to your cleaning and/or presents a stripping problem. Also, there are few natural products as beautiful as marble, and shoe leather is one of the finest polishers or abrasives you can use. If you maintain a surface coating over marble, you will not allow the floor to take its natural sheen, so long identified with terrazzo.

Neutral Cleansers Only

Cleaning: Terrazzo should be cleaned only with neutral liquid cleaners. The cleaning cycle to be set up will be regulated by the amount of traffic. For general cleaning, 1 cup of neutral cleaner is mixed with each 3 gallons of water. For extremely dirty areas, increase the amount of cleaner. Wet mop the solution onto the floor, allow several minutes for the grime-dissolving action to take place, then squeegee, wet vacuum, or mop up the dirt-laden solution. It is important that the floors be kept wet at all times during the cleaning operation to prevent dissolved soil from drying back onto the floor. Also, if the solution is mopped up, it is important that the custodian change his rinse water regularly so that complete removal is assured and unsightly "moplines" are eliminated.

Electric scrubbing machines used periodically with a solution of neutral cleaner will loosen dirt that is hard to remove during normal daily wet-mop cleaning.

Cleaning Materials: The liquid cleaner selected *must* be neutral with a pH of less than 10.0 and free from any harmful alkali, acid, etc. that may ruin the floor. The N.T.M.A. specifically warns that soaps and scrubbing powders containing water solubles, inorganic salts, or crystalizing salts should *never* be used in the maintenance of terrazzo. Many terrazzo, quarry, and ceramic tile floors have been destroyed by improper selection of cleaning materials.

Nonoily Dressings

If a mop dressing is used for daily sweeping, be sure is it nonoily. Sweeping compounds containing oil are a fire hazard and most of them contain sand, which is hard to sweep up and abrades if left on the floor. Floor oils are not only a fire hazard but they will penetrate and permanently discolor terrazzo.

Stain Removal: We don't know of any flooring material that, when under constant use, will not begin to show some staining from daily use and abuse. Fruit, chewing gum, soda pop, and cigarettes are typical floor stainers; on and on the list could go. In a carefully maintained building, the custodians should be instructed to treat stains as soon as possible, as they become more difficult to remove after they have dried. However, *no one* should attempt to remove a stain until he knows what the stain is and why a certain type of remover is being used. Only as a last resort should chemicals be used to remove stains. Stain removers either dissolve the substance that causes the stain; absorb the stain; or act as a bleaching agent.

Removal of Stains

Environment has a definite effect upon the quality of an employee's work as well as his reaction to his supervisor and the general public. The appearance of your floors, free from stains, will do much to achieve this.

Stains should be treated as soon as possible, as they become more difficult to remove when they have dried. But, no one should attempt to remove stains unless he knows what the stain is and why a certain type of remover is being used. *Only as a last resort should chemicals be used to remove stains.* When removing stains with chemicals, directions should be carefully followed (i.e., if the procedure specifies treatment with a solvent prior to cleaning, it may be that if reversed, the alkali in the soap would set the stain and make it impossible to remove).

Stain removers dissolve the substance that causes the stain; absorb the stain; or act as a bleaching agent. Thus stain removers fall into three general classes:

1. Solvents such as carbon tetrachloride which dissolve grease, chewing gum, lipstick, etc.

2. Absorbents such as chalk, talcum powder, blotting paper, or cotton which absorbs fresh grease or moist stains.

3. Bleaches such as household ammonia, hydrogen peroxide, acetic acid, or lemon juice which discolor stains.

Realizing the importance of knowing the nature of the stain, the custodian should ask himself the following questions before trying to remove it:

1. Is it a water base stain? . . . then water will remove it.

2. Is the stain alcohol? . . . then alcohol will remove it (i.e., tincture of iodine).

3. Is the stain acid? . . . then use an alkali to neutralize it.

4. Is the stain alkali? . . . then use an acid to neutralize it.

5. Is the stain grease? . . . then use soap.

6. Does the stain contain albumin (milk or blood)? . . . do not use a hot solution (it will cook the albumin).

For common stains in your building, prepare a chart listing the kind of re- mover to use on specific stains or specific surfaces. It is a constant challenge to produce effective results and highly rewarding. Remember: Floors (any floors) that are deeply embedded with sand or soil deteriorate quickly and, once they have started wearing, no amount of cleaning will bring them back. Proper and regular maintenance creates environment. Your terrazzo floors should and will add aesthetically as well as economically. When purchasing your janitorial sup- plies, it is wise to evaluate not only how far or how many square feet the product will cover but most important how long and how well it will serve you . . . before you consider price. More often than not, the 'expensive' quality products will outserve and outlast inferior ones so that ultimately you save . . . not only on materials, but also your terrazzo floors will be protected properly and will look their very best.

Suggested Methods for Removing Stains from Terrazzo

Ink Stains: Different inks require different treatments. Ordinary writing inks may etch concrete due to acid content. To remove a stain of this type, make a strong solution of sodium perborate in hot water. Mix with whiting to a thick paste, apply in ¼-inch layer, and leave until dry. If some of the blue color is visible after the poultice is removed, repeat. If only a brown stain remains, treat it by Method 1 for iron rust stains. Sodium perborate can be obtained from any druggist.

Many red, green, violet, and other bright-colored inks are water solutions of synthetic dyes. Stains made by this type of ink can usually be removed by the sodium perborate poultice described above. Often the stain can be re- moved by applying ammonia water on cotton batting. Javelle water is also ef- fective, used the same as ammonia water, or mixed to a paste with whiting and applied as a poultice. A mixture of equal parts of chlorinated lime and whiting reduced to a paste with water may also be used as a poulticing material.

Some blue inks contain Prussian blue, a ferrocyanide of iron. These stains cannot be removed by the perborate poultice, Javelle (calcium or sodium hypo- chlorite) water, or chlorinated lime poultice. Such stains yield to treatment of ammonia water applied on a layer of cotton batting. Strong soap solution applied the same way may also be effective.

Indelible ink often consists entirely of synthetic dyes. Stains may be treated as outlined above for that type. However, some indelible inks contain silver salts which cause a black stain. This may be removed with ammonia water applied by bandage. Usually several applications are necessary.

Iron Stains: Method 1—Dissolve one part sodium citrate in six parts water. Mix thoroughly with equal volume of glycerin. Mix part of this liquid with whiting to form a paste just stiff enough to adhere to the surface in a thick coat. Apply with putty knife or trowel. This will dry in a few days. It should then be replaced with a new layer or softened by addition of more liquid. While this treatment has no injurious effects, its action may be too slow to be practical with bad stains. Ammonium citrate may produce quicker results than sodium citrate, but may injure polished surfaces slightly.

Method 2—For deep and intense iron stains, it is more satisfactory to use sodium hydrosulphite ($Na_2S_2O_4$). The surface should be first soaked with a solution made by dissolving one part of sodium citrate crystals in six parts water. Dip white cloth or cotton batting in this solution and paste over the stains for ten or 15 minutes. On horizontal surfaces, sprinkle over with thin layer of hydrosulphite crystals, moisten with water, and cover with stiff paste of whiting and water. On a vertical surface, place whiting paste on a plasterer's trowel, sprinkle on layer of hydrosulphite, moisten slightly, and apply to the stains. Remove after one hour. Do not leave longer or a black stain may develop. If stain is not completely removed, repeat the operation with fresh materials. When the stains disappear, rinse surface thoroughly with water.

Lubricating Oil Stains: Lubricating oils penetrate some concrete readily. It should be mopped off immediately, covering the spot with Fuller's earth or dry powdered material such as hydrated lime, whiting, or dry portland cement. If treated soon enough, there will be no stain. However, when the oil has remained for some time, other methods will be necessary.

Saturate white Canton flannel in a mixture of equal parts of acetone and amyl acetate and place over stain. Cover with slab of concrete or pane of glass. If stain is on a vertical surface, improvise means to hold cloth and its covering in place. Keep the cloth saturated until stain is removed. If the solvent tends to spread the stain, a larger cloth should be used. Covering saturated cloth with glass drives the stain into the concrete, while dry slab of concrete draws some oil into it. Scrubbing with gasoline or benzine will often remove oil stains.

Tobacco Stains: The following formula is generally effective. Dissolve 2 pounds of trisodium phosphate crystals in 1 gallon of hot water. Mix 12 ounces of chlorinated lime to a paste in a shallow enameled pan by adding water slowly and mashing the lumps. Pour this and the trisodium phosphate solution into a 2-gallon stoneware jar and add water until full. Stir well, cover the jar, and allow lime to settle. To use, add some of the liquid to powdered talc until a thick paste is obtained. Apply with trowel as a ¼-inch poultice. To apply with a brush, add about one teaspoon of sugar to each pound of powdered talc.

When dry, scrape off with wooden paddle or trowel. This mixture is a strong bleaching agent and is corrosive to metals. Care should be taken not to drop it on colored fabrics or metal fixtures.

This method is valuable for treating other stains. Trisodium phosphate may be purchased at drugstores, chemical supply, or laundry supply houses. If the stain is not bad, grit scrubbing powders, commonly used on marble, terrazzo, and tile floors are often satisfactory as a poulticing material. Stir powder into hot water until mortar consistency is obtained. Mix thoroughly then apply to stained surface in a ½-inch layer. Leave until dry. In most cases, two or more applications will be necessary.

Urine Stains: Use method as outlined for tobacco stains. Should the stain prove stubborn, saturate cotton batting in the liquids and paste over remaining stain. Resaturate the cotton if necessary.

Coffee Stains: Coffee stains can be removed by applying a cloth saturated in glycerin diluted with four times its volume of water. Javelle water, or the solution used on iron stains is also effective.

Iodine Stains: An iodine stain will gradually disappear of its own accord. It may be removed quickly by applying alcohol and covering with whiting or talcum powder. If on a vertical wall, mix talcum to a paste with alcohol, apply some alcohol to the stain, then cover with paste.

Caution: These treatments should be used by trained and experienced personnel. Improper use may result in bleaching the terrazzo matrix, if a color dye had been added.

Wood floors. Wood floors can be scrubbed to death with harsh cleaners. They will slip, appear dried out and shrunken. These conditions are caused by fast, harsh cleaners, with powerful wetting agents which have carried strong alkalies deep into the celled structure of the wood, causing crystallization, splitting, and drying out of the natural oils.

When this occurs the wood cell structure is opened up and bacteria, molds, and dirt are carried into these voids by cleaners, floor oils, and harmful wetting. Serious damage and deterioration of wood fibers results, necessitating costly reconditioning, increased maintenance, or even expensive replacement.

To insure against such damage, a treatment must be used to fill and close all wood cells, binding them together so that even moisture cannot penetrate the surface. Because of the lateral sawing of lumber, the surface is the factor of concern (1/64 inch in depth). Proper sealing of this surface will protect the whole floor.

The following nine points can be used as minimum standards of a good floor finish product:

1. The seal must penetrate the top surface of the wood.
2. It must seal the pores so as to keep dirt out and resist soil stains.
3. The finish must reflect light so as to improve illumination.
4. The finish with its penetrating quality must not darken the wood.
5. The finish must be nonslippery.
6. The finish must not mar, scratch, nor flake off.

7. The finish must be of such quality so that if it becomes necessary to touch up worn spots, it can be accomplished without complete refinishing.

8. The finish (sealer) should be resistant to water.

9. The finish, after application, must not present a maintenance problem. It must insure economy and maintenance as to eliminate constant resanding and complete refinishing.

Concrete floors. Concrete floors are made of a mixture of portland cement, sand, stone, and water. If the materials are properly mixed and skillfully troweled, a concrete floor will give many years of service—depending on traffic conditions and maintenance. Proper sealing will greatly lengthen concrete floor life, prevent dusting, and make maintenance easier and more economical.

Unsealed concrete floors create a number of problems. In the first place, they encourage, constant dusting which leads in time to gradual floor surface disintegration. Secondly, since dusting never stops once started—it is never entirely cleaned up, it is an unending maintenance headache, it gets into equipment and products, and it is a menace to the health of employees. And spillage seeps into the pores when there is no seal to stop it. Floors rapidly become unsightly because stain removal is difficult and often impossible. A sealed concrete floor helps eliminate these disadvantages.

However, free lime on the surface of new concrete will either attack or interfere with a bond of paints and seals. Therefore, in many cases the floor should be neutralized with a solution of muriatic acid before sealing. The recommended solution is one part of 30 per cent commercial muriatic acid added to fourteen parts of water. This solution should be mopped freely on the floor, and remain until bubbling or fizzing action is completed. The floor should then be rinsed and allowed at least six to eight hours of drying time. Only after thorough drying is the seal applied.

When the floor needs both cleaning and neutralizing, select a liquid detergent which when diluted with water and muriatic acid will both clean and neutralize in one operation.

Selection of the proper seal is important. Heavy traffic floors are best treated with a modified phenolic or epoxy resin floor seal. A first coat should be applied for maximum penetration so as to form a firm bond right in the pores of the concrete. A second coat should be applied to produce maximum wear resistance, and a smooth, easily maintained surface.

When a modified phenolic seal is used, there is no need to purchase a penetrating seal and a surface seal. A good modified phenolic seal is both because proper dilution with inexpensive mineral spirits or turpentine will produce the exact penetration required. Phenolic seals are available in several colored versions as well as in clear.

Where concrete floors are exposed to mineral oils and greases, mild acids and alkalies, use a chlorinated rubber-base seal. Where concrete is already saturated with oil, use a resin modified nitrocellulose-base primer before sealing. This locks in old oil, locks new spillage out, to provide a more easily maintained surface.

To clean sealed concrete floors use a neutral, synthetic, free-rinse liquid detergent. This will quickly emulsify oil and grease from stained surfaces without injuring the seal. Regular sweeping with a yarn broom treated with a liquid dust-control agent is all that is necessary to maintain sealed concrete floors. Such maintenance is easy—yet highly effective, efficient and economical.

Cleaning Materials

Regardless of the method used for floor maintenance, the end results will, to a great degree, depend on the quality of materials used. For example, the use of the wrong cleaner on resilient flooring can cause serious damage after only a few applications. There are a few general rules which can be of great assistance in choosing and applying the proper floor cleaners.

Always use a product specially designed for resilient floors. Most cases of serious floor damage are the result of using furniture polish, concrete cleaner, kitchen cleaners, soap and washing powders containing abrasives and free alkalies, rather than *floor* cleaners. There are hundreds of cleaners which have been formulated especially for resilient flooring. They are available from flooring manufacturers and their dealers, or by companies which specialize in floor maintenance materials. Both the Rubber Manufacturers' Association and the Asphalt and Vinyl Asbestos Institute have set up standards for resilient floor solutions and most floor cleaner manufacturers follow their recommendations.

Avoid excessive use of water or cleaning solution. Excessive use of water, especially hot water, may penetrate between tiles or seams to the adhesive and cause a loosening of the adhesive and eventually warp the floor. The use of excessive cleaning solution is not only uneconomical, but requires additional rinsing to prevent leaving a residue.

Do not use abrasive cleaners. Resilient floorings are relatively soft and may be permanently scratched by abrasives if used routinely. Abrasives may be used with care for removing deep-seated stains.

Avoid harsh alkaline cleaners. Strong alkalies may attack the binder, filler or color pigments used in resilient flooring to cause embrittlement, fading or roughness of the surface. In addition to these general rules, solvents and oils should not be used on rubber or asphalt tile floors. On other resilient floorings, solvents may be used sparingly for removing spots or stains. However, excessive solvent may penetrate between tiles or seams and cause adhesive bleeding or loosening of the floor.

Types of Coatings

There are several types of coatings for floors which can be used to protect and beautify clean floors. The proper choice of material depends on the type of floor and the type of traffic to which it is subjected. The following finishes are available at local dealers:

1. *Liquid water emulsion wax.* A dispersion of wax and other modifying materials in water. This is the most widely used maintenance finish. It is suitable for use on all types of resilient flooring except natural cork. Products with excellent slip resistance are available. Water emulsion waxes dry with a gloss which

may be improved by buffing with a floor machine. Scuffs and traffic marks may be buffed out. Worn areas are easily patched.

2. *Paste water emulsion wax.* Similar to the liquid water emulsion wax except in paste form. The floor must be polished for luster. Suitable for use on all types of resilient floors except natural cork. Water emulsion paste wax is usually applied with a thin steel wool pad on a floor polishing machine. The wax is spread with the machine and allowed to dry. It is then polished to a high luster with a second, thick steel wool pad. The lubricating effect of the wax prevents the steel wool pad from scratching the floor.

3. *Liquid solvent wax.* A mixture of waxes and other ingredients in a solvent base. The floor must be polished for luster. Liquid solvent wax should not be used on rubber or asphalt tile, but may be used on other resilient floors. Care should be taken on thinner gauge floors since excess solvent may penetrate the seams and cause bleeding of the adhesive.

4. *Paste solvent wax.* Similar to liquid solvent wax except in paste form. Should not be used on asphalt tile or rubber floors, but may be used on other resilient floors and is recommended for natural and pre-waxed cork floors.

5. *Water emulsion resins.* A dispersion of natural or synthetic resins and modifying materials in water. They are usually plasticized to impart some buffing. Water emulsion resins may be used on all resilient floors with the exception of natural and pre-waxed cork. They show poor adhesion when applied over waxed surfaces and will rapidly wear off under this condition. Generally, the water emulsion resins produce hard or tough coatings with a high luster. Because of their hardness, they tend to scratch rather than scuff and the scratches are more difficult to buff out than is the case with water emulsion waxes. For this reason, these finishes should be avoided in areas subjected to large amounts of sand or gritty soil. Some of the water emulsion resins have excellent resistance to oil and grease, and most of them are better in this respect than the waxes. Their high scuff resistance and luster make them well suited for prestige floors where good appearance is of paramount importance.

10B / FLOOR MOPPING

Job Description

The job of floor mopping pertains to the daily and/or periodic maintenance of floor surfaces in buildings by means of either damp mopping or wet mopping or a combination of both.

Damp mopping. Damp mopping is performed during inclement weather, to remove dirt and water from floors of building entrances and at intervals for cleaning hard floors such as marble, tile, and terrazzo in lobbies, halls and washrooms; and in other areas where lightly imbedded dirt is not adequately removed by the sweep mop and where a cleaning solution is not required.

Wet mopping. Wet mopping involves the use of a cleaning solution for thorough cleaning of hard floors where the damp mop procedures are inadequate, and where the building area and/or soil is insufficient to warrant the use of power scrubbing equipment on a regular basis. Wet mopping when performed as an adjunct to other jobs such as floor waxing is considered in the production rates for such jobs. The method used however is the same as described in this section.

Procedure. Floors should be swept with a treated sweep mop before wet mopping and before damp mopping as required. This work may be performed by the mop man or by room cleaners where assigned in the area to be mopped.

Employees assigned to this work are responsible for reporting any defective building items noticed in the work area during the performance of their duties.

Job Methods

Operation Steps	*Explanation*

A. *Damp Mopping:*

Step 1—Prepare for Work

1. Obtain equipment needed.	a. Bucket (12 qt.)
	b. Mop
	c. Mop unit (bucket and wringer, small mopping unit or mop tank)
	d. Putty knife (1¼″)
	e. Sweep mop, treated (as required)
	f. Warning signs
	g. Water hose (approx. 6 feet)
2. Proceed to work area.	Proceed to the work area where damp mopping is to start.

Step 2—Damp Mop the Floor

1. Prepare the work area.	a. Generally, it is not required that the floor area be cleared of movable furniture and equipment for this work.
	b. The area should be swept before damp mopping, if necessary, to remove visible debris and dirt.
2. Place warning signs.	Warning signs of wet floors should be placed in conspicuous locations.
3. Remove tar, gum and other substances.	Remove wads of gum, tar, and similar substances from the floor surface with putty knife.

Operation Steps	*Explanation*
4. Damp mop floor.	Immerse the mop in clean water and wring as dry as possible. Draw the mop parallel to baseboards and furnishings to avoid splashing these items before mopping the open areas. Turn the mophead over every four or five strokes to facilitate the cleaning. Rinse and wring the mop frequently. Change the water often enough to maintain a clean mop.
5. Mop strokes.	The mopping stroke should be one which gives the greatest coverage and speed with the least amount of fatigue. Where possible, use an approximately nine foot side-to-side sculling stroke, stopping about four inches of the baseboards and furnishings. Random or forward and backward strokes can be used under and/or around furnishings.

NOTE

1. New mops should be rinsed before use.
2. Mopping equipment should be kept ahead of the work in the area next to be mopped.

Step 3—Return Equipment to Storage

1. Replace moved pieces of furniture.	Replace any pieces of furniture moved during the damp mopping operation.
2. Warning signs.	When floor has dried remove the warning signs.
3. Final inspection.	Before leaving the work area, make a final inspection to see that the work has been performed as required.
4. Return to gear room location.	Gather up all equipment items and return to the gear room where items are to be stored.
5. Clean and store equipment.	Empty the mop unit and clean all equipment items before placing in their respective storage positions. (See "Equipment Care".) Hang wet mops to facilitate drying.

Operation Steps   *Explanation*

B. *Wet Mopping:*

Step 1—Prepare for Work

1. Obtain equipment and materials needed.

    a. Bucket (12 qt.)

    b. Deck scrub brush

    c. Detergent (may be prepackaged)

    d. Floor squeegee and water pickup equipment (optional)

    e. Measuring cup (if detergent is not prepackaged)

    f. Mop tank (three-compartment or mop tank, two-compartment and mop bucket or small mop units)

    g. Mops, cotton (2 pounds)

    h. Overshoes, rubber (1 pair per man)

    i. Putty knife (1¼")

    j. Scouring powder

    k. Sweep mop, treated (as required)

    l. Warning signs

    m. Water hose (approx. 6 feet)

    n. Wiping cloths

2. Prepare cleaning solution.

    Prepare the cleaning solution in one compartment of the mop tank using the type of detergent and in the proportions specified by the supervisor.

3. Fill rinse water compartment of mop tank.

    Add clean rinse water, as required, to another compartment of the mop tank, if a three-compartment tank is used, otherwise use a separate mop bucket. The wringer side is left empty to receive the waste liquids picked up from the floor.

4. Proceed to work area.

    Move the equipment and materials to the area where the work is to be done.

NOTE

Parts 2 and 3 of this step may be postponed until the equipment is moved to the work area.

Operation Steps	*Explanation*

Step 2—Prepare the Area

1. Prepare space to be mopped.	The area to be mopped should have been swept and any items located in the immediate path of the cleaning operation moved.
2. Place warning signs.	Caution signs warning of slippery floors should be placed in conspicuous locations.
3. Remove tar, gum, and other substances.	Use the putty knife to remove gum, spots of tar and other adhesive materials from the floor surface before mopping.

Step 3—Apply Cleaning Solution

1. Wet the mop.	Saturate the mop with cleaning solution in the mop tank. Let surplus solution drain from the mop back into the tank to prevent splashing.
2. Put down cleaning solution.	Draw the mop parallel to baseboards and furnishings to avoid splashing these items with the solution. Then spread the solution over a portion of the floor to be cleaned using side-to-side strokes where possible. Apply cleaning solution to an area approximately nine by twenty feet in sufficient quantity so that the floor does not dry before solution is picked up but not so much as to run under doors, and on floor coverings.
3. Mop strokes.	The mopping stroke should be one which gives the greatest coverage and speed with the least amount of fatigue. Where possible, use an approximately nine foot side-to-side sculling stroke, stopping about four inches short of the baseboards and furnishings. Forward and backward strokes can be used under and around furnishings. Turn the mophead over every four or five strokes to further distribute the solution.

Operation Steps   *Explanation*

Step 4—Clean the Floor

1. Agitate cleaning solution.

After a portion of the floor area has been wet with the solution, swing the mop over the covered area putting a downward pressure on the heel of the mop to remove heavily imbedded dirt.

2. Clean inaccessible areas.

Areas inaccessible to the mop such as corners may be cleaned by means of mop strands grasped in the hand.

3. Use scrub brush for hard-to-clean spots.

Use the deck scrub brush and scouring powder to clean areas that do not respond to the mop cleaning.

4. Pick up cleaning solution.

Wring the cleaning solution mop as dry as possible and pick up the used cleaning solution. Pass the mop over the wet floor area in side-to-side and/or forward and backward strokes turning the mophead over from time to time until it is saturated. During this pickup work continually wring the mop into the empty compartment of the mop tank provided for this purpose. Dry the floor of cleaning solution as thoroughly as possible.

NOTE

1. Optional equipment such as floor squeegee and pickup pan or wet vacuum equipment may be used for this water pickup work.
2. New mops should be rinsed before use.
3. Mopping equipment should be kept ahead of the work in the area next to be mopped.

Step 5—Rinse Floor (optional or as required)

1. Wet the floor.

Use a clean rinse mop, fully wet, to apply rinse water to the mopped area before the floor surface dries.

2. Pick up rinse water and dry mop floor.

Rinse and wring the mop as dry as possible and mop the floor dry. Change the rinse water as often as required to maintain a clean mop.

PRECAUTIONS

1. Take care not to splash baseboards, walls, or any articles nearby with cleaning solution and/or rinse water.

2. Be sure that corners and other hard to reach areas are properly cleaned.

3. Do not allow the mop wet with cleaning solution to stand in contact with the floor either in the cleaned areas or in those portions still to be cleaned as the concentration of solution may spot the floor surface at that point.

4. Keep the wet mops and solutions away from electric floor outlets to avoid possible short circuits.

5. Use a damp cloth to wipe spatterings or other water marks from walls, furnishings, and wherever necessary.

6. Avoid using an excessive amount of cleaning solution or rinse water during mopping or rinsing.

Operation Steps   *Explanation*

Step 6—Return Equipment and Materials to Storage

1. Replace moved furniture.	Replace any furnishings moved during the cleaning operation. Remove the caution signs that warned of slippery floors. Before leaving the work area, make a final inspection to see that the work has been performed as required.
2. Return to gear room location.	Gather up all equipment and materials and return to the gear room where items are to be stored.
3. Clean and store equipment.	Empty compartments of mop tank and clean all equipment items before placing in their respective storage positions. (See "Care of Equipment".) Hang all wet mops, cloths and brushes to facilitate drying.
4. Store usable materials.	Replace all unused materials in proper storage location.

NOTE

In small buildings without elevators where there is a need to carry the equipment up and down stairs, mop buckets or small portable mopping units should be furnished.

Care of Equipment

A. *Deck Scrub Brush:*

1. Rinse in clean water after each day's use and shake dry of excess water.

2. Place for drying and storage in a position so that the bristles do not bear the weight of the brush.

3. Avoid knocking the brush against objects as such action tends to split the brush block.

B. *Floor Squeegee (optional equipment):*

1. Rinse the squeegee blade in clean water after each day's use.

2. Wipe squeegee dry and store in a clean dry location.

C. *Measuring Cup:*

1. Wipe the cup free of detergent, and dry off all moisture.

2. Store so that the cup will not be damaged.

D. *Mop, Cotton:*

1. After each day's use and before storage, rinse the mop in clean water and squeeze or wring to remove excess water.

2. As required, wash dirty mops in a warm detergent solution, rinse in clean water and wring or squeeze to remove excess water.

3. Fluff the mop strands by shaking.

4. Store with the mophead up in a well-ventilated place to allow for better drying.

E. *Mop Tank, Bucket and/or Mop Unit:*

1. Rinse tank compartments, buckets, and wringer thoroughly and wipe them dry after each day's use.

2. Remove any remaining mop strands, strings, etc., from the compartments and wringer.

3. Store in a clean dry location with the wringer in a released position and otherwise maintain in a clean and orderly condition.

4. Report any defects or necessary repairs to the supervisor.

F. *Overshoes, Rubber:*

1. Rinse off dirt and other matter and wipe dry.

2. Store in such a manner that overshoes will not be damaged.

G. *Putty Knife:*

1. Wipe the knife free from moisture after each day's use.

2. Store in a clean dry place.

3. Use only for the purpose specified. Do not use as a screw driver or hammer.

H. *Sweep Mop, Treated:*

1. Shake out or vacuum the treated sweep mop during and at the close of operations each day. Keep free of grime or oily material.

2. Every two weeks or as required, remove the mophead, vacuum or shake out as much dust as possible and wash mop in a warm detergent solution. Rinse thoroughly in clear water and wring dry.

3. Treat the mop with prescribed mop treating material as directed by supervisor.

4. Replace the mophead on the frame and hang mop so that a good air circulation will be obtained to assist in drying the mop.

I. *Wet Vacuum Equipment (optional equipment):*

1. Rinse and dry water compartment after each day's use.

2. Wipe equipment free of dirt and moisture.

3. Coil hose and electric cord loosely and hang in proper storage positions.

4. Store equipment in clean dry location.

5. Report any defects or necessary repairs to the supervisor.

J. *Warning Signs:*

1. Wipe the signs free of dirt and/or moisture.

2. Store in assigned storage position so that they will not be damaged in any way.

K. *Water Hose:*

1. Drain the hose of water.

2. Wipe outside free of dirt and/or moisture.

3. Store in a dry space and in such a manner that the hose will not be damaged.

L. *Wiping Cloths:*

1. Rinse the wiping cloths in clean water after each day's use.

2. Wash dirty cloths as required in a warm detergent solution, rinse thoroughly and wring or squeeze out excess water before spreading them to dry.

Performance Inspection Guide

A. *Preparation for Mopping:*

1. Cleaning solutions, where used, have been mixed thoroughly and in the proportions specified without undue spillage of either solution or rinse water.

2. Proper precautions should have been taken to advise building occupants of wet and/or slippery floor conditions.

3. Taking care to prevent damage to furnishings, etc., the space to be mopped should have been properly prepared for the mopping operation by sweeping the floor area as necessary and otherwise clearing it of visible debris.

B. *Floor Mopping:*

1. The mopping work should have been performed in such manner as to properly clean the floor surface, care having been taken to see that the correct type and mixture of cleaning solution, if required, has been used.

2. All mopped areas should be clean and free from dirt, streaks, mop marks and strands, etc.; properly rinsed, if required, and dry mopped to present an overall appearance of cleanliness.

3. Wall, baseboards, and other surfaces should be free of watermarks, scars or marks from the cleaning equipment striking the surfaces and splashings from the cleaning solution and rinse water.

4. Care should have been taken throughout the mopping operation to prevent the liquids and equipment from coming into contact with electric outlets located in the floor areas or baseboards.

C. *Equipment and Materials:*

1. All equipment items used in the floor mopping work should be in a clean and well cared for condition and properly placed in a neat and orderly manner in assigned storage locations.

2. All unused materials should be properly stored in the assigned location.

Frequency and Production

A. *Frequency and Production Schedule:*

	Main Corridors	*Secondary Corridors*
Frequency	Daily	Weekly
Production Rate per Man-day	20,000 sq. ft.	

B. *Qualifying Factors:*

1. *Frequency*

   a. The frequencies given in the Frequency and Production Schedule are for a normal standard of cleaning under average conditions. These frequencies may be modified to meet local conditions.

   b. Main corridors are those located on the first or ground floor of a building or those that serve as primary access and/or exit areas. Secondary corridors are those less frequently used and usually located on upper floors and in the basement.

   c. Frequencies for other types of areas mopped should be established as required locally.

2. *Production Rate per Man-day*

   The production rate per man-day as established may be subject to modification at the various locations due to type of occupancy, floor condition, and traffic. They provide, however, a standard for certain buildings from which such deviation, if any, should be made.

Staffing

The man-day requirements of labor for floor mopping can be determined by use of the following formula:

$$\text{Man-days (of labor)} = \frac{\text{Quantity}}{\text{Production}} \times \frac{1}{\text{Frequency}}$$

where a quantity represents the square feet of floor area to be mopped at a given frequency; production is the square feet of floor area mopped per man-day, and frequency is the number of work days between operations.

If the frequency and/or production rate values as established in the Frequency and Production Schedule (under Floor Mopping) have been modified to meet existing conditions at a particular location, then those modified values should be used in the above formula for the computation of the labor requirement.

SAMPLE CALCULATION

Assume a building having a total of 5,000 square feet of main corridor area plus 25,000 square feet of secondary area, and with such existing conditions that the established values from the frequency and production schedule may be used; then compute the man-days of labor as follows:

$$\text{Man-days} = \overset{\textit{Main Area}}{\underbrace{\frac{\text{Quantity}}{\text{Production}} \times \frac{1}{\text{Frequency}}}} + \overset{\textit{Secondary Area}}{\underbrace{\frac{\text{Quantity}}{\text{Production}} \times \frac{1}{\text{Frequency}}}}$$

$$= \frac{5,000}{20,000} \times \frac{1}{1} + \frac{25,000}{20,000} \times \frac{1}{5}$$

$$= .250 + .250$$

$$= .500 \text{ man-days of labor required (jobs)}$$

To this total requirement must be added a factor to compensate for leave in order to arrive at the total number of positions needed to accomplish the work.

Conversion factor including compensation for leave = 1.15
Conversion factor × jobs = total positions required
1.15 × .500 = .575 positions required

• Where the floor mopping duty does not require the full time of the employee, other duties should be assigned to secure 100 per cent utilization of his time.

Equipment and Materials Allocation and Replacement

Items	Allotment per Cleaner	Replacement or Reissue
Bucket (12 qt.)	1	1 per year
Deck scrubbing brush	1	1 per year
Deck scrubbing brush handle	1	1 per 2 years
Detergent, synthetic or powdered	As required*	As required
Floor squeegee (18") (optional)	1	2 per year
Measuring cup	1	1 per year
Mop handle	2	2 per year
Mop, cotton (2 pounds, four-ply yarn)	2	6 per year
Mop tank or unit	1	1 per 5 years
Overshoes, rubber	1 pair	1 pair per year

Putty knife (1¼")	1	1 per year
Scouring powder	1 pound	As required
Sweep mop frame and handle, and mophead (optional)	1 1	1 per 3 years 1 per year
Vacuum equipment (optional)	1 unit	1 per 10 years
Warning signs	As required	As required
Water hose (approx. 6 feet)	1	1 per year
Wiping cloths	2	6 per year

* Recommended as a general purpose cleaner and in the proportions of one ounce per gallon of water or as instructed.

The allotment per cleaner and replacement or reissue quantities are based on one cleaner assigned floor mopping duty 100 per cent of his time.

• Where the situation exists that other related jobs are assigned to the cleaner to secure 100 per cent utilization of his time, consideration should be given to reduction in the replacement quantities in accordance with the actual amount of floor mopping duty assigned.

pH SCALES
OF CLEANING PRODUCTS FOR SAFETY

This pH is safe for these surfaces (at normal temperatures and for normal exposure times):

For glass, glazed china, glazed ceramic tiles	0.0–13.0
(Ditto, in very hot solutions)	0.0–12.0
(*Exception:* chemicals containing hydrofluoric acid)	
For rubber, both natural and synthetic, and	
polyethylene and most plastics	0.0–13.0
For stainless steel	3.0–14.0
For bronze and brass	5.0–13.0
For aluminum and its alloys, magnesium	5.0–10.0
For common steel and cast iron	8.0–14.0
For stainless steel	3.0–14.0
For marble and unfinished concrete	7.0–12.0
For sealed or varnished concrete	2.0–10.0
For unfinished wood floors	3.0–11.0
For sealed, varnished or well waxed floors	2.0–10.0
For linoleum (battleship type)	2.0– 9.5
For asphalt tile, vinyl tile, rubber tile	2.0–10.5

10C / FLOOR WAXING AND BUFFING

Job Description

The job of floor waxing and buffing pertains to the scheduled periodic maintenance of floor surfaces by such procedures as cleaning to remove dirt and old wax, application of a thin wax film and polishing. Spot refinishing of heavy traffic areas, damp mopping and separate buffing work, as required, are also included as part of this job.

The floor waxing and buffing methods as outlined in this chapter are broken down into several component parts such as:

1. *Preparation for Work:* Consisting of preparation of the equipment and materials for the job, including cleaning solutions, and movement to the work area to commence work.

2. *Space Preparation:* Preparing the area to be waxed and buffed by movement of furniture, turning back edges of rugs, etc.

3. *Wet Cleaning* (as applicable): Including sweeping the floor, as required, removal of gum and other adhesive substances, removal of dirt and old wax film, scrubbing with machine if required, pickup of used cleaning solution, rinsing and dry mopping.

4. *Waxing:* Application of a thin film of the proper wax to the floor surface in the correct manner and allowing to dry.

5. *Buffing:* Polishing the floor area to an acceptable degree of luster either after the wax application or as a separate job after damp mopping in lieu of rewaxing.

6. *Space Readjustment:* Turning down all rug edges previously turned back and replacement of all moved furniture to its original location without damage to the furniture or wall areas.

7. *Return of Equipment and Materials to Storage:* Return of all equipment and unused materials to storage locations, cleaning and proper storage of equipment used and the replacement of unused material to storage.

Employees assigned to this work are responsible for reporting any defective operating equipment. They should also report burned out lights, floors in need of repairs and other defective items noticed in the work area during the performance of their duties.

Job Methods

Operation Steps	*Explanation*

A. *Using Water Emulsion, Spirit Wax* * *and Rotary Type Polishing Machine*

Step 1—Preparation for Work

1. Obtain necessary equipment and materials.	a. Bucket (12 qt.) b. Mops, cotton c. Mop tank, three compartment; or mop tank, two compartment and mop bucket d. Mop unit (2 bucket with wringers) e. Polishing machine and accessories f. Putty knife g. Water hose (approx. 6 feet)

Operation Steps	*Explanation*
	h. Detergent (may be prepackaged) or floor cleaning solvent **
	i. Steel wool No. 00
	j. Warning signs
	k. Wax (water emulsion or spirit)
	l. Wax applicator (optional)
	m. Wiping cloth

* Spirit waxes should be used on wood floors or linoleum floors. Spirit waxes and solvents should never be used on asphalt, rubber and similar type floors.

** Approved wax strippers may be substituted for use in lieu of prescribed cleaning solutions as authorized by supervisors.

2. Prepare floor cleaning solution.	Prepare the floor cleaning solution in one compartment of the mop tank using the type cleaning material and proportions as specified by the supervisor. Add clear rinse water, if required, to another compartment of the mop tank. If rinse water is not required in the cleaning operation then the two compartment tank is adequate for this work. The wringer compartment of the tank, or the mop bucket with wringer is left empty to receive the waste liquids picked up from the floor.
3. Proceed to the work area.	Move the required equipment and material to the area where work is to start.

Step 2—Space Preparation

1. Move furniture.	a. Clear the floor area that is to be waxed of all wastebaskets, chairs and other light movable pieces of furniture. Move these items in an orderly fashion so that they can be replaced easily in their original positions.
	b. Duplicate numbered tags may be utilized to assist in replacement of the moved equipment.
2. Turn back edges of rugs.	Where floors are covered with rugs, turn back the edges of rugs so that

Operation Steps	*Explanation*
	the floors for a distance of approximately one foot beneath the rug edges may be cleaned and waxed.
3. Place warning signs.	Erect warning signs in the areas that are being cleaned and/or waxed to caution occupants about the wet floors.
4. Remove tar, gum, etc.	Use the putty knife to take up wads of gum, spots of tar, tape and other adhesive materials from the floor surface.

PRECAUTIONS

1. Take care not to scar the furniture or walls.
2. Replace all moved equipment items in their proper locations upon completion of the waxing work.
3. Do not place chairs, wastebaskets, and other items on top of desks and tables.

Step 3—Wet Cleaning

| 1. Wet the mop. | The area to be cleaned and waxed should have been swept to remove loose dirt and visible debris. (This floor sweeping work is usually performed by the room cleaner.) Saturate the mop in the cleaning solution in the mop tank. Let surplus solution drain from the mop back into the tank to prevent splashing. |

NOTE

New mops should be rinsed before being put into use.

| 2. Put down cleaning solution. | With first strokes, cut in section to be cleaned. Draw the mop parallel to baseboards and furnishings to avoid splashing these items with solution. Then outline an approximate area of nine by twelve feet, and mop inside this area with side to side strokes. Use enough cleaning solution so that the floor stays damp during the in- |

Operation Steps   *Explanation*

terval it remains on the floor surface to permit maximum cleaning action. (Solution should not be allowed to remain on floor longer than ten minutes.)

3. Mop strokes.

The mopping stroke should be one which gives the greatest coverage and speed with the least amount of fatigue. Where possible, use an approximate nine foot side-to-side stroke, stopping about four inches short of the baseboards and furnishings. Forward and backward strokes can be used under and around furnishings. Turn the mop over every four or five strokes to further distribute the solution.

4. Agitate the cleaning solution.

Starting at the point where the cleaner was first applied, agitate the solution on the floor surface with a saturated mop to loosen the dirt and old wax film. Clean small hard to reach areas by means of several mop strands grasped in the hand.

NOTE

Where the use of mechanized scrubbing equipment will result in more efficient performance for this agitation then such a machine should be used where possible.

5. Removal of spots and stains.

Spots and stains may require the application of No. 00 steel wool and/or mild abrasive powder.

6. Very dirty floors.

In cases of very dirty floors, agitation may be performed by means of the polishing machine and scrubbing brushes or fine grade steel wool pads.

7. Pick up the cleaning solution.

Immediately after agitation wring the mop as dry as possible and proceed with picking up the used cleaning solution. Pass the mop over the wet floor area in side-to-side and back-

Operation Steps   *Explanation*

ward and forward strokes, turning it over from time to time until it is saturated. Continually wring the mop during this pickup work. Dry the floor of cleaning solution as thoroughly as possible.

8. Rinse and dry mop the floor.

a. On floors where rinsing is required, use a clean rinse mop and rinse water to thoroughly rinse the cleaned floor area before the floor surface dries.

b. Wring the mop as dry as possible, and go over the floor removing all rinse water. Change the rinse water often enough to maintain a clean mop during this operation.

NOTE

1. This step applies only to those types of floors that are not harmed by the application of water.

2. Avoid use of water solutions on wood flooring.

Step 4—Waxing

1. Arrange equipment for waxing.

Pour an adequate amount of wax into the mop bucket. Place equipment in best location to start the waxing operation. Where practical, start at a point farthest from the exit, and work toward the exit.

2. Wetting the mop with wax.

a. Although the instructions relate specifically to the use of a mop for this waxing work, an applicator with pad may be substituted for the mop with equally satisfactory results.

b. Thoroughly saturate the lower two-thirds of a clean mop with wax. If water emulsion wax is used, wet the mop with water and wring before saturating with wax. Wring excess wax from the mop leaving just sufficient wax in the mop to

Operation Steps	*Explanation*
	deposit the desired wax film upon the floor.
3. Apply wax to floor.	Using long sweeping strokes from side-to-side with slight overlap on each stroke, spread a thin film of wax evenly over the cleaned floor area. If polishing is to follow the waxing operation, an approximately four-inch space at the baseboard area is not waxed as buffing will spread over enough wax to cover this area of little or no traffic. Repeat wetting the mop with wax as needed to spread the desired wax film onto the floor.
4. Rinsing the wax mop (when water emulsion wax is used).	When water emulsion wax is used, rinse the wax mop periodically to remove wax accumulations from the mop.
5. Keep traffic from waxed area.	Where possible, keep all freshly waxed areas blocked off from traffic to allow proper drying.

NOTE

Where paste wax is used in place of liquid wax use pads of folded cloth to spread the wax onto the floor area. This type of wax application should be limited to very small areas where liquid wax is undesirable.

PRECAUTIONS

1. Be sure that the buckets and mops are clean before using.
2. The floor area should be clean and dry to touch before applying wax.
3. Take care to avoid splashing baseboards and furniture with the mop or wax.
4. Be careful not to leave puddles or streaks of excess wax on the waxed floor during the wax application.
5. Keep only enough wax in the wax bucket to sufficiently wet the mop.

Step 5—Buffing

1. Set up machine.	Inspect the brushes to see that they are in satisfactory operating condi-

Operation Steps	*Explanation*
	tion. Snap the brush plate onto the machine.
2. Connect machine.	See that the OFF-ON switch on the machine is in the OFF position. Insert the cable plug into the electric outlet.
3. Start the machine.	Raise the brushes from the floor before starting or stopping the machine. Place the switch in the ON position and lower the brushes to the floor to start the polishing operation.
4. Polish the floor.	Starting in an area farthest from the room exit, polish the waxed floor. Where possible, guide the machine from side-to-side, covering the greatest area with the fewest possible steps by the operator.

NOTE

If necessary, the use of a felt carpet, or lamb's wool pad fitted beneath the brush is effective in removing highlights from the floor after the tampico brush buffing is completed.

PRECAUTIONS

1. Where possible, allow the wax or floor surface to dry to the touch before starting to buff.
2. Avoid damaging or scarring the furniture and baseboards with the buffing machine.
3. Erect signs or take other precautionary measures to warn occupants against possibilities of slipping on the wet floors.

Step 6—Space Readjustment

1. Turn down edges of rugs.	When the waxing and buffing work is completed in the space, turn down all folded up edges of rugs.
2. Replace furniture.	Replace all furniture to its original location taking care to avoid damage to furniture and walls.
3. Final inspection.	When all work has been completed in an area, make a final inspection to see that the work has been performed

Operation Steps	*Explanation*
	as required and that all moved articles have been replaced correctly.

Step 7—Return Equipment and Materials to Storage

1. Gather up equipment and materials.	Gather together all pieces of equipment and materials being careful to leave the floor areas in a clean condition.
2. Return to gear room.	Upon completion of work, return equipment and unused materials to their respective storage locations.
3. Clean and store equipment and usable materials.	Clean all equipment (mops, mop tank, buckets, etc.) before placing in their respective storage locations. Hang all wet mops and brushes to facilitate drying. Replace all unused cleaning materials in their proper storage locations.

PRECAUTIONS

Any wax remaining in the wax bucket upon completion of the waxing operation *should not* be poured back into the new wax container but stored in a separate container. If poured back into a new wax container it will contaminate the new wax and generally results in spoilage of the wax supply.

B. *Damp Mopping and Buffing:*

NOTE

Damp mopping and buffing may sometimes be used in place of or in between the regular waxing operation.

Step 1—Preparation for Work

1. Obtain necessary equipment.	a. Bucket (12 qt.) b. Mops, cotton c. Mop tank or mop bucket unit d. Polishing machine and accessories e. Putty knife f. Warning signs g. Wiping cloths
2. Proceed to work area.	Move the required equipment to the area where the damp mopping and buffing work is to start.

Operation Steps   *Explanation*

Step 2—Space Preparation

1. Move furniture.

Clear the floor area that is to be damp mopped and buffed of wastebaskets, chairs and other light movable pieces of furniture. Turn back edges of rugs. Move the furnishings in an orderly fashion so that they can be replaced easily in their original positions.

2. Place warning signs.

If necessary, place caution signs in conspicuous locations to warn occupants of slipping or tripping hazards.

NOTE

The area to be buffed should have been swept to remove loose dirt and other waste material. This floor sweeping work is usually performed by the room cleaner.

Step 3—Damp Mopping and Buffing

1. Damp mop the floor.

Using the clean mop dampened in clean cold rinse water and wrung as dry as possible in the wringer, go over the floor area to be buffed employing the same basic method as that used for wet mopping. Rinse and wring the mop frequently. Change the water often enough to maintain a clean mop. Use putty knife to remove gum, tape, etc. from floor surface.

2. Polish the floor.

Follow the same procedure in buffing after damp mopping as that outlined for buffing after waxing.

PRECAUTION

Avoid scarring furniture, walls or baseboards with the polishing equipment.

Step 4—Space Readjustment

1. Replace furniture.

Replace all moved pieces of furniture taking care to avoid damage to furniture and walls. Turn down all folded up rug edges. Arrange these replaced articles so that each is placed in its original location as nearly as possible.

Operation Steps	*Explanation*
2. Final inspection.	When all work has been completed in an area, make a final inspection to see that the work has been performed as required.

Step 5—Return Equipment to Storage

1. Return to gear room.	Upon completion of work, return all equipment to their respective storage locations.
2. Clean and store equipment.	Clean all items of equipment before placing in their respective storage positions. Hang all wet mops and brushes to facilitate drying.

Care of Equipment

A. *Applicator, Wax:*

1. Remove and thoroughly wash applicator pad in detergent solution, or in floor cleaning solvent if spirit wax is used.
2. Rinse thoroughly and shake free of excess liquid.
3. Replace pad on block and store applicator where there is free circulation of air and where there will be no weight against it.

B. *Cloths, Wiping:*

1. Rinse wiping cloths in clean water after each day's use.
2. Wash dirty cloths, as required, in warm detergent solution and rinse thoroughly.
3. Wring or squeeze excess water from the cloths before spreading to dry.
4. *Do not* leave wet dirty cloths piled together in the storage location.

C. *Cup, Measuring:*

1. Wipe cup free of detergent and other matter. Dry off all moisture.
2. Store so that it will not be damaged.

D. *Hose, Water:*

1. Wipe free of dirt and moisture.
2. Store in a dry place and in such a manner that the hose will not be damaged.

E. *Mops, Cotton:*

1. After each use and before storage, rinse mops in clean water and wring or squeeze to remove excess water.
2. Mops used for waxing should be cleaned within a few minutes after their use before the wax remaining in the mop has set.
3. Wash dirty mops in warm detergent solution; rinse in clean water and wring or squeeze dry.

4. Fluff the mop strands by shaking.

5. Store mops with mopheads up in well-ventilated area to allow for better drying.

F. *Mop Tanks, G. Mop Unit, Buckets and Wringers:*

1. Rinse tank compartments, buckets and wringers thoroughly and dry them after each use.

2. Remove any remaining mop strands and other foreign matter from compartments, buckets and wringers.

3. Avoid any damage to the equipment such as would tend to shorten its life.

4. Store in a dry place and otherwise maintain in a clean and orderly condition.

5. Report any defects or equipment failure to the supervisor.

H. *Polishing Machine and Accessories:*

1. Remove the brush or brushes from the machine. Wash dirty brushes in a warm detergent solution, rinse thoroughly and shake free of excess water. Hang brushes or lay with bristles up to facilitate drying.

2. After each day's use, wipe the machine and cord free of dirt and dust with a damp cloth. A cloth dampened with detergent solution may be used to remove spots otherwise not removed. Wipe the machine and cord dry and rewind the cord loosely on the hooks provided.

3. If cylindrical type polishing machine is used, remove dust bag, empty and replace. Clean the fan at the direction of the supervisor.

4. Wash and rinse the machine wheels and wipe dry.

5. Use metal polish as directed to clean and polish the machine.

6. Store the machine and accessories with the brush removed in assigned storage space.

7. Report needed repairs to the supervisor.

I. *Putty Knife:*

1. Wipe the knife free of dirt or moisture.

2. Store in clean dry place.

3. Use only for purpose specified. Do not use as screw driver or hammer.

J. *Warning Signs:*

1. Wipe the signs free of dirt and other matter.

2. Store in assigned storage positions so that they will not be damaged in any way.

Performance Inspection Guide

A. *Preparation of Floor Area for Waxing:*

1. The floor area should be free of dirt and dissolved wax particles, cleaning material residue, streaks, mop strands, and otherwise be thoroughly cleaned.

2. Walls, baseboards, furniture bases and other surfaces should be free of

watermarks, marks from the cleaning equipment, and splashings from the floor cleaning solutions.

3. All cleaned surfaces should be wiped dry and the floor ready for application of the wax.

B. *Waxing:*

1. The surface waxed should have had the proper type of wax applied in accordance with best operating practices.

2. The wax should have been applied thinly, uniformly and evenly in such a manner as to avoid skipping of areas, and has been allowed to properly dry before being polished.

3. Walls, baseboards, furniture and other surfaces should be free of wax residue and marks from the equipment.

4. The waxed area should be free of streaks, mop strand marks, skipped areas and other evidence of improper wax application.

C. *Buffing:*

1. The waxed or damp mopped surface shall have dried to the touch before being buffed.

2. Baseboards, furniture and equipment should not be disfigured or damaged during the buffing work.

3. The finished area should be polished to an acceptable, uniform lustre and free of extreme highlights from the brushes of the machine.

D. *Furniture Arrangement in Waxed Area:*

1. All rug edges should be replaced to their proper position.

2. All moved items of furniture and office equipment should be returned to their original positions.

3. Care should have been exercised to avoid damage to building and/or office equipment during movement of the furniture, etc.

E. *Equipment and Materials:*

1. All items of equipment used in the waxing and/or buffing work should be clean and properly placed in a neat and orderly manner in their assigned storage locations.

2. All remaining materials should be returned to their assigned storage locations.

3. Left-over wax, if kept, should not be returned to the original wax container with the new wax, but stored in a separate container securely covered.

Frequency and Production

A. *Frequency and Production Schedule:*

1. Clean floor, apply wax, polish, using rotary type polishing machine:

	Office Area	Open Area
Frequency	Every 66 days	
Production Rate per Man-day	3,000 sq. ft.	5,000 sq. ft.

2. Damp mopping and buffing; using rotary type polishing machine:

	Office Area	Open Area
Frequency	As required	
Production Rate per Man-day	30,000 sq. ft.	40,000 sq. ft.

NOTE

Polish floor without applying additional wax.

B. *Qualifying Factors:*

1. *Frequency*

a. The frequencies given in the Frequency and Production Schedule are for a normal standard of cleaning under average conditions. These frequencies may be modified to meet local conditions.

b. Frequencies for other types of areas should be established as required locally.

2. *Production Rate per Man-day*

The production rates as established may be subject to modification at the various locations due to type of occupancy, floor condition, etc. They provide, however a standard, from which such deviation, if any, can be made.

Staffing

The man-day requirements of labor for floor waxing and buffing can be determined by use of the following formula:

$$\text{Man-days (of labor)} = \frac{\text{Quantity}}{\text{Production}} \times \frac{1}{\text{Frequency}}$$

where quantity represents the square feet of floor area waxed and/or buffed at a given frequency; production is the square feet of floor area waxed and/or buffed per man-day for the type of job method utilized; and frequency is the number of work days between operations.

If the frequency and/or production rate values as established in the Frequency and Production Schedule of this section have been modified to meet existing conditions at a particular location, then those modified values should be used in the above formula for the labor requirement computation.

SAMPLE CALCULATION

Assume a building having a total waxing and/or buffing area of 239,400 square feet of office space and 30,200 square feet of open space with existing conditions such that the established values from the Frequency and Production Schedule for the rotary type polishing machine may be used.

Assume also that the 30,200 square feet of open area is to be damp mopped and buffed two times in-between the regular every 66 day waxing operation, or eight times per year. The following chart shows a typical schedule for this work:

SCHEDULE FOR WAXING, DAMP MOPPING
AND BUFFING

Waxing Frequency (Days)		66			66	
Damp Mopping and Buffing Frequency Interval (Days)	22	22	22	22	22	22
Operation to Be Performed	Wax	Buff	Buff	Wax	Buff	Buff

For purpose of computing the man-day requirements for this damp mopping and buffing work in the open area, the frequency value to be used is 33 days as the area would be buffed twice in the 66 day interval between waxing operations.

Then the total man-day requirement is computed as follows:

$$\text{For normal waxing} = \underset{\textit{Office Area}}{\frac{239,400}{3,000} \times \frac{1}{66}} + \underset{\textit{Open Area}}{\frac{30,200}{5,000} \times \frac{1}{66}} = 1.301$$

$$\text{For damp mopping and buffing} = \frac{30,200}{40,000} \times \frac{1}{33} = \frac{.023}{1.324}$$

Total man-days
of labor required
(jobs)

To this total requirement must be added a factor to compensate for leave in order to arrive at the total number of positions needed to accomplish this work.

Conversion factor including compensation for leave = 1.15
Conversion factor × jobs = total positions required
1.15 × 1.324 = 1.52 positions required

• Where the floor waxing and buffing duty does not require the full time of the employee, other duties should be assigned to secure 100 per cent utilization of his time.

Equipment and Materials Allocation and Replacement

Items	Allotment per Wax Crew	Replacement or Reissue
Applicator, complete wax (optional)	1	1 per 2 years (refill pads as required)
Basket (12 qt.)	1	1 per year

Items	*Allotment per Wax Crew*	*Replacement or Reissue*
Cloth, wiping	3	12 per year
Cup, measuring	1	1 per year
Detergent, synthetic	As required	As required
Handle, mop	3	3 per year
Hose, water (6 feet)	1	6 feet per year
Mop, cotton, wet (2 pounds)	3	12 per year
Mop unit (2 buckets with wringer)	1	1 per 5 years
Polishing machine (with accessories)	type and quantity As required	As required As required
Putty knife (1¼″)	1	1 per year
Solvent, floor cleaning	As required	As required
Steel wool No. 00	As required	As required
Tank, mop	1	1 per 5 years
Warning sign	As required	As required
Wax, liquid	As required	As required
Wax, water emulsion (16 per cent)	As required	As required

10D / FLOOR SCRUBBING

Job Description

The daily or periodic maintenance of marble, terrazzo, and other floor surfaces in buildings is accomplished by scrubbing with mechanized scrubbing equipment.

The job method involves preparation for the scrubbing work by readying the equipment and preparing the floor cleaning solution; preparation of the area to be scrubbed; scrubbing the floor area; and care, maintenance and proper storage of the equipment and materials used for this work.

Employees assigned to this work are responsible for reporting any defective operating equipment. They should also report burned out lights and other defective items noticed in the work area during the performance of their duties.

Job Methods

A. *Electric Floor Polishing Machine (Used for Scrubbing):*

Operation Steps	*Explanation*

Step 1—Prepare for Work

1. Obtain necessary equipment and materials.	a. Buckets (12 qt.)
	b. Detergent (may be prepackaged)
	c. Floor polishing machine, electric, with scrub brushes. (Check

Operation Steps	*Explanation*
	brushes for wear; if bristles are less than one-half inch long, install new brushes.)
	d. Measuring cup (if detergent is not prepackaged)
	e. Mop tank, three compartment; or mop tank, two compartment and mop bucket.
	f. Mops, cotton (2 pounds)
	g. Overshoes, rubber
	h. Putty knife (1¼")
	i. Warning signs
	j. Water hose (approx. 6 feet)
	k. Wiping cloths
2. Prepare cleaning solution.	Prepare the cleaning solution in one compartment of the mop tank using the type of detergent and the proportions specified by the supervisor.
3. Proceed to work area.	Move the equipment and materials to the area where the work is to start.

NOTE

Parts 1 and 2 of this step may be postponed until the equipment is moved to the work area.

Step 2—Apply Cleaning Solution

1. Prepare space.	The area to be scrubbed should have been swept if necessary and any items located in the immediate path of the cleaning operation moved. Caution signs warning of slippery floors should be placed in conspicuous locations.
2. Wet the mop.	Saturate the mop in the cleaning solution in the mop tank. Let surplus solution drain from the mop back into tank to prevent splashing.
3. Put down cleaning solution.	Spread the solution over the area to be scrubbed using side-to-side strokes where possible. Draw the mop parrallel to the baseboards and furnishings to avoid splashing these items

Operation Steps	*Explanation*
	with solution. Apply sufficient solution so that the floor does not dry before the solution is picked up.
4. Mop strokes.	The mopping stroke should be one which gives the greatest coverage and speed with the least amount of fatigue. Where possible, use an approximately nine foot side-to-side stroke, stopping about four inches short of the baseboards and furnishings. Forward and backward strokes can be used under and around furnishings. Turn the mophead over every four or five strokes to further distribute the solution.
5. Remove gum, tar and similar substances.	The putty knife is used during this step to remove wads of gum, tar and similar foreign substances from the floor surface.

NOTE

Solution tanks are also available for attaching to the polishing machine so that the cleaning solution may be put down at the same time the floor is scrubbed.

Step 3—Scrub Floor

1. Attach brush.	Check the brush for wear and put on new one as required. Attach the brush or brush plate by placing the polisher on its side and securing the brush on the plate or shaft by hand. Do not attach brush by running machine over it and allowing it to lock in place by starting the motor.
2. Connect machine.	Make sure the OFF-ON switch on the machine is in the OFF position. Insert the cable plug into the electric outlet and secure cable at this point to prevent damage by pulling on plug and/or outlet.
3. Operate scrubbing machine.	Most scrubbing machine brushes rotate in a counterclockwise direction.

Operation Steps	*Explanation*
	The movement of the concentrated weight machine is controlled by raising the handle to move to the right and lowering the handle to move to the left. The divided weight machine is operated by moving the machine in a straight forward or backward direction. The scrubbing operation should start with the machine directly in front of the operator.
4. Proceed with scrubbing.	Press down on the machine handle to start with the brush off the floor. Place the OFF-ON switch in the ON position and begin scrubbing. Use machine to provide maximum coverage and speed with the least amount of fatigue.
5. Mop areas inaccessible to machine.	Areas inaccessible to the machine may be scrubbed by hand with the deck scrub brush. In all cases the machine or brush is directed in such a manner that the cleaning solution is moved in the direction in which the scrubbing is being done.
6. Pick up cleaning solution.	Immediately after scrubbing, wring the mop as dry as possible and pick up the used cleaning solution. Pass the mop over the wet floor area in side-to-side and/or forward and backward strokes, turning the mophead over from time to time until it is saturated. During this pickup work continually wring the mop into the empty compartment of the mop tank provided for this purpose. Dry the floor of cleaning solution as thoroughly as possible.

PRECAUTIONS

1. Take care not to splash baseboards, walls, and other property with cleaning solution.
2. Do not permit the machine to strike and scar the furnishings during the scrubbing operation.

3. Do not let the machine stand on the scrubbing brushes except when in use as this tends to flatten the bristles and make the brushes useless for the purpose intended.

4. Do not allow the wet mops and/or scrub brushes to rest on the floor either in cleaned area or in those portions still to be cleaned.

5. Be sure that corners and other hard to reach areas are cleaned during this operation.

Operation Steps	*Explanation*

Step 4—Rinse Floor (*optional or as required*)

1. Wet the floor.	Use a clean rinse mop, fully wet, to apply warm rinse water to the scrubbed area before the floor surface dries.
2. Pick up rinse water and dry mop floor.	Rinse and wring the mop as dry as possible and go over the floor removing all moisture. Change the rinse water as often as required to maintain a clean mop.

Step 5—Return Equipment and Materials to Storage

1. Replace moved furniture.	Replace any pieces of furniture moved during the scrubbing operation. Remove the caution signs that warned of slippery floors. Before leaving the work area, make a final inspection to see that the work has been performed as required.
2. Return to gear room location.	Gather up all pieces of equipment and materials and return to gear room where items are to be stored.
3. Clean and store equipment and usable materials.	Empty compartments of mop tank and clean all equipment items before placing in their respective storage positions. (*See* "Equipment Care".) Hang all wet mops, cloths and brushes to facilitate drying. Replace all unused materials in proper storage locations.

Care of Equipment

A. *Bucket:*
   1. Rinse and dry the inside and outside after each day's use.
   2. In storing, place the buckets so that they will not be damaged.

B. *Floor Polishing Machine:*

1. Remove the scrubbing brush from the machine and wash as specified under H.

2. After each day's use, wipe the machine and cord free of dirt and dust with a damp cloth. A cloth dampened with detergent solution may be used to remove spots not otherwise removed. Wipe the machine and cord dry and rewind the cord loosely on the hooks provided.

3. Wash and rinse the machine wheels and wipe dry.

4. Use metal polish as directed to clean and polish the machine surfaces.

5. Store the machine and brush (with brush removed) in assigned storage space.

6. Report needed repairs to supervisor.

C. *Measuring Cup:*

1. Wipe cup free of detergent and other matter. Dry of all moisture.

2. Store so that it will not be damaged.

D. *Mops, Cotton:*

1. After each use and before storage, rinse mops in clean water and wring or squeeze to remove excess water.

2. Wash dirty mops in warm detergent solution. Rinse in clean water and wring or squeeze to remove excess water.

3. Fluff the mop strands by shaking.

4. Store mops, with mopheads up, in well-ventilated place to allow better drying.

E. *Mop Tank and/or Mop Bucket:*

1. Rinse tank compartments, buckets and wringers thoroughly and wipe dry after each day's use.

2. Remove any mop strands or other foreign matter from compartments, buckets and wringers.

3. Always leave wringer in released position.

4. Avoid damage to the equipment.

5. Store in a dry place and otherwise maintain in a clean and orderly condition.

6. Report any defects or equipment failure to the supervisor.

F. *Overshoes, Rubber:*

1. Rinse off dirt and other matter and wipe dry.

2. Store in such a manner that they will not be damaged.

G. *Putty Knife:*

1. Wipe the knife free of dirt and moisture.

2. Store in clean dry place.

3. Use only for purpose specified. Do not use as screw driver or hammer.

H. *Scrubbing Brushes:*

1. Remove scrubbing brushes from floor polishing machine after each day's use.

2. Wash all scrubbing brushes in warm detergent solution. Rinse thoroughly and shake free of excess water.

3. Store brushes so that the bristles are free to straighten out.

I. *Scrubber Vacuum Machine:*

1. Drain and flush both the cleaning solution and the dirty water tanks.

2. Remove vacuum screen and clean.

3. Clean and reverse brushes. Install new brushes if required.

4. Clean guide wheel and under carriage.

5. Clean and replace brush shields.

6. If electric scrubber is used, clean and dry wipe electric cable.

7. Wipe dirt and other matter from all machine surfaces except gasoline engine.

8. Store machine in assigned storage position.

9. Report any needed repairs to supervisors. Do not attempt to repair the machine or accessories.

J. *Warning Signs:*

1. Wipe the signs free of dirt and other matter.

2. Store in assigned storage positions so that they will not be damaged.

K. *Water Hose:*

1. Wipe free of dirt and/or moisture.

2. Store in a dry space and in such a manner that it will not be damaged.

L. *Wiping Cloths:*

1. Rinse wiping cloths in clean water after each day's use.

2. Wash dirty cloths as required in warm detergent solution and rinse thoroughly.

3. Wring or squeeze the excess water from cloths before spreading to dry.

4. Hang up cloths to dry. Do not leave wet dirty cloths piled together in the storage location.

Performance Inspection Guide

A. *Preparation for Work:*

1. The machine and other equipment should have been checked and readied for work in a careful and thorough manner.

2. Additions of motor oil, where required, should have been accomplished in a safe and careful way so as to avoid spillage and overfilling.

3. Cleaning solutions should have been mixed thoroughly and in proportions specified without undue spillage of either solution or rinse water.

B. *Operation of Machine:*

1. The mechanized equipment should have been operated only by authorized personnel having sufficient instruction as to its proper and efficient operation.

2. The scrubbing machine should have been started and operated in a safe and reasonable manner.

3. Care of the mechanized equipment should have been exercised at all times during its operation to avoid damage to personnel, the building and equipment.

C. *Floor Scrubbing and Rinsing:*

1. Proper precautions should have been utilized to inform the building occupants of wet and/or slippery conditions during the scrubbing operation.

2. The scrubbing work should have been performed in such manner as to properly clean the floor surface with care taken to see that the proper cleaning solution has been used.

3. All areas, including areas inaccessible to the machine and which are cleaned by means of deck scrubbing brushes and/or mops, should be clean and free of dirt, water streaks, mop marks, and string; properly rinsed; and dry mopped to present an over-all appearance of cleanliness.

4. Walls, baseboards, and other surfaces should be free of watermarks, scars from the cleaning equipment striking the surfaces, and splashings from the cleaning solution and rinse water.

D. *Equipment and Materials:*

1. All items of equipment used in the floor scrubbing work should be in a clean and well cared for condition and properly placed in a neat and orderly manner in their assigned storage locations.

2. All supplies of motor oil should be stored in approved safety containers in a safe and orderly manner.

3. All unused materials should be properly stored in the assigned locations.

Frequency and Production

A. *Frequency and Production Schedule:*

	Main Corridors	Secondary Corridors
Frequency	Daily	Weekly
Production Rate per Man-day	20,000 sq. ft.	

B. *Qualifying Factors:*

1. *Frequency*

   a. The frequencies given in the Frequency and Production Schedule are for a normal standard of cleaning under average conditions. These frequencies may be modified to meet local conditions.

   b. Main corridors are those located on the first or ground floors of a building or those that serve as primary entrance and/or exit areas. Secondary corridors are those less frequently used and usually located on upper floors and in the basements.

   c. Frequencies for other types of areas scrubbed should be established as required locally.

2. *Production Rate per Man-day*

The production rate per man-day as established may be subject to modification at the various locations due to narrow or angular corridors, difficult access to water and drains, and very dirty floors that must be gone over more than once. They provide, however, a standard for normal operated buildings.

Staffing

The man-day requirements of labor for floor scrubbing can be determined by use of the following formula:

$$\text{Man-days (of labor)} = \frac{\text{Quantity}}{\text{Production}} \times \frac{1}{\text{Frequency}}$$

where quantity represents the square feet of floor area scrubbed at a given frequency; production is the square feet of floor area scrubbed per man-day for the type of job method utilized; and frequency is the number of work days between operations.

If the frequency and/or production rate values as established in the Frequency and Production Schedule have been modified to meet existing conditions at a particular location, then those modified values should be used in the above formula for the labor requirement computation.

SAMPLE CALCULATION

Assume a building having a total of 45,000 square feet main corridor area plus 225,000 square feet of secondary area and with such existing conditions that the established values for the electric floor polishing machine may be used.

Then the man-day requirement is computed as follows:

$$
\begin{array}{cc}
\textit{Main Area} & \textit{Secondary Area} \\
\end{array}
$$

$$
\text{Man-days} = \frac{\text{Quantity}}{\text{Production}} \times \frac{1}{\text{Frequency}} + \frac{\text{Quantity}}{\text{Production}} \times \frac{1}{\text{Frequency}}
$$

$$
= \frac{45,000}{20,000} \times \frac{1}{1} + \frac{225,000}{20,000} \times \frac{1}{5}
$$

$$
= 2.25 \qquad\qquad + 2.25
$$

$$
= 4.5 \text{ man-days of labor required (jobs)}
$$

Since this is normally performed by a two-man crew (machine operator and mop man), this represents a requirement of 2.25 man-days for the machine operator and 2.25 man-days for the mop man.

To this total requirement must be added a factor to compensate for leave in order to arrive at the total number of positions needed to accomplish the work.

Conversion factor including compensation for leave = 1.15
Conversion factor × jobs = total positions required
1.15 × 4.5 = 6.75 positions required

• Where the floor scrubbing duty does not require the full time of the employee, other duties should be assigned to secure 100 per cent utilization of his time.

Equipment and Materials Allocation and Replacement

Items	Allotment per Crew	Replacement or Reissue per Crew
Bucket (12 qt.)	2	1 per year
Detergent, synthetic	As required	As required
Measuring cup	1	1 per year
Mop, cotton	3	12 per year
Mop handle	3	4 per year
Mop tank	1	1 per 5 years
Overshoes, rubber	2 pair	2 pair per year
Putty knife (1¼")	2	1 per year
Scrub brush, deck	1	2 per year
Scrubbing machine or floor polishing machine with accessories	1	1 per 10 years (brushes as required)
Warning signs	*	As required
Water hose (approx. 6 feet)	1	1 per year
Wiping cloths (waffleweave)	2	12 per year

* As many as necessary.

The allotment and replacement or reissue quantities in this standard as established are based on a two-man floor scrubbing crew having continuous floor scrubbing duty.

10E / **FLOATING FLOORS**

Floating floors are a new flooring concept for proposed or existing structures that provide:

1. Subfloor access for power lines and other services.
2. Unlimited flexibility for rearrangement of machine layout.
3. Subfloor space for heating and ventilating.

The floor is laid several inches (six to ten) above existing or new subfloor; it is easy to lay, rearrange, or to remove with a flexibility heretofore impractical with conventional techniques. The installation generally consists of 36½" x 36½" modules, which are carried on an adjustable pedestal. Pedestal heads are so designed that the modules are self-locking as they fall into place, without need for adjustment or tightening. The screw flange base of the pedestals rest freely on the subfloor and do not have to be bolted down. The pedestal is readily adjustable to raise or lower the floating floor to compensate for unevenness in the subfloor. This adjustment makes it possible to provide either more or less space between the two floors. Access to the subfloor is obtained by lifting out a panel.

Cutouts in individual floor plates permit entry of electrical wiring or pipe lines from the subfloor area for connection to machinery. Plates can be cut for installation of air registers or duct work whenever needed. The subfloor area can be utilized as a pressurized plenum chamber for heating or cooling. A radiant heating or cooling effect can be achieved in this manner. Floating floor installations have been made in offices that have computing machines which require a maze of power and feeder cables for operation.

11 / HIGH CLEANING

(*Source:* Office of Buildings Management, General Services Administration, Washington, D. C.)

Job Description

High cleaning is a specialized job and consists of planned and scheduled periodic cleaning of lights, transoms, pipes, high files; dusting venetian blinds and other objects high enough to require a ladder (or scaffolding), and too high for the room cleaner to reach while standing on the floor. High cleaning is performed in all areas in buildings where such work is required; and the assignment also includes proper maintenance and care of equipment and materials used for this type of work.

Employees assigned to these jobs should be instructed to turn out lights not in the immediate area being cleaned, close any windows found open, and should report any defective items noticed such as: burned out lights, broken venetian blind cords, slats, and other items to their supervisor.

Job Methods

Operation Steps   *Explanation*

Step 1—Prepare for Work

1. Obtain necessary equipment, materials and supplies.
   a. Ammonia
   b. Brush, wall, dust
   c. Buckets (12 qt.)
   d. Chamois
   e. Cloths, dust
   f. Cloth, wiping
   g. Detergent
   h. Gloves, rubber
   i. Ladder platform, safety (of proper height), or scaffolding
   j. Soap, white, cake
   k. Sponges, cellulose

Operation Steps	*Explanation*
	l. Sweep mop
	m. Truck, platform (push type)
	n. Vacuum cleaner with attachment for cleaning venetian blinds and overhead pipes
2. Proceed to assigned work area.	All equipment, materials and supplies should be taken to the starting point of the work area, to be used as needed.

Step 2—Dust Surfaces

1. Use vacuum cleaner, sweep mop and/or dustcloth to dust.	a. Wall surfaces
	b. High moldings around ceilings and walls, doors and windows
	c. Tops of high partitions
	d. Overhead piping
	e. Venetian blind slats and tapes
	f. Clocks, pictures, and plaques
	g. Wall or ceiling ventilators
	h. Tops of high bookcases and lockers
	i. Other high surfaces as directed
2. Dust wall and overhead fans and door checks.	a. Wall and overhead fans and door checks requiring ladders to reach should be dusted with cloths that are kept separate, and used for this purpose only, to avoid oil smears on other surfaces.
	b. Use ladder or scaffolding as required in a safe manner in the performance of this work.

Step 3

Damp wipe surfaces as required.	Use a water and ammonia solution (mixed as directed), sponge and chamois as required to damp wipe and dry high surfaces, such as transoms, clock glass, picture frames and glass, smudged areas surrounding air grilles and ventilators.

NOTE

Do not damp wipe picture surfaces not covered by glass.

Operation Steps   *Explanation*

Step 4

1. Clean light fixtures.

Due to the many variables connected with this phase of high cleaning, specialized job methods to cope with the type of fixture, degree of dirt accumulation, fixture height from the floor, and any other fixture problem, will need to be formulated and instituted at the various locations for the accomplishment of this work.

2. Group light replacement.

Where a group light replacement procedure is followed in a particular location, then the scheduled cleaning of these fixtures should be incorporated as a part of the replacement schedule.

3. Formulation of specialized methods.

a. Set up work platform (ladder, scaffolding, and other equipment) at point most convenient to clean fixture.

b. Place cleaning material and equipment on platform.

c. Remove light fixture parts as required, clean and set aside while remainder of fixture is cleaned.

d. Replace fixture parts and move work platform to next fixture location.

PRECAUTIONS

1. Lights should be turned off in time for the fixture to cool before being cleaned.
2. Liquids or damp rags should be used with care around fixtures to avoid shock hazards.
3. Make sure that fixtures are dry before turning on lights.
4. Care should be exercised to see that the fixture is properly secured in place after cleaning so there will be no danger of it falling.

Step 5

Return equipment and unused materials to storage.

a. Replace any furniture items moved during the high cleaning operation.

Operation Steps	*Explanation*
	Before leaving the work area, make a final inspection to see that the work has been performed as required. Gather up all equipment and materials and return to the gear room where items are to be stored.
	b. Clean equipment and store equipment and unused materials in proper storage position.

PRECAUTIONS

1. Caution should be taken that liquids are not dripped on furniture and/or rugs.

2. Any dirt and dust resulting from the work performance should be cleaned up and removed.

3. Defective items such as burned out lights, broken venetian blinds, cords, and slats, broken plaster, and others should be noted and reported to the supervisor.

Care of Equipment

A. *Brush, Wall, Dust:*
   1. Comb bristles and use vacuum cleaner to remove remaining dust.
   2. Store in proper manner.

B. *Bucket (12 qt.):*
   1. Rinse and dry the inside and outside of the bucket.
   2. In storing do not place heavy objects on the bucket.

C. *Chamois:*
   1. Rinse chamois in clean water and squeeze out excess water after each use.
   2. Stretch chamois to original shape before placing to dry.
   3. Avoid contaminating chamois with oils, polish or cleaning materials.

D. *Cloths, Dust and Wiping:*
   1. Keep the cloths clean and free from grime or oily material.
   2. Wash the cloths once a week or as required in a warm synthetic detergent solution. Rinse in clear warm water.
   3. Hang the cloths to dry where there is good circulation of air.

E. *Gloves, Rubber:*
   1. After each use, wash gloves in a warm detergent solution, rinse in clear warm water and wipe dry.

2. Always maintain gloves in first-class condition. When holes develop the gloves should be discarded.

3. Be careful not to snag gloves on nails or other sharp objects.

4. Gloves should be stored in a clean dry place.

F. *Ladder or Scaffolding:*

1. Ladder or scaffolding should be wiped free of dirt and moisture and returned to proper storage area.

2. Equipment should be cleaned and inspected frequently and that with defects withdrawn from use.

G. *Sponge:*

1. Rinse the sponge in clean warm water after each day's use. Wash a very dirty sponge in a warm detergent solution, rinse in clean water and squeeze out excess. Store in clean dry place.

H. *Sweep Mop:*

1. Sweep mop should be shaken or vacuum cleaned both during use and at the close of operations each day. Keep clean and free from grime or oily material.

2. Wash mophead frequently enough to maintain a clean and usable mop.

3. Replace mophead on the frame and hang to dry where there is a good circulation of air.

I. *Truck, Platform:*

1. Do not allow truck to strike wall surfaces, radiators, etc.

2. Clean truck daily of dirt and dust.

3. All defects and damage to the truck should be reported immediately to the supervisor. Arrangements should be made to have the wheels oiled regularly.

J. *Vacuum Cleaner and/or Equipment:*

1. Empty dirt receptacle. Wipe machine after each day's use with a dry cloth. Wind cord loosely on the hooks provided.

2. Do not run machine over electrical cords or pull against cord when it is plugged into receptacle.

3. Do not allow machine to strike walls, furniture, radiators, or other objects.

4. After use, return hose, and attachments, to assigned space in storage room.

Performance Inspection Guide

A. *Transoms:*

1. Both sides of glass should be clean, and surrounding woodwork should be free of dust.

B. *Clock:*

1. Face should be clean and free of streaks.

2. Edge should be wiped free of dust.

C. *Molding Around Ceiling:*

1. Should be free of dust accumulations.

D. *Tops of Partitions:*
   1. Should be free of dust.

E. *Overhead Pipes:*
   1. Should be free of dust. If very dirty, pipes should be washed.

F. *Wall Fans:*
   1. Should be free of dust. (Thorough cleaning should be done by electrician.)

G. *Venetian Blinds:*
   1. Slats should be free of dust, both sides.
   2. Window frame and adjoining area should be free of dust.

H. *Pictures, Placques, and Similar Objects:*
   1. Glass should be clean and free of streaks.
   2. Frames should be free of dust.

I. *Wall or Ceiling Ventilators:*
   1. Should be free of dust.
   2. Framework around ventilator should be wiped clean.

J. *File Cases and Bookcases:*
   1. Tops of file cases and bookcases should be free of dust.

K. *Lockers:*
   1. Tops should be free of dust.

L. *Walls:*
   1. Should not be streaked and surfaces free of dust.
   2. Cobwebs should have been removed.

M. *Light Fixtures:*
   1. Should be free of dust.
   2. No dirt should be left on furniture underneath fixture or on rug, or floor.
   3. If washed, should be clean and wiped dry.
   4. Fixtures should be properly reassembled.
   5. No water drippings should appear on rugs, floors, or furniture.

N. *Equipment Care:*
   1. All pieces of equipment should be clean and properly placed in their assigned storage locations.
   2. Soaps, detergents and other materials left over should be returned to their storage location.

Frequency and Production

A. *Frequency and Production Schedule:*

	Office Building Area
Frequency	Every four months
Production Rate per Man-day	10,000 sq. ft.

B. *Qualifying Factors:*

The job of high cleaning is not limited to offices but covers all areas in the building where such work is needed. Due to the many variables connected with high cleaning work, the development and application of specialized methods and work zones at the individual locations may be required for its efficient accomplishment. These detailed procedures should follow as closely as possible the general guidelines presented herein.

Staffing

The man-day labor requirements for high cleaning can be determined by use of the following formula:

$$\text{Man-days (of labor)} = \frac{\text{Quantity}}{\text{Production}} \times \frac{1}{\text{Frequency}}$$

Where quantity represents the square feet of floor area in which the work is to be performed, production is the square feet of floor area high cleaned per man-day; and frequency is the number of work days between cleanings.

If the frequency and/or production rate values as established have been modified to meet existing conditions at a particular location, then those modified values should be used in the above formula for the labor requirement computation.

SAMPLE CALCULATION

Assume a building having a total of 400,000 square feet of office building area plus 30,000 square feet of post office workroom area with existing conditions such that the established frequency and production rate values may be used.

Then the man-days requirement is computed as follows:

$$
\begin{array}{ll}
& \textit{Office Building Area} \qquad\quad \textit{Post Office Workroom Area} \\
\text{Man-day} = & \dfrac{\text{Quantity}}{\text{Production}} \times \dfrac{1}{\text{Frequency}} + \dfrac{\text{Quantity}}{\text{Production}} \times \dfrac{1}{\text{Frequency}} \\[2ex]
= & \dfrac{400,000}{10,000} \times \dfrac{1}{66} + \dfrac{30,000}{7,000} \times \dfrac{1}{21} \\[2ex]
= & 0.61 \qquad\qquad\qquad + 0.21 \\[1ex]
= & 0.82 \text{ man-days of labor required (jobs)}
\end{array}
$$

To this figure must be added a factor to compensate for leave in order to arrive at the total number of positions needed to accomplish the work.

Conversion factor including compensation, for leave = 1.15
Conversion factor × jobs = total positions required
1.15 × 0.82 = 0.94 positions required

• Where the high cleaning duty does not require the full time of the employee, other duties should be assigned to secure 100 per cent utilization of his time.

Equipment and Materials Allocation and Replacement

Items	Allotment per Cleaner	Replacement or Reissue
Ammonia, household	As required	As required
Brush, wall, dust	1	1 per year
Brush, handle, wall, dust	1	1 per year
Bucket (12 qt.)	2	1 per year
Chamois	1	2 per year
Cloth, dust	2	24 per year
Cloth, waffleweave	2	24 per year
Detergent cleaner, synthetic	As required	As required
Gloves, rubber	1 pair	6 pair per year
Ladder, safety platform (with shoes or scaffolding)	1	As required
Polyethylene glycol or other dust mop and cloth treating compound	1 pint solution	1 quart compound per year
Soap, white (cake)	1	6 per year
Sponge, Cellulose	As required	As required
Sweep mophead	1	4 per year
Sweep mop, handle and frame	1	1 per year
Truck, platform (push type)	1	1 per 5 years
Vacuum cleaner and attachments	As required	1 per 10 years (Does not apply to attachments)

12 / TREES AND SHRUBS

Planting Seasons

Off-season planting should be avoided whenever possible. Great care should be taken to move trees or shrubs successfully at seasons ordinarily regarded as unfavorable. Often when they are set out during the wrong season, they will survive only after a heavy cutting back, and will remain in a very depleted condition, giving little or no leaf effect until the following natural spring growing season. Therefore, nothing is gained, and the expense is increased due to the necessity of utilizing more careful, laborious moving methods, and by the lengthening of the period of artificial watering.

Trees

A good maintenance program will include: (1) pruning, (2) spraying, (3) fertilizing, (4) mulching, (5) leaf disposal, and (6) drainage.

Pruning. The shaping of a tree, and the removal of excess growth is generally called pruning. This is only done to the degree necessary to balance the top of the tree to the root system. The roots must have the ability to supply moisture to the leaf area, or it may die or drop its leaves. A tree should be kept to its natural shape as much as possible. All pruning should be done vertically to the direction of growth.

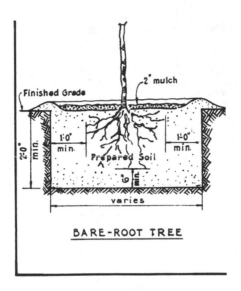

Figure 12-1

Pruning equipment should be kept sharp and clean. When moving from one tree to another, the equipment should be cleaned with alcohol to prevent the transmittal of possible disease.

Spraying. In the early spring, a dormant spray should be applied before the tree is in bud, and when the temperature is higher than 40°. Generally one or two additional sprays are required during the growing season for the control of insects.

Fertilizing. Either a liquid or solid fertilizer is used in the spring of each year. Good leafage and rapid wooded growth are the results of fertile soil. Where tree roots are located under lawns, they should be fertilized. Fertilizers containing a large proportion of nitrogen are desirable because nitrogen is the element contributing most to the leaf growth.

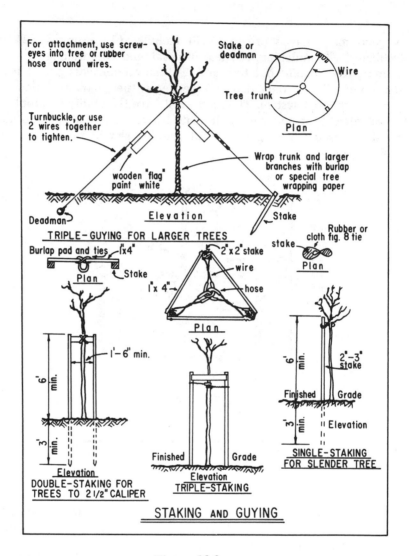

Figure 12-2

Mulching. Mulching is the practice of covering the ground around the tree trunk with a substance to prevent weed growth, washing away of soil, and to retain moisture. Materials used for mulching are peat moss, peanut shells, and leaves. *Leaves* contain plant food where the others do not, and they are known to furnish some nutritive elements.

Drainage. Most plants will die if the soil about them remains saturated for any great length of time. If plants are not capable of enduring such conditions, drainage must be provided.

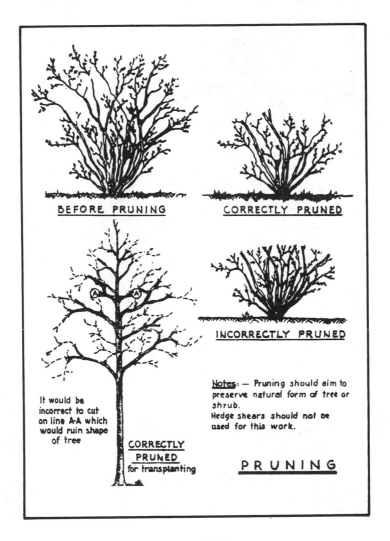

Figure 12-3

Shrubs

A good maintenance program will include: (1) spraying, (2) fertilizing, (3) mulching, (4) trimming, and (5) snow protection.

As indicated in the recommendations for trees, a dormant spray should be used, and either a liquid or solid fertilizer applied in the early spring. The amount of fertilizer used is based on the size and kind of shrub. The ground around the shrub should have leaf mulch or peat moss to prevent weed growth.

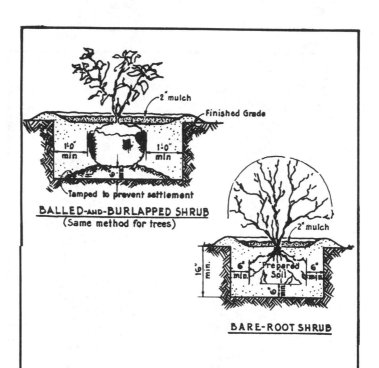

Figure 12-4

Depending upon the type of shrub, from one to three trimmings may be required each growing season. Trimming equipment should be kept sharp and clean. When moving from one shrub to another, the cutters should be cleaned with alcohol to prevent the transmittal of possible disease.

In areas that receive snow, shrubs should be bound with chicken wire or similar material in the late fall of the year. This will prevent the shrubs from becoming broken or spread apart by the snow.

13 / LANDSCAPE JOB ESTIMATING

(Reprinted with permission from January, 1968, issue of
Grounds Maintenance. Copyright 1968 by Implement &
Tractor Publications, Inc.)

This guide contains the average times needed to perform landscape maintenance and construction jobs. As a reference source, you may call upon this guide throughout the year for various uses in your operation: (1) to prepare bids for contracting maintenance and construction work; (2) to determine by comparison of times the efficiency of your present manpower and equipment; (3) to program a work schedule, drawing maximum output from labor and machinery; (4) finally, to see weaknesses in your organization that might require reinforcements with more men or additional equipment.

For maintenance work, you will find mowing timetables (following) and an overall maintenance timetable. The comprehensive maintenance table includes upkeep information for: Turf Areas, Shrub Areas, Trees, Flower Beds, Paved Areas. To properly use the table for Estimating Maintenance Jobs, note the instructions.

Landscape construction tables provide you with helpful times for planting evergreens, shrubs, shade trees, ground covers, as well as average times for performing different tree-care jobs.

Useful Tables and Rules

(Reprinted with permission from January, 1968, issue of *Grounds Maintenance.* Copyright 1968 by Implement & Tractor Publications, Inc.)

MOWING TIME POTENTIALS

Cutting Width	Area*			
	4 acres		6 acres	
	**Walking	Riding	Walking	Riding
Rotary				
48-inch	3 hrs. 30 min.	2 hrs. 45 min.	5 hrs. 15 min.	4 hrs. 10 min.
60-inch	2 hrs. 50 min.	2 hrs. 15 min.	4 hrs. 15 min.	3 hrs. 20 min.
Reel				
76-inch	2 hrs. 15 min.	1 hr. 45 min.	3 hrs. 20 min.	2 hrs. 40 min.
84-inch	2 hrs. 2 min.	1 hr. 35 min.	3 hrs. 2 min.	2 hrs. 20 min.

*Area cut was calculated at 85 per cent efficiency, allowing 15 per cent time for stops and turns and overlap considerations.

**Walking times were calculated at 2¾ m.p.h. and riding times at 3½ m.p.h.

Source: Information from the Institute of Maintenance Research, Salt Lake City, Utah, and Planning Calculator, Toro Mfg. Co.

(Reprinted with permission from January, 1968, issue of *Grounds Maintenance*. Copyright 1968 by Implement & Tractor Publications, Inc.)

MOWING POTENTIALS* (Commercial Reel Type)

Sq. Ft. Lawn	16" Hand	18" Power Sp	25" Power Walk	25" Ride	58" Ride
1000	10 min.	5 min.	3 min.	2 min.	—
2000	20 min.	10 min.	6 min.	4 min.	—
3000	30 min.	15 min.	9 min.	6 min.	—
4000	40 min.	20 min.	12 min.	8 min.	—
5000	55 min.	25 min.	15 min.	10 min.	5 min.
6000	1 hr. 20 min.	35 min.	18 min.	12 min.	6 min.
7000	1 hr. 40 min.	45 min.	21 min.	14 min.	7 min.
8000	2 hrs.	55 min.	24 min.	16 min.	8 min.
9000	2 hrs. 25 min.	65 min.	27 min.	18 min.	9 min.
10,000	—	1 hr. 25 min.	32 min.	20 min.	10 min.
20,000	—	3 hrs.	1 hr. 10 min.	45 min.	20 min.
30,000	—	5 hrs.	1 hr. 45 min.	1 hr. 10 min.	30 min.
1 Acre	—	—	2 hrs. 20 min.	1 hr. 30 min.	40 min.
2 Acres	—	—	5 hrs.	3 hrs.	1 hr. 20 min.
5 Acres	—	—	—	7 hrs.	3 hrs. 20 min.

*This table assumes that after you get over 30 to 40 minutes of mowing time on an industrial area, you are reaching a critical point—time and costwise. This is especially true if your area is being used exclusively for mood and setting purposes.

(Reprinted with permission from January, 1968, issue of *Grounds Maintenance.* Copyright 1968 by Implement & Tractor Publications, Inc.)

EVERGREEN PLANTING CHART

	Dwarf and Spreading Type (Actual Planting on Jobs)			Upright Type (Actual Planting on Jobs)	
Size	In prepared beds— water, clean up, outline	Spade or remove sod, excavate subsoil, add top soil, water, clean up, outline	Size	In prepared beds— water, clean up, outline	Spade or remove sod, excavate subsoil, add top soil, water, clean up, outline
18–24"	.3 hr.	.42 hr.	18–24"	.25 hr.	.4 hr.
2–2½'	.4 hr.	.6 hr.	2–2½'	.3 hr.	.5 hr.
2½–3'	.5 hr.	.85 hr.	2½–3'	.4 hr.	.7 hr.
3–3½'	.7 hr.	1.0 hr.	3–3½'	.5 hr.	.85 hr.
3½–4'	.9 hr.	1.5 hrs.	3½–4'	.65 hr.	1.0 hr.
4–5'	1.5 hrs.	2.3 hrs.	4–5'	.9 hr.	1.5 hrs.
5–6'	2.3 hrs.	3.8 hrs.	5–6'	1.25 hrs.	2.3 hrs.

(Reprinted with permission from January, 1968, issue of *Grounds Maintenance.* Copyright 1968 by Implement & Tractor Publications, Inc.)

SHRUB PLANTING CHART

	Shrubs and Dwarf Trees Bare Root (Actual Planting on Jobs)			Shrubs and Dwarf Trees Balled and Burlapped (Actual Planting on Jobs)	
Size	Prepared Beds	Spade, etc.	Size	Prepared Beds	Spade, etc.
18–24"	.2 hr.	.3 hr.	18–24"	.3 hr.	.42 hr.
2–3'	.25 hr.	.42 hr.	2–3'	.42 hr.	.6 hr.
3–4'	.33 hr.	.5 hr.	3–4'	.66 hr.	.85 hr.
4–5'	.5 hr.	.7 hr.	4–5'	.8 hr.	1.0 hr.
5–6'	.66 hr.	.85 hr.	5–6'	1.0 hr.	1.25 hrs.

(From *Grounds Maintenance Handbook,* Second Edition, by H. S. Conover. Copyright 1958 by McGraw-Hill Book Company. Used by permission of McGraw-Hill Book Company.)

AVERAGE TIME FOR PLANTING 1000 SQ. FT. WITH GROUND COVERS

Number and kind of plants	Labor to dig bed 10 in. deep	Labor (spread manure and peat and prepare bed)*	Time to plant	Total labor (time)
500 Ivy	9 hrs.	9 hrs.	5 hrs.	23 hrs.
4000 Pachysandra	9	9	18	36
750 Sarococca	9	9	7½	25½
4000 Ajuga	9	9	18	36
334 Roses	9	9	18	36
334 Honeysuckle	9	9	9	27
1000 Vinca minor	9	9	10	28

*On each bed was spread four bales of peat moss 1 in. deep and 1 ton of rotted manure, also 1 in. deep.

ESTIMATING MAINTENANCE JOBS

How to use this table:

1. Figure the square footage for your maintenance job. Fill this number in the first column, "Your Total Area."

2. In the "Frequency per Year" column fill in the number of times you plan to perform this job annually.

3. Divide your total square footage by 1000 and multiply this answer by average time figure from "Minutes per 1000 Square Feet." Then, multiply this answer by "Frequency per Year" determined from Step 2. Fill this answer in "Your Total" under "Total Annual Time in Minutes."

4. Determine your "Material Cost per 1000 Square Feet" and put this figure in that column.

5. Multiply your material cost per 1000 square feet by the total number of square feet. Then multiply this answer by the figure in the "Frequency per Year" column and fill this in "Total Material Cost."

6. Take your hourly local labor rate (include tax, insurance, etc.) and fill this in "Wage Rate" column.

7. Divide "Total Annual Time in Minutes" by 60, which should be multiplied by hourly wage rate to determine "Total Labor Cost."

Area and Operation	Your Total Area	Frequency Job Performed per Year — Average	Frequency — Your Area	Minutes per 1,000 Sq. Ft.	Total Annual Time in Minutes — Average Min/1000	Total Annual Time — Your Total	Average Material Cost per 1000 Sq. Ft.	Your Material Cost per 1000 Sq. Ft.	Total Material Cost	Average Equipment Retail Purchase Price	Wage Rates Your Area Include Tax, Insurance, etc.	Total Labor Cost	Your Total Costs per Year — Labor and Materials (No Equipment or Management, Etc., Included)
LAWNS													
Mow													
16" hand		30		10	300					$ 30			
18" power		30		5	150					150			
25" power		30		3	90					350			
58" power rider		30		1	30					500			
(See more time potentials in Mowing Time Potentials Table)													
Fertilize		2		3	6		$3			15			
Crabgrass Control (24" spreader) annual		1		15	15		3/yr.						
3 year		1/3		15	5		4/3 yrs.						
Weed Control (2, 4-D+)		2		15	30		.20			15			
3 gal. hand pump power rig–30" boom		2		4	8					120			
Rake													
hand (cavex)		1		60	60					4			
power (20"–24")		1		10	10					200			
Sweep													
leaf rake		3		25	75					5			
25" sweep rake		3		5	15					60			
30" power rake		3		2	6					180			
Edge—Trim—Clean-up (Min./1000 Lin. Ft.)													
hand (1 wheel mow) walks		30		25	750								
shrub edge		10		60	600					25			
gas power walks		30		5	150								
shrub edge		10		20	200					100			
Vacuum (Min./1000 Sq. Ft.)													
Shrub hose		3		5						550			
SHRUB AREAS													
Weed													
hand hoe		15		60	900		.35			15			
spray out		10		15	150		.07			15			
spray after mulch		5		10	50								

8. Add "Total Material Cost" and "Total Labor Cost." This will give you "Total Costs per Year."

9. For bidding purposes you can add your profit percentage, overhead, and contingency fund for a final estimate.

(NOTE: Pay special attention to columns showing "Frequency per Year," "Average Material Cost per 1000 Square Feet," and "Average Equipment Costs." These averages are meant to give you a comparative guideline for your operation. If your data vary greatly from these averages, you may want to check to be sure you are employing optimum materials, equipment, and labor.)

Task					
Police-Up					
hand pick-up	30	15	450		550
vacuum	30	7	210		10
Prune	5	60	300		10
Fertilize	1	5	5	1	
Mulch					
ferti-mulch	2	30	60	50	
coarse sawdust	2	30	60	15	
Spray (pests)	2	30	60	.50	15
TREES		Min./Tree for 10 Yr. Old			
Prune					
from ground	2	20	40		25
high work (if needed)	¾	60	15		200
Fertilize	1	30	30		15
Pest Control					
spray	3	20	60		150
systemic	1	10	10		
FLOWER BEDS		Min./1000 Sq. Ft.			
Spring Prepare	1	200	200		
Plants (flats)	1	600	600	10	
Weed—no mulch	25	60	150	40	5
Cultivate—no mulch	25	30	75		5
Mulch	1	30	30	50	5
Weed with mulch no cultivation	15	20	300		
Spray	0	10	30	.35	15
Fertilize	3	5	10	1.50	10
Police-Up					
hand pick-up	2	15	450		550
vacuum	30	10	300		584
Fall Clean-Up					
include pick-up mulch for re-use	1		400		
PAVED AREAS					
Walks					
sweep—hand	30	30	900		10
vacuum	30	4	120		550
snow removal hand	10	80	800		5
power (24")		12	120		350
Drives—Parking					
vacuum	10	3	30		550
snow control	10	10	100		350
Cut strips along curbs, drain center—and over parking marking center lines.					

AVERAGE TIME FOR DIGGING, HANDLING, PLANTING, WATERING, PRUNING, GUYING, AND WRAPPING TREES

Ball size diam. depth (in.)	Cu. ft. of soil in ball	Weight of ball (lb.)	Time to dig and lace (min.)	Time to handle ball (min.)	Size hole required (in.)	Cu. ft. soil hole, to excavate	Time to dig hole (min.)	Cu. ft. soil dis-place-ment	Time to plant and prune (min.)	Time to water, wrap, guy and clean up (min.)	Cu. ft. topsoil handled in moving	Total time in moving
12 x 12	7/10	56	15	10	24	3¾	20	3	15	4	10½	64 min.
18 x 16	2	160	30	20	30	7½	28	5½	21	5	21	1-2/3 hr.
24 x 18	4	320	60	40	36	13	65	9	49	12	38	3-2/3 hr.
30 x 21	7½	600	114	76	48	26½	133	19	100	25	76	7-1/3 hr.
36 x 24	12½	980	189	126	54	38	190	25½	143	36	114	11-1/3 hr.
42 x 27	19	1520	285	190	66	64	320	45	240	60	185	18½ hr.
48 x 30	28	2040	420	280	72	85	360	57	270	68	254	23-1/3 hr.
54 x 33	38½	3060	579	386	84	127	635	88½	476	119	370	36½ hr.
60 x 36	52	4160	780	520	90	159	795	107	596	149	474	47-1/3 hr.
66 x 39	68	5440	1020	680	96	196	905	128	679	168	596	57½ hr.
72 x 42	87	7160	1305	870	108	267	1240	180	930	233	795	76 hr.

(Reprinted with permission from January, 1968, issue of *Grounds Maintenance.* Copyright 1968 by Implement & Tractor Publications, Inc.)

SHADE TREE PLANTING CHART

Bare Root (Actual Planting on Jobs)			Balled and Burlapped (Actual Planting on Jobs)		
Size	Prepared Beds	Spade, etc.	Size	Prepared Beds	Spade, etc.
6–8'	.7 hr	1.0 hr.	6–8'	1.2 hrs.	1.7 hrs.
8–10'	.8 hr.	1.2 hrs.	8–10'	1.5 hrs.	2.2 hrs.
10–12'	.9 hr.	1.3 hrs.	10–12'	2.0 hrs.	2.8 hrs.
1½–2"	1.0 hr.	1.5 hrs.	1½–2"	3.5 hrs.	3.3 hrs.
2–2½"	1.3 hrs.	2.0 hrs.	2–2½"	5.0 hrs.	4.5 hrs.
2½–3"	1.5 hrs.	2.5 hrs.	2½–3"	7.0 hrs.	6.0 hrs.
3–4"	2.5 hrs.	3.7 hrs.	3–4"	7.0 hrs.	8.5 hrs.

Source: Originally prepared for National Landscape Nurserymen's Assn., by Harold Hunziker, Hunzikers, Inc., Landscape Nurserymen, Niles, Mich.

14 / **LAWNS**

(*Source: Plant* magazine) April 1960)

A good lawn is a must to welcome and impress customers or visitors of a building or plant. Fortunately good lawn turf is easy to grow, providing you know what is required. A lawn that is fertilized, mowed, watered, and protected will make a good impression on all clients and visitors.

The choice of grass is governed by the climate and sunlight intensity. Kentucky bluegrass and red fescues are logical choices in the north. Bentgrasses make beautiful lawns, but require specialized management. Kentucky bluegrass or improved merion is the proper choice for full sunlight. Red fescue strains are the best grasses for dry soil either in sun or shade. Creeping red fescue spread rather quickly. Under moist shade conditions, Kentucky bluegrass (*poa pratensis*) should be included.

Some seed mixtures contain redtop or rye grass to give faster initial cover. This practice is questionable since these grasses can persist for several seasons and may crowd out the more desirable types. A mixture of 60 per cent merion or Kentucky bluegrass, and 40 per cent fescue is ideal for most industrial lawns where shade as well as sun can be expected.

In the south, improved Bermudas are preferred in the open. Zoysia and St. Augustine thrive in the shade where Bermuda will fail.

Most lawn soils are amply endowed with potash. Sand and peats may be the exceptions. Phosphorous levels are often low at the time of planting. Both of these elements along with the lime needs, should be checked every three to five years by means of soil testing. Many state experiment stations and some fertilizer firms provide this service. These samples should be taken in the late winter or early spring at a depth of approximately two inches below the surface of the ground.

Upon the completion of the soil analysis, the proper amounts of fertilizer should be applied. If large amounts of lime and fertilizer are required, two or three small feedings are recommended over a single feeding. This will eliminate nitrogen burn and excessive watering. Then lawn fertilizer becomes a matter of supplying enough nitrogen to maintain color, density and freedom from weeds.

There are many good fertilizers to choose from. Melorganite is a common choice for supplying nitrogen, phosphorous and the vital trace elements needed by grass. The amount of fertilizer which should be used depends upon the length of the growing season and the kind of grass. For the cool season, bluegrasses and fescues, 17 to 20 pounds for each 1,000 square feet for every month of good growing weather, are required. Double this rate is advocated for the ravenous Bermuda in the south. Lesser amounts will do some good, but the rate suggested is for the ultimate in turf quality.

Lawns must be protected from insects, diseases, weeds, and improper mowing practices. They should be mowed whenever the growth of the grass justifies cutting. For best results, no more than one-third of the total leaf length should be removed in any one cutting. Bermuda and bentgrass demand close one-half to three-quarter inch cutting. All others should be mowed one inch or better. The matting of clippings can do damage to a lawn. This can be eliminated by the collection of clippings, generally accomplished by rakings. It is a good practice to vary the cutting pattern with each mowing. The mower should be kept sharp at all times.

Some insects feed on grass roots, others attack the above ground leaves. Both can be controlled with chlordane, aldrin, dieldrin, BHC, and lead arsenate.

Root zone control of fungus, grubs, crabgrass, will last three years or longer. This should be incorporated with fertilizing to reduce labor cost. Surface feeding insects like cutworms and webworms seldom contact poisons in the soil. Several surface applications of insecticide each season may be needed to keep these caterpillars in check, especially in the south.

Broadleaf dandelion and plantain control is standard with 2,4-D. The propionic form of 2,4,5-T is good on yarrow, chickweed and clover. Crabgrass can be controlled with disodium methyl arsenate or potassium cyanate after it emerges or before germination with lead arsenate, calcium arsenate or chlordane. All weed authorities agree that the best control of all is prevention by growing a thick stand of grass.

Disease attacks can be bad on lawn turf. Merion bluegrass resistance to leaf spot is one reason it is preferred over common Kentucky bluegrass in the north.

Best defense in the battle against weeds is a lush, well-fertilized stand of grass. But, until a good, thick lawn has been established, the following chemical controls are recommended:

Use 2, 4-D on:

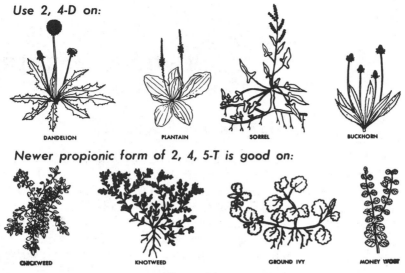

DANDELION PLANTAIN SORREL BUCKHORN

Newer propionic form of 2, 4, 5-T is good on:

CHICKWEED KNOTWEED GROUND IVY MONEY WORT

Figure 14-1

Everybody's pest:

CRAB GRASS

Figure 14-2

Lawn grasses can also be hit by brownpatch, dollar-spot, snowmold, and other diseases. There are several good fungicides available to control them.

Lawn grass must have an ample supply of moisture. However, the tendency has been to use too much rather than too little water. This creates conditions favorable for diseases and weeds, and wastes fertilizers. Lawns will grow well with an application of one to one and one-half inches of water per week during dry summer periods. A properly designed and installed irrigation system takes this into account. The secret of good watering is a thorough soaking applied as infrequently as possible.

Even on the best turf, there will usually be some spots where the grass is too thin, and which requires some patching. In the latitude of the Middle Atlantic States, the best time for doing this is during the month of September. This operation involves the use of a top dressing, consisting of top soil, peat moss, and sand, which is applied to the area to be patched at the time the seed is sown.

15 / LIGHTING MAINTENANCE

(*Source:* The Port of New York Authority, New York)

Methods and Procedures

The in-service foot candle levels that are to be maintained in any lighting installation are determined by a study of the conditions under which the system will operate. This study generally includes the foot candles required, the system and luminaire to be used, the reflection of the walls, ceiling, floors, and the resulting coefficient of utilization. The maintenance cost of a lighting system is a prime factor in annual operating charges, which must be given equal consideration along with initial cost and life expectancy of the various types of luminaires.

In maintenance of a lighting system, original design should give consideration to accessibility of the light source, to facilitate cleaning and lamp replacement. The fixture itself should be designed for easy removal of dust and dirt from its lenses, louvers, lamps, and reflectors. In special problem areas, consideration should be given to the feasibility of providing disconnecting and lowering hangers, which would permit lowering the light fixture for maintenance. No matter how efficiently the lighting system may have been designed initially, its adequacy in operation can be assured only by a well-planned maintenance program. A maintenance program must be established with the goal of keeping illumination up to original design.

In the operation of any lighting system, three maintenance factors effect the amount of light obtained:

1. The decrease in the output of light from the lamp as it ages. The average lumen output throughout the life of a lamp is approximately 12 to 27 per cent lower than initial output.

2. Loss of light through accumulated dust and dirt on the transmitting or reflecting surfaces of the fixture or lamp itself. Fixtures should be washed once per year. If a large amount of dust accumulates during interim periods, additional dusting may be required.

3. Loss of light through the accumulation of dirt on walls and ceilings. Walls and ceilings should be cleaned once per year.

When a lighting system is installed, a fixed amount of capital is committed for payment of installation and fixtures. Depending upon the hours of usage, the following costs are also committed: power costs, lamp costs, labor charges to replace the lamps, and the expense of cleaning the fixture and lamp in order to obtain the maximum amount of light.

When instituting a maintenance program, it must be decided who is going to do the work. Employees selected should be able to properly identify different types of lamps. They should be trained in methods of washing and dusting fixtures, safely replacing lamps from a ladder or ladder cart, and changing lamps in explosion proof fixtures. It is recommended that the repair of any troubles such as faulty ballasts or damaged sockets, be handled by trained electrical personnel. The plant or building should be divided into areas, considering similar fixtures, height, and types of lamps to be installed or cleaned. This will aid in securing maximum performance from employees, and in setting work standards.

Records and Forms

A signal system designed to cover the frequency of operation should be established. This will include the design of a work card with the designation of each month across its top.

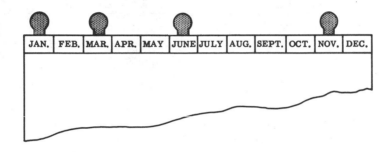

Figure 15-1

A tab system should be initiated, whereby a signal is placed on the month during which certain work is to be performed. For example in Figure 15-1, after the work which was designated for March has been accomplished, the signal would be moved to the next scheduled maintenance date, possibly June. Varied colored tabs may be used to designate frequency such as red for bimonthly, inspect for burnouts and replacements; yellow for semiannually, dust fixture; blue for annually, wash fixtures and lamps.

Records can also be maintained on punch cards. Special reporting cards can be marked with an electrographic pencil, as work is performed, or as additions or deletions are made in the system. These are forwarded to the data processing department for keeping the system up to date and accurate for the next report. Reports and schedules are submitted to the building superintendent or plant manager on predetermined dates by the data processing department, indicating work to be scheduled during that month, quarter, or year.

An area lighting maintenance card should be maintained by the building superintendent or plant manager, in his master card file. This card should cover information on the fixtures, lamps, area, and work accomplished. It should also cover other pertinent data essential to the operation of a good maintenance program. This will be an accurately posted record of the history and performance of each lamp and fixture. It will also denote the frequency of certain types of inspection, and will alert him to pull the area work card. (See Figure 15-2.)

JAN.	FEB.	MAR.	APR.	MAY	JUNE	JULY	AUG.	SEPT.	OCT.	NOV.	DEC.

AREA LIGHTING MAINTENANCE   Card No. *L-12*

Bldg. *OFFICE*   Area *ILLUSTRATING – DRAFTING*

Type Fixture *RECESSED FLUORESCENT, 2-40 WATT, ALBALITE GLASS*

CHROME FRAME   Rated Lamp Life *7600 HRS.*

Manufacture *WILHELM*   Location *SCRANTON*

No. of Fixtures *98*   No. of Lamps *196*   Type Lamp *F-40 T12 WW*

Ballast Cat. No. *89 P707*   Area Time Standards *16 HRS – 20 MIN.*   Group Cleaning *16 HRS 20 MIN.*   Group Cleaning and Relamping *22 HRS 52 MIN*

Average Daily Useage *14 HRS.*   Average Lamp Cost *$.60*   Average Hourly Rate *$1.65*

INSTRUCTIONS

Type Relamping *GROUP-REPLACE LAMP – WASH FIXTURES – INSPECT AND REPLACE MONTHLY*

Cleaning *ANNUAL – DUST TUBE, WASH FIXTURE*

Remarks *GROUP REPLACE AFTER 39 BURNOUTS OR 80% OF RATED LIFE, WHICHEVER COMES FIRST*

Date	Lamps Chg'd.	Fixture Cleaned	No.Lamps Used	Time	Total Cost
3/30/71	196	98	196	21 HRS.	$193.45

MAINTENANCE REPAIR and REPLACEMENT RECORD

Date	Mechanic	Clock #	Description of repairs-part replaced	Hrs.	Cost
1/14/71	MATTHEWS	76	LEAKING BALLAST – NEW BALLAST $8.59	3/4	$9.34

Figure 15-2 / Area lighting maintenance card. Information on the fixtures, lamps, area and work done. Kept in the office Master File.

The area work card should be numbered to correspond with the area lighting maintenance card; as shown in the illustrations, both cards are numbered L-12. The area work card should be pulled from the file and given to the foreman in charge of lighting maintenance, or to the person in charge. Every card which schedules work for a particular month should be given to the foreman at the beginning of the month, in order to aid him in scheduling and laying out his work in a proper and efficient manner. It may be required in some areas, that an inspection for replacements be held once a month, while in other areas where peak lighting is not compulsory, inspection for replacement may be held on a two or three months' basis. The practice of going over three months for inspection is not recommended under any circumstances. Since burnouts will be fewer immediately after relamping, inspections may be held every two months for the first four months after the relamping. After that, it would be advantageous to revert to monthly inspections. By using the signal system, a record can be maintained without continually changing the context of the card. A study should be made of operations covered on the area work card by the foreman, and notes made from it. The employee should be informed of the work standard, and a demonstration made of the method of performance desired to obtain the standard. (See Figure 15-3.)

AREA WORK CARD				Card No. L-12 Schedule-Monthly			
Bldg. OFFICE			Area or Room ILLUSTRATING-DRAFTING				
Type Fixture RECESSED FLUORESCENT, 2-40 WATT							
ALBALITE GLASS, CHROME FRAME							
No. of Fixtures 98		No. of Lamps 196		Rated Lamp Life 7600 HRS.			
Type Lamp F-40 T-12 WW			Ballast Cat. No. 89P707				
INSTRUCTIONS							
Group Relamping X and wash Fixture, Inspect & Replace Monthly							
Spot Replacement ___							
Group Replace after 39 Burnouts or 80% of Rated Life							
Group Cleaning- Annual wash fixture, dust tube							
Spot Cleaning							
Remarks: Fixtures to be washed when lamps are changed							
Standards Per. Fixture	Replace Only 4 MIN.		Group Cleaning 4 MIN.			Group Cleaning and Relamping 14 MIN.	
Lamps Changed	Date	Time Required	Fixtures Cleaned	Date	Time Required	Total Burnouts	Date
196	3/30/71	13 HRS.	98	3/30/71	16½ HRS.	2	3/30/71
6	12/1/71	24 (MIN)	—	—	—	8	12/1/71
8	2/6/71	32 (MIN)	—	—	—	16	2/6/71
4	3/21/71		98	3/21/71	16 HRS.	20	3/21/71

Figure 15-3 / Area work card. Work cards are given to the lamp changer foreman or person doing the work, to show the work to be done in the area.

Each person who is changing lamps will record the amount of time spent in each area, and the work done, on a daily lighting maintenance card. This information is transferred to the area work card, and from the area work card to the area lighting maintenance card in the master file. (See Figure 15-4.)

	DAILY LIGHTING MAINTENANCE CARD				
Name _B. Matthews_			Clock No. _76_		Date _3/30/71_
Area	Lamps Replaced	Fixtures Washed	Starters Replaced	Time Start	Time Finish
ILLUSTRATING	36	18	————	8:00 AM	12:00 AM
ILLUSTRATING	34	17	————	1:00 AM	5:00 PM

Figure 15-4 / Daily lighting maintenance card. Lamp changers record time spent and the work done on this card. This information is transferred to the Area Work Card, then to the Area Lighting Maintenance Card (Master File).

Whether a system of spot relamping and cleaning, group relamping and cleaning, or a combination of both, is the best method for lighting maintenance, depends upon the particular installation. The *type* of lamp replacement used in each area, depends upon the *economics* of maintaining the area, weighed against the *requirements*.

Spot Replacements

Spot replacement is the replacing of burned out lamps as they fail. This can prove to be a tedious process, as time spent in locating such burned out lamps makes this type of replacement very costly. Where appearance is not a factor, a modified spot replacement program is used, with maintenance crew inspecting and replacing burnouts weekly. With a spot relamping program, the continual occurrence of burned out lamps, and new lamps installed next to old ones, results in lower lighting levels, interferes with work performance, and creates a poor appearance. A formula for calculating spot replacement cost vs. group replacement is shown in Figure 15-5. If a spot replacement program is used, it is recommended that the fixtures be washed a minimum of once per year. Suggested procedure is covered in this chapter under Standard Work Methods.

In most operations, large or small, costs can be reduced by using a program of group relamping, over a spot replacement program.

Group Relamping

Group relamping is the systematic replacement of lamps before they burn out. This results in higher lighting levels with no extra cost for electricity. Better lighting reduces fatigue and increases productivity. The numerous visual operations of

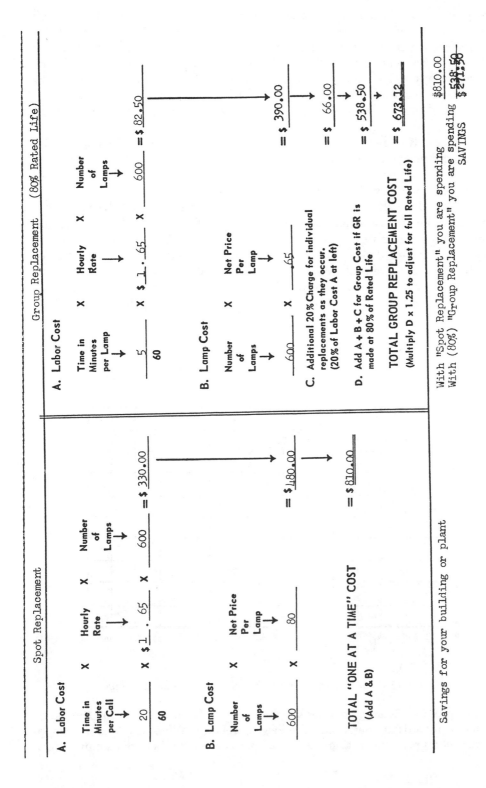

Figure 15-5 / Spot replacement cost vs. group replacement.

employees are quickened, made easier and more accurate. Quality of work will improve with less waste, fewer accidents and better morale. Maintenance calls for lamp replacements during working hours are eliminated. Group relamping can be done at a convenient time—during vacation shutdowns, after working hours, or on Saturdays. Programs can be varied slightly, to fit into schedules when they will result in less work interruptions. This prevents wasted time, as employees will not be distracted by the maintenance crew while they are changing light bulbs.

Another advantage of group replacement is that savings are realized by eliminating damage and burning out of starters and ballasts caused by blinking and slow start in aged lamps. It has been estimated that the life of a ballast can be reduced as much as 45 per cent from conditions such as these.

In establishing the proper time interval for group relamping, the main consideration is the life of the fluorescent lamps. One of the methods for determining group replacement, is to use the percentage of failures. The number of failures in a group indicates the portion of average life delivered by the group. Figure 15-6 shows that fluorescent lamps have reached 70 per cent of their average life when 13 per cent have burned out, and 80 per cent of average life when 21 per cent have burned out. After 80 per cent of average life, the rate of failure

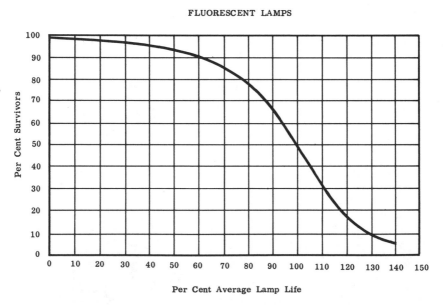

Figure 15-6 / **Average mortality curve—fluorescent lamps.**

increases rapidly. Relamping after 20 per cent of failures is one of the most accepted standards in determining when to relamp. When an area is group relamped at 80 per cent of average life, 20 per cent of the best remaining lamps are retained as individual replacements during the interim period, before the next relamping. These should be lamps with the best appearance from the standpoint of brightness and clean ends. When this supply is exhausted, it is a signal that group relamping is again required. The cycle is then repeated. If it is decided that the lamps will be used beyond 80 per cent of their average life, it is not recommended that these be used for interim replacement. Only new lamps should be used for replacements under these conditions. This would also be true of filament lamps after 85 per cent of their rated life. (See Figure 15-7.)

Another method is to determine the minimum amount of foot-candles recommended for a particular area. When this minimum is reached, it is time to replace lamps. Replacement inspections should be scheduled with this type of program, as previously described.

If previous experience does not include a group relamping program, the following criteria may determine the frequency of group relamping in your plant or building:

Single-shift operation   *Relamp every two years*
Two-shift operations   *Relamp every year and a half*
Three-shift operations   *Relamp once every year*

In most installations the group replacement program will prove to be more efficient and economical over the spot replacement program. Tables covering a particular fluorescent lamp mortality curve can be secured from the lamp manufacturer.

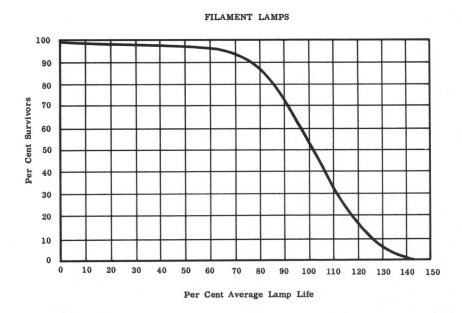

FILAMENT LAMPS

Figure 15-7 / **Average mortality curve—filament lamps.**

Planned Fixture Cleaning

Planned fixture cleaning is a necessity. The considerable expenditure of money in new lighting installations for maintenance, fixtures, and installation are sometimes soon forgotten. Dust continually accumulates on fixtures, and after a period of approximately twelve months without maintenance, installations are known to produce less than 50 per cent of their initial output. From this fact, it would be more practical and less expensive to install only half as many fixtures, and to keep them clean. Another example would be a plant or building designed for 40 foot-candles, but due to dirt they are producing 20 to 25 foot-candles. This is almost similar to hiring forty men when twenty-five can do the job.

Equipment

The purchase of special equipment, or the modification of a regular ladder, will increase the efficiency and justify the cost in a short time. Some special types of equipment available are:

1. Electric fork lift trucks.
2. Platform ladder with cart attached.
3. Pressure containers for spray water will eliminate the need for dropcloths and buckets.
4. Floor cart with two tanks, one holding cleaning solution and the other rinse water, for washing louvers.
5. Special louver trucks designed so that the louvers have space between them.
6. Movable scaffolds (telescopic or sectional type).

Greater efficiency is achieved if such equipment is designed to pass over obstructions such as desks, machines, or other equipment. Filament lamps can sometimes be changed from the floor with a lamp changing device. Equipment should be purchased with its versatility kept in mind, for other operations between lighting maintenance programs.

Relamping and Cleaning

1. Have the employee start the job completely equipped. *For example*, this should include a ladder cart combination of light construction, with platform to hold one pail of water with detergent, and one pail of clear rinse water. One box of new lamps and one empty carton to hold the old lamps should be attached to the front of the cart. Hooks on both sides of the cart will hold a clean louver on one side, and the dirty one on the other side. He will also need the following items: two flat sponges, one louver truck, if louvers are to be replaced, dropcloths if required, a supply of Electrical Repairs Needed tags (Figure 15-8), and

Electrical Repairs Needed slips (Figure 15-9). The ladder cart combination should be of such construction that it be movable on rubber casters, and so that the two casters on the step portion of the ladder will be retractable when the ladder is positioned.

2. Turn off lamp in the area to be serviced.

3. Check to see that all equipment is in a safe position, and accurately position the ladder truck.

4. Glass Type (A): Unlatch the glass and remove the lamps from the fixture and place in spare, empty box. Put on rubber gloves, and thoroughly clean the entire fixture with a detergent solution. Rinse the fixture with sponge and clear water, wipe dry, and remove gloves.

Louver Type (B): Unlatch the louver and hang along the side of the ladder cart. Remove lamps from the fixture and place in spare empty box. Put on rubber gloves, and thoroughly clean the entire fixture with a detergent solution. Rinse the fixture with a sponge and clear water, wipe dry, and remove gloves.

5. Install new lamps taken from carton attached to truck.

6. Latch glass into position, or replace clean louver taken from side of truck.

7. Dismount ladder and reposition it. If louvers are being replaced, remove dirty louver from the side of the ladder and replace it with a clean one. If the louvers are being washed in conjunction with the lamp changing, it may now be washed in a special portable tank.

If spare louvers are maintained, or if several are taken down prior to starting a large area, from a remote section, which is to be serviced last, group washing will aid your efficiency.

8. Repeat the above operation for the duration of the shift, leaving sufficient time to wash dirty louvers and other equipment, replacing the equipment to its proper location for storage. Remove discarded lamps to proper location, and secure a new supply of lamps for the next shift. Make out required reports.

A red Out of Order tag should be used to coordinate the work between the person changing lamps and the electricians. The tag should be of sufficient size to be easily seen. (See Figure 15-8.) The tag should be numbered and have a corresponding notice of electrical repairs needed slip, which should be filled out by the person changing the lamps, with the corresponding number of the red tag recorded on it, and the location. These should be printed to cover the required information, date, number, building, floor, room, lamps changed, type of lamps, brief description, and your signature. This slip should be turned in to the foreman at the end of the shift. (See Figure 15-9.) This will result in a maintenance order and is a method of keeping the system in good repair.

Maintenance of Luminous Ceilings

The maintenance of luminous ceilings starts with preliminary recommendations prior to installation. Specifications should include the following:

1. All surface areas above the ceiling should be sealed. Use aluminum foil or white paint. This will aid in light reflection.

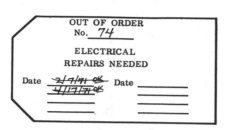

Figure 15-8 / **Electrical repairs needed (tag).**

ELECTRIC REPAIRS NEEDED
Date 4/17/71 — Tag# 74
Bldg. Main Office Floor 2nd Room or area 207
Tubes Changed Yes — Type F40T12WW
Brief Description 2-Busted Sockets
return to foreman Jo Sims
at the end of your shift — SIGNATURE

Figure 15-9 / **Electrical repairs needed (slip).**

2. Loose mortar should be chipped off the ceiling and walls.

3. All windows or holes above the ceiling must be completely sealed off.

4. Do not install ceiling panels until immediately prior to occupancy. This will prevent dust problems.

The average luminous ceiling should be washed once a year, along with the fixture and dusting of lamps. To clean and destaticize plastic panels or fixtures, use a normal household detergent in warm water. *Do not wipe dry.* Allow plastic to air dry in order to avoid accumulating static charge. Two tanks and a rack cart are recommended for cleaning plastic ceiling panels.

Tank #1—Warm water for washing with detergent, and brushes installed on both sides of the tank, and a drain valve at the bottom.

Tank #2—Warm water for rinsing with a waxing agent which will destaticize the plastic, so that it will repel dust.

Rack Cart—Multi-sectioned cart so that the panels can stand on edge to drain properly, and to air dry. A trough should catch all water and a drain valve attached.

During the past several years, many cities are now served with lighting maintenance companies. Where the value of group replacement and planned maintenance programs are recognized, lighting maintenance companies are employed to do the work. Due to their specialized work, they should be equipped with the most efficient equipment and best methods. Through their efficiency they should be able to perform the work at a lower cost. These firms are usually willing to provide free estimates on the costs of their services.

Voltage Line Check—Incandescent Lamps

A voltage check should be made on line voltage at various locations throughout the building on 120-volt incandescent lamps. In purchasing 120-volt lamps, the voltage should not exceed this rating. It has been discovered recently, that in some installations the voltage has been exceeding 120 volts. A 120-volt lamp operated at 122½ volts loses approximately 25 per cent of its rated life. If a 125-volt lamp is operated at 122½ volts, it gains approximately 35 per cent of its rated life. Lamp life burning at a lower than rated voltage does not materially affect the light output or efficiency relative to power consumption. By changing to 125-volt lamps, life expectancy would be about 170 per cent of that obtained from 120-volt lamps. With extended life of incandescent lamps as explained, this would decrease the number of lamps to be replaced by 40 per cent, thus reducing the work required. The 120-volt lamp is considered standard, and most manufacturers will furnish a 115-volt or 125-volt lamp at the same price as the 120-volt lamp.

New Developments a Challenge

New developments are constantly taking place in the lighting field which will require constant attention, in order to qualify one in planning a maintenance program. Research studies are now being made to combine a single panel element that will provide control of temperature and light at the same time. Increased use of lighting design to accommodate air conditioning systems with the use of a capacitive ballast is being employed, which would eliminate most of the wattage which loads an air conditioning system and much of the luminaire weight. Designs in lighting are being made to be included in the heating system in order to take the heat from the fixtures and pipe it to the heating ducts. An ultraviolet lamp is being developed for the annihilation of dangerous bacteria in heating and air conditioning ducts. Flat electroluminescent light panels that will emit practically no heat and consume a minimum amount of current will completely diffuse the light without special fixtures or baffles. It is also reported that a light control system has been developed to measure daylight, and will balance the electric light to assure constant room illumination.

With these developments in mind, maintenance responsibility increases, demanding constant development of know-how in the field of lighting maintenance methods and procedures.

A lighting maintenance program should be installed in every plant or building, or improved upon. All methods of education and reading must be explored to keep one qualified as a building superintendent or plant engineer. It has been an established practice to spend huge sums of money for research and development of new equipment, with only token offerings for maintenance. It is recommended that in the purchase of new equipment, the purchase order contain the requirement of a parts catalog, operating instructions, and all maintenance manuals. A building superintendent or plant manager is charged with the responsibility of protecting and getting full utilization of capital expended for fixtures and equipment. These catalogs and manuals will assist him in performing his job in a more efficient and economical manner for his employer.

16 / LOBBY AND ENTRANCE CLEANING

(*Source:* Buildings Management Division, General Services Administration, Washington, D. C.)

Job Description

Lobby and entrance cleaning includes all cleaning work in building lobbies and entrances, except floor scrubbing which is done by a scrubbing crew, and entrance door and glass cleaning which is done as a part of sidewalk cleaning and grounds maintenance.

The job of lobby and entrance cleaning consists of sweeping, dusting, polishing metal and wood surfaces, wall spotting, cleaning drinking fountains, mopping, and also the proper care and maintenance of the cleaning equipment and materials. Employees assigned to this job are instructed to notice lights out, defective door equipment and similar service defects, and to report them to their supervisor.

Equipment and Materials

A. *Equipment:*

    1. For sweeping and dusting.

        a. Dustpan
        b. Floor sweep mop
        c. Radiator dust brush
        d. Putty knife
        e. Dust cloths (polyethylene glycol treated cheesecloth)

    2. For polishing, wall spotting and mopping.

        a. Cheesecloth
        b. Cellulose sponge
        c. Cotton mop
        d. Buckets with wringer

    3. For equipment and materials transportation.

        Equipment carrier

B. *Materials:*

    1. For polishing, wall spotting and mopping.

        a. Detergent (mopping) solution
        b. Lemon oil polish
        c. Liquid metal polish
        d. White soap (cake)

2. For treating dust cloths.

    Polyethylene glycol solution

Job Methods

Operation Steps	*Explanation*

Step 1

1. Obtain equipment and materials.	Equipment and materials needed are listed under "Equipment and Materials".
2. Proceed to work area.	After placing equipment and materials on equipment carrier, move to area where work is to start.

PRECAUTION

Do not allow the equipment carrier to strike the walls, radiators or furniture.

Step 2

1. Dust radiators.	Use the radiator dust brush to clean radiators and grilles located within the assigned area as required.
2. Sweep lobby and/or entrance.	Use the floor sweep mop to sweep the lobby and/or entrance floor area, collecting the sweepings from such area into the dustpan. Dispose of the sweepings as directed by the supervisor.
3. Remove gum, tar and other foreign substances.	Use the putty knife to remove any gum, tar or similar substances from the floor areas during the sweeping.

PRECAUTIONS

1. Make sure all loose dirt is removed from baseboard areas and corners in the process of sweeping.
2. Be careful when using the putty knife not to gouge the floor surfaces.

Step 3

1. Dust with treated dust cloths.	Use the polyethylene glycol treated cloths to dust radiator tops, grilles,

Operation Steps   *Explanation*

door frames and mouldings, ledges and all other surfaces that are to be dusted.

Start the dusting at the highest surfaces that can be reached while standing on the floor and work downward toward the floor.

PRECAUTION

Avoid using the treated dust cloths on painted wall surfaces.

Step 4

1. Clean and polish metal surfaces.

Use clean cheesecloth, and metal polish as directed, to clean and polish metal doorknobs, push bars, kickplates, railings and other metal surfaces.

2. Clean and polish wood surfaces.

Use clean cheesecloth and lemon oil polish to clean and polish wooden handrails, doors and other wood surfaces as directed.

3. Clean drinking fountains.

Use warm water, white soap and a sponge to clean the bowls of all drinking fountains. Wipe remainder of fountain with a clean damp cloth. Wipe all chromium fixtures with a clean dry cloth.

PRECAUTION

Metal polish is not to be used on painted, lacquered, chrome-plated or other metal surfaces that have been specially treated.

Step 5

1. Place mop buckets for mopping.

Place the buckets of detergent (mopping) solution and rinse water in convenient positions for the beginning of the mopping operation.

2. Mop the lobby and/or entrance floors.

Use mop and detergent solution to thoroughly clean the floor areas. After dipping the mop into the solution, wring out enough excess solution so that the mop does not drip. Push the

Operation Steps	*Explanation*
	mop close to the baseboards, taking care to remove all dust from corners. In open areas swing the mop from side to side in long arcs, being careful to cover all the floor and to keep from splashing the baseboards or walls. After wringing the mop, go back over the mopped area to pick up the excess cleaning solution.
3. Rinse and dry the mopped areas.	Use a second mop and clean rinse water to thoroughly rinse and dry the mopped floor. Frequently rinse and wring the mop during this operation.

PRECAUTIONS

1. Avoid using an excessive amount of cleaning solution or rinse water during the mopping or rinsing.

2. Change the rinse water frequently.

3. Be careful not to splash or scar the walls or baseboards with the equipment.

4. As may be appropriate, suitable warning should be given to the building occupants against the danger of slipping on the wet floor.

Step 6

1. Return all equipment and unused materials to storage location.	Gather all pieces of equipment and unused materials, place them in the equipment carrier and proceed to storage location.
2. Clean and store equipment.	a. Shake cloths to remove excess dust and dirt; rinse out if required, and when dry, treat with polyethylene glycol solution. Place for drying and storage.
	b. Wipe the pan and knife free of dust and dirt and place for proper storage.
	c. Clean the dust and dirt from the brush, comb the bristles and place in its proper storage position.
	d. Rinse the mops, sponge and wiping cloths in clean water, squeeze or

Operation Steps	*Explanation*
	shake dry of excess water, and place for drying and storage.
	e. Empty buckets, rinse, wipe dry and return buckets and wringer to their proper storage positions.
	f. Wipe the carrier free of any moisture, dust and dirt, and place it in its proper storage location.
3. Store unused materials.	Return all unused materials to their proper storage places.

PRECAUTION

Make sure that lids are replaced securely on the containers for the cleaning and polishing materials.

Care of Equipment

A. *Dust Pan:*

1. Clean the dust pan of loose dirt and dust before storing.

2. When dumping the dust pan, avoid banging it with force against the waste container as such action tends to damage the sides and edges of the pan.

3. Hang by the handle so as not to allow any damaging weight to be put on the pan during storage.

B. *Floor Sweep Mop:*

1. Treated sweep mops should be shaken or vacuum cleaned, both during use and at the close of operations each day. They should be kept clean and free from grime or oily material.

2. Once every two weeks, remove mop head, shake out as much dust as possible and wash mop in a warm solution of synthetic detergent. Rinse in clear warm water.

3. Wring out as much water as possible and treat with polyethylene glycol (mix thoroughly three parts of water with one part of glycol). Spray mixture on mophead. Quantities required are approximately one ounce for three-quarter pound mop; one and one-half ounces for one pound mop; two ounces for one and one-quarter pound mop.

4. Replace the mophead on the handle and hang the mop so that a good air circulation will be obtained to assist in drying the mop. Let the mop dry for at least eight hours before using.

C. *Radiator Dust Brush:*

1. Comb the bristles periodically to clean the brush.

2. Store the brush either by laying it flat or by hanging it so that the bristles stand free of surrounding objects.

3. Do not use the radiator dust brush for any other purpose than for dusting radiators and grilles.

D. *Putty Knife:*

1. Use the putty knife only for the removal of gum, tar and other foreign substances during sweeping. *Do not* use as a screw driver or hammer.

2. Keep the putty knife clean and in good condition at all times.

3. Wipe free of dust and dirt after each use and store in a clean, dry place.

E. *Dust Cloths (polyethylene glycol treated cheesecloth):*

1. Thoroughly shake, or, if vacuum facilities are available, vacuum the dust cloths to remove loose dust.

2. Wash dirty cloths in warm detergent solution, (mix one ounce of detergent per gallon of warm water); rinse in clean water and wring or squeeze to remove excess water.

3. To treat new or washed cheesecloth, spray or sprinkle approximately one fluid ounce of polyethylene glycol solution (mixed in proportion of three parts of water to one part of polyethylene glycol) per square yard of cheesecloth. The cloth should be treated at least eight hours before using.

F. *Cheesecloth:*

1. Rinse all cloths after each use.

2. Wash dirty cloths in a warm detergent solution, (mix one ounce of detergent per gallon of water), and rinse in clean water.

3. Wring or squeeze to remove excess water and spread or hang the cloths to dry in their proper storage places.

4. Avoid leaving wet, dirty cloths piled together in the storage location.

G. *Cellulose Sponge:*

1. Rinse the sponge after each use.

2. Wash the sponge in detergent solution, as necessary, to remove dirt, and rinse in clean water.

3. Squeeze, *do not* wring, the sponge to remove excess water and place in proper storage position.

H. *Cotton Mops:*

1. After each use and before storage, rinse mops in clean water and wring or squeeze to remove excess water.

2. Wash dirty mops in warm detergent solution, (mix one ounce of detergent per gallon of warm water); rinse in clean water and wring or squeeze to remove excess water.

3. Fluff the mop strands by shaking.

4. Store the mops with the mopheads up to allow better drying.

I. *Buckets and Mop Wringer:*

1. Rinse and dry the buckets and wringer thoroughly after each use.

2. Remove any mop strands and other foreign matter remaining in the wringer.

3. Wipe the outside of the buckets free of dirt.

4. Avoid striking the buckets and wringer against objects as such action tends to create leaks in buckets and shorten the life of the equipment.

5. Store in a dry place and in such positions as to have no damaging weight bearing on the buckets or wringer.

J. *Equipment Carrier:*

1. Do not allow the carrier to strike the walls, furniture, or other equipment.

2. Wipe the carrier clean of dirt and dust at least once a week; maintain in a clean and orderly condition at all times.

3. Report promptly any defects or equipment failure to the supervisor.

Performance Inspection Guide

A. *Sweeping and Dusting:*

1. Lobby and entrance floors should be clean and free of dirt streaks and there should be no dirt remaining under radiators, in corners, behind doors or where the dirt was picked up with the dustpan after the sweeping operation.

2. Wads of gum, tar, and other sticky substances should have been removed from the area.

3. Radiators, grilles and woodwork should be dust-free after dusting. Dust should have been removed rather than merely pushed around.

4. There should not be any spots or smudges on the painted wall surfaces, caused by touching the wall with the treated dustcloth.

B. *Polishing and Wall Spotting:*

1. Doorknobs, push bars, kick plates, railings, doors and other surfaces should be clean and polished to an acceptable lustre.

2. Any drinking fountains located within the assigned area should be clean and free of stains. The wall surfaces around the drinking fountains should be free of water spots and streaks.

3. Wall surfaces up to a standing height should be free of finger marks, smudges and other dirt spots of any kind.

C. *Mopping:*

1. Lobby and entrance floors should be free of loose and/or caked dirt particles and should present an over-all appearance of cleanliness after the mopping operation.

2. Walls, baseboards and other surfaces should be free of watermarks, scars from the cleaning equipment striking the surfaces, and splashings from the cleaning solution and rinse water.

3. All surfaces should be dry and the corners and crevices clean after mopping.

D. *Equipment and Materials:*

1. All items of equipment should be clean and properly placed in a neat and orderly manner in their assigned storage locations.

2. All metal and furniture polish, soap and other materials left over from the cleaning operation should be returned to their assigned storage locations.

Frequency and Production

A. *Frequency and Production Schedule:*

	Sweeping and Dusting	*Polishing and Wall Spotting*	*Mopping*
Frequency	Daily *	Weekly *	Daily *
Production Rate per Man-day	16 main (large) lobbies or entrances 32 secondary lobbies or entrances		

* Or as directed by the supervisor

B. *Qualifying Factors:*

1. *Frequency*

   a. The frequency of lobby and entrance cleaning is dependent upon weather conditions, type of building and location, type of occupancy, and amount of traffic through the lobby and/or entrance.

   b. The frequencies given in the Frequency and Production Schedule are for a normal standard of cleaning under average conditions.

2. *Production Rate per Man-day*

   a. The main (large) lobbies or entrances are those that serve as the principal access to the buildings.

   b. The secondary lobbies or entrances are those of lesser importance, smaller in size and usually less frequently used.

   c. Due to the various types and sizes of lobbies and entrances, the production rate as established may be subject to modification at the various locations.

Staffing

The man-day requirements of labor for lobby and entrance cleaning can be determined by use of the following formula:

$$\text{Man-days (of labor)} = \frac{\text{Quantity}}{\text{Production}} \times \frac{1}{\text{Frequency}}$$

where quantity is the actual number of lobbies and entrances to be cleaned in the building; production is the number of lobbies and entrances cleaned per man-day; and frequency is the number of work days between operations.

If the frequency and/or production rate values as given in the Frequency and Production Schedule in this chapter have been modified to meet the existing conditions at a particular location, those modified values should be used in the above formula for the computation of the labor requirement.

SAMPLE CALCULATION

Assume a building with two main lobbies and/or entrances, plus ten secondary lobbies and/or entrances, and subject to such conditions that the average cleaning values from the Frequency and Production Schedule may be used.

$$\text{Man-days} = \frac{2}{16} \times \frac{1}{1} + \frac{10}{32} \times \frac{1}{1}$$
$$= 0.13 \qquad + 0.31$$
$$= 0.44 \text{ man-days of labor required (jobs)}$$

In order to determine the total number of positions needed to accomplish the work, this figure is multiplied by a conversion factor to compensate for leave allowances.

Conversion factor including compensation for leave = 1.15
Conversion factor × jobs = total positions required
1.15 × 0.44 = 0.51 positions required

Where the normal cleaning duty does not require the full time of the employee, other duties such as toilet cleaning, stairway cleaning, or some other job should be assigned to the employee to secure 100 per cent utilization of his time.

17 / CARE OF OUTSIDE AREAS

(*Source:* Buildings Management Division, General Services Administration, Washington, D. C.)

The care and maintenance of outside building and garage areas as outlined in this chapter has been broken down into the following duties: (A) Area policing; (B) sweeping; (C) outside entrance cleaning; (D) hosing sidewalks; (E) area scrubbing; (F) lawn maintenance; and (G) snow removal. These duties pertain to areas outside of the buildings except where garages and driveways are located within the building confines.

Job Description

In general, this job consists of keeping the specified areas policed of paper, trash and other debris; periodic sweeping of sidewalks and other paved areas as scheduled; wet cleaning sidewalks, steps, exterior portions of garages, and other required areas either by hosing down and spot scrubbing with a deck brush where necessary by mopping, or by machine scrubbing; cleaning exterior portions of building entrances; maintaining the lawn areas in a presentable condition by mowing and trimming the grass, trimming shrubs, raking leaves, watering grass and shrubbery; and, assisting in the removal of snow and ice as necessary.

It is suggested that the advice and experience of local county agents and other recognized experts on lawn and shrubbery care for your area be utilized as an aid in the achievement of satisfactory results.

All employees assigned to this work are responsible for reporting defective operating equipment and building items encountered during their tour of duty.

Job Methods

Operation Steps	*Explanation*

A. *Area Policing:*

Step 1—Prepare for Work

1. Obtain equipment needed.	a. Canvas shoulder bag or other approved trash container
	b. Dust pan, long handle
	c. Gloves
	d. Sidewalk brush (24")
	e. Litterstick with sharp point (for picking up paper)
2. Proceed to work area.	Proceed to the assigned location where area policing is to start.

Step 2—Wastepaper and Trash

1. Pick up wastepaper and trash.	Go over the assigned outside areas and pick up loose wastepaper and/or trash. Place collected paper and trash in the trash room, or dispose of it as directed by the supervisor.
2. Litter sweeping.	During the policing operation use the brush to sweep up litter, loose accumulations of dirt, leaves, etc. from the hard surfaced areas as required. This litter sweeping should not substitute for the regularly scheduled sweeping job.

Step 3—Return Equipment to Storage

1. Return to storeroom location.	Gather all policing equipment and return it to the storage location.
2. Clean and store equipment.	Clean the equipment as directed and place in the assigned storage location.

Operation Steps   *Explanation*

B. *Sweeping:*

Step 1—Prepare for Work

1. Obtain equipment needed.

    a. Dust pan, long handle
    b. Gloves
    c. Sidewalk brush (24")
    d. Trash basket (3 bushel oak splint) or other suitable trash container

2. Proceed to work area.

    Proceed to the assigned location where sweeping is to start.

Step 2—Sweep the Assigned Area

1. Sweep with brush.

    Sweep the assigned area using short push strokes of two and one-half to three feet, and pressing down slightly on the brush bristles when sweeping rough surfaces. Overlap the brush strokes as necessary to avoid leaving fine dirt behind.

2. Pick up string, pieces of paper, etc.

    Pick up pieces of paper, string, and other trash which cannot be pushed forward with the brush and place them in the trash container.

3. Dispose of trash.

    Empty the collected sweepings and trash into designated containers in the trash room, or dispose of as otherwise directed by the supervisor.

Step 3—Return Equipment to Storage

1. Return to storeroom location.

    Gather all pieces of equipment used during the sweeping operation and return them to the storage location.

2. Clean and store equipment.

    Clean the equipment as directed. Place all items in their proper storage positions.

C. *Outside Entrance Cleaning:·*

Step 1—Prepare for Work

1. Obtain equipment and materials needed.

    a. Ammonia
    b. Bucket (15 qt.)

Operation Steps	*Explanation*
	c. Cheesecloth
	d. Deck scrub brush
	e. Detergent
	f. Dustpan
	g. Lemon oil
	h. Metal polish
	i. Mop unit (2-compartment mop tank or mop buckets with wringer)
	j. Mop, cotton
	k. Putty knife (1¼″)
	l. Sidewalk brush (24″)
	m. Sponge
	n. Squeegee, floor
	o. Squeegee, window
	p. Water hose (length as required)
2. Proceed to work area.	Proceed with the equipment and materials for work to be performed to the area where the entrance cleaning is to start.

Step 2—Sweep the Entrance

1. Sweep with brush.	Sweep the entrance area, using short push strokes, and pressing down slightly on the brush bristles when sweeping rough surfaces. Overlap the brush strokes as necessary to avoid leaving fine dirt behind.
2. Pick up string, pieces of paper, and other waste material.	Pick up by hand pieces of paper, string and other trash which cannot be pushed forward with the brush, and dispose of same with the other sweepings.
3. Remove gum and other foreign substances.	Use putty knife to remove wads of gum, spots of tar, or other foreign substances from the floor area.

Step 3—Wet Clean the Entrance

1. Hose down the entrance.	If practical, use a hose to wet clean the entrance steps, doors and areaway. As necessary, use the mop and/or deck brush and detergent solution (mix one ounce per gallon water) to remove grime and dirt.

Operation Steps	*Explanation*
	Thoroughly rinse after detergent solution is used.
2. Pick up excess water.	Squeeze off or mop up all excess water remaining after the cleaning operation. Rinse and wring the mop frequently in clean water.

NOTE

This work should be performed during periods of minimum traffic through the building entrances to avoid inconveniencing the building occupants and visitors. As necessary, warning signs should be conspicuously placed near the work area.

Step 4—Polish Glass, Wood and Metal

1. Clean entrance glass.	Use a sponge, clean water and window squeegee to thoroughly clean both sides (inside and out) of glass surfaces in the entrance. As necessary, ammonia may be added to the water to loosen oil and grime adhering to the glass. Wipe up water drippings from the glass cleaning operation with the sponge.
2. Clean and polish metal surfaces.	Use clean cheesecloth, and metal polish as directed, to clean and polish metal doorknobs, push bars, kick plates, railings and other metal surfaces.
3. Clean and polish wood surfaces.	Use clean cheesecloth and lemon oil polish to clean and polish wooden handrails, doors and other wood surfaces as directed.

PRECAUTIONS

Metal polish is not to be used on painted, lacquered chrome-plated or other metal surfaces that have been specially treated.

Step 5—Return Equipment and Materials to Storage

1. Return to storeroom location.	Gather up all equipment and unused materials and return them to the storage area.
2. Clean equipment and store equip-	Clean all equipment items as directed.

Operation Steps	*Explanation*
ment materials.	Store equipment and unused materials in their assigned storage locations.

D. *Hosing Sidewalks:*

Step 1—Prepare for Work

1. Obtain equipment and materials needed.	a. Rubber boots or overshoes b. Squeegee, floor c. Water hose (length as required) d. Water hose nozzle e. Wiping cloths
2. Proceed to work area.	Proceed with the equipment to the assigned area where work is to start.

Step 2—Clean the Assigned Area

1. Hose down the area.	Connect the water hose to a convenient water outlet and hose down the sidewalk and other designated areas. Play the water stream from side to side as required to clean the hard-surfaced areas.
2. Remove excess water.	Use the squeegee to remove water remaining on the cleaned surfaces.
3. Wipe up spatterings.	Use the wiping cloth to clean water marks or spatterings from nearby wall surfaces, door, hand railings, and other places, if necessary.

PRECAUTION

1. Use as little water as possible to properly clean the area.
2. Avoid directing the water stream on surrounding plants, shrubs and buildings.
3. This work should be performed at times of least inconvenience to occupants.

Step 3—Return Equipment to Storage

1. Return to storeroom location.	Disconnect water hose from water outlet, drain water from hose, and return it and other equipment items to the storage area.
2. Clean and store equipment.	Clean the equipment items as directed. Store the items in assigned storage locations.

Operation Steps   *Explanation*

E. *Area Scrubbing:*

Step 1—Prepare for Work

1. Obtain equipment and materials needed.

    a. Bucket (15 qt.)
    b. Deck scrub brush
    c. Detergent
    d. Mop unit (2-compartment mop tank or mop buckets with wringer)
    e. Mops, cotton
    f. Putty knife
    g. Rubber boots or overshoes
    h. Scouring powder
    i. Squeegee, floor
    j. Water hose (length as required)

2. Proceed to work area.

Proceed with the necessary equipment and materials to the assigned area where work is to start.

Step 2—Wet Cleaning

1. Mop or scrub the assigned area.

Wet clean assigned areas, such as garage, driveway, and portico by the most practical of the following prescribed methods:

2. Clean with water hose and deck scrub brush.

    a. Use water hose to wet the area to be cleaned. Scrub with deck scrub brush and detergent solution (mix one ounce detergent per gallon water). Use scouring powder as necessary to remove embedded dirt and grime.

    b. Thoroughly rinse cleaned area with water hose, and use floor squeegee to remove excess water. As necessary, a mop may be used to remove excess water and to dry the floor.

3. Clean with mop and detergent solution.

Job methods for wet cleaning floors as outlined in Section 10B, Floor Mopping, may be followed in the performance of this work.

4. Clean with power scrubbing ma-

Where using a power scrubbing ma-

Operation Steps	Explanation
chine.	chine will result in more efficient performance, one should be used for this work.
5. Remove grease and oil spots, wads of gum and tar, noticed in area.	Use the putty knife and/or scouring powder and deck scrub brush to remove grease and oil spots and stains, particles of tar and gum, and other embedded dirt not otherwise removed.
6. Rinse the cleaned area.	Use clean rinse water and mop, or water hose, to thoroughly rinse the cleaned area. As applicable, the floor squeegee may be used to remove excess rinse water.

PRECAUTIONS

1. Before hosing down areas, check the drains to see that they are open to carry off the wash water.

2. Avoid splashing dirty water on surrounding equipment, walls, etc.

3. Thoroughly rinse all areas where detergent solution or scouring powder has been used.

Step 3—Return Equipment and Materials to Storage

1. Return to storeroom location.	Gather all equipment items and unused materials and return them to storage area.
2. Clean equipment, and store equipment and materials.	Clean all equipment items as directed. Store equipment and materials in their assigned storage locations.

F. *Lawn Maintenance:*

Step 1—Prepare for Work

1. Obtain necessary equipment.	a. Edger
	b. Gloves
	c. Hand truck (as necessary)
	d. Lawn mower (hand or power machine)
	e. Lawn rake (as necessary)
	f. Power equipment accessories (if power equipment is used)

Operation Steps   *Explanation*

g. Shrubbery tools (shears, knife, and others)

h. Trash basket (3 bushel oak splint), or suitable containers for trimmings and leaves.

i. Water hose (length as required)

j. Wiping cloths

2. Proceed to work area.

After obtaining the necessary equipment required for the particular job to be done, proceed to assigned area where work is to start.

Step 2—Grass Cutting

1. Grass cutting season.

In general, the first spring grass cutting should be delayed until the grass has had an opportunity to grow, to allow for greater root development. The last cutting in the fall should take place at such time to allow the grass to grow back to a length of two to three inches as a protection against the winter season.

2. Cutting frequency.

For purposes of staffing the cleaning standard establishes the frequency for grass cutting as once every five days. However, this work should be performed as the growth of the grass requires. A good general schedule to follow is to cut the grass whenever it reaches a height of three inches, one and one-half inches since the last cutting.

3. Grass clippings.

Usually grass clippings may be left on the lawn. They act as a mulch, reduce moisture evaporation from the soil, help keep the soil cool, and contribute to the feeding of the soil. However, long clippings that lie on top of the grass rather than filtering down onto the soil should be raked up and disposed of to prevent smothering or killing the grass, and to eliminate the unsightly appearance of

Operation Steps	*Explanation*
	dried, discolored clippings on the lawn.
4. Cutting height.	Generally a minimum cutting height of one and one-half inches is recommended. At this height the turf is less likely to suffer, even when the grass is frequently cut, and the grass tips can be kept uniformly trimmed so that the lawn has a well-groomed appearance.
5. Cutting procedure.	Cut up and down the lawn or in a circular direction, overlapping the previous swath by about two inches. When using power equipment, use the longest uninterrupted run practicable. When cutting the crown of a terrace mow up and down the slope, as cutting crosswise may result in scalping the crown. Trim the grass areas adjoining sidewalks, shrubs, and other areas, as required during the lawn mowing operation.
6. Mower blade maintenance.	Well set, sharp mower blades provide a fast, clean cutting action. This prevents the bruising of grass blades, which occurs when dull blades chew off the ends rather than shearing them evenly. For safety, police the lawns before cutting the grass to remove stones, broken glass and other debris. This helps keep the mower blades from becoming nicked and avoids damage and serious breakdowns of the equipment.
7. Rolling, aeration, feeding, weed and insect control.	Various other duties pertinent to lawn maintenance, such as lawn rolling, soil aeration, feeding, elimination of weeds, and pest control should be handled on a local basis as required.

GENERAL GRASS CUTTING RULES FOR PROPER LAWN MAINTENANCE:

1. Delay the first spring cutting until the grass has made some new growth.

2. Cut the grass at a minimum height of one and one-half to two inches, using well-set sharp mower blades.

3. Remove the clippings if extra long and during wet weather when the clippings are heavy enough to mat.

4. Stop cutting the lawn early enough in the fall to insure a growth of two to three inches before winter.

Step 3—Pruning Shrubbery

NOTE

The pruning information outlined in this step is of a general nature. The advice of recognized pruning experts should be followed in the performance of this work in the various localities.

Operation Steps	*Explanation*
1. When to prune.	Generally, early blossoming shrubs which bloom on the previous year's growth should be pruned just after they bloom. Late blossoming shrubs which bloom on the same year's growth should be pruned in late winter or early spring. Late summer pruning is unsafe because the new growth of wood that occurs rarely becomes hardened enough to withstand severe winter weather.
2. Pruning tips.	Pruning is performed to remove surplus or undesirable growth at the proper time of the year. All cuts must be made cleanly with a sharp knife, saw or shears, leaving no projecting stubs or branches. Cut surfaces should be treated with creosote or other suitable recommended materials and coated to prevent decay. Branches should never be broken off and left untreated.

Step 4—Watering

1. Watering frequency.	Water lawn only often enough to keep it alive. Loam and clay soils hold more water and longer than sandy and gravel soils and so need not be watered as frequently. A method for determining the dryness of the soil

Operation Steps	*Explanation*
	in some areas is to examine it to a depth of two to three inches. Dryness of the upper part of the soil generally indicates need for watering.
2. Amount of water.	If the lawn must be watered, thorough soaking to a depth of six inches is much better than light sprinklings. Light sprinklings in quantities that do not allow thorough penetration of the soil tends to draw the grass roots to the surface of the ground. Because of surface runoff and greater moisture loss, special consideration should be given to steep slopes and slopes facing in a southerly direction. Apply the water only as fast as the soil can absorb it to prevent surface runoff and consequent water loss.

NOTE

When watering lawns, care should be taken to prevent muddy water splashing from bare ground areas onto nearby building walls, foundations, and walks.

Step 5—Return Equipment to Storage

1. Return to storeroom location.	Gather all equipment items and return them to the storage location.
2. Clean and store equipment.	Clean all pieces of equipment as directed, and store in their assigned storage positions.

G. *Snow Removal:*

Step 1—Prepare for Work

1. Obtain equipment and materials needed.
   - a. Bucket (15 qt.)
   - b. Gloves
   - c. Hand truck
   - d. Ice scraper
   - e. Plow attachments for power equipment (type and quantity as required)
   - f. Raincoat

Operation Steps	*Explanation*
	g. Rubber boots (or overshoes)
	h. Sand and/or chemical compound
	i. Snow shovel
2. Proceed to work area.	Move the required equipment and materials to the assigned area where work is to start.

Step 2—Removal of Snow and Ice

1. Clean sidewalks, entrances, and driveways of snow and ice.	Use shovels, ice scrapers, and/or plows attached to power equipment to clear areas of snow and ice. Pile the snow and ice on lawns and around trees rather than pushing into streets and gutters.
2. Sand sidewalks, walks and driveways.	Use an approved chemical snow and ice remover compound or sand on areas as required to control hazards to occupants and visitors.

NOTE

Chemical snow and ice remover compound is preferable to sand for use at building entrances and on adjoining sidewalk areas where possible. Use of sand in such areas results in trackage of the sand into buildings, which necessitates additional cleaning. Where sand is used, provision should be made for its pickup and proper disposal, as directed, as soon as is practical after it has served its purpose.

Step 3—Return Equipment and Materials to Storage

1. Return to storeroom location.	Gather all equipment and unused materials and return them to the storage location.
2. Clean and store equipment; store materials.	Clean all pieces of equipment as directed. Store equipment and unused materials in their proper storage positions.

Care of Equipment

A. *Canvas Shoulder Trash Bag or Other Approved Receptacle:*
   1. Thoroughly empty bag or receptacle of trash and dirt after each day's use.
   2. Avoid snagging or tearing the bag.
   3. Store in a clean, dry location.

B. *Cheesecloth:*

1. Rinse all cloths after each day's use.

2. Wash dirty cloths as required in a warm detergent solution, rinse thoroughly and wring or squeeze out excess water before placing them to dry.

C. *Deck Scrub Brush:*

1. Rinse in clean water after each day's use and shake dry of excess water.

2. Place for drying and storage in such position that the bristles do not bear the weight of the brush.

3. Avoid knocking the brush against objects for this tends to split the brush block.

D. *Dustpan:*

1. Clean dustpan of dirt and dust before storing.

2. Hang dustpan by the handle so that it will not become bent or otherwise damaged.

E. *Gloves:*

1. Shake or wipe free of dirt and dust after each day's use.

2. Avoid snagging gloves on nails or other sharp objects.

3. Store gloves in a clean, dry location.

F. *Hand Lawn Mower:*

1. Wipe all parts free of dirt, moisture and grass cuttings after each day's use.

2. Store the mower in its assigned location and in such a position as to prevent its being damaged.

3. The mower should be kept properly lubricated and sharpened. Report all defects and/or necessary repairs to the supervisor.

G. *Hand Truck:*

1. Clean the truck of dirt and trash after each day's use.

2. The wheels of the truck should be kept properly lubricated. Report all necessary repairs or defects to the supervisor.

H. *Ice Scraping Tool:*

1. Wipe the tool free of moisture and dirt after each day's use.

2. Store in a safe manner in its assigned location

I. *Lawn Rake:*

1. Wipe the rake free of moisture, dirt and other debris after each day's use.

2. Store by hanging in a safe and secure manner so it will not be damaged.

J. *Litterstick:*

1. Wipe the stick free of moisture and dirt after each day's use.

2. Place a cork or similar protective object over the pointed end before storage.

K. *Mop, Cotton (2 lb.):*

1. Before storing after each day's use rinse the mop in clean water and squeeze or wring to remove excess water.

2. Wash dirty mops in a warm detergent solution, rinse in clean water, and wring or squeeze to remove excess water.

3. Fluff the mop strands by shaking.

4. Store, with the mophead up, in a well-ventilated place to allow for better drying.

L. *Mop Unit (2-compartment mop tank or mop buckets with wringer) and Bucket (15 qt.):*

1. Rinse bucket, mop unit compartments and wringer thoroughly and wipe dry after each day's use.

2. Remove any remaining mop strands and other matter from the bucket, mop unit and wringer.

3. Always leave mop wringer in released position.

4. Store in a clean dry location and otherwise maintain a clean and orderly condition.

5. Report any defects or necessary repairs to the supervisor.

M. *Power Sidewalk Sweeper:*

See instructions for equipment care under *H* in the following section "Performance Inspection Guide."

N. *Putty Knife:*

1. Wipe knife free of dirt and moisture after each day's use.

2. Store in a clean dry place.

3. Use only for the purpose specified. Do not use as a screw driver or hammer.

O. *Raincoat:*

1. Wipe raincoat free of dirt and excess moisture after each day's use.

2. Store by hanging on a hanger in a clean and well-ventilated place.

3. Avoid snagging on nails and other sharp objects.

P. *Rubber Boots:*

1. Rinse off dirt and other matter, and wipe dry after each day's use.

2. Store in a clean, well-ventilated place in such a way that the boots will not be damaged.

Q. *Shrubbery Tools:*

1. Wipe tools free of dirt and moisture after each day's use.

2. Maintain tools in a well-sharpened and lubricated condition as necessary.

3. Hang tools for storage in such a way that they will not be damaged or endanger personnel.

R. *Sidewalk Brush (24"):*

1. Wash an oily or dirty brush in a warm detergent solution and rinse in clean water. Shake free of excess water and hang to dry with bristles in a downward position and free of any weight.

2. Store in a clean, dry place so that the bristles do not bear the weight of the brush.

3. Reverse the position of the handle in the brush block weekly to assure longer brush life.

S. *Snow Plow Attachments:*

1. Wipe equipment free of dirt and moisture after each use.

2. Upon storing, coat bare metal portions of the equipment with a rust preventive compound as directed by the supervisor, and maintain in a well lubricated, clean condition.

3. Report necessary repairs and defects to the supervisor.

T. *Sponge:*

1. Rinse sponge in clean water after each day's use.

2. When dirty, wash in warm detergent solution, rinse thoroughly, and squeeze out excess water.

3. Store in a clean, dry location.

U. *Squeegees:*

1. Rinse squeegee blade in clean water after each day's use.

2. Wipe dry and store in a clean, dry location.

V. *Trash Basket:*

1. Clean trash basket of trash and dirt after each day's use.

2. Store in such a way that it will not be damaged.

W. *Water Hose:*

1. Drain hose of any remaining water and wipe outside free of dirt and excess moisture.

2. Coil and properly store hose in its assigned location.

X. *Wiping Cloths:*

1. Rinse wiping cloths in clean water after each day's use.

2. Wash dirty cloths as required in a warm detergent solution, rinse thoroughly, and wring or squeeze out excess water before spreading to dry.

Performance Inspection Guide

A. *Area Policing:*

1. Lawn areas, sidewalks, driveways and other assigned areas should be free of trash, paper and other debris after the policing work has been performed.

2. Sidewalks and building entrances should be kept free of cigarette butts, loose sand, string, paper, and other debris, and be presentable in appearance at all times.

B. *Sweeping:*

1. Sidewalks, entrances, garages and other portions of the assigned areas should have been swept clean, as scheduled.

2. Loose trash, such as paper, string, sticks, and other litter should have been cleared away in the sweeping process.

C. *Entrances:*

1. Entrance steps and areas should be policed, and kept free of dirt and other debris.

2. Entrance glass, metal and wood surfaces should be clean and, where applicable, polished to an acceptable lustre.

D. *Hosing Sidewalks:*

1. Sidewalks and other designated paved areas should have been hosed down and cleaned as scheduled.

2. Excess water should have been squeegeed off the cleaned areas.

3. Water spatterings should have been wiped from railings, entrances, and other surrounding surfaces as required.

E. *Area Scrubbing:*

1. Paved sidewalks, entrances, garages, and other paved areas should have been wet cleaned as scheduled.

2. Grease spots and stains should have been removed and the areas left in a clean and presentable condition.

F. *Lawn Maintenance:*

1. All grass and shrubbery should present a well-groomed appearance at all times.

2. Grass should not have been cut to a height of less than one and one-half inches except for those plant varieties that may be mowed closer.

3. In season, leaves should be raked up and removed from the lawn areas as necessary.

G. *Snow Removal:*

1. Sidewalks, entrances, drives and other such areas should be as free as possible from accumulations of snow and ice.

2. As required, such areas should be kept sanded or chemically treated to provide safe footing or traction.

H. *Equipment Care:*

1. All equipment used in performing this work should be clean and properly cared for.

2. All equipment and materials should be stored in assigned storage locations when not in use.

Frequency and Production

A. *Frequency and Production Schedule:*

	Frequency	Production per Man-day
1. *Area policing.* Includes entrances, parking areas, garages, ramps, sidewalks, lawn areas, driveways.	Daily	300,000 sq. ft.

	Frequency	Production per Man-day
2. *Sweeping*. Includes parking areas (hard-surfaced), garages, ramps, driveways, and sidewalks.	Every 5 days	Hand— 50,000 sq. ft. Machine— 100,000 sq. ft.
3. *Outside entrance cleaning*. Includes daily sweeping, wet cleaning weekly, and cleaning and polishing glass, metal and wood surfaces as required.	Daily	Main—16 Secondary—32
4. *Hosing sidewalks*. Includes hosing down sidewalk and other designated paved areas periodically and removing resulting water puddles from such surfaces.	As required	30,000 sq. ft.
5. *Area scrubbing*. Includes wet mopping or scrubbing of exterior portions of garages, ramps, driveways, and other specified paved areas.	As required	Mopping— 20,000 sq. ft. Scrubbing Machine 25,000 to 50,000 sq. ft.
6. *Lawn maintenance*. Includes cutting and edging of grass areas, watering, care of shrubbery, raking leaves, etc., as required.	Every 5 days	100,000 sq. ft.
7. *Snow removal*. Includes removing snow and ice from entrances, sidewalks, parking areas, driveways, and ramps, and spreading sand or snow and ice remover compound to reduce safety hazards.	As required	Performed by special crews from maintenance force.

B. *Qualifying Factors:*

The values for frequency and production given in the above schedule are for a normal standard under average conditions. These values may be subject to modification at certain locations due to climatic conditions, type of occupancy, terrain, and other reasons. However, they provide a standard for buildings from which such deviations, if any, should be made.

Staffing

The man-day requirements of labor for this work can be determined by use of the following formula:

$$\text{Man-days (of labor)} = \frac{\text{Quantity}}{\text{Production}} \times \frac{1}{\text{Frequency}}$$

Where quantity represents the square feet of area maintained at the given frequency; production is the square feet of area maintained per man-day for the type of equipment and the job method utilized; and frequency is the number of work days between operations.

If the frequency and/or production rate values established in the Frequency and Production Schedule, have been modified to meet existing conditions at a particular location, then those modified values should be used in the above formula for the labor requirement computation.

SAMPLE CALCULATION

Assume a building location having the following areas to be maintained, under such conditions that the established values from the Frequency and Production Schedule may be used:

Sidewalks	=	16,700 sq. ft.
Driveways, garages and parking areas	=	22,900 sq. ft.
Lawn area	=	23,200 sq. ft.
Main entrances	=	4

Then the area to be policed is 16,700 sq. ft. + 22,900 sq. ft. + 23,200 sq. ft., or 62,800 sq. ft.; sweeping area is 16,700 sq. ft. + 22,900 sq. ft., or 39,600 sq. ft.; sidewalk hosing area is 16,700 sq. ft. based on a weekly basis; area scrubbing is 39,600 sq. ft. based on a quarterly basis; and lawn maintenance area is 23,200 sq. ft. Therefore, the man-days (of labor) are computed as follows:

$$\text{For Area Policing} = \frac{62,800}{300,000} \times \frac{1}{1} = .21$$

$$\text{For Sweeping} = \frac{39,600}{50,000} \times \frac{1}{5} = .16$$

$$\text{For Sidewalk Hosing} = \frac{16,700}{30,000} \times \frac{1}{5} = .11$$

$$\text{For Entrance Cleaning} = 4/16 \times \frac{1}{1} = .25$$

$$\text{For Area Scrubbing} = \frac{39,600}{25,000} \times \frac{1}{66} = .02$$

$$\text{For Lawn Maintenance} = \frac{23,200}{100,000} \times \frac{1}{5} = .05$$

$$\text{Total Jobs} \qquad\qquad\qquad .80$$

To this total requirement must be added a factor to compensate for leave in order to arrive at the total number of positions needed to accomplish this work.

Conversion factor including compensation for leave = 1.15

Conversion factor × jobs = total positions required

1.15 × .80 = positions required—.92

Where this work does not require the full time of the employee, other duties should be assigned to secure 100 per cent utilization of his time.

Equipment and Materials Allocation and Replacement

Items	Allotment per Cleaner	Replacement or Reissue
Boots, rubber	1 pair	1 per 5 years
Brush, deck scrub	1	1 per year
Brush handle, deck scrub	1	1 per 2 years
Brush, radiator	1	1 per year
Brush, sidewalk (24")	1	2 per year
Brush handle, sidewalk	1	1 per year
Bucket (15 qt.)	1	1 per year
Cheesecloth	2 yards	8 yards per year
Cloth, wiping	2	6 per year
Detergent *	1 pound	50 pounds per year
Dustpan	1	1 per 2 years
Gloves	1 pair	2 pair per year
Hose, water	As required	As required
Ice scraping tool	1	1 per 5 years
Lawn edger, sprinkler, and Lawn equipment (miscellaneous)	As required	As required 1 per 5 years
Lawn mower, hand	1 (as required)	
Lawn mower, power †	As required	As required
Litterstick	1	2 per year
Mop, cotton	1	6 per year
Mop handle	1	1 per year
Mop unit	1	1 per 5 years
Putty knife (1¼")	1	1 per 2 years
Raincoat	1	1 per 5 years
Rake, lawn	1	1 per 2 years
Sand or chemical compound	As required	As required
Scouring powder	1 pound	25 pounds per year
Shrubbery tools	As required	As required
Sidewalk sweeper, power †	As required	As required
Snow plow attachments	As required	As required
Snow shovel	1	As required
Sponge	1	2 per year
Squeegee, floor (18")	1	2 per year
Squeegee, window	1	1 per year
Trash bag, canvas shoulder or other approved receptacle	1	As required
Trash basket (3 bushels)	As required	As required
Truck, hand	1	1 per 5 years

* Recommended as a general purpose cleaner. Has proven satisfactory for general floor cleaning in proportions one ounce per gallon of water.

† Mechanized equipment should be provided with safety features, such as fire extinguisher, fill and vent fitting, and others as prescribed to insure compliance with safe operating practices and regulations.

The allotment per cleaner and the replacement or reissue quantities in this section are based on one cleaner spending 100 per cent of his time on this work.

• Where the situation exists that other duties are assigned to the cleaner to secure 100 per cent utilization of his time, consideration should be given to a reduction in the replacement quantities in accordance with the actual amount of this work assigned.

Operation of Gasoline-Powered Grass Cutting Machine

Operation Steps   *Explanation*

NOTE

Although these instructions pertain specifically to the operation of power lawn mowers, they also have application to power trimmers, edgers and leaf mulchers.

Step 1—Preparation for Work

1. Obtain equipment and materials needed.

    a. Approved safety gasoline can
    b. Engine oil (S.A.E. 30)
    c. Gasoline (white, unleaded)
    d. Lawn mower, gasoline-powered
    e. Radiator brush
    f. Wiping cloths

2. Check oil level in crankcase.

    a. Remove oil filler plug and check to see if crankcase is full of oil.
    b. If oil is low, add enough oil to bring it up to proper level.

3. Check gasoline tank.

    a. Remove filler cap of gasoline tank.
    b. When necessary fill gasoline tank with white (unleaded) gasoline *after* machine has been taken outdoors.
    c. Be sure that vent hole in filler cap is open.

PRECAUTIONS

1. Always wipe any accumulated dirt from gasoline filler cap, oil filler cap, oil filler plug, and the area around them before they are removed for checking or refilling.
2. Be sure dirt and water do not enter the gasoline tank or engine crankcase.
3. Be sure there is no lighted cigarette, cigar, pipe, or match nearby when handling gasoline.
4. Do not fill gasoline tank while engine is running.

5. Avoid spilling gasoline on hot engine.

6. Always fill gasoline tank outdoors.

7. When using a rotary blade mover, be sure that guard is in place to prevent possible injury from the blade, flying stones, and other debris.

Operation Steps	*Explanation*
4. Procedure to start engine.	a. Open gasoline shut-off valve located near bottom of gasoline tank.
	b. Open gasoline throttle slightly.
	c. Close choke on carburetor.
	d. Crank engine with rope starter, foot pedal starter, or hand crank provided.
	e. When engine starts gradually open choke valve until engine runs smoothly with choke open.

NOTE

1. When engine is difficult to start, operator should check the following:

   a. No fuel in gasoline tank.
   b. Improper choking.
   c. Throttle valve stuck or improperly adjusted.
   d. Throttle rod loose.
   e. Loose or defective wiring.
   f. Cracked spark plug.

2. If operator cannot start engine he should notify his supervisor.

5. Proceed to work area.	a. Open throttle to speed engine up slightly.
	b. Engage the propelling clutch.
	c. Control forward speed of mower by opening or closing throttle.

PRECAUTIONS

1. Be sure all clutches are disengaged before starting engine.

2. The engine exhaust contains carbon monoxide gas which although colorless and odorless is a deadly poison. Be sure that engine is permitted to run only in well-ventilated areas.

3. Do not wrap starting cord around hand or wrist, or place thumb on opposite side of crank handle from palm of hand when cranking engine. This will avoid injuries which might result if the engine should backfire.

Operation Steps   *Explanation*

Step 2—Cut Grass

1. Start machine and operate.

    a. Engage reel or rotating blade clutch, if mower is equipped with one.

    b. Engage propelling clutch.

    c. Control forward speed of mower by opening or closing gasoline throttle.

    d. Overlap previous swath about two inches.

2. Stop machine.

    a. Disengage propelling clutch.

    b. Disengage reel clutch, if any.

    c. Close gasoline throttle.

    d. Press grounding spring against top of spark plug and hold in this position until engine has stopped.

PRECAUTIONS

1. Be sure to pick up stones or other solid objects in the path of the mower before mowing grass.

2. Always stop engine before cleaning or working near the reel or rotating blade.

3. Be careful to avoid letting foot slip under guard of rotating blade while blade is in motion.

4. Always stop engine and leave propelling clutch engaged if there is any possibility of mower moving when it is out of gear and operator is away from the controls.

5. Do not permit mower to run into or strike against any object.

6. Check oil and gasoline every four hours.

7. Be sure to remove wire from spark plug before attempting to make adjustments or haul the lawn mower.

Step 3—Returning Equipment and Materials to Storage

Return all equipment and unused materials to storage location.

    a. Gather all pieces of equipment and materials and proceed to designated gear room.

    b. Brush off accumulated grass and wipe power lawn mower, except the engine, with a dry cloth.

    c. Use radiator brush to remove any dirt or grass that has accumulated

Operation Steps   *Explanation*

on fly-wheel housing or between the cylinder fins of engine.

d. Comb tufts of radiator brush out straight. Store on a flat surface so that tufts will be flat.

e. Rinse out wiping cloths used (or, if dirty, wash in warm detergent solution), wring and place for drying.

f. Return all unused oil and gasoline, and other materials to their proper storage.

PRECAUTIONS

1. Make sure gasoline is kept and stored in approved safety cans.
2. Oil must be stored in closed metal containers with covers securely in place.
3. Be sure to close gasoline shut-off valve.
4. Notify supervisor when lawn mower, or engine, is in need of adjustment, servicing or repair.
5. When not in use, park machine in a well-ventilated area that is clear of combustibles.

18 / **PAINTS**

(*Sources: Buildings* magazine, "Guide to Paint Selection and Application" by J. E. Spector, *American Machinist* magazine.)

Color and light can do much to build efficiency in an office or plant. Good color and lighting can relax tensions, improve morale, stimulate output, decrease absenteeism. Colors should not be distracting or disturbing. Such colors will pull the employee's attention from his work and leave him staring at the wall. It usually costs no more to use a color correctly.

Paint Formulation

The two basic ingredients in paint are pigment and vehicle. The pigment consists of small particles of several types of opaque materials which perform a number of functions including color, hiding power, and protection from deterioration from the sunlight.

The binder or vehicle is the liquid portion of the paint and contains two basic components, the thinner and solvent. The thinner is the volatile part of the paint. After application, it evaporates, leaving behind the solid paint film. The purpose of the thinner is to reduce the viscosity of the paint so that it can be spread easily and uniformly. In solvent based paints, the thinners are organic materials, such as turpentine, mineral spirits or naphtha. These materials are largely responsible for the odor associated with paint.

The binder has the important function of binding the pigment particles into a uniform paint film. The nature and amount of binder determines the important properties of the paint, such as washability, toughness, adhesion, aging stability and resistance to abrasion, scrubbing, weathering and corrosive chemicals. Linseed oils, soya oil, dehydrated castor oil, and alkyd resins or varnishes are commonly used binders in solvent thinned paints. Some of the binders used in latex paints are butadiene-styrene, polyvinyl acetate and acrylic resins. These binders are not actually dissolved in the thinner, but are dispersed in the form of tiny particles.

Emulsion Paints

Water emulsion paints are easy to apply and are quick drying. Of these, acrylic and polyvinyl acetate are probably the most widely used.

Drying Time: Both acrylic and polyvinyl acetate paints are dry to the touch and ready for the second coat in less than an hour. Therefore, two coats can be applied to a large room in less than a day.

Color Stability: Color retention is good for both indoor and outdoor applications. Much of this color stability is due to the light resistant nature of both binders. Since the binder is colorless, the true pigment colors come through with exceptional clarity.

Odor Characteristics: Both of these paints have a mild odor which dissipates as soon as the film is dry.

Moisture Permeability: This property enables the paint film to breathe. Therefore moisture, one of the common causes of failure due to blistering and peeling, can pass through the paint film.

Application Conditions: Both paints can be applied without difficulty under unfavorable atmospheric conditions, including high humidity. A freshly dried coat will be impervious to water spotting in case of rain.

Surfaces

Preparation. Any material that will interfere with the adhesion of paint to the surface should be removed. On new work, the dirt should be brushed off, then dusted. To prepare a surface for repainting, all dirt, grease, loose or peeling paint and other foreign matter should be removed. The surfaces should be clean and dry. Rust must be removed from metal and a proper primer used. Wax must be removed from floors, as it interferes with the adhesion of the paint, and will delay the drying. Peeling paint on exterior wood surfaces is generally due to moisture getting behind the paint. Proper treatment requires the removal of all loose paint, and correction of the moisture condition, or the peeling will likely recur.

Wood. The first coat of paint for wood should be chosen primarily for adhesion and for a base for further coats. For interior use, enamel under-coaters are very satisfactory, giving an excellent surface for finish coats. For exterior use, where wood may be expected to swell and shrink with moisture changes to a greater extent than inside, house paint primers should be used.

Metal. The first coat of paint for metal should act to prevent corrosion of the metal, as well as furnish a base for future paint coats. When repainting, remove any rust with steel wool or a wire brush. For iron and steel, anti-corrosive primers containing red lead or zinc yellow, acts to prevent the spread of corrosion, if the paint is damaged mechanically. Other primers, usually containing iron oxide are used when corrosive conditions are less severe.

Old aluminum should be cleaned of any dirt and loose oxide by brushing or sanding. Follow this with a coat of zinc chromate metal primer and then with exterior metal primer followed by exterior metal enamel. To keep the original metal color, use a clear varnish or one of the special metal lacquers. *New aluminum* should first be treated with a phosphoric acid compound containing grease solvents.

Copper, brass and *bronze* do not rust, but rain water running down from these metals will stain the masonry or paint below. It is easier to give the metals a protective coat of lacquer or varnish than to remove the stains later.

Zinc dust-zinc oxide paints do best on galvanized metal. These primers are also very satisfactory on iron and steel in special cases and are often used on the interior of water tanks.

Masonry. On brick, concrete, stucco, plaster, and similar masonry surfaces, the principal problem is the free alkali which attacks and softens oil base coatings. A suitable primer which seals the alkali primer permits the use of any desired finished coat. Certain masonry paints are made of an oil or varnish base, fortified with resin, and these give excellent results where conditions are not too severe. However, when the maximum alkali resistance is called for, paints based on a solution of a synthetic rubber resin, or portland cement based paints are preferred.

Bituminous coatings. Asphalt or coal-tar pitch tend to dissolve in most paint solvents, "bleeding" into the succeeding coats and discoloring them. For this reason, bituminous coatings are preferred for such surfaces unless the desired color cannot be obtained with bitumens. In this case, a water dispersed primer, usually a latex, is necessary, since the solvent does not dissolve the bitumen.

Asbestos cement. Painting asbestos-cement shingles, siding and sheeting presents a problem because the alkali in the cement often combines with oil paint to make a soluble soap that harms the paint film. The following methods are recommended:

1. Solvent thinned resin coating with the resin being a derivative or modification of rubber. Chlorinated rubber paints are an example of this type. Solvent thinned resin coatings have the advantages of extreme water resistance,

stain resistance, protection of metal fittings from corrosion, good abrasion resistance and excellent hiding power.

2. The use of specially formulated resin coating as a primer, followed by conventional oil or alkyd resin base topcoats. They are sold as asbestos shingle primer, alkali resistant primer, etc.

3. The use of two coats of exterior latex paint. The principal advantages of the latex paints are ease of application, rapid dry satisfactory application on damp surfaces, easy cleaning of painting equipment in water, and transmission of interior water vapor so that blistering of the paint film is uncommon.

There are oil vehicle masonry paints containing modified linseed or other oils that reduce the alkali sensitivity of the oils. On weathering asbestos-cement siding, they will give satisfactory performance, but if there is doubt as to the degree of weathering, a solvent thinned primer should be used as a base.

Walls and ceilings. For walls and ceilings flat paints are usually chosen because of better hiding power and more uniform distribution of reflected light. Flat paints are usually based on alkyd resins, which are thinned by paint solvents, or on emulsions (styrene butadiene, polyvinyl acetate, or acrylic) which are water thinned. The choice between alkyd and latex is largely a matter of personal preference. The alkyd paints usually have slightly better hiding power and are more washable, especially early in the life of the film. Latex paints have slightly less odor during painting and, being water thinned, cleanup of tools, drips, etc. is easier.

Where a high degree of washability and resistance to abrasion and marring is required, semigloss finishes are usually used. Diet kitchens and washrooms are best finished in such coatings. Gloss finishes have the best washability and resistance to abrasion and water. Semigloss finishes have a higher hiding power and less tendency to show up defects in the underlying surfaces.

The color of walls, ceilings and trim is a matter of personal preference. The light reflection of such surfaces is part of the lighting of the room and must be taken into consideration. The lighting has usually been designed for surfaces of a certain reflection and, unless repainting is kept within the same range, the light distribution of the room will be thrown out of balance.

Wall coatings are sometimes done in multicolor paints. One or more colors can be applied in a single operation. Most of these are sprayed on, but some formulations can be applied with rollers. These multicolor paints give an extremely tough chip and scratch resistant textured finish, and because of the random distribution of color flecks, do not show dirt and other marks. They can withstand repeated scrubbings.

Ceilings (acoustical). Emulsion type coatings are recommended. They should be brushed so as not to fill the small diameter holes. Since they are flat, they do not add any glare for the lighting.

Exteriors. Where long life and little or no exposure to abrasion or standing water is likely to be encountered, the conventional house paints, based on mixed pigments, are the best selection. Where industrial fumes are prevalent, a lead free paint will avoid staining from hydrogen sulfide.

Paints to be used above or next to brick surfaces should be of a nonchalky type to prevent staining of the brick.

Trim paints are used where high gloss and extreme color retention are required, and exterior enamels are preferred for outdoor furniture and for metal surfaces, over a good primer.

If a suitable sealer is used on masonry, any good exterior paint will usually give satisfactory service as a finish coat, but it is usually safer to use a finish coat, latex or synthetic rubber, designed for exterior masonry. Cement water paints are also excellent.

Where a clear coating for brick or stone is required, water repellent solutions, usually silicones are available.

Natural coatings for exterior wood are usually varnishes, oils or stains. None of these have a life comparable with conventional pigmented coatings, but, if refinished at suitable intervals, will give satisfactory service.

Special insecticides in powdered form are available for adding to paint, making it effective for killing insects such as ants, flies, roaches, mosquitoes, etc., that come in contact with the surface.

Roofs. Roofs of most general concern will be either of galvanized metal or of an asphalt-felt composition. Galvanized metal requires a special zinc dust-zinc oxide primer, which can be followed by any tough paint. It should be remembered that any pastel colors or natural aluminum will help reduce under-roof temperature by reflecting sun heat.

Since the solvent in conventional paints will often attack the asphalt roofing, they should not be used. There are a number of special roofing paints formulated for this purpose which will give protection to roofing material from the sun, heat, and moisture. Colored aluminum roofing paints can also be used. Many of the exterior latex products can be used over this type of roof. Light colors, which will have to be renewed more frequently, will also reduce heat absorption.

Interior surfaces. Interior surfaces to be painted must be clean and free of any foreign matter that will prevent the paint film from bonding with the surface.

Procedures for preparing an interior surface for painting and for applying the paint itself vary with the type of paint. The best recommendations are those provided by the manufacturer for using his product. The following tips are applicable for most interior paint types under normal conditions.

Interior Paints and Their Qualities

1. *Cumar:* Alkali and acid resisting, this paint will not "burn" aluminum powder flakes, and is often preferred as an aluminum heat-resisting and interior coating.

2. *Ester Gum:* This is a good water and fade resistant paint.

3. *Alkyd:* Color and gloss retention are two of alkyd's outstanding qualities. It is also weather and water resistant.

4. *Phenolic:* Phenolic is fast drying, and has good weather, water and chemical resistance, although it usually chalks easily and often yellows in white formulation.

5. *Epoxy:* Outstanding hardness and chemical resistance are attributed to epoxy. However, it is often low on elasticity and weather resistance.

6. *Maleic:* This fast drying paint is usually harder than ester gum and holds color better than cumar.

7. *Silicone:* Silicone has outstanding water resistance and slickness, but it is often difficult to make subsequent paint coats adhere.

8. *Latex:* This paint has outstanding washability and color stability but usually needs about thirty days for complete curing.

9. *Synthetic Rubber:* Excellent resistance to chemicals, alkali, acids and abrasion are the desired qualities in this paint, but usually has little resistance to petroleum solvents.

High temperatures. Where temperatures in the 200 degree range are encountered, the acrylic enamels should be used. They should be applied over a clean, non-primed surface since no moisture will ever be present and because acrylics, due to the hot solvent in them, will not tolerate a conventional primer under them.

For temperatures above 400 degrees, such as engine exhaust manifolds, mufflers and boiler stacks, a silicone type coating is recommended. The service life of such material is sometimes considered in direct proportion to the cleanliness of the surface to which it is applied. Generally where such surfaces are sandblasted, they average two years of service life.

Water storage tanks. The tanks should be sandblasted and coated with vinyls or red lead primer in a phenolic resin with a compatible finish coat of aluminum. If this painting can be accomplished before the tank is put into service, the effective life will, in most cases, exceed fifteen years. Although the expense is high initially, it more than pays for itself over the years due to reduced maintenance cost.

Cooling towers. Cooling tower piping should be sandblasted and coated prior to installation. Assuming that the cooling tower piping has been sandblasted, several coatings can be used. The best is zinc enriched inorganic coating. Next to the zinc enriched coatings are the polyamide epoxies which, when properly applied with a film thickness of eight to fourteen mils, will insure a performance life of several years. Where a tower cannot be shut down, merely get the pipe as dry as possible, then rig a tarpaulin as a wind and spray breaker and apply a red lead primer followed by an asphalt mastic. If desired, apply a finish coat of aluminum or any other color, after the mastic has had sufficient time for the entrapped solvent to evaporate.

Primers. The use of primers in buildings for maintenance painting is generally regulated to equipment located in the basement or on the roof, with very little other metal being involved. In general, wherever a highly corrosive condition is encountered, such as moisture and constant dripping, a red lead or slow drying primer is recommended, preferably two coats, followed by the desired finish color on alkyd resin. Otherwise the corrosion is not a problem and a quick drying primer is usually satisfactory when applied in two coats with a finish coat of the desired color.

Thinners. Usually no difficulty is experienced with the use of thinners, except when too much is used in a given quantity of paint. Paint containers invariably show the thinner to be used. Contract painters should not be allowed to over-thin a material in order to stretch out the square footage, subsequently giving a thinner film and a shorter satisfactory service life.

Application. Paint may be applied by brush, spray, or roller. The method to be used depends on the type of surface, the area involved, the type of paint, and the problems of cleanup. Certain types of paints including lacquers and the fast drying coatings are designed for spray application only. This is also true of most multicolor finishes. Small areas and irregular surfaces are usually done best with a brush. Rollers work best on large and fairly uniform surfaces.

Storage. The storage of paints is not difficult, especially if the building is air-conditioned. Extremes of heat and cold should be avoided, and paints should not be stored near fires or open flames. When paint is stored, it should be maintained in temperatures of 60° F. to 90° F. Emulsion type paints will take several freezing cycles before they are unfit for use. Some types of coatings, particularly those with a phenolic resin, will have chemical reaction take place in the container when they are stored in temperatures exceeding 115° F.

Unopened containers should be inverted periodically, to avoid breaking up any skin which may have formed.

Some paints, particularly dark colors, tend to lose drying properties in storage, and material which has been on hand for some time should be tested for drying properties before use. If the paint does not dry readily, the addition of a small amount of drier will correct the difficulty.

Brushes and rollers should be cleaned with a suitable solvent after use. If they are not to be used within a short period of time, they should be washed out with soap and water, wrapped in paper and stored, either laying on their sides or hanging up. If brushes or rollers are to be used presently, they may be suspended in a suitable solvent. Under no circumstances should brushes stand on the bristles.

Rags and drop cloths should be spread out to dry, particularly if they are soaked with oil, or stored in a tightly closed container. Wadded up rags will heat up, from oxidation of the oils, and under favorable conditions, may ignite.

Glossary of Paint Terms

Acid. Any chemical compound containing one or more hydrogen atoms available for reaction with active metals or alkaline solutions.

Acid number. The number of milligrams of potassium hydroxide required to neutralize the free fatty acid in one gram of fat, oil, wax, or resin.

Acrylic acid. Colorless, organic liquid which polymerizes to a hard solid when heated.

Acrylic resins. Family of synthetic resins made by polymerizing esters of acrylic acid.

Alkali. Substance that neutralizes acids; highly destructive to oil paint films.

Alkyd. Generic name indicating a family of synthetic materials formed from a complex organic acid and an alcohol.

Alligatoring. Condition of a painted surface caused by improper buildup.

Barium sulphate. Heavy, white crystalline extender pigment.

Binder. The non-volatile portion of a paint vehicle which serves to bind or cement the pigment particles together.

Bleeding. When coloring material from either the wood or undercoat works up into succeeding coats and imparts to them a certain amount of color.

Blistering. Formation of bubbles on the surface of paint film, usually caused by moisture. Also caused by excessive heat.

Bloom. Condition of clouding or fogging of paint film, usually caused by reactive materials in paint film coming into contact with dust, oil, deposits from gases in the air or soluble matter in rain.

Blushing. Applied to lacquers when they become flat or opaque and white on drying. Usually occurs when applied in a humid atmosphere.

Brushability. Adaptability of paint to application with a brush.

Body coat. Intermediate coat of paint between priming and finishing.

Calcimine. A paint composed essentially of calcium carbonate or clay, and glue.

Casein. Protein obtained from milk, soluble in alkaline water solution.

Casein paints. Paint in which casein solution has replaced binder.

Chalking. Presence of loose powder on surface of paint.

Checking. Formation of short narrow cracks in surface of film.

Coagulation. Precipitation of colloids into single mass; usually caused by excessive heat or catalytic agents.

Coating. All material which has been spread as film over surface while in fluid condition and which remains upon surface.

Cohesion. Attractive force between polymers of similar nature which tends to hold them together.

Colloid. Composed of ultramicroscopic particles of solid, liquid or gas dispersed in a different medium.

Color-in-oil. A paste formed by mixing a color pigment in linseed or other oil. Used for tinting.

Copolymer. Product obtained when two different monomers are polymerized or linked together chemically to yield a resin.

Copolymerization. Simultaneous polymerization of two compounds which have properties different from polymer obtained with either monomer separately.

Cracking. Form of paint failure in which breaks in film extend through all coats down to building material.

Crazing. Hairline cracks in paint film.

Cure. Toughening or hardening of paint film.

Dimer. Formed by chemical combination of two similar molecules.

Dispersed. Finely divided or colloidal in nature.

Dispersing agent. Material used to aid in holding finely divided matter in dispersed state.

Dispersion. Generic term describing any heterogeneous system of solids, gases or liquids.

Drier. Composition of certain metals that accelerates drying action of oil when added to paint or varnish.

Drying oils. Vegetable or animal oils which, when applied as film, absorb oxygen and dry to a relatively tough, elastic coating.

Efflorescence. A deposit of water soluble salts on the surface of masonry or plaster caused by the dissolving of salts present in the masonry.

Emulsifier. Material which, when added to mixture of dissimilar materials will produce stable emulsion.

Emulsion. A preparation in which very fine particles of liquid such as oil, are suspended in another, such as water.

Enamel. Special type of paint made with varnish or lacquer as the vehicle.

Extender. Pigment of low hiding power; may contribute desirable properties to paint products, such as improved flow and sheen control.

Fade-o-Meter. Mechanism used to artificially reproduce effect of sunlight on paint.

Failure. Commonly used to describe condition of paint film at end of useful life.

False body. Describes characteristic of paint which becomes viscous on standing but thins down on stirring.

Flaking. Detachment of small pieces of paint film.

Flashing. Nonuniform appearance on surfaces in which coating dries with spotty differences of color or gloss.

Flash point. Lowest temperature at which mixture in an open vessel gives off enough vapors to produce momentary flash of fire when small flame is passed near surface.

Flatting agent. In oil paints, a soap added to paints to reduce gloss.

Floating. Separation or layering of pigment in mixture of pigment.

Flocculate. Agglomeration of undispersed pigment particles.

Ghosting. Patches of lighter color showing in dry coat.

Hiding power. Ability of paint to obscure surface.

Holidays. Unintentional missing of areas of surface being painted.

Incompatible. Describes material which cannot be mixed with another material without damaging original properties.

Intumesce. To foam, swell, froth or bubble up as a result of heat, liquid, air or chemical action.

Lacquer. A material that dries by the evaporation of the thinner or solvent.

Latex. Generic term describing any of many relatively stable dispersions of submicroscopic, insoluble resin particles in a water system.

Lifting. Buckling of finish coat when applied over previous coat which is not yet dry, or when solvents in second coat are too strong.

Mil. Unit of thickness, 1/1000 inch.

Monomer. Organic compound capable of polymerizing or linking together.

Pigment Volume Concentration (P.V.C.). This represents value obtained by dividing volume of pigment by total volume of solids in dry film.

Polymer. High molecular weight material composed of large number of repeating units of monomer linked together.

Polymerization. Type of reaction resulting in linkage of monomers to form high molecular weight polymers.

Polyvinyl Acetate Latex (PVA). Latex composed primarily of vinyl acetate.

Rain spots. Describes defects on paint film caused by rain.

Resin. A solid or semi-solid organic material of either natural or synthetic origin.

Sagging. Wet paint film which tends to flow downward and become thicker in some areas.

Sheen. Special type of gloss measured in terms of reflected light at an angle of 30 degrees or less.

Solvents. Defines products used to dissolve film-forming constituents; volatilize during drying and do not become part of film.

Spreading rate: Area of surface over which unit volume of paint will spread, usually expressed in square feet per gallon.

Varnish. Any homogeneous, clear liquid which forms film on curing.

Vehicle. Liquid portion of paint.

Water-based paints. Misnomer; no paint employs water as a permanent, film-forming base.

Applying Paint to Metal

The accompanying chart is applicable to most of all problems of applying paint to metal, and can be used as a basic guide.

Special problems, especially those involving high production painting, should be carefully worked out with a paint manufacturer's representative to insure maximum surface life, ease of application and minimum cost.

APPLICATION OF PAINT

Items	Surface Preparation	Primer Coat	Type of Finish Coat
Constant Exposure—Outdoors			
1. *Top Abuse.* Cutting, pushing, lifting, scuffing, etc.	Remove all loose scale, rust, and foreign matter.	Polyamide Epoxy	Polyamide Epoxy
2. *Medium Abuse.* Guard rails, covers, housings, etc.	Remove all loose scale, rust, and foreign matter.	Epoxy	Epoxy
3. *Occasional Abuse.* Switch panel doors, light standards, etc.	Wire brush, remove all loose matter.	Containing Keytoxine compounds	Industrial enamel
4. *Weathering only.* Iron grillwork, metal canopies, light hoods, etc.	Wire brush, remove all loose matter.	Containing Keytoxine compounds	Industrial enamel
5. *Construction equipment.*	Wire brush or sandblast.	Enamel undercoat	Alkyd enamel
Marine Equipment			
1. Metal parts on shore, exposed to salt spray.	Remove all loose scale, rust, and foreign matter.		

APPLICATION OF PAINT (*continued*)

Items	Surface Preparation	Primer Coat	Type of Finish Coat
2. Metal parts exposed to fresh water.	Wire brush, and remove all loose material.	Rubber base	Rubber base
3. *Boats.* Submerged or exposed parts.	Wire brush surface as clean as possible.	Chlorinated rubber	Chlorinated rubber

Number of Coats		Method of Application	Drying Time (Hours)	Approx. Life Expectancy (Years)
Prime	Finish			
1	1	Brush, roller, spray	2-4	5
1	1	Brush, roller, spray	2-4	5
1	1	Brush, spray	4-6	2-3
1	1	Brush, spray	4-6	2-3
1	2	Brush, spray	4-6	1-2
1	2	Brush, spray	4-6	1-2
1	1	Brush, spray	1-2	1-2

Items	Surface Preparation	Primer Coat	Type of Finish Coat
Indoor Plant and Office Equipment and Furnishings			
1. Storage facilities, lockers, closets.	Wire brush, remove loose particles.	Enamel undercoat	Alkyd enamel
2. Materials handling equipment, rough use.	Wire brush, remove loose particles.	Enamel undercoat	Alkyd enamel
3. Floor trucks, cranes, other powered equipment.	Wire brush, remove loose particles.	Enamel undercoat	Alkyd enamel
4. Components subject to scuffing—elevator doors, kick plates, shelf surfaces, metal stair treads.	Remove loose matter, wash with mineral spirits.	Epoxy	Epoxy
5. Machine tools.	Wire brush, remove loose particles.	Enamel undercoat	Alkyd enamel
6. *Large surfaces.* Grillwork, metal partitions, canopies.	Remove loose matter, wash with mineral spirits.	Epoxy	Epoxy

Number of Coats		Method of Application	Drying Time (Hours)	Approx. Life Expectancy (Years)
Prime	Finish			
1	1	Brush, spray	4-6	3-5
1	2	Brush, spray	4-6	2-3

APPLICATION OF PAINT (*continued*)

| Number of Coats | | | | |
Prime	Finish	Method of Application	Drying Time (Hours)	Approx. Life Expectancy (Years)
1	1	Brush, spray	4-6	3-5
1	1	Brush, roller, spray	2-4	2-3
1	1	Brush, spray	4-6	3-5
1	1	Brush, roller, spray	2-4	2-3

Items	Surface Preparation	Primer Coat	Type of Finish Coat
Atmospheric Abuse			
1. *General acid.* Plating rooms, textile finishing, electrolytic processes.	If possible, sandblast.	100 per cent Solids epoxy	Cut top coat 20 per cent
2. *Alkali.* Salt spray.	If possible, sandblast.	100 per cent Solids epoxy	Cut top coat 20 per cent
3. *Fine abrasive particles.* Grinding, milling.	If possible, sandblast.	100 per cent Solids epoxy	Cut top coat 20 per cent
4. *Petroleum.* Petro-chemicals, coal and coke products, organic residues.	If possible, sandblast.	100 per cent Solids epoxy	Cut top coat 20 per cent
Fabricated Metal Products			
1. Severe outdoor exposure.	Degrease and chemical etch.	Rust inhibitor, zinc chromate or chromate-oxide	Alkyd or alkyd-melamine
2. Expendable containers.	Remove visible dirt and grease.	None	Alkyd or alkyd-urea
3. Flat surfaces.	Degrease	None	Alkyd-urea or alkyd-melamine
4. Deep-drawn and curved surfaces, precoated. Deep-drawn and curved surfaces, after coated.	Degrease	None	Alkyd-urea or alkyd-melamine

| Number of Coats | | | | |
Prime	Finish	Method of Application	Drying Time (Hours)	Approx. Life Expectancy (Years)
1	1	Brush, spray	2-4	3-4
1	1	Brush, spray	2-4	3-4
1	1	Brush, spray	2-4	3-4
1	1	Brush, spray	2-4	3-4

APPLICATION OF PAINT (*continued*)

Number of Coats Prime	Finish	Method of Application	Baking Time (Temperatures or Hours)	Approx. Life Expectancy (Years)
1	1	Spray, flow, dip, roller, curtain	275-350°F.	2-5
1		Spray, flow, dip, roller, curtain	275-350°F.	1 plus
1		Spray, flow, dip, roller, curtain	275-350°F.	Indefinite
1		Spray, flow, dip, roller, curtain	275-350°F.	Indefinite

Items	Surface Preparation	Primer Coat	Type of Finish Coat
Fabricated Metal Products (*continued*)			
5. *Sanitary products.* Often-cleaned surfaces in restaurants, kitchens, laboratories.	Degrease and chemical etch.	Epoxy or phenolic rust inhibiting chromate-oxide	Epoxy, vinyl, polyurethane, polyester
6. *Rigid metal goods.* Castings, forgings.	Degrease and chemical etch.	Rust inhibiting chromate-oxide	Alkyd-melamine, alkyd-urea, or epoxy
7. High temperature exposure.	Degrease and chemical etch.	Silicone-alkyd or straight silicone	Silicone-alkyd or straight silicone
8. Low temperature exposure.	Degrease and chemical etch.	Vinyl rust inhibiting type	Vinyl
9. High-gloss surface.	Degrease and chemical etch.	Sanding type rust inhibiting chromate-oxide	Alkyd-melamine or alkyd-urea
Consumer Goods			
1. *Toys,* for non-toxic finish.	Degrease	None	Lead free alkyd or lacquer
2. *Toys,* for scuff and bruise resistance. *Expendables:* pails, mops, inexpensive tools.	Wash and clean as well as possible.	None	Lead free alkyd enamel
3. *Metal furniture,* indoor, outdoor.	Wire brush. Remove rust, and loose matter.	None	Lead free alkyd enamel

APPLICATION OF PAINT *(continued)*

Number of Coats		Method of Application	Baking Time (Temperatures or Hours)	Approx. Life Expectancy (Years)
Prime	Finish			
1	1	Spray, flow, dip, roller, curtain	275-350°F.	2-5
1 (Spot Fill)		Spray, dip	275-300°F.	Indefinite
(Intermediate)	1 or 2	Spray, flow, dip, roller, curtain	375-450°F.	Indefinite
1	1	Spray, dip, roller, curtain	250-275°F.	Indefinite
1 or 2	1	Spray, curtain	250-300°F.	Indefinite
			Bake or Force Dry	
1		Spray, dip, roller, curtain	2-4	Indefinite
2		Brush, spray	2-4	1-2
2		Brush, spray	2-4	1-2

Fire Retardant Paint

The ability to reduce the flame-spread rate of ordinarily combustible material and tests indicate that they have a low rate of smoke emission. When a flame is applied, the film forms a foam that is many times thicker than the original paint film. The foam chars and becomes ridged. For a time, it effectively insulates the combustible surface from flame and heat, helps delay the fire, helps save lives and reduces property damage. For this reason, it is sometimes recommended for schools, hospitals, office buildings, nursing homes, churches, or for any building with combustible interior walls.

Fire retardant properties which should be established when choosing a fire retardant paint or coating for the surface to be covered include: *Flame Spread, Fuel Contributed, Smoke Developed, Rate per Coat (Coverage), Total Number of Coats, Fire Hazard Classification, Rating, Flame Spread on Noncombustible Substrates, Heat Insulating Properties.*

Quantitative and Qualitative Properties

Glass
Light Reflectance
Viscosity Range
Odor
Toxic Properties
Stability
Wash and Scrub Resistance
Solvent/Maximum Thinning
Recommended Application
Recommended Coverage
Maximum Coverage

Resistance to Moisture
Hiding Properties
Color Compatibility and Tinting
Film Flexibility
Impact Resistance
Freeze, Then Stability
Dry Time
Recoat Time
Working Properties
Packaging

19 / **PARKING LOT LAYOUT**

Several factors must be taken into consideration in laying out a parking area. Some of these are:

1. Dimensions of the parking area.
2. The route by which the cars will enter and exit.
3. The average length of time each car is parked in the lot.
4. The number of times per day of car "turnover."

The installation of fences, gates, light standards, and over-the-walk driveways must all conform to the parking pattern selected.

Lots that are 76 feet and wider are usually laid out for angular parking. This is preferred because it is easier to park at an angle, and the circulation of cars are easier to control. The width of the driveway will determine the parking angle, which can vary from a 90° angle to a sharp 45° angle.

To design a good functional parking area, there should be little or no waste of pavement. The proper type of fences, gates, lighting, signs, reflective parking lines, directional arrows, and locations of entrances and exits should be given careful consideration. To eliminate damages by a car to any of these items, concrete stops are recommended to be located at the end of each space. These stops also aid as a guide in keeping the parking uniform.

The schematic drawing illustrates 60-degree diagonal parking on a lot approximately 100 feet by 150 feet. The dimensions shown in the table for 45, 55, 60 and 90-degree parking are the ones most used.

DIMENSION TABLE

Dimensions when parking at any of these angles:

		45°	50°	55°	60°	90°
Offset	A	18′	15′8″	13′4″	11′	1′6″
Car Space	B	12′	11′4″	10′8″	10′	8′7″
Stall Depth	C	16′	16′8″	17′4″	18′	18′6″
Stall Depth	D	18′	18′4″	18′8″	19′	19′
Overhang	E	2′	2′1″	2′2″	2′3″	2′9″
Driveway	F	13′	14′6″	16′	17′6″	25′
Turnaround	G	17′	16′	15′	14′	14′
Extra	H	6′	5′	4′	3′	0

To determine the number of cars that can be parked in each of the four banks:

1. Deduct the area lost in parking the first car. (Dimension "A" in the illustration.)

2. Then divide the car space (dimension "B") into the total length of each bank of cars, plus the extra factor "H", if there is any extra space.

Lot Layout
Economical Use of Parking Blocks

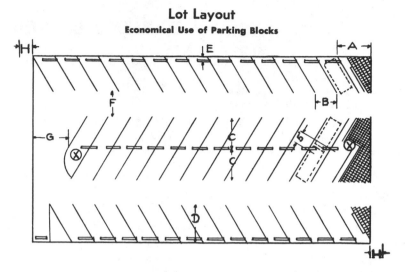

Figure 19-1

The shaded space of the parking lot may be used for planting. Dotted lines illustrate two large cars in parking position with alternate wheels against block—bumpers separated. Light standards are shown in drawing by "X".

20 / PEST MANAGEMENT

(Robert Snetsinger and David E. Schneider, Department of Entomology, The Pennsylvania State University, University Park, Pennsylvania.)

Pest management means considerably more than the use of pesticides to kill insects and other pests; it also means the manipulation of the environment so as to prevent or discourage pest problems from developing. Pesticides are only one means used to control pests. The term sanitation is often used for the preventive aspects of pest management. Pest management experts usually divide control into two phases: (a) "clean-up" or "clean-out", during which an outbreak is reduced to a manageable level; (b) contracted or preventive services where, through regular chemical applications and sanitation, reinfestation is forestalled.

Control of pests, in most instances, is the responsibility of management. Through various federal, state, and local laws, apartment owners, food services, and other businesses are forced to maintain pest control programs. Also, the very destructiveness of the pests may require building supervisors to institute pest control.

The Federal Food, Drug, and Cosmetic Act of 1938, as amended, was established to regulate interstate shipment of adulterated and misbranded foods and other products, including foods containing excessive pesticide residues and insect fragments. Fines of $1000 and/or one year imprisonment may be the punishment for conviction under certain provisions of the act. Other legislation regulates pesticide usage, pesticide residues, interstate movement of various products, sanitation in food plants, methods of control, and many other aspects of pest control.

Pest Control Services

Professional pest management is the science and business of protecting the health and comfort of man, his domesticated animals, his plants and plant products, and buildings and structures from insects, rodents, birds, weeds, wood-destroying fungi, and related pests. There are more than 6000 different firms in in the United States which provide professional pest control services. Pest control operator, PCO, exterminator, sanitation specialist, and insect control or pest control expert are titles commonly used by these specialists. In general, a pest control operator must (a) be able to recognize a wide variety of animals, plants, and kinds of damage; (b) have a respectable understanding of chemical pesticides and their safe use; (c) be able to operate and repair a variety of sprayers and other equipment; (d) have the ability to understand and work with people; and (e) be a businessman.

Most apartment owners, hospitals, and business institutions employ a professional pest control operator on a contract basis to keep premises free of pests. Building superintendents and plant engineers are usually responsible for hiring and inspecting the services of a pest control expert. In order to effectively evaluate pest control services, work with a pest control operator in preventive control programs, and in some circumstances carry out company programs, it is important to have an understanding of types of pest control services, some common pest problems, and the safe use of pesticides.

Selection of a pest control operator can be based on several criteria: appearance of representatives and service employees, standing in the community, experience or length of time in business, trade association affiliations, municipal or state licenses (where required), references or recommendations from previous satisfied customers, and many other points which management generally uses to hire its own employees. Firms usually offer several types of pest control services, including general pest control, termite control, rodent control, bird management, and ornamental pest control and vegetation maintenance.

20A / GENERAL PEST CONTROL—
SOME COMMON PESTS

Domestic Fly Problems

The term housefly, as used by most people, refers to a considerable number of species of two-winged insects found in and about dwellings. Under the name "housefly" the layman ignores the differences between bottle flies, flesh flies, fruit flies, moth flies, cluster flies, and other kinds of flies of diverse habits. Flies have egg, larval, pupal, and adult stages. It is usually the adults that are nuisances or carry disease organisms; for best results, however, it is the larval or maggot stage that should be controlled.

Larval cluster flies are parasites on earthworms. The adult flies invade homes in great numbers during the fall where they are found in windows, attics, etc., and are attempting to overwinter. Larval moth flies develop in traps in sink drains, basement drains, and in the filter beds of sewage treatment plants; adults may swarm in buildings. Larval bottle flies develop in the feces of pets, garbage, and other organic matter; they also enter homes. Other species of flies have other habits; unless the species of fly is recognized, preventive control may be difficult.

Preventive control involves keeping garbage cans clean, regular garbage removal, at least once weekly during the cool season and twice weekly during the summer, removal of animal wastes, elimination or proper covering of privies, disposal of dead animals, use of screening, etc.

Chemical control of flies involves treatment of the breeding sites with larvicides or residual sprays applied to surfaces where adult flies land. The use of aerosols or mist sprays quickly kills flying adults; baits which contain attraction substances mixed with a poison, and the use of various devices including DDVP resin strips are also useful. A variety of insecticides including DDVP, dimethoate, Malathion, and pyrethrin are used in control.

Cockroaches

Cockroaches (water bugs) feed upon a wide variety of foodstuff, including fats, oils, pastes, glues, soap, cereals, beer, soft drinks, cheese, meat products, and garbage. There are more than a dozen species of cockroaches that infest buildings; however, the German, American, Oriental, and brown-banded species are the common nuisance species in the U.S.A. In addition to the nuisance aspect of cockroach infestations, these insects play a role in the transmission of certain diseases, such as *Salmonella* food poisoning.

During construction, a new building may become infested with cockroaches; also these insects are carried into houses from infested stores in food, food packages, on furniture, etc. In crowded urban areas, cockroaches can travel from building to building and apartment to apartment with relative ease.

In general, cockroach control is best done by a professional on a contract basis. The first step is called the "clean-out." This involves application of pesticides to reduce the population to a reasonable level; in most cases, the clean-out is followed by contracted control which includes twice monthly, monthly, or bimonthly service calls. The cost of clean-out is about five times more than the regular service call. Cockroach control is necessary on a regular basis because of the mobility, reproductivity, longevity, and behavior of these pests.

Control procedures involve the application of residual pesticides, such as 2% chlordane, 0.5% diazinon, 1% baygon, and combinations of 0.5% dichlorvos and 0.5% diazinon. The technique of flushing (chasing them out) is sometimes used to determine where residual applications should be made. Pyrethrins combined with piperonyl butoxide are often used for this purpose. In some situations, phosphorus paste or kepone baits are used. Different species of cockroaches have different habits and require different techniques of control. Application of pesticides to various surfaces requires the adjustment of the sprayer nozzle to get the most effective control and no discoloration or damage to wallpaper, paint, etc. The development of resistance to chemical pesticides is a serious problem with cockroaches. Sanitation practices, such as garbage removal and maintaining clean food serving areas, are also important in cockroach control.

Ants

Ants become a problem in dwellings during the summer, particularly when foods are left on counters or crumbs and debris remain on the floor. Pavement ants, larger yellow ants, thief ants, and carpenter ants are commonly found in homes. About 15 thief ants laid end to end measure an inch, while only 2–4 carpenter ants will make an inch. Besides size, ants also vary in color—yellow, reddish, brown, and black. With the exception of the carpenter ant, which may damage wooden structures, these insects are more a nuisance problem than anything else.

Ant control involves knowing their habits and where they enter from the outside. Oil-based sprays of 2 to 5% chlordane are commonly used against ants. At times, baits, dusts, or syrups are used. Ant swarms resemble termite swarms and sometimes the two are confused. Ants have narrow "waists," "elbowed" antennae, and four developmental stages (egg, larva, pupa, adult), while termites have no "waist," straight antennae, and three developmental stages (egg, nymph, adult).

Fleas

The cat flea, associated with both cats and dogs, is a common pest in homes. The presence of fleas is most commonly reported by pet owners returning to an apartment or house following a trip; this is because the pet has been removed for some time and the fleas are hungry. Fleas are also associated with rats, mice, and other mammals. Rat control programs usually include the use of DDT dust or "tracking powders" to remove the possibility of fleas and other ectoparasites from leaving the dead rats and attacking humans. Adult fleas feed on the blood of man and other mammals; these pests also transmit a number of diseases to man, including plague.

Control of fleas involves treatment of the area where the dog or cat spends its time. This includes vacuum-cleaning, removal, and disposal or treatment of the pets' bedding materials, and treating the infested area with a residual pesticide such as DDT, lindane, or diazinon. Pets should not be treated with these materials, but rather with commercially available flea powders.

Bedbugs and Other Biting Bugs

Bedbugs are most common in crowded tenements, rundown hotels and motels, and in apartment buildings with a rapid turnover of transient residents. During the daylight hours these insects hide in cracks and crevices in walls, beneath mattress covers, and under loose wallpaper in bedrooms. They are also found in couches, sofas, and chairs in dwellings; seats on street cars, buses, trains; and in movie theaters. When it is dark, bedbugs attack humans, feeding on their blood. Bloodsucking conenoses (kissing bugs), masked hunters, bat bugs, and bird bugs have similar biting habits, but occur under somewhat different conditions.

Application of residual pesticides, including Malathion and DDT into cracks, crevices, walls, and other hiding places, is essential to achieve control; sometimes fumigation methods are used.

Carpet Beetles and Clothes Moths

In most cases when fabric is destroyed by insect pests, it is the work of carpet beetles rather than of clothes moths. Leaving rags, old clothing, and old rugs stored undisturbed for long periods of time may create breeding sites for carpet beetles and clothes moths. Keeping rugs and other fabrics clean, dry cleaning, and exposing clothing, etc. to sun and fresh air reduce likelihood of infestation. Considerable care is necessary in treating rugs and other fabrics with insecticides because of the chance of chemical spotting and soiling. A specialist should be called for carpet beetle control and mothproofing.

Spiders

For the most part, spiders are beneficial, eating insects and other pests. However, the black widow and the brown recluse spiders have bites that are poisonous to man. During the fall, many species of spiders become nuisances by invading buildings. Spiders have eight legs, usually eight eyes, and two body regions connected by a narrow "waist."

Several black widow species occur in the United States. These species are quite similar in appearance; the adult females are about ½ inch long and black with a reddish hourglass marking on their undersides; males are much smaller and of little concern. The bite of the female black widow is intensely painful for the first six to 24 hours. Pain normally subsides by the second day and the bite is usually not fatal. However, the victim of a black widow bite should be taken to a physician for treatment as soon as possible. Black widows are locally common and found in outdoor privies, under boards, in rock piles, etc.

The brown recluse spider and related species are not common but are fairly widespread in the United States. The adult female is about 1½ inches long, brownish-yellow, and with a black fiddle-shaped marking on the upper surface of the head region. The bite of this spider causes a necrotic skin lesion which is

COMMON HOUSEHOLD PESTS

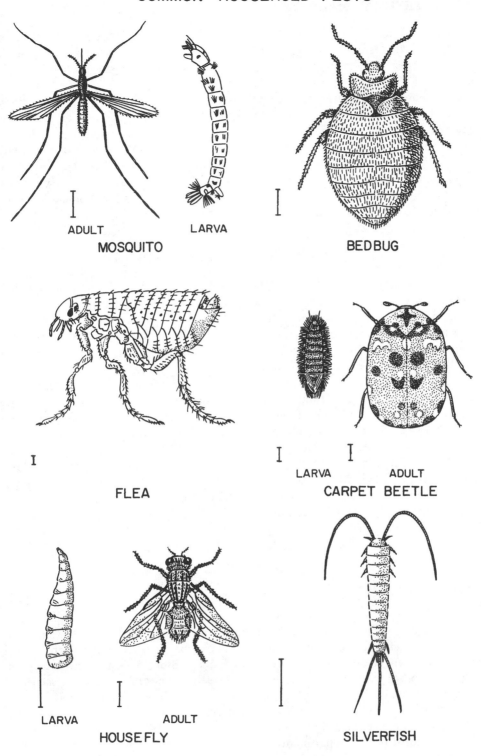

ADULT   LARVA
MOSQUITO

BEDBUG

FLEA

LARVA   ADULT
CARPET BEETLE

LARVA   ADULT
HOUSEFLY

SILVERFISH

THE LINES INDICATE THE ACTUAL SIZES OF THE PESTS.

Figure 20-1

slow to heal and leaves a sunken scar. Several deaths have been reported from the bite of this spider. In case of a bite, treatment by a physician is required. Brown recluse spiders may be found on undersides of furniture and in storage areas.

In general, it is unusual to encounter black widows and brown recluse spiders. In most cases, people seek the assurance that they do not have one of these two spiders. Sheet-web, wolf, jumping, and house spiders, are commonly found in buildings and do no harm. Tarantulas are groups of spiders that people fear; however, tarantulas do not have bites poisonous to man. Also, they are only common in certain areas of southwestern United States.

Residual treatments of 2% chlordane or 0.5% diazinon solutions are recommended where spiders are active and control is desired.

Centipedes and Millipedes

Centipedes are sometimes called 100-leggers and millipedes are called 1000-leggers; centipedes have one pair of legs per body segment and millipedes have two pairs of legs per body segment. Centipedes are predators and feed on insects, while millipedes feed on vegetable matter. The house centipede may regularly live in cellars and other areas of a building, while millipedes invade buildings during the late summer or fall.

The house centipede has long, slender legs and moves rapidly but awkwardly; it has 15 pairs of legs. Millipedes may aggregate in great numbers in cellars, stairwells, garages, and along foundations in autumn. They are active during the early hours of the night, but are hidden beneath soil debris during the day. By midwinter millipedes are not a problem.

Residual applications of 2% chlordane or 4% Malathion in basements, along baseboards, etc. where centipedes are active will provide control. In general, a program of cockroach control is also effective for house centipedes. Diazinon applied around the exterior foundation and at entry points is effective against millipedes.

Mosquitoes

Control of mosquitoes is an entire field of pest control by itself. These insects are relatives of flies and have only one pair of wings. The larval stage is found in quiet waters—ponds, flooded areas, birdbaths, roof gutters, tin cans and other containers, etc. Control programs should be directed against the larval stage and then the adults. Only adult female mosquitoes take a blood meal from man and other vertebrate animals. It is at this time that certain species of mosquitoes may transmit mosquito-borne encephalitis, malaria, and certain tropical diseases including yellow fever and dengue. Mosquito control is achieved through community programs and is not usually successful by individual approach. State and local health departments may direct mosquito control programs. Their aid should be sought during the planning stage of large construction projects to

COMMON HOUSEHOLD PESTS

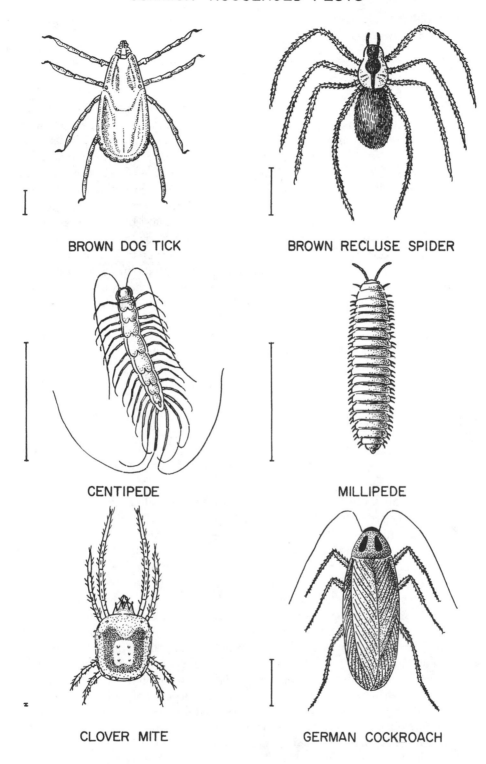

BROWN DOG TICK

BROWN RECLUSE SPIDER

CENTIPEDE

MILLIPEDE

CLOVER MITE

GERMAN COCKROACH

THE LINES INDICATE THE ACTUAL SIZES OF THE PESTS.

Figure 20-2

avoid problems. The cost of equipment necessary for mosquito control programs is beyond the means of those not engaged full-time in this business.

Ticks

Ticks feed on the blood of mammals, birds, and reptiles. Dogs play an important role in bringing man into contact with ticks. Also, increased interest in outdoor recreation brings more people into tick-infested areas. These eight-legged arthropods transmit a number of diseases to man, including spotted fever.

The brown dog tick has become a serious household pest in recent years. This species feeds on dogs, but will attack people, usually children. The entire life cycle of the brown dog tick is completed in the home. During part of its life, the tick drops off the dog; at this time it may be found behind picture frames, baseboards, etc. Various stages may live for six months or more without feeding.

The American dog tick is one of a number of other tick species that commonly attack man and dogs. This tick does not complete its life cycle in dwellings. It is rather carried into dwellings by dogs and man from fields and woody areas. The American dog tick and related western species are the most important vectors of spotted fever and other tick-borne diseases in the United States.

Control of brown dog ticks is very difficult and usually requires the services of a pest control specialist. Treatment involves the use of residual materials such as 1% diazinon spray. Satisfactory control is based upon a knowledge of tick behavior and repeated chemical treatments. Control of American dog ticks and related species requires extensive treatment of infested fields, etc. The use of some of the standard flea-tick compounds on dogs, usually gives temporary relief to the dog.

Clover Mites

During the spring, and sometimes during the fall and winter, reddish-orange or dark-greenish mites may invade homes, apartments, and other structures in great numbers. Usually these are clover mites; they feed on grass, clover, and lawn weeds growing near the foundation of a building. These mites do not bite people or damage property, but do cause considerable alarm to housewives, etc. The life cycle of the mite consists of egg, larval, two nymphal, and adult female stages; males are not known in the U.S. Eggs are laid in exterior cracks and crevices in a building and other stages hide in these cracks when not feeding.

Clover mites are dormant during the summer, and chemical control measures applied during June to early September are of no value; however, on bright, sunny days during the winter, the mites may be active. Spring and fall applications of kelthane or chlorobenzilate sprays applied to the lawn at a distance of 15 feet from the structure and to the foundation is recommended. The removal of grass plantings up to 18 inches from the foundation will also check the problem. Ornamental plantings may be used to replace the grass. Generally, the services of a pest control expert is needed for satisfactory control.

Chiggers

Chiggers are larval mites, less than 1/150 inch in length, which attack people on legs, abdomen, etc. and cause reddish welts on the skin which are intensely irritating. The nymphal and adult stages of chiggers are called harvest mites and feed on insect eggs, etc. People are attacked by chiggers when they walk or sit in fields where there are populations of these pests, and small mammals on which they normally feed are also susceptible to them.

Chiggers are not normally found in houses. The use of repellent materials such as "6-12" applied to skin and clothing prior to outdoor hiking, etc. and the treatment of lawns, certain park areas, campsites, etc. with lindane or chlordane are recommended.

20B / TERMITES AND THEIR CONTROL

Termites of many species are distributed throughout the United States. Infestations have been reported at one time or another in all of the 48 continuous states. Distribution of infestations follows a general east-west band across the United States. Florida, the Gulf Coast, and California have the highest incidence of infestation from all species. A broad band from east central United States, to the southwest, and northern California have "moderate to heavy" infestations. The more northern tier of states have a "slight" infestation index. Northern Maine, northern Minnesota, most of the Dakotas, and a large part of Montana have limited problems with termites.

Termites have been called "white ants" or "flying ants" by uninformed people. Termites are not ants. They are not related even though both live in societies. Ants can always be distinguished from termites by their narrow constricted "waist"; swarming ants have four wings of uneven length (the forward ones extending over the abdomen). Termites, on the other hand, have no waist at all; both the front and hind wings of swarming termites are of approximately equal length. Unlike the white termite workers, swarmers are dark. Ant swarmers are dark, but they are the same color as the other ants in a nest. As a general rule, ants more often swarm in the fall, and termites more often swarm in the early spring. Careful observation will avoid mixups.

There are two major groups of termites which cause the greatest damage over most of the United States. These are the subterranean termite group and the dry-wood termite group; the latter is the less frequent of the two.

Subterranean termites dwell in soil and use the cellulose content of wood as food. They are common beneath the surface of any woodland. They live in nests, sometimes very deeply in the soil, and are social in habit. There are four social castes: winged reproductives (king and queen), soldiers, workers, and secondary reproductives. The winged reproductives lose their wings after the mating flight; the other three castes are wingless. Through a system of tunnels which maintain humidities of about 95%, termites are able to forage for wood some distance from their nest. When food is found on the surface of the ground, workers may extend tunnels above the soil in the form of tubes. The tubes serve to block out sunlight and maintain the needed high humidity inside the tube-tunnel system. Termites shun light as a rule. They may enter a structure at a protected spot where wood touches soil or is very close. Then they proceed to eat the softer portions of the wood, leaving the more hardened "rings." Termites will seldom eat through the surface of wood. This is why infestations are difficult to find until the wood has been weakened so greatly that it no longer will support weight.

Another way infestations are discovered is through swarming. In the spring of each year, a strong colony will produce an excess of winged reproductives. Shortly following the first warm spring rains, winged reproductives swarm from infested wood or soil. In the ensuing flight, dispersion and mating takes place. Mated pairs retire to soil or wood and start a new colony, which may mature in about five years.

If termites are already working within the wood of a structure, or beneath the soil of a basement containing a furnace, swarming times may be altered because of the unusual conditions created by a local heat source. Under these conditions, termites can swarm at any time of the year. Swarming and weakened wood are two major ways termites may be discovered by accident; it is much safer and less spectacular to find them by a careful inspection.

Inspection

Inspection includes finding the conditions favorable for termite activity or damage. An inspector should look for several conditions:

(1) *Soil-wood contact.* Since termites can enter wood under concealment directly from the ground, it is essential that no structural wood touch the soil. While there are some obvious flaws in construction which can often be seen and corrected, there are many ways which builders use to cover over, disguise, or otherwise hide such soil-wood contact. During an inspection, the question should always be asked: "Does this wooden member touch soil even though I cannot see it?"

(2) *Wood near soil.* Wood may be near soil and still be attacked, for termites are able to construct tubes without external support for distances up to 3 feet. Also termites can move through incredibly small cracks in concrete (1/32 inch) to reach wood. Therefore, cracks in basement floors must be considered suspect and checked. Foundation walls may also have cracks; this is a termite hazard if the wall is below the soil surface.

(3) *Moisture conditions.* Excess moisture can be conducive to termites. Dripping spigots, defective plumbing, defective eavespouting, poor ventilation in underbuilding areas resulting in condensation of soil moisture—all can contribute to excess dampness under which termites thrive.

(4) *Warmth.* Furnaces or heat ducts in or on soil or near concrete walls will warm the soil also. Termites in temperate regions move deeper into soil in winter

to avoid cold. They will increase their activity in soil warmed by furnaces or heat ducts, provided moisture conditions are at least minimal. If settling or expansion cracks are present near these heat sources, the structural wood nearby is also vulnerable.

These four factors can be observed with no special tools; only a keen and suspicious eye is required. A termite inspection where the object is to find termites (not just the conditions under which they can survive) is rather different. Minimal tools for this job are a probe (such as an ice pick), powerful flashlight, coveralls, and the same keen eye, and again a suspicious nature. Commercial termite control operators may also carry a variety of equipment to help them measure and estimate costs of treatment and repair.

Several things may be found on such an inspection. The ice pick is used to probe wood at suspected places to penetrate into termite-riddled wood. Tapping is preferred to picking if the wood under inspection is decorative and would be marred by poking it. If termites are found active in wood, the probe can be used to determine the extent of damage by prying apart the infested wood.

Tubes may be found extending from the soil over walls or from soil directly to overhead wood members. Simply breaking these tubes does little good, since the persistent termites can reconstruct them in a matter of hours or days.

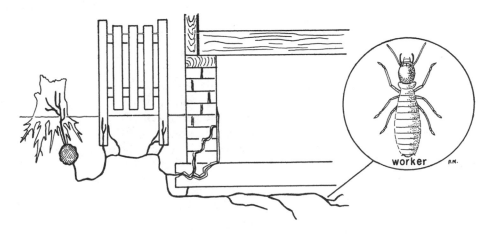

CROSS SECTION ILLUSTRATION SHOWING TERMITE ENTRY POINTS

Figure 20-3

Prevention of Termite Damage

Prevention of termite damage is a matter of periodic inspection to insure that there is no soil-wood contact, no vulnerable wood near cracks in concrete of foundation walls, no undue moisture accumulation about susceptible wood, and no possible entry points near sources of warmth.

Though the idea has been proposed in the past, there is probably no such thing as a structure built to be termite proof. Concrete may be solid when freshly cured, but cracks develop eventually.

Mechanical methods of preventing termites from entering buildings have taken the form of shields, which are supposed to fit around all members which touch ground and prevent termites from tubing directly into other wood. Termites must "come out in the open" to get around a termite shield. Shields have, however, been shown to be, at their maximum, only partially effective. They must be maintained so that there is no bending of the shield. Inspection to check the efficiency of the shields must be regular. Of course, the shield must be installed carefully and properly; this is seldom the case.

One method of prevention of termites' damage which has been more successful than any other method is that of pre-construction treatment of the soil on which the building is constructed.

Modern chlorinated hydrocarbon insecticides have been tested by the U. S. Forest Service and found to have an effective life in the soil of 20 years (to date); this figure has been increasing year by year, for some of the chemicals, since the tests began in 1949.

These tests consist of treating a small plot of soil, then driving stakes into that soil or laying a board on the treated soil. Success of the test is determined by the number of years that the stakes are left undamaged. That the soil in this area of the southern United States is thoroughly infested with termites is confirmed by the fact that termites make short work of untreated wood stakes driven into untreated plots of soil within the test areas.

The object of treating the soil beneath a structure is to provide a barrier of chemical through which termites will not pass. The necessity for pressure-treated wood being incorporated into the lower portion of the structure is thus removed. Treating soil in this way does not make the soil unfavorable for plants, flowers, shrubs, or trees. Some of the older termite control chemicals were toxic to vegetation as well as to termites, and they did not last nearly as long.

Chemicals designed for pre-construction treatment are the same ones which are commonly used for corrective treatment (or post-construction treatment). These chemicals are:

> aldrin
> chlordane
> dieldrin
> heptachlor
> lindane

Chemicals are applied in a water emulsion of 0.5 to 2.0% with a quantity of water so they soak into the ground. Even though it would appear to be a simple job, it is not, but the care and upkeep of equipment and bulk buying of chemicals enables commercial pest control firms to perform such pre-construction treatment on a volume basis at relatively low unit cost.

Correction of Termite Damage

Termite control after damage has been done is a more expensive process. It involves basically four steps: (1) removal and replacement with treated lumber of all wood destroyed by termites; (2) correction of structural faults which may encourage termites (or fungus rot); (3) treatment of wood in the more vulnerable lower portion of the structure; (4) treatment of soil around the inside and outside perimeter of the structure, and insofar as possible, beneath the structure.

Corrective termite control is not a job which can be done by an untrained layman. It is best left to specialists in termite control who are associated with reputable firms, have adequate insurance and performance bonds, and will stand behind their work with a warranty on work performed. Such firms are often members of a trade association which supports the warranty, as does the National Pest Control Association.

In corrective termite control, strategic placement of the chemical is the goal of the termite control specialist. To apply the chemical properly may require that portions of the existing structure be drilled so as to reach otherwise un-accessible areas. Treatment is designed to place a chemical barrier around the whole of the structure as in pre-construction control. Depending upon the amount of damaged wood and the nature of the damage, state regulations govern-ing commercial pest control, and the customer's wishes, termite operators may remove and replace structural members themselves, or they may subcontract this job to carpenters, or leave the job for the customer to complete. Replace-ment costs of damaged wood may contribute heavily to the cost of a termite job.

Most termite control operators use many laborsaving devices, such as electric or pneumatic drills, enabling them to bore through concrete rapidly and neatly. Their trucks are usually equipped with tanks into which carefully measured doses of chemical are placed, mixed thoroughly, and applied by pumps using pressures up to 200 pounds per square inch.

Nonsubterranean Termites

In the southwestern USA, dry-wood termites are an important pest. The nests of dry-wood termites do not require a connection with the soil. Colonies are started when mated pairs enter a crack or crevice in wood. Dry-wood termite infestations may be discovered because they make a small hole to the exterior surface of wood. They dispose of their waste products through such holes. These wastes form small piles of what appears to be "sawdust." Careful inspection can reveal the source as dry-wood termites. (Other insects, such as powder-post beetles in wood, may produce "sawdust" piles.)

Even though dry-wood termites are found in structures in the more southern and southwestern portions of the country, it is not unusual for a single infested piece of furniture to be discovered far beyond the normal range. Chairs, pianos, and tables are common items found infested. Such isolated colonies can survive within heated buildings even in the north.

Control of dry-wood termites often involves the use of fumigating gases, commonly methyl bromide. If the infestation is widespread, the whole structure is usually enclosed in a tarpaulin cocoon and gas introduced for an extended period before ventilation. Small lots of infested goods may be fumigated in special chambers or under small tarpaulins. Fumigation is definitely a job for specialists and should not be attempted by inexperienced persons.

Other Wood-Destroying Organisms

Fungi, known as wood-rot fungi, may attack building timbers in areas where moisture conditions are high and favorable to their growth. These fungi will not be recognized as mushrooms or toadstools. Rather the fungus penetrates the wood with tiny living threads which extract the stored nutrients on which the fungus grows. After the fungi have done their work and died out, the wood begins to dry and crack in a typical cuboidal pattern. Certain fungi are capable of transporting water from a particularly moist place to another less moist place where they are growing.

Most termite inspectors are trained to recognize fungi and, more importantly, the conditions under which fungi flourish.

It is therefore suggested that plant management make a practice of having periodic (yearly or biennially) surveys of susceptible buildings. This is particularly important if the building is made largely of wood or has areas which might be vulnerable to termite or fungus attack.

20C / ORNAMENTAL AND TURF PESTS

Shade trees, shrubs, flowering plants, and turf are subject to attack by insects and mites. Recognition of pests of plantings is essential to develop programs of prevention and control. Aphids or plant lice suck sap; mites rasp leaves causing discoloration, reduced vigor, and leaf-drop; caterpillars and grubs of various types defoliate; scale insects encrust branches and twigs and suck the sap; borers of many kinds enter bark and wood, weakening and killing; diverse gall-making pests cause malformations of leaves, fruits, and branches. Certain root-feeding insects and mites transmit fungal and viral diseases of ornamental plantings.

It is essential for good ornamental maintenance to inspect plant specimens regularly. Within a period of a few weeks, insect and mite populations can rapidly increase and cause serious injury. Proper timing for spray application is essential for good control. There are a multitude of pest species, but recognition of a relatively few species and their damage will enable maintenance personnel to deal with about 70% of their pest problems. Also, County Agricultural Agents and entomological specialists at a state university will assist in identifying pest problems and provide control recommendations, usually at no charge. The following pests are some of the more common ones encountered on ornamental plants.

Gall-Forming Pests

Malformations in leaf growth (tumors, warts, etc.) prompt a great deal of concern by plant growers, particularly during the early summer months. Galls are overgrowths which are caused by plant reaction to the presence of certain mites, aphids, wasps, tiny flies, and other pests. Maple bladder gall, caused by mites, is one type that raises much concern. In May and June wartlike greenish and reddish growths on the upper surfaces of soft maple leaves are an indication of this problem. It is too late to control the problem once the galls are present.

Species of oak and elm are also subject to attack by many types of gall-producing pests. In general, galls do not cause serious injury to plants and cannot be controlled at all or can be controlled only by preventive application of pesticides early in the spring. Most people are unwilling to use a preventive spray for gall-producing insects that do not cause serious damage anyway.

Scale Insects

Scale insects cause serious damage to ornamental plants because most people do not recognize the problem. These pests cover their bodies with a whitish, grayish, or brownish waxy substance, so that they resemble bark on twigs and branches. Only the tiny crawlers (nymphs) are able to move; it is this stage that is most easily controlled because it is not protected by the waxy layer. However, crawlers are only present for about two weeks; therefore, proper timing of the spray application is essential to good control. Malathion or methoxychlor are recommended for the control of the crawler stage; on deciduous ornamentals, dormant oils may be applied during the winter season.

Aphids

There are many species of aphids or plant lice which occur on ornamentals. Some aphids are found only on certain plant species, while others, such as the green peach tree aphid, feed on many plant species. Aphids tend to attack new plant growth and undersides of leaves by inserting their beaklike mouthparts into plant tissues and sucking plant juices. Heavy infestations of aphids produce quantities of a sugary liquid called honeydew. This material may drip from trees onto cars, etc. leaving spots; also when the honeydew remains on the leaves, a blackish, sooty mold may grow on it. Applications of lindane, Malathion, or diazinon are effective against aphids.

Spider Mites

Small white flecks of dead leaf tissue, silking, yellowing or bronzing of the leaves, and finally leaf-drop are characteristics of spider mite damage. Spider mites are barely visible to the eye and require the use of a special microscope or suitable hand lens for positive identification. Bedding plants purchased from greenhouses and other sources are often infested with spider mites which rapidly increase after the plants are set out in floral gardens. Also spruce, hemlock, boxwood, elm, roses, and other plantings have mite problems. Application of kelthane, chlorobenzilate, or morestan are recommended for control.

Borers

Most borers are larvae of some type of beetle or moth. The adult female beetle or moth lays eggs under bark of trees and shrubs; eggs hatch and the young enter the plant tissue causing serious damage. Newly planted and poorly growing older trees and shrubs are highly attractive to these pests. The smaller European elm bark beetle is a borer-type insect, which in addition to causing direct damage, also transmits the fungus which causes Dutch elm disease. Methoxychlor and other insecticides applied against the adult beetle in the spring, along with sanitation practices (remove infected trees promptly) and fertilization, is used to discourage the spread of this disease.

COMMON ORNAMENTAL PESTS

SCALE INSECTS FEED BY SUCKING JUICES FROM PLANT STEMS.

MAPLE BLADDER GALLS ARE CAUSED BY GALL-PRODUCING MITES.

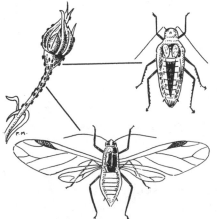

Both winged and wingless forms are shown. APHIDS SUCK PLANT JUICES.

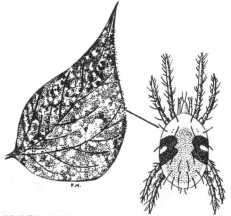

SPIDER MITES DAMAGE PLANTS BY RASPING TISSUE AND SUCKING JUICES.

Figure 20-4

Defoliators

The larvae of many moths (caterpillars) and of sawflies (larvae), and the adult or larval (grub) stages of many beetles cause serious damage by eating leaves of trees, shrubs, and other plants. Yellow-necked caterpillars on apples and ornamental crabapples, walnut caterpillars on walnut, redheaded pine sawfly larvae on pine, and the elm leaf beetle on Chinese elm are a few common defoliators. Applications of methoxychlor or similar insecticides, applied in the spring or when the damage starts, will provide control; however, large-scale control programs against defoliators need to be carefully planned to avoid adverse community reaction and misapplication of pesticides.

Root-Feeding or Soil Insects

The most serious pests of turf attack the roots. The Japanese, May or June, and Oriental beetles are some of the more serious problem species. The larvae or grubs feed from ½ to 2 inches below soil level. Here they feed on the fibrous roots, causing large areas of turf to die. Sod webworms are another type of turf pest; these caterpillars eat grass blades during the night. Armyworms and cutworms may cause damage to turf during certain years. The most commonly used material to control turf pests is chlordane.

20D / PESTICIDE USAGE

The term pesticide means a chemical or other agent that will kill or otherwise protect something (in which man has a special interest) from destruction or loss from a pest species. There are various types of pests including insects, mites, rodents, birds, weeds, etc.; against these, man uses various types of pesticides, insecticides, miticides, rodenticides, avicides, herbicides, etc. Under insecticides and each other type, there are a variety of chemical and biological substances which have different residual properties, degrees of effectiveness, formulations, and levels of safety. Some 800 individual active chemical pesticide ingredients and 60,000 pesticide formulations have been registered with the federal government (U. S. Department of Agriculture).

To obtain a pesticide registration, the chemical manufacturer must provide information to U.S.D.A. to show that the material is effective and safe to use. If the material is applied on food items, the U. S. Food and Drug Administration must approve its use and establish tolerances. Many years of costly research and development take place before a new pesticide can be marketed.

Safety in Using Pesticides

When using pesticides, it is essential to avoid property damage and injury to humans. Each year there are about 1,500 fatal and 16,000 nonfatal cases of poisoning (from all causes) in the U.S.A. About 10% of the cases are caused by pesticides. These poisonings result from mishandling; when pesticides are handled properly and according to manufacturers' recommendations, they pose little hazard.

(a) Accidents may happen when instructions on the label of the pesticide container are not followed.

(b) Accidents may happen when pesticide storage areas are not locked.

(c) Accidents may happen when pesticides are placed in other than their original containers (pop bottles, paper bags, etc.).

(d) Accidents may happen when protective clothing or devices are not worn.

(e) Poisoning may occur when pesticides are spilled on the body or clothing and not washed off immediately.

(f) Accidents may happen when pesticide containers are not disposed of safely.

If a supply of pesticides is to be maintained by a building plant supervisor, he should:

(a) Have a locked, well-ventilated storage room which is used only for pesticide storage;

(b) Use only those pesticides which are recommended for general use; these should be correctly measured for use and applied only according to recommendations on the label of the container;

(c) Allow only careful and experienced personnel to handle and apply pesticides, or employ a professional pest control expert;

(d) Make certain that empty pesticide containers are disposed of safely so as not to become available to children, pets, and uninformed people.

Types of Pesticides

Pesticides may be classified in a number of ways. They may be grouped according to their effectiveness against certain kinds of pests: insecticides versus insects, herbicides versus weeds, etc.; according to how they are formulated and applied: dusts, fogging oils, granulars, wettable powders, etc.; and according to the chemistry of the pesticide: chlorinated hydrocarbons (chlordane), organophosphates (Malathion), natural organic insecticides (pyrethrum).

Effectiveness against a particular pest species, safety, chemical hazard to property, type of formulations available, equipment required, and cost of material must all be taken into account when choosing a pesticide for a particular job. Recommendations change with experience, the development of new materials, and new governmental regulations. However, there is a degree of stability and most recommendations last over a period of years. The pesticides recommended in this chapter are likely to remain for many years; it should be kept in mind that these recommendations are not the only ones that will work, but are standard ones that have and will work if properly executed.

Chlordane is chlorinated hydrocarbon. It is a wide spectrum, long residual insecticide widely used against household pests, for termite control, and against turf pests. It is regarded as moderately toxic; however, certain formulations commonly used for termite control have a high percentage of the active compound and should be regarded as quite hazardous to nonprofessionals. Preformulated 2–3% chlordane oil solutions are available to the nonprofessional for cockroach control. Generally the nonprofessional lacks the equipment and knowledge to do a satisfactory job of cockroach control.

Diazinon (spectracide) is an organophosphate type, broad spectrum insecticide which has a rather long residual and is fairly toxic to man. It is widely used in household pest control against cockroaches, ticks, ants, silverfish, spiders, and many others. Diazinon is formulated as a 50% wettable powder or 25% emulsion. When used by the nonprofessional, considerable care should be exercised and directions followed precisely.

DDVP (vapona, dichlorvos) is an organophosphate, volatile insecticide-acaricide which is used under special conditions. While quite toxic, DDVP breaks down rapidly. It is used in greenhouse and ornamental pest control to some extent; also it is used in cockroach control programs by professional pest control operators and widely used against flies, formulated as a resin strip which is hung from the ceiling. In many cases, these resin strips are used in such a manner as to be ineffective. One or two strips cannot possibly protect a huge room which has a constant source of fresh air entering from the outside. Follow the formulators' recommendations when using this device.

Kelthane (dicofol) is a chlorinated hydrocarbon type miticide which is relatively safe when used according to directions. It is widely used for the control of mites, including both the clover mite and the two-spotted or red spider mite. It is available as a 35% wettable powder. Recommended for use by nonprofessionals.

Malathion is an organophosphate type, broad spectrum insecticide which has a very low hazard when used according to directions. While only slightly toxic to man and other mammals, it is highly toxic to fish and birds. It is widely used in ornamental pest control of aphids, whiteflies, scale insects, and others; however, it is not very effective against the two-spotted spider mite. Premium grade, 2–3% Malathion residual sprays can be used against most household pests; there is less chance of an odor problem with the premium rather than the regular grade. Malathion may be purchased as a 57% emulsion concentrate or a 25% wettable powder. Recommended for use by nonprofessionals.

Methoxychlor (marlate) is a chlorinated hydrocarbon type, slightly toxic insecticide that is being used as a replacement for DDT. Methoxychlor is not accumulated in human body fat and does not contaminate the environment as does DDT. However, methoxychlor is less effective and does not have the residual life of DDT. For certain ornamental pest control applications, methoxychlor is satisfactory. It is available as a 50% wettable powder and is commonly sold as marlate. Recommended for use by nonprofessionals.

Pyrethrins are botanical insecticides derived from chrysanthemum flower buds. Pyrethrins are relatively nontoxic to humans and are used with other insecticides to give a quick knockdown or flush insects from hiding places. Commonly found as ingredients in some aerosol preparations.

Aldrin, dieldrin, and **heptachlor** are persistent chlorinated hydrocarbon type insecticides. Their major use about structures is in termite control. These chemicals are not recommended for use by nonprofessionals.

Chlorobenzilate is a chlorinated hydrocarbon type miticide which is effective against clover mites and other species of spider mites. It is available as a 25% wettable powder. Recommended for use by nonprofessionals, but is sometimes difficult to obtain.

DDT is a chlorinated hydrocarbon type, broad spectrum insecticide which is very stable and persistent. It is only moderately toxic to man. However, because of its cumulative and persistent qualities, it is no longer widely used in areas where there is any hazard of environmental contamination. In some states, it is no longer legal to use DDT. However, where legal, under some circumstances such as the control of lice, fleas, bedbugs, and bats, pest control operators and other health workers still use DDT because it gives the best results and is the safest material available. Not recommended for use by the nonprofessional.

Dimethoate (cygon) is a moderately toxic, organophosphate type of insecticide which is used for fly control. Not recommended for use by nonprofessionals.

Lindane is a chlorinated hydrocarbon type, moderately persistent, and toxic insecticide now being replaced by other pesticides less likely to contaminate the environment. Lindane is sometimes sold with vaporizer devices. Such sales are illegal because these devices are not federally registered. The prevention of the sale of these devices is difficult, but there appears to be a definite hazard to humans breathing the vapors produced by them. Lindane spray still has a limited use in household pest control and is effective against aphids and certain other ornamental pests. Not recommended for use by nonprofessionals.

Other pesticides. There are many pesticides available. Those listed here are ones most likely to be encountered by nonprofessionals. It is undesirable for nonprofessionals (for safety and economic reasons) to invest in large quantities of pesticides listed here and others, because they may decompose with age.

20E / **RODENT CONTROL**

Rats and mice have the ability to adapt to many situations, even some which on the surface appear to be unfavorable for them. They can live in sewers, on dumps, in hospitals, in homes, on islands, on ships, and on farms. Unlike some living organisms which have rather narrow tolerances of living requirements, rats and mice have rather wide tolerances. A family of mice were once found living "comfortably" inside a large food cooler; their adaptation was a thick coat of shaggy hair to protect them from the cold. There is no question that rats and mice are very flexible in their behavior.

The living requirements of rats and mice are not great. They can get by on relatively little food and water. Some is required, but surprisingly little. Though rats and mice have preferences for certain foods (or qualities) when given choices, they can subsist and reproduce on what we might consider very vile substances.

Rats and mice have a very high reproductive potential; they bear offspring regularly and in large numbers. A small rat population, if left unchecked with plenty of food, water, and shelter can within a few months produce hundreds of rats. Limitations of food, water, and shelter keep populations from growing. Even a population of 50 rats near an industrial plant can damage raw materials, contaminate processed or unprocessed food, or lower employee morale. Those in charge of building maintenance should keep in mind that rat problems can be avoided or forestalled.

In some plants, workers eat their lunches from sacks or lunch pails near the spot where they work. Sometimes they are careless and they leave papers, crumbs, and uneaten food about. This waste is attractive to rats and indeed will support a small colony. Cleaning up will increase overall sanitation and avoid the possibility of rats.

Weeds in a field next to an office building or manufacturing plant can harbor a colony of rats. Under the weed cover, few if any rats may be noticed. Debris such as scrap lumber or metal form the same type of rat harborage. Removal of weeds and debris will permit building supervisors to inspect for rats quite quickly. Since rats do not prefer to traverse open spaces, there is a lowered chance that rats will enter a building surrounded by such open spaces.

A periodic check of exterior conditions of a building will indicate if and how rats might enter the building. First, check the regular openings of the building to see that they are in good repair. Doors should close tightly, and have automatic closing devices to insure that they are not accidentally left open. Screens and windows, especially those near ground level, should be intact. Window frames, if made of wood, should be checked to see that no gnawing marks are present which might indicate that rats are attempting to make an entry point. Ventilation louvres should be in good repair and screened if openings would permit rodent entry. Holes in soil along foundation walls should be checked to find if they are rodent holes. Interior foundation walls should have no large cracks which would enable a rat or mouse to burrow through soil and enter. Entry points of all utilities (gas, electricity, water, etc.) should be sealed securely. Holes made for the insertion of pipes are sometimes left in a ragged condition which will permit rodent entry along the pipes. Signs of rats may also be seen. These include worn paths in grass, burrows, fecal pellets, gnawed wood, damaged products, or rubmarks made by a rat's greasy body as it rounds a corner or goes up or down a step. Vigilance is a major key to preventing rat troubles in any structure.

A large part of a rat problem can be solved by making the environment unfit for rats. Ratproofing consists of either complete replacement of a damaged wooden part or covering the damaged portion with a sheet metal "shoe," which makes the wood inaccessible to rats' gnawing. Metal, covering exposed susceptible wood, will prevent gnawing because of its hardness and smoothness.

Most universal of the modern rat toxicants are the anticoagulant rodenticides. These tend to prevent blood from clotting. Small doses taken each day for approximately one week will cause rats to develop many small internal hemorrhages (bleeding spots) and die of internal bleeding. Similar chemicals in different doses are used in human heart disease therapy to dissolve dangerous clots in the circulatory system.

The active ingredients of anticoagulant toxicants are administered in very small concentrations. If a pet or a child accidentally ingests exposed rodent baits, little chemical is present to do permanent harm. Also, if larger doses are taken over longer periods, an antidote is readily available from hospitals in the form of Vitamin K, which promotes clotting.

Several active ingredients of the anticoagulant type are available in ready-to-use form from retailers or wholesalers. These are warfarin, fumarin, diphacin, and pival and are sold as a cereal grain bait with approximately .025% active ingredients.

Such baits should be placed about the area to be treated. Bait cups or stations should be plentiful. Saturation baiting is a better method than placing just a couple of bait stations and hoping that rats find them. It is also advisable to find and remove (or make inaccessible) the food on which the rats have been regularly feeding. Otherwise, there is competition from the alternate food sources. It is better to have just one food source, and that the rodenticide bait.

Traps can be used in place of bait stations or as supplements to them. If the manufacturing plant processes food, generally no toxicants are permitted within the food plant. In this case, saturation trapping would be advised. Place snap type rat or mousetraps just off observed runways (if baited) or directly on runways, if it is decided to make use of the expanded trigger footfall trap. In this type, one places a piece of cardboard over the treadle of the trap, then sets the trap; when a rat steps on the treadle, the trap will spring. Expanded treadle traps should not be baited.

Baited traps can be set with peanut butter, partly sizzled bacon, half gum drops, or bits of meat or fish. Traps with perishable food as bait should be checked often and have the baits replaced at the first sign of staleness. With a small population, use at least six traps for each estimated rat. For a larger one, of course, the economic consideration of trap costs will limit the number used. But the general rule is use plenty.

Traps and anticoagulants will serve the nonprofessional controller well in combating most simple rat problems. For more complicated problems, it is advised that building engineers and managers seek the counsel of a professional pest control operator. He has a variety of other tools which are generally reserved for professional use only. Occasionally, these are needed to get rid of a difficult rat problem. He has highly toxic calcium cyanide, a fumigant dust for use in rat burrows. He also carries a stock of more toxic rat control compounds such as thallium sulfate, red squill, sodium fluoroacetate, and others, and has men trained in the proper use of these dangerous chemicals. Sometimes baits must be formulated from highly concentrated chemicals. These are best left to men accustomed to handling such chemicals who will take proper precautions against personal and environmental contamination. The work of a reputable pest control operator is generally "satisfaction guaranteed," and PCOs are insured for the use of the toxic chemicals they employ.

Recommendations for control given in this section should be considered first aid, or what to do until the professional arrives. Many simple control problems can be managed with the simple techniques listed. If problems are not easily solved, the plant manager should then consult with a pest control operator to achieve the desired results. Some of the more difficult problems which professionals are better equipped to handle involve rat populations in sewers where cooperation with municipal authorities is required. Widespread infestations, both indoors and outdoors, require a many-pronged attack which only the pest control operator has the time and materials to cope with. Rat problems, where regular food is sufficiently attractive to cause rats to avoid the baits placed for them, require the special talents of pest control operators. Jobs where the plant engineer sees only the superficial problems involving rats may require the specially trained eyes and mind of a PCO who is used to dealing with deep-seated rat problems.

20F / BIRD MANAGEMENT

Three species of birds are likely to be pests in or about structures, whether dwellings, offices, production, or storage. They are the house sparrow, European starling, and common pigeon. All three are imported species; they are not native to North America and have expanded their range relatively free of natural enemies. These birds are well adapted to life near humans, and make full use of human structures for both nesting and roosting.

House sparrows (sometimes called English sparrows) are rather small birds; males exhibit a small black "bib" on their throat, and a white shoulder stripe. A male's head has a gray cap and the nape of the neck is chestnut colored. In birds frequenting dirty buildings in a city, some of these colors may be obscured with dirt. Females are pale brown all over and have a light, dusky line over each eye. House sparrows are somewhat social birds. They will nest near another pair of the same species, if nest space is available. Often these sparrows usurp nest holes in a purple martin house. Nest spaces will usually be under a protected eave, or in a hole (defect) in a structure. Nests are rather bulky, sloppily built masses of grass, string, paper, etc.

European starlings nest in almost any hole, whether it is in a tree, fence post, or in a building. Starlings during the breeding season are black; the sexes look alike. Juvenile birds are uniformly brown. After molting in late summer, starlings are black with white spots and iridescent speckles. These spots and speckles wear off during the winter months. Starlings have long, spear-shaped, pointed beaks which are yellow in fall and winter, becoming black during spring and summer. Starlings in flight exhibit a "delta-wing" silhouette. Careful observation will avoid confusing starlings with other birds which are black, such as grackles, cowbirds, and red-winged blackbirds.

The common pigeon was imported to North America by fanciers to breed for racing, plumage, and for food. Wild or feral pigeons, which make up the large urban populations that are pests, descend from those which escaped from domestic flocks. Pigeons are colored blue-gray, with a gray breast, pink feet, and a collar of dark gray feathers which is often attractively iridescent with green and purple. Other color variations may be seen in an urban flock: reddish-brown, white, and black; these colors result from hybridization of color strains.

Pigeons commonly nest on protected ledges and build a loose nest of rough twigs. They will soil such ledges with excrement. They roost or loaf on roof peaks, eaves, and on structures such as statues. These, too, are defaced with water-insoluble excrement.

Urban pest birds about structures cause a mess with their droppings and nesting litter. But possible damage and discomfort can go beyond that. House sparrows have been incriminated several times with carrying live cigarette butts into their nests. Many burned-out cellulose filters have been found as nest linings; some fires have been credited to house sparrows.

Birds which gain access to interiors of large buildings (warehouses, hangars, production lines, etc.) may do considerable damage, if their excrement or feathers fall into raw or processed food products. Pigeon excrement has corrosive effects on certain metals, especially those used to make the "skin" of aircraft. Any bird litter or filth falling onto machinery or personnel can have a damaging effect on production and morale.

Nests of birds contain insects and mites; some are parasitic upon the birds, some only live in the nest because it affords them food and protection. At certain seasons these organisms leave the bird's nest and invade human quarters; they may bite or crawl on humans. People affected by such tiny organisms are often thought to be "imagining" itches which are actually very real.

Control Procedures

Control of pest birds consists of basically a birdproofing technique. Seal all defects in buildings so that there are no nesting sites on the outside or entrance holes in the building. Old nests should be removed before sealing. Decorative architecture or ledges which afford protection can be blocked or screened with inconspicuous wire or hardware cloth. Ledges can be made inhospitable by installing slanted boards at a 45 degree angle so that birds can gain no foothold.

There are several products which can be installed, especially against birds. These involve a system of pointed wires or slender spikes on a base. Birds alighting on a ledge with such wires installed have the sensation of sitting on a cactus. Such wires have also been installed inconspicuously on statues.

Chemicals can be used to repel birds from alighting or roosting on buildings. There are several proprietary products which employ a "sticky-foot" principle. When such material is applied to a ledge, etc. with a caulking gun, birds landing become uncomfortable because of sticky, slippery footing, and they fly away. Such preparations are not toxic and will not kill birds. These sticky repellents are prepared with ingredients which will not run in hot weather and remain rather sticky even in cold weather. Reputable pest control operators can apply such chemicals skillfully. Some nationally known products of this type are: Roost No More, No Roost, 4 the Birds, Go Birds Go, and others. Grossly underpriced service estimates should be met with suspicion for some people have applied "homemade" concoctions to repel birds, and such materials can deface buildings far worse than birds can.

One product (Rid-A-Bird) is available for killing birds by absorption of a toxicant through their feet. This product can only be used inside structures, for the material used is highly toxic to other birdlife and to humans.

A variety of toxic chemicals can be formulated into baits for use in poisoning pest birds. Certain toxicants can only be used during the winter months so as to avoid exposing poisons to desirable birds. Formulation of toxicants is best left to knowledgeable pest control operators who are familiar with birds, chemicals, techniques, and state and federal laws.

An additional variety of chemical materials, which can be used by a progressive pest control operator, include such things as anesthetics so birds can be knocked out quickly, retrieved, and later released elsewhere or disposed of humanely, and distress inducers (Avitrol) which cause birds eating a toxicant-treated bait to emit cries of distress, scaring others of a flock away (this technique may not work well on certain species).

In certain circumstances, pest birds may be shot with shotgun loads using fine pellets or dust. Pyrotechnic (fireworks) devices may be useful in certain situations to frighten birds away, but these should be left to persons trained in their use.

An electronic adaptation of the sticky-repellent technique involves installation of wires to ledges where birds roost. With the landing of birds on the two-part wire and the completion of a circuit, birds get a "hot foot." This technique usually is effective only against pigeons. Other pest birds have smaller feet.

In an attempt to develop a "one-shot cure-all" for bird problems, various manufacturers have developed devices which make noise, produce lights, or create some mechanical movement supposed to be frightening or feared by birds. Most of these devices will produce the initial scaring reaction, but they do not produce the final desired result—no birds in the area. Birds, like humans, become accustomed to certain discomforts, and learn not to fear that which does them no real harm. Such devices may be effective for approximately one week, then birds tend to ignore the noise, light, or movement.

Bird control does not offer any "one-shot cure-all" for problems. Experts have a variety of tools which they apply to a given situation. Anyone claiming to have a single device, chemical, or technique which will do the whole job of building protection and bird removal or repelling, should be questioned carefully. Bird control is not easy, nor is it cheap, but the job, when done by a person who knows what he is doing, will pay off in the satisfaction that the problems of contamination, building protection, and personal comfort are solved.

Federal, state, and local laws protect birds. No bird control should be undertaken without knowledge of these laws.

20G / VEGETATION MAINTENANCE

Weeds are plants which grow where they are not wanted. Even cultivated plants which grow larger than intended, or begin to spread out of bounds, become "weeds." An example of the latter is ivy-covered walls of buildings. Young or controlled ivy lends an attractive appearance to a building, but when ivy begins to cover windows or becomes too thick, control becomes imperative.

Weeds produce many seeds which increase the chances of survival of the species. Weeds are very persistent and may grow in odd places: cracks in pavements, in soil where no other plants can compete, etc. Many species of weeds are not readily killed by mechanical cutting, or will quickly recover and continue to grow. Some weeds grow rapidly during times when desirable plants are dormant, and thus they may spread and "shade out" the desirable species. Other weeds may be toxic to livestock or have noxious properties (thorns, irritating nettles). All of these factors must be taken into account when planning a grounds maintenance program for vegetation control.

There are several types of vegetation maintenance programs related to human needs and to the vegetation itself.

(1) *No or Limited Vegetation.* In or about parking lots near buildings, no weeds are desired. Weeds, however, constantly encroach on such open spaces where there is little competition. Cracks in asphalt, or bare places in gravel lots, provide the minimum soil necessary for some types of unsightly weeds.

With ornamental plantings about buildings, generally bare soil or soil covered with a mulch (wood chips, etc.) is desired. Too often lawn grasses or weed grasses may begin growth in these bare areas. These unwanted plants must be removed.

Some manufacturing plants store raw materials near the main building. Weeds may gain a foothold in such areas, because regular mowing is difficult or impossible. Control of weeds is especially important in this situation because:

(a) There is a fire hazard in the autumn when the weeds die and dry out. This is especially important when some of the stored materials are flammable products.

(b) Weed growth retains moisture near the ground. Metal materials stored in a weedy area will rust or corrode from excess moisture.

(c) Weeds may make finding stored raw materials more difficult.

(d) Moist weeds cause employees to get wet when working in a weedy storage lot. Indeed, wet weeds are slippery and serious falls may result.

(e) Mice, rats, and other animal pests may breed in certain weedy areas.

In such areas where mowing is difficult, chemicals offer the best method for reducing or eliminating weeds. Chemicals to assure no vegetation are called soil sterilants.

(2) *No Broad-leaved Weeds.* On lawn areas grasses are desired, but the fast growing, nongrassy (broad-leaved) weeds are not. Mowing will generally help control broad-leaved weeds; however, certain weeds persist under mowing and grow more flatly; plantain and dandelion are two examples. Some chemicals are specific for broad-leaved weeds; these, applied to a lawn, will remove most broad-leaved weeds and do not harm grasses. Such chemicals are called selective broad-leaf herbicides.

(3) *No Grassy Weeds.* This is the point where control becomes most difficult, for in the category of grassy weeds are crabgrass, orchard grass, foxtails, and various other grasses, most of which grow in a bunch or cluster. Chemical removal of one species of a grass from turf of another species has formerly been impossible. Chemicals have been developed to prevent crabgrass from growing

in newly seeded turf, but the best method for getting rid of crabgrass is by developing a strong, healthy lawn of desirable grasses.

(4) *No Woody Plants (Brush)*. When fence rows or out-of-the-way areas are neglected, eventually some woody plants (trees, shrubs, and vines) will grow and continue growing each year. They become very difficult to control physically by cutting because the roots persist and soon send up even more shoots. Chemicals have been developed to kill such brush and the use of these chemicals on young, bushy growth insures against difficult jobs of brush removal later.

(5) *Weed-Free Water Bodies*. Some buildings have decorative ponds on premises which must be maintained. Some vegetation in ponds is desirable as decoration; however, overgrowth is undesirable. Proper construction of ponds will help prevent overgrowth by undesirable plants. Periodic inspection and maintenance by specialists will keep a decorative pond pleasing to viewers and minimize insect and odor problems associated with neglected, slow flowing, or stagnant water.

Some General Principles

The best prevention of weeds in a lawn is a healthy, thriving lawn. In other situations, controlling problems early is easier and more economical than trying to control them later.

A major problem in weed control is understanding the idea of early or preventive control. To those alert to weed growth, tiny sprouts seen in the spring or early summer are the prime targets of their efforts. Unfortunately for the uninformed, they do not see a weed problem until a 4- or 5-foot weed is "staring them in the face." Only then do they think that they need control service. It is not really "too late" at this point but the effort and expense needed to gain desirable control is greater.

(1) Strive to keep exterior grounds of buildings looking presentable. Make maintenance a regular schedule; that way little problems will not become large problems.

(2) Survey periodically. Common sense and alertness will help you recognize incipient weed problems. A grounds keeper alert to what a good lawn or storage area should look like will be able to spot an incipient infestation of broad-leaved weeds.

(3) Schedule treatments at the most economical time, biologically. Preventive treatments for crabgrass in temperate North America should be applied by April 1, while crabgrass is just sprouting. Most turf grasses are "cool season" grasses and grow most actively in the spring and fall months; their growth tapers off during the hot summer. At this season crabgrass grows most actively. By treating crabgrass early, the desirable turf will tend to "crowd out" the weed grasses.

Dandelion and plantains should likewise be controlled in the spring. It is obvious that control before flowering and seed production lowers the number of weeds which will grow in subsequent years. Too often control is delayed until flowering; then the economical chemical dosage is too little for complete kill.

(4) Consult with a professional weed expert who is knowledgeable about proper maintenance of desired vegetation. Contracts for certain phases of grounds maintenance (fertilizing, chemical weed control, etc.) by reputable custom sprayers can be more economical for building management than stopgap attempts by grounds personnel who must share their time doing many other maintenance jobs.

Weed Control Uses a Variety of Chemical Types

Soil sterilants are used where no vegetation is desired. Older formulations depended upon arsenical compounds (sodium arsenite) to prevent growth of any vegetation. These products are still in use and, because they are very poisonous, should be handled with the utmost care, preferably by specialists.

Modern advances in herbicide chemistry have provided a variety of less hazardous organic products which can be used as soil sterilants. One group which chemists have supplied are known generally as substituted urea compounds, because urea is the basic chemical. Some of these chemicals are diuron (telvar), monuron (karmex), and uracil (hyvar-X). Another group of organics (triazines) which can be used as soil sterilants are simizine and atrazine. Several compounds utilize the herbicidal abilities of certain borate compounds. These are occasionally used in combination with the strong herbicidal properties of sodium chlorate. This material is a flammable product and addition of borates reduces fire hazard. Chemical manipulation of the basic formula of 2,4-D or 2,4,5-TP (silvex) produces chemicals with soil sterilant properties not possessed by 2,4-D. Such is the case with erbon (baron), a soil sterilant classed as a phenoxy compound along with 2,4-D and silvex. Most soil sterilants will persist in the soil for periods up to a year (possibly longer, depending upon the chemical and the dose used).

For control of broad-leaved weeds in established turf, few chemicals surpass the utility of 2,4-D in its selective effectiveness. Basically, 2,4-D is a growth hormone which, when absorbed by broad-leaved plants (not just weeds), will cause the plants to grow excessively and out of control. This pattern results in plants looking curled and contorted; they "grow themselves to death."

The chemical 2,4-D is a tricky one to use. A single sprayer should be set aside for use with 2,4-D and not used for any other purpose; 2,4-D is very difficult to clean completely from any spray apparatus. Cases are known where a sprayer was used for weed control one week and used for rose spraying later; all the roses were killed because of the traces of 2,4-D in the sprayer. Soybeans, roses, tomatoes, and grapes are some highly susceptible plants to 2,4-D damage.

Hazards of drift must also be mentioned in connection with 2,4-D use. Drift may result from too much pressure or too small a nozzle, causing a mist to form. Such mist may be carried on breezes to damage nontarget plants. One form of 2,4-D, the ester formulation, is highly volatile; it forms vapor. This vapor can rise from a sprayed area on warm days and be carried on convection currents to contaminate nontarget plants. The amine form of 2,4-D is less volatile, but its use near susceptible plants is still questionable. While most broad-leaf weeds in a lawn can be controlled with 2,4-D, some troublesome ones cannot be controlled with this herbicide. Proper dosages of 2,4,5-T can be used against such troublesome broad-leaved weeds. Since 2,4,5-T is effective against woody plants and shrubs, its use on lawn areas landscaped with woody plantings is limited. Under no circumstances should 2,4,5-T be used on or near food crops, about the home or recreational areas, or near water areas.

Dicamba (banvel D) is a product which can be used for control of otherwise difficult-to-kill perennial, broad-leaved weeds in turf. Another relative in the 2,4-D family of chemicals is mecoprop (mecopex), which is effective against some of the broad-leaved weeds that 2,4-D cannot kill.

For control of grassy weeds, the problem becomes more touchy, because desirable lawns are composed of grasses. Since there are few chemicals so selective as to kill one kind of grass and leave another unharmed, researchers have devised some schemes whereby removal of weed grasses is a matter of timing and dosage of application. Several chemicals have been developed to be applied in the early spring to established turf before crabgrass has sprouted. These products are called pre-emergence crabgrass controls. Some of them are dacthal, bandane, and DMPA (zytron). One interesting chemical which has been developed for pre-emergence crabgrass control has the very special use of being applied to bare soil when grass seed is sown. To prevent unwanted invasion by crabgrass (the seeds of which are probably already in the soil), siduron (tupersan) kills germinating crabgrass seedlings but leaves desirable grass seedlings unharmed. Tupersan, however, will not kill annual bluegrass (*Poa annua*), which is considered a weed grass by those who want pure stands of Kentucky or Merion bluegrass. DSMA (disodium methanearsonate) is a selective contact herbicide used for crabgrass control on established turf.

For grass control on nonlawn areas, chemicals such as amitrole and atrazine can be used. To completely remove all grasses from an area so that other plants can be planted later, dalapon can be used; however, it persists in the soil. A specialized product, cacodylic acid, is used to give complete kill of all grasses in a lawn so that the lawn can be reseeded a few days later with better grasses; tilling or cultivation is unnecessary with the use of this chemical. The old dead grass serves as a straw mulch.

For control of brush found along fence lines, etc., 2,4,5-T or 2,4,5-TP (silvex) can be used. Some chemicals with a broader effectiveness against brush are picloram (tordon) and ammonium sulfamate (ammate X). For control of a specific bush or shrub, granules of a material such as fenuron (dybar), a substituted urea compound, will be absorbed by the roots and kill that individual bush or tree.

Aquatic weed control is probably the most touchy and difficult of the many types of weed control and should be left to the professional. Control is difficult because the plants are growing in a different environment (water) and this alone presents many problems. There are surface floating weeds, rooted shoreline and shallow-water vegetation, rooted deep-water vegetation, all of which require a different approach. Measurement of the volume of water to be treated must be relatively precise, so that the dosage can be regulated properly. The chosen chemical must be applied in such a manner that fish in the water will not be harmed. Local laws regarding potable water supplies must be adhered to and restrictions on chemicals which can be used must be obeyed. Techniques differ widely, depending upon whether the water is impounded or flowing.

A commonly used chemical for aquatic weed control is 2,4-D. Special formulations designed for aquatic work are available, but these are used for certain types of aquatic problems, generally still water. Endothall is effective in flowing streams; here the rate of flow must be accounted for and dosage regulated accordingly so that the chemical will be dissipated before coming into contact with desirable vegetation or animals downstream.

Acrolein is an especially dangerous chemical and has a utility for only certain types of lake-weed control work. Its sale is restricted to professional applicators. Sodium arsenite is used in small dosages as an aquatic weed control herbicide. Diquat can be used in fish-inhabited lakes.

Some chemicals are used for more than one type of weed control. Differing dosages determine the effectiveness of a chemical for a certain type of work. At low doses, a product is effective as a pre-emergence grassy weed controller; at higher doses, the same chemical is effective as a soil sterilant, all the more reason for reading and understanding label directions before proceeding with any chemical application.

The chemicals mentioned here are not the only ones which can be used for weed control. This listing is not meant to exclude from consideration any which were not mentioned. Directions for use were not mentioned; for proper methods of application and safe use, consult the label. Read the label and heed it carefully. Labels of pesticides generally are the best guide for achieving the desired results with use of any chemical. (Sales brochures are not labels; guard against unwarranted claims in such brochures. Base buying decisions on label directions alone.) Each compound will have specific limitations which will be detailed on the label directions for use. For the greatest satisfaction in the control of weeds or the cultivation of fine turf, experts should be consulted and/or retained. Their results are generally performance guaranteed.

Cultural Methods of Vegetation Maintenance

In most cases, chemicals alone will not completely solve a weed problem. Cultural methods alone also will not completely solve weed probems, but used together, both methods can help to prevent recurrence of problems.

To prevent weed growth on a vacant area, it can be covered with asphalt or concrete. Noncrop areas can also be tilled periodically to eliminate weeds, but this generally leads to problems with erosion and mud during rains. Mowing is, of course, a cultural method and mowers designed for weedy areas are available. With regular mowing, generally, the most unsightly, fast growing, tall weeds will be eliminated.

Cultural methods for eliminating broad-leaved weeds from grass areas have already been mentioned: keep a healthy turf. Keeping turf healthy is a matter of periodic fertilization, regular strategic watering (soak as infrequently as possible), insect control, disease prevention with fungicide application, periodic dethatching (thatch is the matted dead grass which surrounds the base of the grass stems; it is removed by a "combing" process), relieving soil compaction by aerifying the soil (by punching or spooning out plugs or cores), and topdressing with a light sand-peat combination.

Cultural methods for keeping grassy weeds out of turf are much the same as for keeping broad-leaved weeds out of turf. Some mechanical removal and replacement with plugs of desirable sod may be necessary when chemical means fail or fall short of expectations.

Cultural control against brush is a matter of periodic cutting to prevent small brush from becoming small trees. A variety of hand-held saws, rotary blades, and knives are available for nonchemical brush removal. These are not completely effective alone, but when used in combination with certain chemicals (such as dybar-fenuron, or ammonium sulfamate) which poison the remaining root systems, control can be complete. Removal of larger trees and their roots is best left to professional tree experts.

21 / PAVEMENT MAINTENANCE (CONCRETE)

The best possible provision for maintenance of a concrete pavement is to use good materials and the best methods of construction, so that repairs will be infrequent. The defects in a concrete road can almost always be traced to some fault in the materials, in preparing or placing the concrete, or to neglect of some important features in preparing the subgrade.

Next in importance to employing good materials and insuring thorough workmanship, is taking care that all imperfections in the road should receive attention as soon as they are discovered. If this is not done, small breaks will become large holes, and expensive repairs will be necessary. The edges of cracks are constantly battered by the tires and in time the cracks are enlarged into chuckholes. Prompt repairs are therefore essential. The road should be inspected annually, and hot bitumen should be poured into all cracks. This repair work is generally done at a small cost, and it constitutes the only charge for upkeep on concrete pavements besides that of taking care of the shoulders along the road.

The major repairs that are likely to be necessary to concrete pavements will fall mostly into three general classes: (1) the replacement of the pavement where it has been cut through to lay or repair underground pipes; (2) the repair of depressions or chuckholes caused either by original faulty construction or by the normal effects of wear; and (3) repairs along expansion joints or self-formed cracks in the pavement. Repair patches that are properly made should prove as durable as the original pavement, and such patches should not be noticeable after being exposed to traffic for a short time.

22 / ROOF MAINTENANCE

(*Source:* "How to Care for Your Built-up Roof," by G. E.
Hann, reprinted by special permission of *Factory Management and Maintenance,* copyright by McGraw-Hill
Publishing Company, Inc.)

Roofs need regular upkeep and care just like any other part of the plant.
But in many plants they get little if any attention—until they leak.

If you want to keep building maintenance costs down, do not wait for the
roof to leak and cause water damage. That means you are too late—and you are
in for some expensive remedies. Routine upkeep of a roof is always cheaper
than more extensive emergency repair.

Your roof contractor may be responsible for the repair of a bonded roof,
but he is not responsible for repair of roof copings and parapets. Many guarantees also do not include roof flashings. And no guarantee ever covers repair of
walls, ceiling, equipment, or product damaged by water. Remember, too, some
guarantees expire when the contractor retires from business or goes bankrupt.
Be sure you read every guarantee clause before you sign the contract.

A reliable roofing contractor can be a big help in planning and carrying
out a good roof maintenance program. But, whether the repairs are made by
the contractor or by your own plant maintenance crew, the prime responsibility
for roof care rests on the plant owner. And it usually falls on the shoulders of
the plant engineer or maintenance superintendent. Therefore, the engineer or
superintendent should acquaint himself with the principles and practices of
proper roof care.

The majority of manufacturing plant buildings are topped with a built-up
roof, therefore, we shall only discuss this type of a roof. Many of the maintenance
techniques, however, are equally applicable to other types of roofs and should
be helpful in inspiring good care of roofs, no matter what their design.

Built-up Roofs

A typical built-up roof consists of a supporting base or deck, and a laminated
surface application or mat.

Deck. The roof deck may be of wood, steel plate, poured concrete, precast
concrete slabs, gypsum slab, or other structural material. The ideal deck is leakproof against common roofing materials such as asphalt or tar. It does not depend
on the first layer of sheathing paper to prevent seepage, and it is structurally firm.

Mat. The mat is composed of one or more thicknesses of roofing felt. This
felt is made of a fabric impregnated with tar or asphalt which provides the waterproof qualities. The fabric acts only as a reinforcement to hold the waterproofing
in place.

Roofing fabrics are normally made of paper, rag, or asbestos fabrics, or combinations of these. Roofing is estimated and sold in squares. A square of roofing

288 / **PHYSICAL MAINTENANCE**

is the amount required to cover 100 square feet. Single-ply felts usually run fifteen pounds per square. Double-ply runs thirty pounds. Special felts may run as high as ninety pounds. Size and area of one roll vary according to brand. The more common widths are thirty-two and thirty-six inches.

In a built-up roof, several layers of roofing felt are applied to the roof with a hot bonding agent. There are two main agents, tar and asphalt. Each has its advantages.

Tar will stand up under water better than asphalt. But it has a low melting point and needs a binder of gravel to keep it from running when softened by hot weather. Hence tar is better for flat than steep grades. Gravel helps to insulate the roof against solar heat and retards drying out. It can take the brunt of foot traffic. It helps protect the roof against fire from sparks and embers. (See Figure 22-1.)

Asphalt can be used on any grade. An asphalt roof has a smooth surface, hence dirt gets washed off easily by wind and rain. You can spot flaws in an asphalt roof more readily. And it is easier to maintain.

Construction and Repair

Laying roof deck. The roof deck must be leakproof against the hot bonding agent. If it is not, apply a preliminary layer or capsheet of 30-pound felt with a two-inch overlap. This layer is nailed on along the seams. The roof plies are then applied over a mopping of tar or asphalt between layers.

The number of plies a roof has is fixed by the width of overlap of each layer of felt. An overlap of half the width of a roll gives two plies. Overlap of two thirds give a three-ply mat.

The felt is laid starting at the lowest point, parallel to the gutter. In a typical five-ply roof using a two-ply capsheet the first gutter strip is cut one-third the width of the roll. Another two-thirds strip is added, then a full width strip, all touching the edge at the gutter. Ensuing full width strips are then laid, with a two-thirds overlap, to the crown of the roof. Here partial width strips are again used, as at the gutter. There are several ways to use different weight capsheets or roof layers with varying overlaps to obtain a given number of plies.

Drainage. A good roof drains well, usually into gutters. It doesn't have any hollows that will hold stagnant water, which rots an asphalt mat. All low points have drains with properly sized downspouts. Valleys between monitors or saw-teeth are pitched to an outlet at either end, or to a center drain.

Parapet and coping. Many buildings have outer walls or parapets that extend above the roof deck. If not properly constructed and joined to the roofing mat, they can be a prolific source of leaks. (See Figure 22-2.)

The outer face of the parapet is usually an extension of the face masonry of the building. The inside face or back-up, however, may be made of common brick, building tile, concrete block, or even poured concrete. In between faces, if they do not meet, you may find loosely filled mortar and rubble, or hollow tile.

Most important part of a parapet in preventing water leakage is the coping, and the cap flashing that seals the top. Copings are made of units of clay tile,

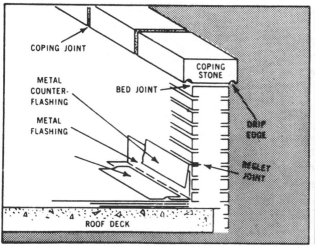

ONE-THIRD WIDTH OF ROLL
TWO-THIRDS BALANCE OF ROLL
FULL WIDTH OF ROLL

30-LB. FELT

2-IN. OVERLAP

ROOF DECK

FULL WIDTH
OF ROLL
TWO-THIRDS

Figure 22-1 / Typical built-up roof is laid with overlapping felt strips so thickness of mat is same at all points. Here a 5-ply mat consists of one 2-ply capsheet and three 1-ply strips. Each layer is mopped over with hot asphalt or tar to make it waterproof.

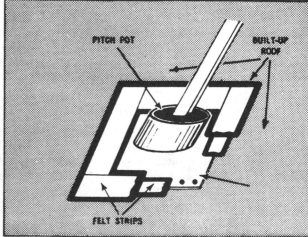

COPING JOINT

METAL COUNTER-FLASHING

METAL FLASHING

COPING STONE

BED JOINT

DRIP EDGE

REGLET JOINT

ROOF DECK

PITCH POT

BUILT-UP ROOF

FELT STRIPS

Figure 22-2 / Good construction of parapet wall and coping adds to life of roof. Flashing keeps water out of corner. Anchors or eyes for cables or signposts will not harm roof if properly sealed off. The pitch pot method is one good way.

metal, cement, brick, or continuous concrete. Joints may be made of cement or mastic. To do a good sealing job, copings themselves must be waterproof, and the joints must stay leakproof.

Flashings. The angle where the roof deck meets the parapet wall is a critical joint that has to be protected against water leakage by a special treatment known as flashing. Different roofing contractors have different ways of installing it. In principle, a good flashing keeps water on the roof from getting under the edge of the roof mat. It also keeps out water that flows down the parapet. This usually requires two kinds of flashing—base flashing and counterflashing.

One common method of *base flashing* is to turn the roof mat up against the vertical surface. Sometimes a separate angle strip is cemented to the roof mat. This is called angle or gutter flashing. In either case, the turned-up portion must extend high enough to exceed the depth of any flow of water along the wall.

Counterflashing hangs downward from a tight joint in the inside parapet wall, known as a reglet joint, and laps over the base flashing. Sometimes it extends through the whole width of the parapet from a point just below or near the coping. Then it is known as a cap flashing or wall flashing.

Flashings are made of various materials—roofing felt, sheet metal, or saturated fabric membrane.

Other roof structures. Every vertical roof structure that interrupts the plane of the roof surface needs joint treatment in the form of flashing.

Some structures are an integral part of the roof, such as elevators penthouses, monitors, skylights, and expansion joints. Other structures—chimneys, stacks, exhaust vents, and other ventilators—pierce through the roof deck. In both types, the aim of the treatment is the same—to keep water out of the joint by use of flashing and counterflashing or their functional equivalents.

Where anchors or eyes for rods or cables are fastened to the roof structure, the anchor is often sealed off by surrounding it with a metal collar known as a pitch pot. The collar is filled with mastic, and treated with flashing around the outside surface. Another way is to use a fabric covering or skirt covered with asphalt or tar.

Elements That Can Ruin Your Roof

Sun. Most roofs do not wear out—they dry out. Hot sunshine on a roof causes the volatile ingredients of tar or asphalt to evaporate. The asphalt oxidizes and becomes brittle. Slowly the roof mat loses its elasticity—the surface coating becomes checked and flakes off, exposing the felt. What you have left is a dried up roof felt that acts like dry paper—brittle, easily torn, and in the last stages, ready to absorb rather than repel moisture. Thus it is susceptible to easy damage from wind, rain, and foot traffic. And it loses the elasticity to respond to temperature change stresses, so it cracks.

Water. When water seeps into a dry roof, or steals through a crack, it causes a leak. But worse damage is yet to come with freezing weather, which turns the moisture into ice. This tears and heaves the roof, leaves it wide open.

Wind. A strong wind takes advantage of any weakness in the roof mat or structure. It drives rain into defective joints in the mat or parapet; it tears at loose seams; and it sways roof structures.

Temperature changes. Expansion and contraction are at work all the time. Often expansion strains in a new wood or concrete structure cause the roof mat to tear. Seasonal changes in temperature expand and contract mortar in the coping joints, and provide an opening wedge for water to do its worst. The various forces of expansion and contraction in the roof deck and building walls wrestle at the angle joint and strain the flashings. Expansion of hot stacks loosens the roof joints.

Settlement. Settlement of roof or wall structures is not unusual. It will cause trouble wherever the horizontal and vertical surfaces meet. As walls settle, or as roof timbers dry and shrink, extra strains are exerted on flashings. Or the roof may settle below the level of the drain pipe. This will cause either a backup of flood water, or a leak through the crack around the drain.

Outside interference. Roof mats are not designed for extra accessories such as signs and electric wires. Nor are they intended for regular foot traffic. Many agencies are not careful, when installing such services, to provide the extra precautions of flashing and mastic waterproofing. Sign supports or braces are spiked or bolted to any convenient spot on the roof, regardless of its structural relationship. Anchorage planks are spiked or lagged to the deck, piercing both the mat and deck, and causing serious leakage.

Inspection

Take frequent looks at your roof. Good care of a roof begins with regular inspection of the roof premises for signs of weakness or repair. The biggest step in holding down maintenance expenses is to catch your roof troubles before they are born, or while they are still young.

You can expect any person to find the bad leaks, but it takes a qualified person to spot the small leaks. Your inspector has to know what to inspect and what to look for. Competent inspection includes many points—copings, parapets, flashings, skylights, ventilators, inside and outside walls, ceilings, and so on.

Semiannual inspections. Inspect the roof every six months, preferably in the early fall to anticipate winter weather, and in the early spring in preparation for the hot summer sun.

Special inspections of the roof should be made after a violent rain or windstorm, or after a nearby fire, or after "outside-interference" workmen have been on the roof.

Use a roof plan. A schematic drawing or plot plan of the roofs of all buildings in your plant can be a big help in making roof inspections.

To be of value, show on this plan the location of each roof, and when installed. Also state who did the job, guarantee (if any) and approximate dimensions. Set up an inspection chart. After each inspection enter brief notes on the condition of the roof and the general nature and cost of repairs. This data will prepare you for the next inspection, and guide you on future repairs and choice of new materials and methods to be used.

One helpful technique is to note any weak points on the roof you feel do not need repair at least until after the next inspection, but must be checked at that time.

How to inspect. Is it better to inspect area by area, or item by item? For instance, is it better to inspect all copings, then all parapets, then all flashings, or do you inspect everything in one round of the roof walls?

It is a matter of choice and convenience. Certainly it seems sensible to complete one roof at a time. The important thing is not to miss any parts of any items.

Use a checklist. Best way to make sure you do not overlook an important detail in roof inspection is to arm yourself with a checklist. You can easily make up your own form, or you might get one from a leading contractor. Use a separate sheet for each roof.

A typical checklist shows all major roof items in the left column, such as copings, parapets, roof deck, roof covering, drains and gutters, skylights, and penthouses. Under each item is a subdivision of the point to be checked. *For example,* under roof covering you might list blisters, loose or torn seams, cracks, old patches, condition of mat. Across the top of the page are headings for vertical columns—condition OK, number of defects, repair at once, repair when convenient, check next inspection, special comments.

A good roof inspector will look at many things besides the roof mat for clues to actual leaks or those in the making. Obviously, the first thing to do is ask the building maintenance staff if there are any known leaks, or any complaints of leaks from employees. Such reports help pinpoint other symptoms later on.

We have grouped these symptoms into convenient parts of the roof structure.

1. *Exterior walls of building.* Look for settlement cracks or other evidences of building movement that might have strained the flashing or displaced drainage facilities. Efflorescence—that whitish, powdery excretion from mortar joints—may mean there are leaks in the coping or parapet up above.

Interior walls. Stains or seepage indicate leaking walls or roof. Damp walls, chipped or discolored paint are clues to roof leakage.

2. *Underside of roof deck.* Look for any signs of deterioration from leakage above. If the deck is of wood, list rotted or warped boards for replacement. Make a note to check roof above any watermarked boards. Sometimes a wooden roof deck is weakened by excessive heat and humidity from processes below it.

If the deck is of concrete or gypsum slab, look for cracks and stains. Spots of rust might come from wet reinforcing rods. In a gypsum deck, leaking water usually causes spalling. In metal decks, rust spots mean trouble with condensation or leaks.

Topside. Only a few defects are apparent in a topside inspection. A sharp indentation or break at right angles to the boards in a wood deck points to a sprung board. If the roof mat gives under you, the board may be rotted. But there are secondary clues. Cracks in roofing felt may be due to cracks in the roof deck. It is always a good idea to check the roof deck wherever you find leaks or holes in the roof mat.

3. *Drainage.* Check for any changes in the original slope of the roof. Has any part of the deck sagged, to produce hollows for standing water?

In a wood deck, you can try to correct or restore the proper level by timber shoring. When this is not feasible, and the hollow is less than four inches, a good remedy is to fill the low area with asphalt mastic either above the old mat or under a new one. Always add a water repellent surface over it.

Deeper hollows can be repaired by a "saddle" or false roof that bridges over the depression. This is covered by a new roof to match the built-up roof. Hollows in concrete roofs can be built up the same way. Sometimes you can clear the trouble by adding or relocating a drain.

Drainage system. Make sure no rain water accumulates on the roof during a rainfall. Slow acting drains and gutters may cause a backup of water above the parapet flashing. See that all drains are sound, flush to the roof, and clear of debris. If settlement or expansion has worked a drain spout above the roof level, re-install it.

4. *Copings.* Coping troubles are mostly of two kinds, either the coping material loses its water repellency and absorbs moisture; or the mortar in the coping joint cracks. In either case, water drains down into the parapet, and where it will finally come out is hard to tell.

Often you can see the effects on the outside facing as you approach the building. More than likely, however, the leak will drain out of the inside facing, because of poor masonry work. Once frost gets in, mortar and bricks may be pushed by ice pressure out of the wall.

Repairing the Roof

Cracked or crumbling coping joints. You can point them up with a mastic that adheres to the coping and adjusts itself to expansion and contraction. First chisel out the old material to a depth of at least one-half inch. If all the mortar is crumbling, remove it and pack all but the outer one-half inch of the joint with oakum. Then force the mastic in the outer groove with a caulking tool or gun, and finish it at surface level. If appearance is not a factor, you can use the less expensive black caulking materials. Otherwise you can get a mastic that will match in color and will not stain.

Porous copings. When copings absorb water, you need to treat the surface to seal them against moisture. There are several ways to do this by brush application of a sealer. Which sealer you use depends upon the need to preserve appearance and the degree of porosity. You can apply an "invisible sealer," but it does not weather as well as other sealers that do have a darkening effect on light copings.

Another method of keeping appearance is to use a pigmented mastic that actually simulates the original appearance. This mastic is available in a variety of colors to suit most conditions. It is especially useful where disintegration of the coping has reached only the early stages, before spalling has begun.

Broken or spalled copings. You can replace broken or spalled copings, but there is a less expensive way of repair. It involves restoration of the shape with suitable mortar and application of the protective coatings as just described. This method of face lifting will work as good as new.

Parapets. Disintegration of a parapet most often starts with a defective coping. But not always. The stormy gusts that sweep along a roof and hit the parapet are a severe test on the backup masonry. Or water may accumulate above the height of the flashing for several days. Thus it permeates the cracked or porous joint. The evidences are efflorescence, spalling, open joints and loose bricks, and leaks to the building interior. The repair of broken parapets is a masonry job.

Sometimes a maintenance superintendent will seek extra security for the inside of his parapet by applying a plaster coat of cement mortar containing a water resistant compound. Where units have been spalled, this procedure gives you a uniform surface. But remember, it does not eliminate water ingress from a defective coping.

There is a question about the advisability of sealing the interior or exterior wall surfaces. If walls are not allowed to breathe, according to some authorities, the moisture in building interiors can not migrate through the walls. This leads to excessive condensation of moisture entrapped in the walls themselves, and weakens them. Vapor barriers should not be applied without careful study of the principles and favorable applications.

Flashings. Flashing flaws are among the commonest causes of roof troubles. Too often flashing defects are overlooked when you inspect the roof for leaks. Look for loose flashings, or for torn or even missing sections.

Be sure to check all flashings for tightness of bond. A metal flashing may be in good condition, but may fail to do its job because of improper bond. This will allow water to seep between the edge of the mat and the base flashing, or between the counterflashing and the wall. Also, a flashing that is not turned up far enough may allow backup water to stand against the wall and leak down to the roof deck.

Once water gets behind a flashing at the parapet angle, frost action follows. The flashing is loosened from its fastenings, thus opening the way to worse damage.

Metal flashings, unless made of copper or other fairly noncorrosive metal, are subject to rust. It pays to give them a good protective coating. Once they get holes, they are poor protection.

You can replace defective metal flashings with so-called membrane flashings at lower cost. There are several types to satisfy various conditions. The "sliding and overlapping" type is most commonly used. Sometimes the "accordion pleat" type is necessary, or the "regleted" type. All are equally effective when properly employed.

Roof mat. Roof mat troubles are not usually traceable to a single cause. When you spot the following symptoms, check carefully for the cause, so you can make the right repairs.

Blisters. Three main causes for blisters are: (1) misapplication has entrapped air, (2) leaks have allowed moisture to form pockets between layers, or (3) volatile ingredients have given off gases that have formed pockets.

Small blisters do not need attention unless there are many of them. In such a case, you must cut them off before you give the roof a general membrane treatment. Cut all large blisters open with a criss-cross cut so you can lay back the roofing felt for examination. If any water is inside, mop it dry. Make sure the felt and deck are unharmed. Replace decayed timber. Recement the felts to the deck and apply a membrane covering over the slits.

Loose or torn seams. This occurs when the original bond of roof felts begins to weaken, putting all the strain on the seams. The seams loosen, lift, and are torn by the wind. When this happens, cement all loose edges back into place. Sometimes you need to reinforce the seams by cementing six-inch widths of membrane over their entire length and covering them with tar or asphalt.

Cracks. Cracks are caused by (1) contraction stains in the mat, and (2) drying out of original saturating oils which leave the mat hard and unelastic.

When an asphalt roof dries up it reveals a dry, powdery brittleness. On a tar and gravel roof you will find bare spots where the gravel has washed or blown away because the tar has lost its adhesiveness.

Contraction in a hardening roof mat often shows up most plainly at the seams. If seams are drawing or curling, it may mean the felts are losing their pliability. Improper application may cause the same effect. In the case of a nailed seam, contraction often withdraws the felt from the nails, so that the nail heads protrude. Unless repaired, this may lead to rusting, then holes and leaks.

Cracks at the ridges of sloping roofs are usually the result of contraction and a movement of the mat down the slope.

Sometimes hardening of the roof mat is abetted by poor results of previous maintenance efforts. Hot or cold applications that do not penetrate the felt, but merely seal the surface, are likely to stiffen and crack in time. Occasionally you will see a roof that has been treated many times this way cracked in an alligator pattern.

Fairly new roofs will also crack. This happens when the roof deck itself cracks. The real solution may be to install an expansion joint, or to make structural repairs.

Repair of local cracks and miscellaneous breaks follows a general pattern. All you need is roof cement and membrane. This membrane is made of cotton, jute, or glass fabrics saturated with asphaltic materials, or unsaturated. It comes in rolls of six, twelve, twenty-seven and thirty-six inches wide. The membrane itself is not waterproof. Its function is to reinforce the cement, which is the real water repellent. The membrane must be closely bonded to a supporting surface—do not use it to bridge over a depression.

To repair a break, first clean it thoroughly, cement any loose roofing felt back into place, and apply a bonding coat of cement. Then cover it with a patch of membrane, and over that apply a second coat.

Wearing spots. Check the causes of abrasions or erosions in the roof surface. Sometimes maintenance employees beat a path on a roof to make regular inspections of superstructures, or to raise the flag every day. For this condition we suggest a superimposed wood walkway. Other worn spots may be caused by water that spills at the same place. Here you need only lay down a splash board or tile to prevent further erosion.

Old patches. The fact that you have had trouble there before is a good reason why you should always check old patches. Make sure they are tight and standing up well.

Old roof. Let us assume that the roof you have inspected is an old built-up one, and that you find one or more of the conditions, just described, is wide-spread: bad cracks, dried up felt, large blisters, gravel surface gone, and open seams.

What are you to do about it? Here are six possibilities—consider which is best for your roof.

1. *Leave the roof alone.* Your roof may be guaranteed, and its condition the responsibility of those who laid it. So you decide to ignore the roof's deterioration, even at the risk of the well-being of the building.

2. *Rip it off.* But would you go to the expense of buying a new roof if the present roof has value in it that can be made to give trouble-free service?

3. *Cement a new covering to the old roof.* This practice of putting on a "cap sheet" seldom proves satisfactory. You are expecting one or two new plies to do the job of a multiple ply roof. You do not get a good bond—heavy winds may tear it. The moisture absorbed by the old dried up roof may cause serious blisters in the new roof.

4. *"Hot mop" the roof.* Such treatment builds up the coating, but it does not resaturate the dry felt, and it prolongs the roof life very little. Very heavy hot applications check and crack readily.

5. *Apply a resaturating agent.* This may preserve the roof, but it is no cure for leaks. It works only if the roof is watertight and you use an agent that insures continued flexibility and water resistance.

6. Cover the entire roof with a flexible membrane embedded into and coated with a roof preservative. Where there are still whole felts left on the old roof, but cracks and breaks occurring, this treatment may give the roof a new lease on life.

A good roof, properly maintained, will last the life of a building. The secret of trouble free, low cost maintenance lies in the regular resaturation of surface treatment of the roofing mat with a proper penetrant or coating while it is still sound and whole. And the early repair of cracks and breaks by the simple membrane treatment is economically sound and safe.

When breaking of felts is widespread, however, and deterioration is well advanced, we recommend you have the roof completely removed and a new roof built up from the base. Do not try to save such a roof by merely coating it with tar or asphalt. Unfortunately many plant maintenance executives have been educated to believe nothing can go wrong with a roof that a few barrels of asphalt or tar, and maybe a cap sheet, will not cure.

Roof Repair Hints

Planning. Know exactly what repairs are to be made and exactly what areas are to be treated. Mark them off with chalk and measure them. List every item. Write out the methods of treatment so that the job will proceed smoothly from beginning to end.

Now you are ready to order the needed materials, and arrange for the necessary equipment and crew. It is a good idea to plan where materials and equipment should be placed when they arrive.

You may need a hoist to lift material to the roof. Sometimes it is wise to store most of the materials on the ground to avoid cluttering up the roof.

You should also decide whether a roofing pump will be of any help. This is a portable power pump with hose attachment for piping cold materials from a container on the ground through a flexible hose to any place on the roof. Such a pump can be especially useful in saving time and labor.

First step. Cleaning all roof surfaces is the first step in major roof repairs. This includes parapets, copings, and flashings. Sometimes you can do this with a long-handled push broom. In the case of a tar and gravel roof, you have to cut the gravel off by hand or use a gravel removing machine.

Marking. Cleaning usually erases some of the original chalk marks. Go over the whole area again with roofer's chalk, indicating where work is to be done.

Local repairs first. It is well to make all minor and local repairs before you undertake roof area treatments. Complete the work on parapets and copings, *for example,* before starting work on the roof proper. By employing this sequence, you can remove the debris resulting from chiseling of mortar joints along with other roof dirt.

Parapet cleaning should include wire brushing of any efflorescence from the brickwork. Do not begin repair of flashings until completion of all pointing or resetting of parapet units and repair of coping joints.

Summary. No manual, however complete, will insure effective roof maintenance or repair. The skill, adaptability, and good judgment of the men on the job are important factors in doing a good job.

Technically there should not be any mystery about how to obtain proper roof maintenance. It really requires three simple procedures: (1) regular inspection to catch defects early, (2) determination of the best remedies, and (3) proper application of corrective measures.

23 / ROOM CLEANING

(*Source:* Buildings Management Division, General Services Administration, Washington, D. C.)

Job Description

The job of room cleaning consists of emptying wastepaper baskets and ash trays, sweeping or vacuum cleaning offices, vacuum cleaning rugs, sweeping adjacent corridors, dusting all parts of rooms and corridor spaces that can be reached while standing on the floor, damp wiping glass and other specified surfaces; also the following duties when they occur in an assigned work area, cleaning washbasins, private toilets, phone booths and drinking fountains; taking proper care of the room cleaning equipment; and performing other work assigned by the supervisor. Employees assigned to room cleaning should be instructed to turn out lights not in the immediate area being cleaned, and close any windows found open. They should also notice burned out lights and other defective items and report these to their supervisor.

Equipment, Materials and Supplies

A. *Equipment:*

For room cleaning:

a. Bags for collecting wastepaper and trash
b. Metal container for collecting cigarette butts and ashes
c. Radiator dust brush
d. Putty knife
e. Dustpan
f. Dustpan broom
g. Dust cloth (cheesecloth treated with polyethylene glycol)
h. Wiping cloths, waffleweave
i. Sponge
j. Bucket
k. Sweep mop (treated with polyethylene glycol)
l. Corn broom
m. Equipment carrier
*n. Toilet brush
†o. Vacuum equipment or vacuum cleaner with hose, bare floor tool and carpet tool

* To be used where needed.
† Alternate equipment.

B. *Materials:*

For room cleaning:

a. White soap cake
b. Sand soap
c. Detergent (soapless synthetic)
d. Liquid metal polish

C. *Supplies:*

For private toilets:

a. Hand soap (liquid, powdered or bar)
b. Towels
c. Toilet tissue

Job Methods

| *Operation Steps* | *Explanation* |

Step 1—Preparation for Work

1. Obtain necessary equipment, materials and supplies.

Obtain necessary equipment, materials and supplies as listed under "Equipment, Materials and Supplies."

Operation Steps	*Explanation*
2. Proceed to assigned work area.	All the equipment, materials and supplies should be taken to the starting point of the assigned work area, to be used as needed.

Step 2—Emptying

1. Empty wastepaper baskets.	Starting at one end of the work area empty each wastepaper basket into the bag for collecting wastepaper and trash. Replace empty basket in original location. When bag is filled place in the corridor for collection. Do not place wastepaper baskets on furniture.
2. Empty ash trays.	While in the area collecting wastepaper, empty each ash tray into the metal container provided for that purpose and replace the ash trays in their original location. Ash trays should not be stacked on furniture.

PRECAUTIONS

1. In order to prevent possible injury, empty wastepaper baskets directly into the bag without handling the contents.
2. Ash trays should not be emptied into wastepaper baskets or into bags for collection of wastepaper and trash.
3. Avoid spilling any paper or ashes on floor.

Step 3—Dusting, Wiping and Polishing

1. Dust rooms and corridor space.	Use a treated cheesecloth to dust horizontal surfaces daily and vertical surfaces as required. Start with highest surfaces that can be reached while standing on the floor and work downward. Surfaces such as radiators, desks, file cabinets, office door panels and louvers and elevator hoistway doors should be dusted.
2. Damp wipe certain surfaces.	Use damp wiping cloth to wipe mirrors, window sills, ash trays, bookcase glass, door glass, door sills, telephone booth glass and all other glass surfaces except windows and transoms.

Operation Steps	*Explanation*
3. Polish metal surfaces.	Use metal polish and dry cloth to polish metal surfaces as directed by supervisor.

PRECAUTIONS

1. Do not disturb papers, books or other material on desks, tables or files.
2. Do not damp wipe desks or steel cabinets.

Step 4—Sweeping

1. Sweep corridor floors.	a. Use the radiator dust brush to clean radiators as required.
	b. During the sweeping operation, use the putty knife to scrape any gum or other foreign material from the floor.
	c. Use the treated sweep mop to sweep all corridor space, including telephone booths, within the area of the zone.
	d. Use the dustpan and dustpan broom to pick up any loose sweepings.
2. Sweep office floors.	a. Use the radiator dust brush to clean radiators as required.
	b. During the sweeping operation, use the putty knife to scrape any gum or other foreign material from the floor.
	c. Use the treated sweep mop to sweep all office floors in the assigned work area. Start at the far side of the room and sweep toward the door.
	d. Shake or vacuum the sweep mop as required during the sweeping operation.
	e. Use the dustpan and the dustpan broom to pick up any loose sweepings.

NOTE

In some instances where corridor scrubbing or floor waxing is scheduled in the assigned room cleaning area it will be necessary to sweep these areas before proceeding with the routine operations.

Operation Steps	*Explanation*

Step 5—(When central vacuum cleaning equipment is available.)

1. Clean corridor floors.	a. Use the radiator dust brush to clean radiators as required.
	b. During the vacuum cleaning operation, use the putty knife to scrape any gum or foreign material from the floors.
2. Clean with vacuum.	Use central vacuum equipment with hose and bare floor tool to clean all corridor floor space, including telephone booths, within the assigned area.
3. Clean office floors.	a. Use the radiator dust brush to clean radiators as required.
	b. During the vacuum cleaning operation, use the putty knife to scrape any gum or foreign material from the floor.
	c. Use central vacuum equipment with hose and bare floor tool to clean all office floor space in the assigned area.

NOTE

In some instances where corridor scrubbing or floor waxing is scheduled in the assigned room cleaning area it will be necessary to sweep these areas before proceeding with the routine operations.

Step 6

Replace furniture and equipment.	Replace in their proper locations any furniture, equipment, wastepaper baskets, and ash trays which have become disarranged during the cleaning operation.

Operation Steps	*Explanation*

Step 7

1. Clean wash basins.	a. Use sponge, warm water, and white soap to clean all wash basins in the assigned work area. If the basin is unusually dirty, use a sand soap in place of the white soap.
	b. Use a damp cloth to wipe all chromium fixtures.
2. Clean private toilets.	a. Use a toilet brush to flush and brushwash the interior of all water closets. Work the brush as far into the trap as it will reach. Use the brush to wash thoroughly under rounded inside rim of bowl.
	b. Wash and rinse water closet seat with a clean wet wiping cloth and white soap as necessary.
	c. Use a wet wiping cloth with white soap to clean all exterior porcelain surfaces. If necessary, use sand soap in place of white soap.
	d. Use a damp wiping cloth to rinse and damp-dry all external porcelain surfaces.
	e. Flush toilet to rinse inside of bowl.
	f. Use a damp wiping cloth to wipe all chromium fixtures.
3. Replenish supplies.	Replenish soap, towels and toilet tissue.

PRECAUTIONS

1. *Do not* use sand soap on chromium finish.
2. Report leaky spigots to supervisor.
3. *Be sure* to keep wiping cloths used in private toilets separate from those used elsewhere.

Step 8

Clean rugs.	Use a vacuum cleaner or central vacuum system to clean all rugs on the assigned work area. Use corn broom to lay nap of rug after vacuuming.

Operation Steps	*Explanation*
	Corn broom may be used to clean rugs in buildings with only a few rugs where a vacuum machine is unavailable.

PRECAUTIONS

1. Empty vacuum machine dirt receptacle after each day's use.

2. Do not strike furniture or baseboards with vacuum equipment.

Step 9

Clean drinking fountains.	Use warm water, white soap and a sponge to clean the bowl of all drinking fountains. Wipe remainder of fountain with a clean wiping cloth. Wipe all chromium fixtures with a clean dry wiping cloth.

PRECAUTION

Do not use sand soap on chromium finish.

Step 10—Return equipment, unused materials and supplies to storage

1. Equipment and unused materials and supplies.	Gather all pieces of equipment and unused materials and supplies, put them in their places on the equipment carrier and return to the gear storage room.
2. Equipment.	Clean and store all articles of equipment. (For details of cleaning and storage see "Care of Equipment.")
3. Unused materials and supplies.	Return all unused materials and supplies to proper storage place.

PRECAUTIONS

1. *Be sure* that covers are replaced securely on the containers of all materials and supplies.

2. *Be sure* to hang wiping cloths so that cloths used on toilets will not be confused with those used elsewhere.

Care of Equipment

A. *Bag for Collecting Wastepaper and Trash:*

   1. Do not empty ash trays into this bag.

   2. Do not allow bag to be caught in door or on sharp objects as this might damage the bag.

B. *Metal Container for Collecting Cigarette Butts and Ashes:*

1. Empty container daily.

2. Use a damp wiping cloth to wipe the inside and outside of container to remove all remaining ashes and other material.

3. Store container so that it will not be damaged.

C. *Radiator Dust Brush:*

1. Comb tufts of brush out straight.

2. Store brush on a flat surface so that tufts will be flat and straight.

3. Do not wet the brush.

4. Do not use a radiator brush for any purpose other than cleaning radiators.

D. *Putty Knife:*

1. Use the putty knife to scrape gum and like substances from the floor.

2. Do not use putty knife for a screw driver or other improper purposes.

3. Store putty knife in a clean dry place.

E. *Dustpan:*

1. Empty the dustpan into waste can and damp wipe it to remove all dirt and dust.

2. Hang up dustpan so that it will not be bent or otherwise damaged.

F. *Dustpan Broom:*

1. When broom becomes soiled, wash it in warm synthetic detergent solution and rinse it in clear water.

2. Store broom so that its weight is not on the bristles. (This is especially necessary when bristles are wet.)

G. *Dust Cloth (cheesecloth treated with polyethylene glycol):*

1. Keep the cloth clean and free from grime or oily material.

2. Wash the cloth once a week or as required in a warm synthetic detergent solution, and rinse in clear warm water.

3. Squeeze out the cloth and treat it with polyethylene glycol as directed.

4. Hang the cloth to dry where there is a good circulation of air.

H. *Wiping Cloths (waffleweave):*

1. Wash the cloths daily with white soap and warm water and rinse them in clear warm water.

2. Wring the water out of the cloths and hang them up to dry where there is a good circulation of air.

I. *Sponge:*

1. Rinse the sponge in clean water and squeeze out excess water after each day's use. Wash a very dirty sponge in a warm synthetic detergent solution, rinse in clean water and squeeze out excess water.

2. Store in a clean dry place.

J. *Bucket (12 qt.):*

1. Rinse and dry the inside and outside of the bucket.
2. Avoid hitting anything with the bucket as this might cause leaks.
3. In storing do not place heavy objects on the bucket which would tend to damage it.

K. *Sweep Mop (treated):*

1. Treated sweep mops should be shaken or vacuum cleaned both during use and at the close of operations each day. They should be kept clean and free from grime or oily material.
2. Once every two weeks remove the mophead, shake out as much dust as possible and wash mop in a warm solution of synthetic detergent. Rinse in clear warm water.
3. Wring out as much water as possible and treat with polyethylene glycol as directed.
4. Replace the mophead on the handle and hang the mop so that a good air circulation will be obtained to assist in drying the mop.

L. *Corn Broom:*

1. Soak a new broom overnight before using it the first time.
2. Never stand a broom on the straws when storing.
3. So that the broom will always be dry, store it where there is free circulation of air.
4. Wash broom as required, using a warm synthetic detergent solution and rinse in clear water and hang broom to dry.
5. Never use the broom when it is wet.
6. Do not use the broom for scrubbing.
7. When using the broom rotate it so that it will wear evenly.

M. *Equipment Carrier:*

1. Do not allow carrier to strike walls, radiators, or other objects.
2. Do not take the carrier into offices; *always* leave it in the corridor.

N. *Toilet Brush:*

1. Wash the brush daily using a warm synthetic detergent solution.
2. Shake out excess water.
3. Hang up where air circulation is good so the brush will dry.

O. *Vacuum Cleaner and/or Equipment:*

1. Empty the dirt receptacle. Wipe machine after each day's use, with a dry wiping cloth. Wind cord loosely on the hooks provided.
2. Do not run machine over electric cord or pull against cord when it is plugged into receptacle.
3. Do not allow machine to strike walls, radiators or other objects.
4. Report needed servicing or repairs to the supervisor.
5. After use, return hose, carpet tools, and extension tubes to a clean dry storage room where they will not be damaged.

6. Do not allow doors to be closed on hose or electric cord.

7. Keep floor tool and carpet tool clean at all times.

P. *Instructions for Treating Mops (all kinds) and Dust Cloths:*

1. Treatment of mops and dust cloths with polyethylene glycol:

a. *Instructions for Mixing solution:*

Polyethylene glycol is not to be used as it is purchased but should be dissolved in water. Use three parts of water to one part of polyethylene glycol. After thorough mixing this solution can be applied to mops and dust cloths.

b. *Instructions for Treating Mops and Cloths:*

The mop or cloth should be hung up and the glycol solution sprayed on the cloth or on the head of the mop. Approximately one fluid ounce should be sprayed on a three-quarter pound mop, approximately one and one-half fluid ounces on a one pound mop, approximately two ounces on a one and one-quarter pound mop, and approximately one fluid ounce should be sprayed on a square yard of dust cloth. For mops of different weights the solution should be applied approximately at the rate specified above. The mop and/or cloth should be treated the day before they are to be used.

Performance Inspection Guide

A. *Trash Removal:*

1. All wastepaper baskets should be empty and in place, clean and ready for use.

2. All ash trays should be empty and in place, clean and ready for use.

3. Ashes and trash should not be left on floor.

4. The trash bags, when filled, should be conveniently and neatly placed for collection.

B. *Sweeping or Vacuum Cleaning:*

1. There should not be any dirt left in corners, behind radiators, under furniture or behind doors.

2. Baseboards, furniture and equipment should not be disfigured or damaged during the cleaning operation.

3. There should not be any dirt left where sweepings were picked up.

4. Furniture and equipment moved during sweeping should be replaced.

5. All radiators should be free of dirt.

6. There should be no trash or foreign matter under desks, tables or chairs.

C. *Dusting:*

1. There should not be any dust streaks on desks or other office equipment.

2. Woodwork, after being properly dusted, should appear bright.

3. Corners and crevices should be free from any dust.

4. There should not be any oily spots or smudges on walls, caused by touching them.

5. When inspected with a flashlight there should be few traces of dust on any surface.

6. Radiators, window sills, door ledges, door frames, door louvers, elevator hoistway doors, window frames, wainscoting, baseboards, columns, and partitions should be free of dust.

D. *Damp-wiping:*

Mirrors, window ventilators, ash trays, door glass, and all other glass that can be reached while standing on the floor, should be clean and free of dirt, dust, streaks and spots. (This job does not include window washing.)

E. *Clean Rugs:*

1. Rugs should be thoroughly clean and free from dust, dirt and other debris.

2. There should be no trash or foreign matter under desks, tables or chairs.

3. Any furniture moved during rug cleaning should be replaced.

F. *Clean Wash Basins:*

1. There should be no stains or spots on surfaces of wash basins.

2. Wash basins should be clean and bright.

3. Wall near wash basins should be free from spots or smears.

4. The floor should be free of water or soap solution.

5. All metal such as spigots and other hardware should be clean and bright.

G. *Clean Private Toilets:*

1. Toilet bowls and other porcelain surfaces should be clean and bright.

2. Towel, toilet paper and soap dispensers should be filled, clean and in good working order.

3. There should be no soil marks on the fixtures or walls.

4. The floor, wainscoting, and partitions should be clean.

H. *Clean Telephone Booths:*

1. Dust, dirt, and other foreign material shall have been cleaned from telephone booths.

2. Glass shall be clean and free of fingerprints, streaks and spots.

I. *Clean Drinking Fountains:*

1. The porcelain or white enamel surfaces of drinking fountains should be clean and free of stains.

2. The wall and floor around the drinking fountain should be free of spots and water marks.

3. All other surfaces of the fountain should be free of spots, stains and streaks.

Frequency and Production

A. *Frequency and Production Schedule:*

Room Cleaning
(Exclusive of rugs)

Frequency	Daily
Production Rate	14,000 sq. ft., plus
per Man-day	adjoining corridor area

B. *Qualifying Factor:*

Frequency rate may be modified as directed by supervisor.

Staffing

The man-day requirements of labor for office cleaning can be determined by use of the following formula:

$$\text{Man-days (of labor)} = \frac{\text{Quantity}}{\text{Production}} \times \frac{1}{\text{Frequency}}$$

where quantity equals the square feet of office area exclusive of corridors and exclusive of special areas, such as file rooms and laboratories which must be considered separately, and where frequency is the number of work days between operations.

If the frequency and/or production rate values as given in the Frequency and Production Schedule above have been modified to meet existing conditions at a particular location, then these modified values should be used in the above formula for the computation of the labor requirement.

For each rug twelve feet by fifteen feet in the assigned area, 175 square feet of office area should be subtracted from the production rate of 14,000 square feet.

SAMPLE CALCULATION

Consider a building that has 110,000 square feet of room area to clean daily which has no rugs, and which is occupied as normal office space so that the average cleaning values from the frequency and production schedules may be used.

$$\text{Man-days} = \frac{110,000}{14,000} \times \frac{1}{1} = 7.86 \text{ jobs}$$

To this figure must be added a factor to compensate for leave in order to arrive at the total number of positions needed to accomplish the work.

Conversion factor including compensation for leave = 1.15
Conversion factor × jobs = total positions required

Nine positions required to clean 110,000 square feet of office space daily, exclusive of rugs, plus adjoining corridor area.

• Where a room cleaning assignment does not require the full time of an employee, other duties should be assigned to secure 100 per cent utilization of his time.

Equipment and Materials Allocation and Replacement

Items	Allotment per Cleaner	Replacement or Reissue
Bags, wastepaper and trash	As required	As required
Broom, corn	1	1 per year
Broom, dustpan (hearth size corn broom)	1	1 per 2 years
Brush, radiator, dust	1	1 per year
Brush, toilet	1	1 per year
Bucket (12 qt.)	1	1 per 2 years
Cheesecloth	2 yards	52 yards per year
Cloth, wiping (waffleweave)	2	24 per year
Container (cigarette butts and ashes)	1	1 per 2 years
Detergent (soapless synthetic)	1 pound	50 pounds per year
Dustpan	1	1 per 2 years
Equipment carrier	1	1 per 10 years
Knife, putty (1¼″)	1	1 per 2 years
Mop, sweep (complete)	1	1 per 3 years
Mop, sweep (refill only)	None	1 per year
Polish, metal (liquid)	1 pint	4 pints per year
Polyethylene glycol	1 pint solution	1 gallon glycol per year
Soap, sand	1 cake	
Soap, white (6 oz. cake)	1 cake	3 cakes per year
Sponge, scrap cellulose	1	25 cakes per year
Vacuum machine and attachments, (Domestic Type II, Class B, Size 2, Heavy Duty)	1	6 per year 1 per 10 years*

* Does not apply to replacement of vacuum cleaner attachments.

The allotment and replacement or reissue quantities are based on one room cleaner working regularly eight hours each working day on the job.

• Where other duties are assigned to a room cleaner to secure 100 per cent utilization of his time, consideration should be given to a reduction in the allotment and replacement quantities in relative proportion to the time spent per day on the other work.

24 / STAIRWAY CLEANING

(*Source:* Buildings Management Division, General Services Administration, Washington, D. C.)

Job Description

Stairway cleaning includes all cleaning work inside the confines of the stairwell that can be reached from the normal walking surface, such as sweeping stair landings and steps; removal of gum or other foreign substances; dusting stair railings, fire apparatus, doors, ledges, radiators and grilles; cleaning and polishing handrails, glass surfaces, metal doorknobs, and other metal and wooden surfaces; wall spotting; mopping or scrubbing stair landings and steps; thoroughly drying all water from such areas as a result of the mopping or scrubbing operation; and proper care and maintenance of the stair cleaning equipment. Employees assigned to this job should be instructed to notice lights out, loose railings and similar service defects, and to report them to their supervisors.

Equipment and Materials

A. *Equipment:*

   1. For sweeping and dusting.

      a. Dustpan broom
      b. Dustpan
      c. Putty knife (1¼")
      d. Dust cloths (polyethylene glycol treated cheesecloth)
      e. Radiator dust brush

   2. For polishing, wall spotting and mopping or scrubbing.

      a. Cotton mop (1¼ lb.)
      b. Deck scrubbing brush
      c. Bucket (with squeeze-type wringer)
      d. Wiping cloths (cheesecloth)
      e. Chamois

B. *Materials:*

   1. For polishing, wall spotting and mopping.

      a. Warm water
      b. Liquid metal polish
      c. Lemon oil polish
      d. Detergent
      e. White soap (cake)

   2. For scrubbing.
      a. Scouring powder
      b. White soap (cake)
      c. Warm water

Job Methods

Operation Steps	Explanation
A. *Sweeping and Dusting:*	
Step 1	
1. Obtain equipment needed.	Obtain equipment as needed for sweeping and dusting. (See list under "Equipment and Materials".)

Operation Steps	*Explanation*
2. Proceed to work zone.	Carry sweeping and dusting equipment to top landing where work is to start.

Step 2

1. Sweep landings and stairs.	Using the janitor set of broom and pan, start top landing and sweep all landings and stairs down to bottom landing, collecting the sweepings at each landing and each step into the dust pan.
2. Remove gum or foreign substances.	Use putty knife to remove any gum or other foreign substances from landings and steps during sweeping operation.
3. Dust radiators.	In the course of the sweeping operation use radiator brush to remove dirt and dust particles from radiators in stairway.

PRECAUTION

Make sure that all loose dirt is removed from the corners of the stair treads in the sweeping operation.

Step 3

1. Dust with treated dustcloths.	Start at bottom landing, and dust stair railings, fire apparatus, doors, ledges, radiator tops and grilles in the stairwell up to top landing. Use treated dustcloth to dust all surfaces that can be reached by standing on landings or steps.

PRECAUTION

Avoid using treated dustcloth on stairway walls.

B. *Polishing, Wall Spotting and Mopping or Scrubbing:*

Step 4

1. Obtain equipment and materials needed.	a. Obtain equipment and materials needed for this operation. (See list under "Equipment and Materials".)

Operation Steps

Explanation

b. Place mopping or scrubbing equipment and materials at top landing clear of passageway until ready for use.

2. Secure clean bucket of warm water.

Obtain a bucket of warm water from the nearest service sink for use in cleaning glass surfaces and walls.

Step 5

1. Clean glass surfaces.

Use a clean, damp chamois to clean glass surfaces in doors and fire cabinets.

2. Clean and polish metal surfaces.

Use clean cloths, and metal polish as directed, to clean and polish metal handrails, metal door knobs and other metal surfaces.

3. Clean and polish wood surfaces.

Use clean cloths and lemon oil polish to clean and polish wooden handrails and other wood surfaces as directed.

PRECAUTION

Metal polish is not to be used on painted, lacquered, or other metal surfaces that have been specially treated.

Step 6

1. Normal wall spotting.

Use a clean, damp cloth to spot clean the wall up to a height that can be reached while standing on the stair landing or steps.

2. Removal of stubborn spots.

Use white soap on the damp cloth as an aid to removing any stubborn spots from the wall that cannot be cleaned off by normal wall spotting.

PRECAUTIONS

1. This step is for spot removal only and not to wash entire wall.

2. Wring cloths out after rinsing in bucket of clear water to avoid an excessive amount of water on the wall. Rinse cloths out often.

3. Be sure to wipe any soap particles from wall with a clean, damp cloth.

Operation Steps	*Explanation*

Step 7 (See Step 8.)

1. Mix warm water and detergent solution for mopping.	Mix one-half to three-quarters of an ounce of detergent per gallon of water.
2. Mop stairs and landings.	Wet mop with solution from bucket, start at top landing and mop stairs down to bottom landing. As may be necessary, grasp a few mop strands in your hand and rub the dirt from the stair tread corners during the mopping operation, and also remove any stubborn spots from the landings and stairs that are not cleaned off by normal mopping.
3. Dry stairs and landings.	Using mop, clean and well wrung out, thoroughly dry all water from the stair landings, steps, risers and base-boards.
4. Keep mop clean.	Rinse and wring out mop frequently in a clean bucket of water in order to properly clean and dry the stairs and landings.

NOTE

There should not be any water traces remaining after this step.

PRECAUTIONS

1. Be sure to rinse stairs and landings thoroughly after the scrubbing operation.

2. Avoid using an excessive amount of solution or water during the mopping, scrubbing or rinsing.

3. Take care not to splash or scar the walls, baseboards or stair risers with the equipment.

4. Do not allow the solution or water to run over the stair tread edges into the stairwell.

5. As may be appropriate, suitable warning should be given to the building occupants against the danger of slipping on the wet landings and stairs.

Operation Steps   *Explanation*

Step 8—When scrubbing is to be substituted for mopping
(See Step 7.)

1. Scrub landing.

Sprinkle scouring powder over landing. Use wet deck scrubbing brush, with handle, to thoroughly scrub landing.

2. Scrub stairs.

Remove handle from deck brush, sprinkle scouring agent on stairs and scrub thoroughly by hand.

C. *Returning Equipment and Materials to Storage:*

Step 9

1. Return all equipment and unused materials to storage location.

Gather all pieces of equipment and materials and proceed to proper storage location.

2. Mop and scrub brush.

Rinse out mop and scrub brush, wring or shake dry of excess water and place in proper storage position.

3. Cloths and chamois.

Rinse out all cloths and chamois in clean water and place for drying and storing.

4. Bucket and wringer.

Empty bucket, rinse and return bucket and wringer to proper storage position.

5. Brushes, dustpan and putty knife.

Clean all articles and return to proper storage position.

6. Unused materials.

Return all unused materials to proper storage place.

PRECAUTION

Make sure that lids are placed securely on the polish, detergent and scouring powder containers.

Care of Equipment

A. *Bucket and Mop Wringer:*
   1. Rinse and dry the bucket and wringer thoroughly after each use.
   2. Wipe the outside of the bucket free of any dirt.

3. Avoid striking the bucket and wringer against objects as such action tends to cause leaks in the bucket and shorten the life of the equipment.

4. Store in dry place so as to have no weight bearing on the bucket or wringer.

B. *Chamois:*

1. Rinse chamois in clean water and squeeze out excess water after each use.
2. Stretch chamois to original shape before placing to dry.
3. Avoid contaminating chamois with oils, polish or cleaning materials.

C. *Dust Cloths (polyethylene glycol treated):*

1. Thoroughly shake or vacuum dust cloths to remove loose dust.
2. When dirty, wash in warm detergent solution, (mix one ounce of detergent per gallon of warm water) rinse in clean water and wring or squeeze to remove excess water.
3. To treat new or washed cheesecloth, spray or sprinkle approximately one fluid ounce of polyethylene glycol solution (mixed in the proportion of three parts of water to one part of polyethylene glycol) to a square yard of cheesecloth. The cloth should be treated at least eight hours before using.

D. *Dustpan:*

1. Clean dustpan of loose dirt and dust before storing.
2. Hang by handle so as not to allow any damaging weight to be put on pan during storage.
3. Avoid getting the dustpan wet at any time.
4. In dumping the dustpan, avoid banging it with force against the waste container as such action tends to damage sides and edges of the pan.

E. *Dustpan Broom:*

1. Comb out broom bristles periodically.
2. Wash oily or dirty broom in warm water and detergent solution. Shake the broom and hang it up to dry. Allow to dry thoroughly before using.
3. When not in use, hang the broom in a clean, dry place so that the weight of the broom is not upon the bristles.

F. *Deck Scrubbing Brush:*

1. Rinse out in clean water after each use and shake dry of excess water.
2. Place for drying and storage in a position so that the bristles do not bear the weight of the brush.
3. Avoid knocking the brush against objects or walls as such action tends to split the brush block.

G. *Mop:*

1. After each use and before storage, rinse mop in clean water and wring or squeeze to remove excess water.
2. When dirty, wash in warm detergent solution, (mix one ounce detergent per gallon of warm water); rinse in clean water and wring or squeeze to remove excess water.

3. Fluff the mop strands by shaking.

4. Store mop with mophead up to allow better drying.

H. *Putty Knife:*

1. Use the putty knife only for removal of gum and other foreign substances during sweeping. Do not use a screw driver or hammer.

2. Keep the knife dry at all times.

3. Store ir a clean, dry place either by placing on a shelf or hanging on a nail.

I. *Radiator Dust Brush:*

1. Comb out brush bristles periodically.

2. Avoid wetting the dust brush at any time.

3. Store brush either by laying flat with bristles out straight or by hanging so bristles stand free of surrounding objects.

4. Do not use the radiator dust brush for any other purpose than for dusting radiators.

J. *Wiping Cloths:*

1. Rinse all cloths used, or when dirty wash them in warm detergent solution, and wring out.

2. Spread or hang cloths to dry in their proper storage places.

3. Avoid leaving wet, dirty cloths piled together in the storage cabinet.

Performance Inspection Guide

A. *Sweeping and Dusting:*

1. Stair landings, steps and all corners of stair treads should be free of loose dirt or dust streaks after sweeping.

2. Stair railings, fire apparatus, door mouldings, ledges, radiators and grilles should be dust free after dusting. The dust should have been removed rather than pushed around.

B. *Cleaning, Polishing and Wall Spotting:*

1. Glass surfaces should be clean and free of any smudges, finger marks, and dirt.

2. Handrails, door knobs and other surfaces should be clean and polished to an acceptable lustre.

3. Walls up to a standing height should be free of finger marks and other dirt spots of any kind.

C. *Mopping and Scrubbing:*

1. Stair landings and steps should be free of loose and/or caked dirt particles and should present an over-all appearance of cleanliness after mopping or scrubbing.

2. Walls, baseboards and stair risers should be free of water marks, scars

from the equipment striking the surfaces and splashing from the cleaning solution.

3. All surfaces should be dry and the corners and cracks clean after dry mopping.

D. *Equipment and Materials:*

1. All pieces of equipment should be clean and properly placed in their assigned storage locations.

2. All metal polish, furniture polish, soap and other similar materials left over from the stairway cleaning operation should be returned to their assigned storage locations.

Frequency and Production

A. *Frequency and Production Schedule:*

	Sweeping and Dusting	*Polishing Fixtures and Wall Spotting*	*Mopping or Scrubbing*
Frequency	Daily*	Weekly*	Weekly*
Production Rate (flights of stairs)	60	20	20

* Or as directed by supervisor.

B. *Qualifying Factors:*

1. *Frequency*

   a. The frequency of stairway cleaning is dependent upon weather conditions, type of building, type of occupancy, and amount of traffic. *For example*, a flight of stairs between the first and second floors that is in daily, constant use by the public as well as by the cleaning occupants would require a greater frequency of cleaning, especially in inclement weather, than would a flight of stairs between the sixth and seventh floors in the same building that mainly serves the building occupants or is used only in case of emergency.

   b. The frequencies given in the schedule are for a normal standard of cleaning under average conditions.

2. *Production Rate per Man-day*

   a. The term "flight of stairs" as used in the schedule is taken to mean an average stairway, approximately four feet in width and with its accompanying fixtures, such as railings, fire apparatus, and other items extending from one floor of a building to the next floor, and including the stair landing at one floor as well as any intermediate landings in between floors.

b. Due to the many and varied types and sizes of stairways in buildings, the production rate as established may be subject to modification at the various locations. It does, however, provide a standard from which such deviation should be figured.

Staffing

The man-day requirements of labor for stairway cleaning can be determined by use of the following formula:

$$\text{Man-days (of labor)} = \frac{\text{Quantity}}{\text{Production}} \times \frac{1}{\text{Frequency}}$$

where quantity is the actual total flights of stairs in the building; production is the number of flights of stairs cleaned per man-day; and frequency is the number of work days between operations.

If the frequency and/or production rate values as given in the Frequency and Production Schedule have been modified to meet the existing conditions at a particular location, then those modified values should be used in the above formula for the computation of the labor requirement.

SAMPLE CALCULATION

Consider a building that has fifteen flights of stairs and is of such construction and subject to such traffic conditions that the average cleaning values from the Frequency and Production Schedule may be used.

$$
\begin{array}{l}
\text{Man-days} = \overset{\text{For Sweeping}}{\underset{\text{and Dusting}}{\frac{\text{Quantity}}{\text{Production}} \times \frac{1}{\text{Frequency}}}} + \overset{\text{For Polishing}}{\underset{\text{and Mopping or Scrubbing}}{\frac{\text{Quantity}}{\text{Production}} \times \frac{1}{\text{Frequency}}}} \\[2ex]
\qquad\quad = \frac{15}{60} \times \frac{1}{1} \quad + \quad \frac{15}{20} \times \frac{1}{5} \\[2ex]
\qquad\quad = \quad 0.25 \qquad\qquad + \qquad 0.15 \\[1ex]
\qquad\quad = \quad 0.40 \text{ man-days of labor required (jobs)}
\end{array}
$$

To this figure must be added a factor to compensate for leave in order to arrive at the total number of positions needed to accomplish the work.

Conversion factor including compensation for leave = 1.15
Conversion factor × jobs = total positions required
1.15 × 0.40 = 0.460 positions required

• Where the normal stairway cleaning duty does not require the full time of the employee, other related duties such as elevator cleaning, toilet cleaning or some similar job should be assigned to the employee to secure 100 per cent utilization of his time.

Equipment and Materials Allocation and Replacement

Items	Allotment per Cleaner	Replacement or Reissue
Bucket	1	1 per year
Chamois	1	1 per year
Cheesecloth	1 yard	18 yards per year
Deck scrubbing brush	1	1 per year
Deck scrubbing brush handle	1	1 per year
Detergent	1 pound	50 pounds per year
Dustpan	1	1 per 2 years
Dustpan broom (hearth size corn broom)	1	1 per year
Lemon oil polish	1 pint	4 pints per year
Metal polish (liquid)	1 pint	4 pints per year
Mop (1¼ lbs.)	1	2 per year
Mop handle	1	1 per year
Mop wringer (squeeze type)	1	1 per 2 years
Putty knife (1¼")	1	1 per 2 years
Radiator dust brush	1	1 per year
Scouring powder (for floors)	1 pound	20 pounds per year
White soap (6 oz. cake)	1 cake	25 cakes per year

The allotment per cleaner and replacement or reissue quantities in this standard are based on one stair cleaner having continuous stairway cleaning duty.

• Where the situation exists that other related duties are assigned to the stairway cleaner to secure 100 per cent utilization of his time, consideration should be given to a reduction in the replacement quantities in accordance with the actual amount of stairway cleaning assigned.

25 / STORAGE CLOSETS

One of the major causes of lost work time in the cleaning operation is the disregard of the fact that even the best supplies, unless easily available will not raise productivity to its maximum. When a worker must travel to the basement or some other out of the way storage area, an average of twenty minutes of work time can be lost. Multiply this figure by the number of employees who daily wander through the building searching for cleaning tools, and you will have a surprising number of valuable man–hours.

The best answer to this problem is a storage closet located on each floor, or strategically located in large areas, to keep travel time to a minimum. Each well-ventilated closet should be designed and laid out properly to include: buckets, wringers, mops, dustpans, counter brush, broom, rubber gloves, toilet brush, and dry mop. It should also include supplies such as: waxes, detergents, hand soap, toilet tissue, hand towels, deodorants, toilet bowl cleaners, cleanser, and polishes or other products to maintain a station or area.

An ideal storage closet should be so arranged that every item has its own assigned space, and can accommodate such equipment as vacuum cleaners, floor machines, or other similar items if required. It should be designed to encourage neatness, and be provided with a drain and low utility sink with a stainless steel rim to protect porcelain finish. Lockers can also be provided for the employees to store their personal property.

Storage Closet Provisions

1. Low utility sink with stainless steel rim to protect porcelain finish.
2. Ample shelving.
3. Mopping outfit in storage position under shelving.
4. Floor machines and vacuums stored away from sink, and in a location easily removable.
5. Fitting and tools mounted on peg board.
6. Mop stored in assigned position.
7. Four-inch space between the mop and the wall to hold it out.
8. Aluminum or ceramic drip tray placed under mop.
9. Twenty-four-inch wide door louver (ventilation).
10. Ceramic tile on concrete floor with drain.
11. Bibb (threaded) faucet, with brace.
12. Length of hose for washing equipment.
13. Minimum of 45 foot candles.
14. Small bulletin board.

26 / TOILET CLEANING AND SERVICING

(*Source:* Buildings Management Division, General Services Administration, Washington, D. C.)

Job Description

The job of toilet cleaning and servicing includes all the cleaning work inside toilet rooms which can be reached while standing on the floor, including emptying the waste receptacles; servicing the soap dispensers, toilet paper holders, seat cover dispensers if provided, and paper towel dispensers; dusting window sills, ledges, grilles and similar items; cleaning the tile walls, mirrors, shelves, dispensers, receptacles; stall partitions and doors, wash basins, water closets and urinals; polishing metalwork as directed; sweeping and mopping or scrubbing the floors; and the proper care and maintenance of the toilet cleaning equipment. This work is easier if performed at night while the tenants are out of the building. Sufficient supply dispensers should be provided to reduce daytime servicing to a minimum. Each employee assigned to this job should be instructed to notice burned out lights, leaky fixtures, and other defective items and report them to his supervisor.

Equipment, Materials and Supplies

A. *Equipment:*

For toilet room.

a. Bag for collecting wastepaper
b. Bucket
c. Cellulose sponges
d. Chamois
e. Cheesecloth
f. Dustpan
g. Dustpan broom
h. Equipment carrier
i. Floor sweep
j. Mops, cotton
k. Mop tank
l. Polishing machine with scrubbing brushes
m. Putty knife
n. Radiator dust brush
o. Rubber gloves
p. Toilet brush or toilet bowl mop
q. Wiping cloths, waffleweave

B. *Materials:*

For cleaning toilet room and fixtures:

a. Grit soap
b. Metal polish
c. Multipurpose cleaner
d. Paper bags (large and small)
e. Scouring powder
f. White soap
g. Bowl cleaner

C. *Supplies:*

For servicing.

a. Hand soap (liquid, powdered, or bar)
b. Paper towels
c. Seat covers *
d. Toilet tissue

* Optional

Job Methods

Operation Steps	*Explanation*

Step 1

1. Obtain necessary equipment, materials and supplies.	Obtain necessary equipment, materials and supplies. (See list under "Equipment, Materials and Supplies.")
2. Proceed to assigned work area.	All the equipment, materials, and supplies should be taken to the first toilet room where work is to begin.

NOTE

Items for cleaning toilet room fixtures must not be intermixed. The following tabulation shows the items this section prescribes for cleaning the various fixtures.

Fixtures	*To Clean* (Item)	*To Dry and/or Polish* (Item)
A. Mirrors and shelves	Green sponge	Chamois
Seat cover dispenser	Green sponge	Chamois
Soap dispenser	Green sponge	Chamois
Towel dispenser	Green sponge	Chamois
Wash basin	Green sponge	None
Wash basin chrome	Green sponge	Chamois
Waste towel receptacle	Green sponge	None
B. Urinal exterior porcelain	Yellow sponge	Waffleweave cloth
Water closet exterior porcelain	Yellow sponge	Waffleweave cloth
Water closet seat and hinges	Yellow sponge	None
C. Urinal interior	Toilet brush, straight head or mop	None
Water closet interior	Toilet brush, straight head or mop	None
D. Sanitary receptacle	Toilet brush, curved head or mop	Cheesecloth
E. Walls and stalls	Pink colored sponge	Damp dry with sponge

Step 2—Collect Used Towels and Other Trash

1. Empty waste towel containers.	Remove contents of waste towel receptacles.
2. Pick up wastepaper and trash.	Pick up all wastepaper and trash from floor and place in bag for collecting

Operation Steps	*Explanation*
	wastepaper. When the bag is full, tie securely and place it on the equipment carrier outside the toilet room.

PRECAUTIONS

1. Be careful of broken glass, razor blades, or other sharp articles that may be in the waste towel container.
2. Place glass and hard or sharp objects in a separate container, *not* in the bag for collecting wastepaper.
3. Wait for toilet room occupants to leave, and do not let others enter until your work is completed and you are ready to leave.

Step 3—Service Dispensers

Check and service dispensers of supplies.	a. Check soap dispensers and fill with soap of the type used (liquid, powder or bar).
	b. Check toilet tissue and seat cover dispensers and add supplies needed.
	c. Check the paper towel dispensers and put in towels needed to fill them.

PRECAUTION

Do not put supplies anywhere other than in the dispensers provided.

Step 4—Dust and Damp Wipe

1. Dust radiators, sills, and other items.	Use radiator dust brush and/or dust cloth to dust radiators, window sills, ledges, grilles, and other items located in the toilet room.
2. Damp wipe dispensers and waste towel receptacle.	a. Wipe towel, tissue, napkin and seat cover dispensers, and the waste towel receptacle, with a damp sponge (green colored) as required.
	b. Use a chamois if necessary to dry after damp wiping.

Step 5—Sweep Floor

Remove trash from floor.	a. Use floor sweep to sweep the floor.

Operation Steps   *Explanation*

b. Remove gum, tar or similar substances from floor areas with putty knife during sweeping.

c. Use dustpan and dustpan broom to pick up sweepings.

Do not sweep trash into the corridor or other adjoining area.

Step 6—Clean and Polish

1. Clean mirrors and shelves.

a. Use a damp sponge (green colored) to wipe all mirrors.

b. Use sponge (green colored), white soap and water to wash and rinse shelves.

c. Dry and polish mirrors and shelves with chamois.

2. Clean wash basins and soap dispensers.

a. Clean wash basins with a clean sponge (green colored), white soap and water. If basins are unusually dirty, use grit soap.

b. Use sponge (green colored) to clean the outside of the soap dispensers.

c. Dry and polish soap dispensers and chromium fixtures with a chamois as necessary.

3. Polish metal.

As directed use metal polish and cheesecloth to polish metalwork.

PRECAUTION

Grit soap should not be used on chromium finish.

Step 7—Clean Sanitary Receptacles

Empty and clean sanitary receptacles.

a. Remove paper bags, together with the contents, from the sanitary receptacles daily. As each bag is removed, close it by folding over the top, and place it in a clean paper bag, ready for disposal. Fasten the collection bag securely to prevent spillage.

Operation Steps   *Explanation*

b. Clean sanitary receptacle daily, both inside and out, with a curved toilet brush (or toilet bowl mop) and a solution of multipurpose cleaner.

c. Wipe receptacle with dry cheese-cloth.

d. Put in new paper bag.

NOTE

Prepare fresh cleaning solution for each toilet room. Mix cleaner and warm water in proportions recommended by manufacturer's instructions on container label.

PRECAUTIONS

1. Wear rubber gloves when cleaning sanitary receptacles.

2. The bags removed from sanitary receptacles, together with their contents, should be disposed of daily by burning or being hauled to the dump.

Step 8—Clean Water Closets and Urinals
(For cleaning with chemicals, see the discussion at the end of this section.)

1. Clean water closets.

a. Flush toilets before starting to clean. Then use a toilet brush (or toilet bowl mop) and multipurpose cleaner solution to brush wash interior of all water closets. Work the brush or mop as far into the trap as it will reach. Wash thoroughly under the rounded inside rim of the toilet bowl. Flush toilet to rinse inside of bowl.

b. Wash water closet seats and hinges with clean wet sponge (yellow colored) and multipurpose cleaner. Rinse thoroughly with clean water. Leave seats in raised position.

c. Use a wet sponge (yellow colored) with multipurpose cleaner to clean all exterior porcelain surfaces. If necessary, use grit soap. Use the damp sponge (yellow co-

Operation Steps	*Explanation*
	lored) to rinse and damp dry all external porcelain surfaces.
	d. Use the damp sponge (yellow colored) to wipe all the chromium fixtures. Polish with a dry cloth as necessary.
2. Clean urinals.	a. Flush urinals. Then use a toilet brush (or toilet bowl mop) and multipurpose cleaner to wash interior of all urinals. Wash thoroughly under the rounded rim of the urinal bowl. Flush urinal to rinse interior of bowl.
	b. Use a wet sponge (yellow colored) with multipurpose cleaner to clean all exterior porcelain surfaces. If necessary, use grit soap. Rinse as required.
	c. Pour approximately a cupful, or as otherwise directed, of multipurpose solution into the urinal. *Do not flush.*
	d. Use a wiping cloth to dry all external porcelain surfaces.
	e. Use the damp sponge (yellow colored) to wipe all chromium fixtures. Polish with a dry cloth as necessary.

PRECAUTIONS

1. Grit soap is not to be used on chromium finish.
2. Report leaky spigots and other defective equipment to the supervisor.
3. Keep sponges, cloths and brushes or mops that are used on water closets, urinals and sanitary receptacles separate from those used in other cleaning operations.
4. Wear rubber gloves when cleaning water closets and urinals.

Step 9—Damp Wipe Walls and Stalls

Clean tile walls, stall partitions and doors.	Use a damp sponge (pink colored) to damp wipe tile walls, stall partitions and doors. As necessary, wash them

Operation Steps   *Explanation*

with a solution of multipurpose cleaner. After washing tile walls, partitions, and doors, wipe them dry.

Step 10—Clean Floors (Mop)
(For additional details see "Floor Mopping".)

Mop floors.

a. In accordance with instructions on the container label, mix multipurpose cleaner and warm water in one bucket of the twin mop tank. Fill the second two-thirds full of clear water.

b. Use mop and the warm solution to thoroughly clean floor area. Dip mop into solution and wring out excess solution. Push mop around the edges of the floor, being careful to remove all dirt from the corners and around pipes. Then mop remainder of floor area. Avoid splashing walls and baseboards.

c. Wring mop and go back over mopped area to pick up excess cleaning solution.

d. Use a second mop and clean water to rinse and dry the mopped floor. Frequently rinse and wring mop during this operation.

Step 11—Clean Floors (Scrub) (See Step 10.)

Scrub floors.

a. Sprinkle floor with scouring powder, and wet with warm water.

b. Use a small polishing machine with scrubbing brushes to scrub floor.

c. Use a mop and clean water to rinse and dry scrubbed floor. Frequently rinse and wring mop during this operation.

PRECAUTIONS

1. Do not use so much water that it tends to splash or run beneath the floor covering.

2. Be careful that equipment does not bump against walls, doors, or stall partitions.

3. Change rinse water frequently.

4. Always start the mopping or scrubbing operation at the wall opposite the door and work toward the door.

5. Pay special attention to corners and other places where dirt accumulates.

Operation Steps   *Explanation*

Step 12—Return Equipment,
Materials and Supplies to Storage

Return equipment and unused materials and supplies.

a. Gather all pieces of equipment and unused materials and supplies and put them in their place on equipment carrier.

b. Return them to gear storage room.

c. Dispose of wastepaper and other waste.

d. Clean and store all items of equipment. (For details of cleaning and storage see "Equipment Care.")

e. Return all unused materials and supplies to proper storage place.

PRECAUTIONS

1. Be sure that covers are replaced securely on all containers of materials and supplies.

2. Be sure to hang wiping cloths so that cloths used on water closets and urinals will not be confused with cloths used elsewhere.

Care of Equipment

A. *Bag for Collecting Wastepaper:*

1. Do not put sharp objects or sweepings into the wastepaper collection bag. Place them in a separate container, such as a cardboard box.

2. Do not allow bag to be caught in door or on sharp objects as this might damage it.

B. *Broom, Dustpan:*

1. When broom becomes soiled, wash it in warm synthetic detergent solution and rinse in clear water.

2. Store broom by hanging in a clean dry place so weight of brush is not on the bristles.

C. *Brush, Radiator Dust:*
1. Comb bristles of brush periodically.
2. Avoid wetting dust brush.
3. Store brush on a flat surface so bristles will be flat and straight.

D. *Brush, Toilet or Toilet Bowl Mop:*
1. Wash brush or mop daily, using a warm multipurpose cleaner solution, and shake out excess water.
2. Hang to dry where air circulation is good.

E. *Bucket (12 qt.):*
1. Rinse and dry bucket, inside and out, after each use.
2. In storing do not place heavy objects on the bucket.

F. *Cloths, Wiping:*
1. Wash wiping cloths daily in a warm multipurpose cleaner solution and rinse them in clear warm water.
2. Wring water out of cloths and hang them up to dry where air circulation is good.

G. *Dustpan:*
1. Empty dustpan into waste can and damp wipe pan to remove all dirt and dust.
2. Hang up dustpan so it will not become bent or otherwise damaged.

H. *Equipment Carrier:*
1. Do not allow carrier to strike walls, radiators, or other objects.
2. Wipe free of moisture and dirt after each day's use.

I. *Floor Sweep (18"):*
1. Comb floor sweep bristles daily.
2. Wash an oily or dirty floor sweep in a warm solution of multipurpose cleaner and rinse in clean warm water. Shake excess water from bristles and hang to dry.
3. Store in a clean dry place so bristles do not bear weight of brush.
4. Once each week reverse brush block of floor sweep by placing handle in the other hole. (This increases the life of the floor sweep.)

J. *Gloves, Rubber:*
1. After each use, wash gloves in a warm multipurpose cleaner solution and rinse in clear warm water.
2. Always maintain gloves in first-class condition. Discard gloves when holes develop.
3. Be careful not to snag gloves on nails or other sharp objects.
4. Store gloves in a clean dry place.

K. *Knife, Putty (1¼"):*
1. Use putty knife to scrape gum and like substances from floor. Do not use it as a substitute screw driver or hammer.
2. Wipe free of dust and dirt after use and store in a clean, dry place.

L. *Mops, Cotton (2 lb.):*

1. After each use and before storage, rinse mops in clean water and wring or squeeze to remove excess water.

2. Wash dirty mops in warm solution of multipurpose cleaner. Rinse in clean water and wring or squeeze to remove excess water.

3. Pluff mop strands by shaking.

4. Store mops with mop heads up to allow better drying.

M. *Mop Tank (two 32-qt. buckets; chassis with casters, one wringer):*

1. Rinse and dry buckets and wringer thoroughly after use.

2. Wipe outside of buckets and the chassis free of any dirt.

3. Avoid striking mop tank outfit against objects, which tends to cause buckets to leak and to shorten life of the equipment.

4. Store in dry place so that no weight bears on any part of mop tank.

N. *Polishing Machine and Scrubbing Brushes:*

1. Remove brush (or brushes) from machine. Wash dirty brushes in multipurpose cleaner solution, rinse thoroughly, and shake free of excess water.

2. After each use, wipe machine and cord free of dirt and dust with a damp cloth. Use multipurpose clean solution to remove spots. Wipe machine and cord dry and rewind cord loosely on the hooks provided.

3. Wash and rinse wheels of machine and wipe with a dry cloth.

4. Use metal polish as directed to clean and polish machine. Store machine and brush, with brush removed, in assigned storage space.

5. Report needed repairs to supervisor.

O. *Sponges (cellulose):*

1. Rinse sponges in clean warm water after use. Wash very dirty sponges in a warm multipurpose cleaner solution, rinse in clear water and squeeze out excess water.

2. Store in a clean dry place.

Performance Inspection Guide

A. *Collection of Used Towels and Other Trash:*

1. All used towel receptacles should be empty.

2. All sanitary receptacles should be clean, both inside and outside, and contain a new paper bag liner.

3. No trash should be on the floor.

4. The large paper bags containing collected contents of sanitary receptacles should have been deposited in a large can pending burning or other approved disposal.

5. All unused towels should be at the designated disposal location.

B. *Replenishment of Supplies:*

All dispensers of supplies should be clean and filled with the proper supplies.

C. *Cleaning of Sanitary Receptacles:*

1. All sanitary receptacles should be empty except for a new paper bag "liner."

2. All sanitary receptacles should be free of spots, stains and finger marks.

3. All sanitary receptacles should be free of odors.

D. *Cleaning of Toilet Room Fixtures:*

1. All porcelain surfaces of wash basins, water closets and urinals should be free of dust, dirt, spots and stains.

2. The wall surfaces should be free of spots and smears.

3. All water closet seats should be left in raised position after cleaning. They should be free of spots and stains, and the seat hinges should be free of green mold.

4. The plumbing fixtures should be free of green mold and water stains.

CENTRAL SUPPLY BUILDING REPORT OF TOILET INSPECTION	Date of inspection
	Inspected by (Signature)

This form is part of Buildings Management's standard inspection procedure.

INSTRUCTIONS

1. Inspection of each toilet will cover all items on check list that apply to room being inspected.
2. Proper abbreviations for items requiring attention will be noted in the remark column.
3. Supervisor performing inspection will give one copy of this report to group foreman of laborers.
4. Foreman of laborers will notify superintendent or proper shop foreman of mechanical repairs required.
5. This report will be retained only as long as required for administrative purposes.

CHECK LIST AND ABBREVIATIONS

CLEANING		SUPPLIES		REPAIRS		
ITEM		ITEM		ITEM		
Floors	C1	Paper	S1	Water Closet	R1	
Stall Partitions	C2	Soap	S2	Urinal	R2	
Walls	C3	Towels	S3	Wash Basin	R3	
Wash Basins	C4	Seat Covers	S4	Soap Containers	R4	
Urinal	C5			Lights	R5	
Water Closet	C6			Drinking Fountain	R6	
General	C7					

Building	Room Number	Attendant	Inspection Time	Remarks and/or Items Requiring Attention
			Actual Size 8 x 10 1/2	

Figure 26-1

E. *Cleaning of Supply Dispensers, Tile Walls, Stall Partitions, Doors, Shelves, Mirrors and Floors:*

1. All supply dispensers should be clean and free of finger marks and water spots.

2. All shelves and shelf brackets should be free of gum, dust, fingerprints, water stains, smudges and other soil.

3. All mirrors should be free of streaks, smudges, water spots, dust and lipstick smudges, and should not be cloudy.

4. Walls, stall partitions and doors should be free of hand marks, dust, pencil marks, lipstick smudges, water streaks, mop marks, and green mold.

5. Floors (especially in corners), should be free of dirt and dust, gum, grease, black marks, loose paper, water, mop stains, and strings.

F. *Care of Equipment, Materials and Supplies:*

1. All pieces of equipment should be clean and properly placed in their assigned storage locations.

2. All materials and supplies left over from the toilet cleaning and servicing operation should be returned and stored neatly in their assigned storage locations.

NOTE

1. A mirror on an offset handle should be used for inspecting underside of rim on toilet fixtures.

2. "Report of Toilet Inspection" should be used as directed.

Frequency and Production

Frequency and Production Schedule:

	Toilet Cleaning	Toilet Servicing
Frequency	Daily	As required
Production Rate per Man-day	80 Fixtures (basins, water closets, urinals)	

The job of toilet cleaning and servicing includes the daily collecting of waste; cleaning dispensers and waste receptacles; cleaning wash basins, urinals, water closets, partitions, walls, shelves, mirrors, doors and floors; and servicing of the dispensers of supplies as required.

Staffing

The necessary staffing can be figured by dividing the number of public toilet room fixtures (wash basins, urinals, and water closets) in the building by the daily production shown in schedule. (Private toilets are cleaned by employees assigned to clean the offices in which the private toilets are located.)

Assume a building with toilet rooms (both men's and women's) containing a total of 76 fixtures (wash basins, water closets and urinals). From the production rate per man-day, each toilet cleaner in an eight-hour period should clean and service toilet rooms containing 80 fixtures (wash basins, water closets and urinals). By dividing the quantity of work to be done by the production rate, the required number of man-days to clean and service toilet rooms with this number of fixtures may be calculated.

Quantity   = 76 fixtures   Man-days $= \dfrac{76}{80} = .95$ jobs
Production = 80 fixtures

To this figure must be added a factor to compensate for leave in order to arrive at the total number of positions needed to accomplish the work.

Conversion factor including compensation for leave = 1.15
Conversion factor × jobs = total positions required
1.15 × .95 = 1.09 positions required

• Where the duty of cleaning and servicing toilets does not require the full time of an employee, other duties should be assigned to secure 100 per cent utilization of his time.

Equipment and Materials Allocation and Replacement

Items	Allotment per Cleaner	Replacement or Reissue
Bags for collecting waste towels	3	6 per year
Bags, paper (25 lb.)	24	6,000 per year
Broom, dustpan (hearth size corn broom)	1	1 per year
Brush, floor, sweep (18")	1	1 per year
Brush, handle (for floor sweep)	1	1 per year
Brush, radiator, dust	1	1 per 3 years
Brush, scrubbing (for use on polishing machine)	1 set	2 sets per year
Brush, toilet (curved head)	1	2 per year
Brush, toilet (straight head)	1	2 per year
Bucket (12 qt.)	1	1 per 2 years
Chamois	1	2 per year
Cleaner (multipurpose)	1 pint	50 gallons per year
Cloth (cheesecloth)	1 yard	52 yards per year
Dustpan	1	1 per 2 years
Equipment carrier	1	1 per 10 years
Gloves, rubber	1	2 pairs per year
Mop, cotton (2 lb.)	1	4 per year
Mop handle	1	2 per year
Mop tank (two 32-quart buckets on twin chassis)	1 set	1 per 2 years

Items	Allotment per Cleaner	Replacement or Reissue
Mop toilet (optional)	1	2 per year
Mop wringer	1	1 per 2 years
Polish metal (liquid)	1 pint	2 pints per year
Polishing machine, scrubbing and (15″)	1	1 per 10 years
Putty knife (1¼″)	1	1 per 2 years
Scouring powder (for floors)	1 pound	25 pounds per year
Soap, grit	1 cake	25 cakes per year
Soap, white (6 oz. cake)	1 cake	25 cakes per year
Sponge, cellulose	2	12 per year
Wiping cloth (waffleweave)	2	12 per year

Supplies for Toilet Servicing *

Items	Unit of Issue
Soap, hand, (cake, V-type)	Case of 144 cakes
Soap, hand, liquid	Drum of 55 gallons
Soap, hand, powdered	Barrel of 175 pounds
Toilet seat covers	5,000 to a case
Toilet tissue (rolls)	Carton of 100 rolls
Toilet tissue (sheets)	Carton of 125 packages
Towels, paper (C-fold)	Carton of 25 packages
Towels, paper (single-fold)	Carton of 15 or 25 packages

* Use supplies required by dispensers.

The allotment of equipment and materials per cleaner and the replacement or reissue quantities in this section are based on one cleaner having continuous toilet cleaning and servicing duty.

• Where the situation exists that other duties are assigned to the cleaner to secure 100 per cent utilization of his time, consideration should be given to a reduction in the replacement quantities in accordance with the actual amount of toilet cleaning and servicing assigned.

Chemical Cleaning of Water Closets and Urinals

Porcelain surfaces of water closets and urinals with stains or incrustations should be cleaned with porcelain cleaner in accordance with the following procedure:

1. Wet hand mop in water to make it absorbent.
2. Pour porcelain cleaner in a nonmetallic container.
3. Saturate hand mop with bowl cleaner and apply to inside of fixture from top to bottom in a swabbing fashion.
4. Check with mirror to see that all incrustation has been removed from under flange at the top or front of the fixtures.
5. When the incrustation is heavy, it takes several minutes for the chemical action of the porcelain cleaner to remove all of the stain. Accordingly, in toilet rooms where heavy incrustations are found, the cleaner should first be applied liberally to the incrusted area of all fixtures, before completing the operation

described in procedure three and four. The removal of stubborn incrustations in traps will be facilitated by pouring hot water into the fixtures.

Although this bowl cleaner, which is an acid, contains an inhibitor that prevents damage to the toilet fixture parts, it is injurious to the skin. Rubber gauntlet gloves and face shields should be worn by personnel performing this job.

As an alternative to cleaning the fixtures as previously described, a sodium bisulfate toilet cleaner may be used every few days to keep stains and incrustation from forming in water closets and urinals. When using this cleaner, first flush the fixture, then sprinkle a small quantity of the cleaner above the water line. Use a toilet brush or mop wet with the solution to wash the area above the water line, and under the flange or rim at top or front of the fixture, and then flush.

Once the porcelain surfaces in water closets and urinals are free from stain and incrustation, the use of sodium bisulfate should keep them free from these deposits. When this is accomplished, subsequent cleaning with porcelain cleaner may not be necessary.

Instructions from the supervisor and/or the manufacturer should be closely followed in performing this job.

27 / URN AND JARDINIERE CLEANING (FLOOR AND WALL TYPE)

(*Source:* General Services Administration, Washington, D. C.)

Job Description

Jardiniere and urn cleaning, floor and wall type, includes the removal of refuse and debris from the jardinieres and/or urns, both floor and wall type, within the assigned area; cleaning all such units, and wiping them dry of moisture as necessary; refilling the units with fresh sand as may be necessary; replacing the units in their original positions; and proper care and maintenance of the jardiniere and urn cleaning equipment and materials. Employees assigned to this duty should be instructed to report to their supervisor any broken or defective jardinieres or urns noticed during the cleaning operation.

Keeping the jardinieres and urns neat and presentable during the day time is a part of the duties assigned to the corridor policing job.

Jardiniere and urn cleaning is assigned to individuals as a regular job, with the exception that cleaning of enclosed wall type urns is assigned as part of the room cleaner's job.

Equipment and Materials

A. *Equipment:*

For collecting refuse from the units.

a. Rubber gloves

b. Scoop or sieve
c. Device for collecting debris
d. Equipment truck (where necessary)

B. *Materials:*

For cleaning, polishing and refilling units.

a. Water
b. Detergent
c. Wiping cloths
d. Sand
e. Counter brush
f. Dustpan
g. Toilet brush (for use as necessary with liquid filled urns)

Job Methods

Operation Steps	*Explanation*

Step 1

1. Obtain necessary equipment and materials.	Obtain necessary equipment and materials as listed under "Equipment and Materials."
2. Proceed to assigned work area.	All equipment and materials should be taken to the starting point of the assigned work area, to be used as needed.

Step 2

A. *Enclosed Wall-Type Urns (no water or sand)*

Start at one end of the work area to clean and refill units.	a. Remove top of receptacle.
	b. Wipe underneath side of top with a damp cloth, using detergent, if necessary, to remove deposits that have formed there.
	c. Lift out bucket type container.
	d. Empty debris into metal collecting device.
	e. Wipe out bucket with a damp cloth, using detergent if necessary.
	f. Before replacing bucket wipe inside of urn with a damp cloth, as necessary, using detergent.

Operation Steps	*Explanation*
	g. Replace bucket.
	h. Wipe exterior with damp cloth, then wipe with a dry cloth.
	i. Use counter brush and dustpan to tidy up around urns as necessary.

NOTE

This type urn should be assigned to room cleaner for cleaning.

B. *Sand Urns, Floor-Type (Jardiniere) and Wall-Type*

Start at one end of the work area to clean and refill units.	a. Screen sand with perforated scoop or sieve.
	b. Place refuse in metal collecting device.
	c. Remove metal pan containing sand, and clean interior of urn with a damp cloth, using detergent as necessary.
	d. Remove soiled sand and replace with fresh sand.
	e. Wipe off exterior of urn with damp cloth, using detergent. Polish with dry cloth.
	f. Use counter brush and dustpan to tidy up around urns as necessary.

Step 3

1. Return equipment for storage.	a. Return equipment to gear room.
	b. Clean and store in area designated by supervisor.

Care of Equipment and Material

A. *Brush:*
1. Wash brush daily, using a warm synthetic detergent solution.
2. Shake out excess water.
3. Hang up where air circulation is good so brush will dry.

B. *Device for Collecting Debris:*
1. Wash with detergent solution and dry.
2. Wipe outside free of dirt.

3. Store so it will not be damaged.

C. *Equipment Truck:*

1. Do not allow truck to strike walls, radiators or other objects.
2. Clean truck daily.
3. Report all defects and damage to the truck to your supervisor immediately. Make arrangements to have the truck wheels oiled regularly. Store truck near the source of sand.

D. *Rubber Gloves:*

1. Wash in soap and water.
2. Rinse.
3. Dry with wiping cloths.
4. Turn gloves inside out to keep them from sticking together.

E. *Scoop (Perforated) or Sieve:*

1. Wipe clean and store in a dry, clean place, either on a shelf or hanging on a nail.
2. Store so that no damaging weight is placed on the scoop or sieve.

F. *Wiping Cloths:*

1. Rinse all cloths after each use.
2. Wash dirty cloths in a warm detergent solution, (mix one ounce of detergent per gallon of water) and rinse in clean water.
3. Wring or squeeze to remove excess water and spread or hang the cloths to dry in their proper storage place.

Performance Inspection Guide

A. *Collecting Refuse from Units:*

Cigarette butts, burnt matches, and other discarded material should have been removed from the jardinieres and/or urns.

B. *Cleaning, Polishing and Refilling Units:*

All jardinieres and/or urns should have been emptied of refuse, and should be clean. Sand should be clean and neat appearing.

C. *Equipment and Materials:*

All pieces of equipment should have been cleaned and returned to their assigned storage locations.

NOTE

1. The use of the old umbrella stand type of jardiniere partly filled with water is not recommended. In cases where some of these still remain in the buildings, the methods used and the production rate should be approximately the same as given herein for the sand type.

2. In addition to nightly cleaning, policing of the jardinieres and/or urns should be done at least twice daily.

Frequency and Production

A. *Frequency and Production Schedule:*

	Urn and Jardiniere Cleaning
Frequency	Daily
Production Rate per Man-day	100

The job of urn and jardiniere cleaning includes the removal of refuse and debris from all types of urns and jardinieres, cleaning and polishing these units, and refilling with clean sand when necessary.

B. *Qualifying Factor:*

Production Rate

The production rate given is related to an average type office building with average density and type of occupancy, equipped with the usual type of jardinieres or urns. The rate as established may be subject to modification due to variations from the normal. It does, however, provide a standard from which such deviations (if any) should be calculated.

Staffing

The necessary staffing can be determined by dividing the number of jardinieres and urns in the building by the daily production shown in the Frequency and Production Schedule.

SAMPLE CALCULATION

Assume an office building containing seventy-five jardinieres. From the production standard, each employee in an eight-hour period should clean one hundred each. Using the formula

$$\frac{\text{Quantity}}{\text{Production}} = \text{Man-days per job performance}$$

calculate the number of employees required to clean the jardinieres in this area as follows:

Quantity $= 75$ Man-days $\frac{75}{100} = .75$ jobs
Production $= 100$

To this figure must be added a factor to compensate for leave in order to arrive at the total number of positions needed to accomplish the work.

Conversion factor including compensation for leave $= 1.15$
Conversion factor $\times$ jobs $=$ total positions required
$1.15 \times .75 = .863$ positions required

• Where the duty of jardiniere and urn cleaning does not require the full time of an employee, other duties should be assigned to secure 100 per cent utilization of his time.

Equipment and Materials Allocation and Replacement

Items	Allotment per Cleaner	Replacement or Reissue
Brush, counter, dust	1	1 per 2 years
Brush, toilet	1	2 per year
Cheesecloth	2 yards	52 yards per year
Cloth, wiping (waffleweave)	2	24 per year
Container, metal (for collecting cigarette butts)	1	1 per 2 years
Detergent (soapless, synthetic)	1 pound	25 pounds per year
Dustpan	1	1 per 2 years
Equipment carrier	1 (where required)	1 per 10 years
Gloves, rubber	1 pair	2 pair per year
Sand, white (for sand urns)	100 pounds	5,000 pounds per year
Scoop or sieve (perforated)	1	1 per 2 years

The allotment per person and replacement or reissue quantities in this section are based on one employee cleaning jardinieres and/or urns full-time, in accordance with the standard.

• Where other duties are assigned to the employee to secure 100 per cent utilization of his time, a reduction in the replacement quantities should be made in accordance with the amount of time consumed at other work.

28 / GENERAL UTILITY WORK

(*Source:* General Services Administration, Washington, D. C.)

Scope of Work

The job of utility work does not consist of the regular performance of a single well-defined duty. It includes, but is not limited to such activities as:

Servicing complaints; performing special cleaning necessitated by the building occupants vacating space, holding special conferences, meetings; performing special cleanup work made necessary by toilet room floods and similar occurrences; cleaning up results of accidents such as coffee and beverage spillages, dropped food items, broken glass, and many others.

Assisting truck drivers in loading, unloading and distribution of building supplies such as toilet room supplies and cleaning materials.

Assisting in laying down and picking up, at required times, mats in building entrances during periods of inclement weather; aiding in the removal of snow and ice from building areas and the placing and removal of sand, chemical compounds or other materials used in reducing slip hazards during such weather conditions.

Assisting in moving furniture as necessary during repainting of office and other types of space; moving light pieces of furniture within the building occupants' space; setting up and/or rearranging furniture items in auditorium and conference rooms to conform with requirements of activities scheduled therein.

Providing occasional help to building forces when necessary such as for digging and filling trenches for repair of sewers and plumbing; cleaning up debris resulting from alterations to the building; assisting the force as necessary in refilling fire extinguishers.

Performing other general duties as directed.

Special Jobs

In certain locations, some of the cleaning work not identifiable as a part of the other individual jobs as defined in the cleaning standards may be of such nature and recurrence that it can be classified as a special job, rather than as a part of the utility work, and so programmed for in a way similar to that used for the regular cleaning jobs.

As a special job the work will have rates established for production and frequency. These rates could be the same as those for related cleaning jobs, such as policing work and room cleaning special areas. In the absence of established rates, observations and studies of the work procedures and needs should provide a basis for the determination of satisfactory values.

Examples of special jobs include: Cleaning escalators; emptying waste crocks located in experimental laboratory space; operation of incinerators; replenishing drinking water supply in individual water coolers, and others.

Job Methods, Equipment and Materials

The job methods, equipment and materials used in the accomplishment of any assigned duty should be in accordance with related work methods, equipment and materials prescribed in other related job descriptions.

Safety regulations and procedures prescribed for proper lifting and other manual effort exerted in the accomplishment of this work should be closely observed to avoid possible injuries to the worker and/or others.

Frequency and Production

Utility work. The frequency for the performance of utility work is generally described as daily.

The production rate, although not specifically defined in terms of individual work units for the varied work performed, is based upon building gross area and is set at 1,000,000 square feet gross per eight hour man-day.

Special jobs. The frequencies and production rates for special jobs should be, where possible, similar to those established for other related cleaning jobs. *For example,* special daytime room cleaning work might be assigned on the standard basis of 14,000 square feet per man-day at whatever frequency local conditions require. Extra policing necessitated by tours, or other means, for instance, might be assigned on the standard basis of 300,000 square feet per man-day at the required frequency.

In those instances where established rates are not pertinent, the work requirements for the particular job should be studied in detail to determine appropriate values.

Staffing

The necessary staffing for the utility job can be figured by dividing the building area by the daily production rate of 1,000,000 square feet.

SAMPLE CALCULATION

Assume an office building with a gross area of 445,000 square feet. Using the formula given below, the required staffing is computed as follows:

$$\text{Man-days (of labor)} = \frac{\text{Quantity}}{\text{Production}} \times \frac{1}{\text{Frequency}}$$

$$= \frac{445,000}{1,000,000} \times \frac{1}{1}$$

$$= .445 \text{ jobs}$$

To this figure must be added a factor to compensate for leave in order to arrive at the total number of positions needed to accomplish this work.

$$\text{Conversion factor including compensation for leave} = 1.15$$
$$\text{Conversion factor} \times \text{jobs} = \text{total positions required}$$
$$1.15 \times .445 = .51 \text{ positions required}$$

Staffing for special jobs, where established, may be determined by means of the same formula using the rate values as required for the particular work involved or through determination by other means of a satisfactory figure. The leave factor, also, has to be applied to the staffing developed for this duty in the manner shown in the formula to arrive at the total positions to accomplish this work.

• Where the utility work and/or special jobs do not require the full time of an employee, other duties should be assigned to secure 100 per cent utilization of his time.

29 / **VENETIAN BLIND CLEANING**

(*Source:* General Services Administration, Washington, D. C.)

Job Description

The job of venetian blind cleaning consists of washing the venetian blind slats and cleaning the tapes and cords by brushing with a stiff bristle brush or by washing.

Employees assigned to this job should be instructed to notice any defects in the blinds and report them to their supervisor.

Job Methods

Venetian blind washing, because of the many variables which affect it (such as the type of blind head, length, width and number of blinds, and availability of contract services), requires closer individual attention, as far as method is concerned, than many of the other cleaning jobs. The method used may vary from location to location to meet the problems in specific buildings. In deciding on the method to be used care must be taken to be sure that accurate costs are determined for each method considered and that similar results are obtained.

Washing in place. Where blinds with complex headers are installed, washing in place will eliminate the cost of labor to remove and replace the blinds, and to transport them to and from the windows, and therefore should be given serious consideration.

The method used when blinds are washed in place is as follows:

1. Before washing tilt all the slats in one direction and wipe down with a greased sweep mop. Then tilt slats in the other direction and repeat the process. If the blinds are extremely dusty, use a vacuum cleaner for this job to avoid spreading dust.

2. Adjust slats to the open position, and "dry-clean" tapes and cords with stiff, short-bristled brush.

3. With the slats in the open position, work from a "safety" stepladder and wash the blind, one slat at a time. Use a quarter section of waffleweave cloth and a bucket of detergent solution (mixed one ounce to one gallon of water). Rinse as necessary with a clean cloth and clear water. Then wipe the slat dry. Repeat this process for each slat, completing each one in turn.

4. When finished, wipe all dust and water spots from the window sill, and sweep the floor around the window.

When this method is used, particular care must be taken to avoid splashing or streaking the window glass.

Removal and washing by hand. When blinds can be removed easily from the windows for washing it is usually more productive to do this. Various types of tanks and other equipment are available to suit the various sizes and types of blinds to be washed and the space available for the washing equipment.

The method used when blinds are removed for washing by hand is as follows:

1. Remove blinds from windows and mark them for location identification before removing them from the individual rooms.

2. When the number of blinds that can be washed in one day have been accumulated on the hand truck, take them to the washing area.

3. Where necessary to soak the blinds before washing, place a few at a time in a tank of detergent solution (mix one ounce to one gallon of water). Limit the soaking time to the absolute minimum required to loosen the dirt film. Be especially careful with wooden slat blinds with worn or damaged finish, as long periods of soaking may result in further damage.

4. Remove the blinds from the soak tank one or more at a time and hang them on a suitable washing rack. Tilt the slats first in one direction and then the other, washing by agitation with a "pullman" brush or a scrub brush dipped in detergent solution. Additional water is supplied as needed, either through the hose connected to the pullman brush, or by a garden hose.

When soaking is not required before washing the blinds, hang the blinds directly on the wash racks and wash them.

5. Rinse blinds, as necessary, with the pullman brush or with the garden hose, and hang to dry.

6. Replace clean blinds in their original location the following day.

To reduce costs to a minimum commensurate with satisfactory results, the use of pullman brushes, drying racks and other labor saving devices should be considered.

Machine washing. In those areas where there are sufficient blinds to justify its purchase, commercially designed venetian blind laundry equipment may be used. The cost of such equipment ranges from approximately $500 upward. Some of the less expensive equipment is sufficiently portable to be moved from one group of buildings to another within a city, and possibly from city to city if desired.

The actual washing production of commercial laundries which use such equipment is reported to be in excess of one hundred blinds, four foot by eight foot, per eight man-hours. Production for the complete job, including removing and replacing the blinds, would of course be less.

The method followed when such equipment is used is as follows:

1. Remove blinds from windows and mark them for location identification before removing them from the individual rooms.

2. When a hand truck load of blinds is accumulated, take them to the venetian blind laundry location, and stack them in a rack to await washing.

3. Hang the blinds, one or two at a time, in the spray booth, and raise them to a sufficient height to permit them to be opened to their full length.

4. Spray blinds that are especially dusty with clear water to remove the

excess dust and dirt. (This rinse water is wasted to the drain.) This conserves the cleaning solution.

5. Adjust the valved connections so that when the cleaning solution is sprayed onto the blind it will run down into the drain through and back into the cleaning solution tank to be recirculated.

6. Spray the cleaning solution, under pressure, onto one side of the slats of the blind. Reverse the slats and spray the other side.

7. Spray the tapes and cords of the blind with the cleaning solution at the same time the slats are being sprayed.

8. Rinse both sides of the blind with clear water, under pressure. (First adjust the valved connections to allow this rinse water to be wasted to the drain.)

9. Remove blinds from the spray booth and hang to dry, in full extended position with the cords slack.

10. After drying, overnight or otherwise, take the blinds down, load them on the hand truck, and return them to the windows from which they were removed.

During the washing operation, inspect each blind, and if any are found in need of repair or repainting, separate them from the others and mark them for such action.

Exercise care to see that the length of time the blinds are removed from the window is kept to a minimum. A suggested procedure would be to remove the blinds one day, wash them the second day, and reinstall them by the third day. This procedure, of course, refers to those blinds not marked for repairs.

Frequency

The frequency for dusting venetian blinds should be once every six months. The frequency for washing is once per year. These frequencies generally will prove satisfactory.

Staffing

The man-day labor requirements for venetian blind cleaning can be determined by use of the following formula:

$$\text{Man-days (of labor)} = \frac{\text{Quantity}}{\text{Production}} \times \frac{1}{\text{Frequency}}$$

where quantity equals the number of venetian blinds to be cleaned; production is the rate of cleaning per man-day; and frequency is the number of work days between washings.

SAMPLE CALCULATION

Consider a building that has 1,500 venetian blinds for which the production rate has been determined to be thirty-five blinds per man-day, and for which the annual washing frequency is satisfactory.

$$\text{Man-days} = \frac{1500}{35} \times \frac{1}{252} = .17$$

To this figure must be added a factor to compensate for leave in order to arrive at the total number of positions needed to accomplish the work.

Conversion factor including compensation for leave = 1.15
Conversion factor × jobs = total positions required
1.15 × .17 = .196 positions required

- Where a venetian blind cleaning assignment does not require the full time of an employee, other duties should be assigned to secure 100 percent utilization of his time.

30 / WALL WASHING

(*Source:* General Services Administration, Washington, D. C.)

Job Description

The job of wall washing is the planned and scheduled periodic cleaning of wall surfaces either by hand or by machine.

The job consists of moving furniture away from the walls, removing pictures, drapes, and other removable articles from walls, covering floors and furniture with drop cloths, dusting walls before washing, as required, washing, rinsing and drying walls as described under "Job Methods", and the proper care and maintenance of equipment used.

Equipment and Materials

Equipment and Material:

For wall washing

a. Buckets (12 qt.)
b. Cloth, turkish toweling
c. Drop cloths (floor and furniture)
d. Gloves, rubber
e. Ladder, safety step (of proper height)
f. Scaffolding (where needed)
g. Sponges
h. Sweep mop
i. Steel wool No. 00
j. Truck, platform, push type
k. Wall washing machine with attachments
l. Detergent (may be prepackaged)

Job Methods

Operation Steps	*Explanation*

Step 1—Prepare for Work

Obtain equipment and materials.

a. Obtain equipment and materials needed for washing walls. (See "Equipment and Materials".)

b. Take all equipment and materials to rooms where walls are to be washed.

Step 2

Prepare walls for washing.

a. Move furniture away from walls toward center of room, except heavy files.

b. Remove pictures, drapes or other furnishings from walls and windows which will interfere with wall washing operation.

c. Spread dropcloths over furniture and floors.

d. Dust walls lightly with a sweep mop to remove loose soil.

e. Remove cobwebs with an upward and outward lifting stroke to avoid smearing the walls.

f. Set up ladder or scaffold and make sure it is firmly anchored.

g. Prepare solution for washing. Concentration of detergent depends upon degree of soil on wall.

Step 3 (See Step 4.)

Wash walls (manually).

a. Dip sponge or turkish toweling into solution. Squeeze out excess solution so it will not drip.

b. Apply cleaning solution to wall in corner next to floor from bottom up, covering an area of approximately ten to twelve square feet,

Operation Steps   *Explanation*

wetting the surface without rubbing to remove the dirt.

c. Repeat the washing operation, applying pressure, or rubbing, to remove the dirt.

d. A straight rubbing motion should be used, either side to side or up and down.

e. Immediately following the washing operation and before the wall becomes dry, rinse the section washed with clean sponges or terry cloth and wipe dry.

f. Lap over into the cleaned area when washing, far enough to prevent border marks.

g. Continue washing lower half of wall until entire lower half is completed.

h. Wash woodwork on windows and doors as reached.

i. Place ladder or scaffolding at starting point and commence washing upper half of wall.

j. Place pails on platform. Be sure ladder or scaffolding is in good condition and safe for this purpose.

k. Apply solution to wall in corner at point where washing of lower half of wall ended and wash as described in this step.

l. If solution spills on wall below, wipe off immediately with clean damp cloth.

m. Wash woodwork around ceiling (if any) as it is reached. (Ceilings washed as directed.)

n. Keep moving ladder or scaffolding as necessary and wash entire upper half of wall.

Operation Steps	*Explanation*

Step 4—Alternate Method (See Step 3.)

Wash walls (by machine).	a. Prepare solution for washing. Concentration of detergent depends upon soil on wall.
	b. Pour solution into tank, fill rinse water compartment with clean water, replace pump unit and pump up to specified pressure.
	c. Fold turkish toweling into the form of pads and fasten on the trowels.
	d. Saturate towels on washing and rinsing trowels. *Be careful* not to flood wall or trowels.
	e. Begin washing wall in lower corner, working up. Hold trowel flat against wall and move it with light pressure in long straight strokes, from side to side, where surface permits.
	f. Wash walls twice rather than apply heavy pressure.
	g. Overlap area just washed in order to avoid border marks.
	h. Stubborn streaks or marks may be removed with the heel of the trowel.
	i. Wash area of approximately five by six feet at a time.
	j. Wipe area just washed with rinsing trowel to remove soiled cleaning solution.
	k. Use drying trowel to pick up moisture still on surface.
	l. Use cloths from trowel, folded, to wash trim, pipes, woodwork, and other places, if necessary.

PRECAUTIONS

1. Use fine steel wool with care on spots.
2. Be careful not to spill any cleaning solution on wall which has not

yet been washed as this will cause streaks which are very hard to remove.

3. Avoid getting sponge or cloth against ceiling as spots will result which cannot be removed unless entire ceiling is washed. (Ceilings washed as directed.)

4. Change towels when dirty.

5. When washing by hand change water and solution frequently to avoid redistribution of dirt over walls.

6. Do not damage walls and furniture with ladders.

7. Care should be taken to avoid snagging water hose over objects.

Operation Steps	*Explanation*

Step 5

1. Remove drop cloths.	Fold up and remove drop cloths from room.
2. Replace furnishings and furniture.	a. Replace pictures, drapes, or other furnishings removed from walls.
	b. Return furniture to proper locations.

Step 6

Return equipment and materials to storage.	a. Empty buckets and machine compartments and clean all equipment items before placing in their respective storage positions (See "Equipment Care".)
	b. Hang all wet cloths or towels, gloves, and mops to facilitate drying.
	c. Replace all unused materials in proper storage locations.

Care of Equipment

A. *Bucket:*

1. Rinse and dry the inside and outside of bucket after each day's use.
2. In storing, place the bucket so that it will not be damaged.

B. *Cloth Wiping:*

1. Wash in warm detergent solution, rinse, squeeze out excess water and hang cloth to dry.

C. *Drop Cloth:*
1. Brush or shake cloth free of dust and dirt after each use.
2. If cloth becomes wet during use, hang to dry.
3. Avoid snagging or tearing.
4. Fold and store cloth in assigned storage location.

D. *Gloves, Rubber:*
1. After each day's use, wash gloves in warm detergent solution, rinse in clear water and shake free of excess water.
2. Store glove in a clean dry place.
3. Always maintain gloves in first-class condition. Discard when holes develop.
4. Be careful not to snag gloves on nails or other sharp-edged objects.

E. *Scaffolding:*
1. Damp wipe scaffolding to remove solution spillage, or any moisture.
2. Store the equipment in its assigned storage location. Avoid creating safety hazards in placing the scaffolding for storage.
3. Report defects or needed repairs to supervisor.

F. *Sponges:*
1. Rinse in clean water after each day's use.
2. When dirty, wash sponges in warm detergent solution, rinse thoroughly and squeeze out excess water.
3. Store in a clean dry location.

G. *Step Ladder:*
1. Damp wipe ladder to remove solution.
2. Store in assigned storage area.

H. *Sweep Mop:*
1. When soiled, wash mop in warm detergent solution, rinse thoroughly, shake free of excess water, hang to dry.

I. *Truck, Hand:*
1. Wipe truck free of dirt and debris after each day's use.
2. Report defects to supervisor.
3. The wheels of the trucks should be properly lubricated.

J. *Wall Washing Machine:*
1. Empty tanks and thoroughly rinse with clean warm water, and wipe machine dry.
2. Remove towels from all trowels and wash.
3. Discard any unused cleaning solution.
4. Force clean water through the hose and trowels to prevent clogging.
5. Drain hose and wind loosely on trowels.
6. Place trowels in proper storage location.
7. Store in clean dry place where it will not be damaged by heavy equipment.

Performance Inspection Guide

A. *Wall Washing:*

1. There should be no streaks or spots remaining on walls or signs of not overlapping.

2. There should be no smudge spots at point where cleaning of the lower and upper halves of the wall overlaps.

3. No water should have been spilled on floor or furnishings.

4. Wall should be uniformly clean all over.

5. Woodwork on doors, windows, and moldings should be clean.

B. *Furniture Replacement:*

1. All furniture, pictures, and other furnishings moved during the wall washing operation should be returned to their original positions.

C. *Equipment:*

1. All equipment used during the wall washing operation should be cleaned and properly placed in its assigned storage location.

2. Wall washing machine or mop tanks should be emptied, rinsed, wiped dry, and placed in proper storage location.

Frequency and Production

Frequency and Production Schedule:

	Wall Washing, Manual or Machine
Frequency	As required
Production Rate per Man-day	2,000 sq. ft. (wall area)

Staffing

The necessary staffing can be figured by use of the formula shown below. The labor supplies, and other items used in this work are chargeable against painting and decorating cost classification.

SAMPLE CALCULATION

Assume a building having a wall washing requirement of 100,000 square feet in an annual frequency. Then the man-day requirement is computed as follows:

$$\text{Man-days} = \frac{\text{Quantity}}{\text{Production}} \times \frac{1}{\text{Frequency (days)}}$$

$$= \frac{100,000}{2,000} \times \frac{1}{252}$$

$$= .199 \text{ labor required (jobs)}$$

To this requirement must be added a factor to compensate for leave in order to arrive at the total number of positions needed to accomplish this work.

Conversion factor including compensation for leave = 1.15
Conversion factor × jobs = total positions required
1.15 × .199 = .229 positions required

Equipment and Materials Allocation and Replacement

Items	Allotment per Cleaner	Replacement or Reissue
Bucket (12 qt.)	2	1 per year
Cloth, wiping (terry cloth toweling)	As required	As required
Detergent	As required	As required
Dropcloth	As required	As required
Gloves, rubber	1 pair	6 pair per year
Ladder (12 ft.)		
Scaffolding (optional equipment)	As required	As required
Sponges, cellulose or natural	As required	As required
Steel wool No. 00	As required	As required
Sweep mops (frame, handle and mophead)	1 or 2	1 or 2 per year
Truck (hand)	1	1 per 5 years
Wall washing machine (optional equipment)	As required	As required

31 / **WASTE DISPOSAL**

(*Source:* General Services Administration, Washington, D. C.)

31A / **PAPER BALING**

Job Description

The job of paper baling includes loading the paper into the baler, adjusting and tamping the paper, operating the baler to compress the paper into a bale, placing the baling wire around the bale, removing the bales to storage, stacking the bales and taking care of the baling equipment. Employees assigned to this job should be instructed to notice lights out, and any other similar service defects, or defects in the operation of equipment and report them to their supervisor.

Equipment and Materials

A. *Equipment:*

1. For baling paper.

    a. Floor sweep (24")
    b. Gloves, gauntlet

    c. Shovel

    d. Goggles, dust

    e. Rake (heavy iron)

    f. Corn broom

    g. Basket (3 bushel, oak)

    h. Tamper

    i. Stevedore truck

    j. Respirator, dust

2. For equipment care.

    Wiping cloths

B. *Materials:*

1. For baling paper.

    Baling wire

2. For equipment care.
    Detergent

NOTE

The baling machine is considered to be included in the building equipment.

Job Methods

Operation Steps	Explanation

A. *Paper Baling:*

Step 1

1. Obtain equipment needed.	Equipment and materials needed for baling paper are listed under "Equipment and Materials."
2. Proceed to assigned work area.	Take equipment and materials not already stored there to the room containing the baling machine.

Step 2—Baling Machine in Pit (See Step 3.)

Bring paper to room where baling machine is kept.	a. Bring paper to baling machine.
	b. Cover the lower platen with cardboard.
	c. Lower the bottom platen to the lower limit.
	d. Remove carbon paper, glass, and

Operation Steps	*Explanation*
	metal from the paper. Further segregate when required by contract.
	e. Using a twenty-four inch floor sweep or metal rake, sweep or rake the paper into the filling chamber.
	f. Using the tamper, tamp down the paper as it is swept into the chamber.
	g. When the level of the paper in the filling chamber reaches the floor, close and lock the rear door.
	h. From the front of the baler fill the remaining space up to the upper platen with cardboard, if available; otherwise fill with paper and place a piece of cardboard next to the upper platen large enough to cover the platen.
	i. When the space up to the upper platen is filled and the cardboard has been placed next to the upper platen, close and lock the front door of the machine.
	j. Push button to raise the lower platen and press bale.
	k. Open both front and rear doors.
	l. Feed in baling wire with the looped end going first from the back to the front, passing under the bale and through the slots in the platen, then up and from the front to the back, passing over the bale and through the slots in the platen. Tie each wire by passing the straight end through the looped end and drawing wire tight.
	m. When the bale is securely tied lower the bottom platen to the desired level to permit easy removal of the bale.
	n. Remove bale to a stevedore truck and store for disposal.

Operation Steps   *Explanation*

Step 3—Baling Machine on Floor (See Step 2.)

Bring paper to room where baling
machine is kept.

a. Bring paper to baling machine.

b. With the upper platen at the highest position place a piece of cardboard large enough to cover the platen on the lower platen.

c. Close both front and rear doors and lock.

d. Swing down the top section of front door.

e. Remove carbon paper, glass, and metal from the paper. Further segregate where required by contract.

f. Using a scoop type shovel load paper into the filling chamber up to the opening.

g. Lower the upper platen and compress paper in filling chamber, then raise the upper platen to its original position.

h. Using the shovel, put in more paper; then lower upper platen and compress the paper again.

i. Raise the upper platen to its original position, load more paper and place a piece of cardboard the size of the platen on the top of the paper and compress again.

j. After the third loading and pressing operation leave the paper compressed and open both front and back doors.

k. Feed in baling wire with the looped end going first from the back to the front, passing under the bale and through the slots in the platen. Tie the wires by passing the straight end through the looped end and drawing wire tight.

Operation Steps	*Explanation*
	l. When the bale is securely tied raise the upper platen to the highest point.
	m. Remove the bale to a stevedore truck and store for disposal.

PRECAUTIONS

1. Be sure to cut the electric power off the machine after completion of the baling operations.

2. Do not smoke while in the trash room or while operating the baling machine.

3. Be careful of broken glass and materials with sharp edges. Always wear gloves.

4. Do not get into the filling chamber during the filling operations.

5. Stack bales at least two feet away from lights, sprinkler heads or steam pipes in the trash room.

6. Always wear dust goggles and dust respirator when loading trash and when tying baling wire around a bale.

B. *Cleaning Baling Machines:*

Step 4—Baling Machine in Pit (See Step 5.)

Clean baling machine.	a. Remove pit cover, *make certain* the light is on and the electric power to the baler is off, and then go down into the pit.
	b. Using corn broom, gather up all paper and trash in the pit.
	c. Place the paper and trash in the basket and haul the basket up to the floor level.
	d. After cleaning out the pit clean all paper and trash from around the machine on the floor level.
	e. Wipe off the entire baler with a damp wiping cloth.

Step 5—Baling Machine on Floor (See Step 4.)

Clean baling machine.	a. Clean all paper and trash from around the baling machine.

Operation Steps   *Explanation*

b. Wipe off the entire baler with a damp wiping cloth.

PRECAUTIONS

1. *Be sure* to wear goggles and gloves when working on a baling machine.
2. *Do not* smoke while in the room with the baling machine.

C. *Returning Equipment and Materials to Storage:*

Step 6

Return equipment and materials to gear storage room.

a. Gather all pieces of equipment and materials that are not normally kept in the baling room and proceed to the proper storage location.

b. Clean the broom and floor sweep, and store them in their proper location.

c. Clean and store the gauntlet gloves.

d. Clean dust particles from shovel, rake, tamper and basket. Store in proper location.

e. Clean off dust from goggles and respirator, and keep in first-class condition.

f. Clean stevedore truck of dust particles and store.

g. Return all unused baling wire to proper storage place.

Care of Equipment

A. *Floor Sweep (24"):*

1. Clean floor sweep daily by combing.
2. Wash oily or very dirty floor sweeps as necessary in a warm solution of synthetic detergent. Rinse in clear warm water. Shake out excess water and hang to dry.
3. Store floor sweep in a clean dry place so that the bristles do not bear the weight of the brush.
4. Once each week reverse the floor sweep by placing the handle

in the other hole of the brush block. This increases the life of the floor sweep.

B. *Gloves, Gauntlet:*

    1. After each use, shake out all dust.

    2. Always maintain gloves in a first-class condition. When holes develop the gloves should be repaired or discarded.

    3. Be careful not to snag gloves on nails or other sharp objects.

    4. Gloves should be stored in a clean dry place.

C. *Shovel:*

    1. Using a dry wiping cloth, clean loose dirt and dust from shovel before storing.

    2. Store shovel so it will not be damaged during storage.

    3. Avoid banging the shovel against the baling machine or any other object as such action tends to damage the shovel.

D. *Goggles, Dust:*

    1. Wipe goggles clean daily.

    2. Care should be taken not to drop or strike goggles.

    3. Goggles should be maintained in first-class condition. When they become chipped or otherwise damaged they should be replaced.

E. *Rake:*

    1. Using a dry wiping cloth clean loose dirt and dust from rake.

    2. Remove any paper that might still be clinging to it.

    3. Store rake so the spikes will not be bent or otherwise damaged during storage.

    4. Avoid hitting the rake against metal or concrete as such action tends to damage it.

F. *Corn Broom:*

    1. Soak a new broom in water before using.

    2. Never stand broom on straws when storing.

    3. Store broom where there is free circulation of air, so that it will always be dry.

    4. Use a synthetic detergent in warm water to wash the broom periodically, and rinse in clear warm water and hang to dry.

    5. Never use broom when it is wet.

G. *Basket (3 bushels, oak):*

    1. Always empty the basket after each use.

    2. Wipe out any dust or dirt in the basket.

    3. Store basket in such a manner that it will not be damaged.

H. *Tamper:*

    1. Wipe tamper off with a damp cloth after each use.

    2. Store the tamper so it cannot be damaged or damage other objects by falling.

I. *Stevedore Truck:*

1. Do not allow truck to strike walls, radiators, or other objects.

2. Clean thoroughly each week using a damp wiping cloth. Wipe off all dust or dirt.

3. All defects or damage to the truck should be reported immediately to the supervisor. Arrangements should be made to have the wheels of the truck oiled regularly.

4. Store truck out of the way of traffic to prevent anyone falling over it.

J. *Respirator, Dust:*

1. Wipe respirator clean daily.

2. Do not snag or strike respirator.

3. Respirator should be maintained in first-class condition. If it becomes damaged in any way it should be replaced.

K. *Wiping Cloths:*

1. Rinse out cloth used, or when dirty wash it in warm detergent solution, and wring out.

2. Spread or hang cloth to dry in the proper storage place.

3. Avoid leaving any wet, dirty cloths piled together in the storage place.

Performance Inspection Guide

A. *Paper Baling:*

1. The trash room should be in first-class condition. Any baled paper and trash should be stacked neatly so it will not fall. The floor should be swept clean.

2. The bales of paper should not contain any large quantities of carbon paper, metal, glass or similar material and should not have an excessive amount of paper protruding.

3. The baling wires should be tied properly and securely on the bales.

B. *Cleaning Baling Machines:*

1. The baling machine and the surrounding area should be free of paper and from dust and dirt.

2. The pit, if any, should be free of any paper, trash or other foreign material. The electric power to the baler should be off and the doors to the baler closed and locked.

C. *Equipment and Materials:*

1. All pieces of equipment should be clean and properly placed in their assigned storage locations.

2. All twine, detergent and similar materials left over from the job should be returned to their assigned storage locations.

Frequency and Production

A. *Frequency and Production Schedule:*

	Paper Baling
	Daily
Frequency	Bales—Total Weight
Production Rate per Man-day	10,000 lbs.

The job of paper baling consists of the daily baling of waste paper, including cardboard and cardboard boxes, and stacking those bales in the trash room for storage, pending disposal.

B. *Qualifying Factor:*

Production Rate

The production rate given relates to power operated balers capable of baling at least five hundred pound bales. The rate as established may be subject to modification due to the use of different types of balers or the degree of segregation of the paper. It does, however, provide a standard from which such deviations (if any) should be figured.

Staffing

Staffing can be figured by dividing the total paper to be baled by the daily production shown in the Frequency and Production Schedule.

SAMPLE CALCULATION

Assume there are 9,900 pounds of paper to be baled daily. From the standard, each baler operator should bale a total weight of 10,000 pounds in an eight hour period. Using the formula as follows, calculate the required number of man-days to bale the paper.

$$\text{Quantity} = 9,900 \text{ lbs.}$$
$$\text{Production} = 10,000 \text{ lbs.}$$
$$\text{Man-days} = \frac{9,900}{10,000} = .99 \text{ jobs}$$

To this figure must be added a factor to compensate for leave in order to arrive at the total number of positions needed to accomplish the work.

Conversion factor including compensation for leave = 1.15
$$1.15 \times .99 = 1.14 \text{ positions required}$$

It is preferable to have two employees working together baling paper.

• Where the duty of baling paper does not require the full time of the employees assigned to this job, other duties should be assigned to them to secure 100 percent utilization of their time.

Equipment and Materials Allocation and Replacement

Items	Allotment per Cleaner	Replacement or Reissue
Baling wire	1 bundle	150 bundles per year
Basket (3 bushel, oak)	1	1 per 2 years
Broom, corn	1	1 per year
Brush, floor sweep (24")	1	1 per year
Brush handle	1	1 per year
Detergent (soapless synthetic, powdered)	1 pound	10 pounds per year
Gloves, gauntlet	1 pair	3 pairs per year
Goggles, dust	1 pair	1 pair per year
Rake (heavy iron)	1	1 per 2 years
Respirator, dust	1	1 per year
Shovel (scoop type)	1	1 per 2 years
Stevedore truck	1	1 per 10 years
Tamper	1	1 per 2 years
Wiping cloths (waffleweave)	1	6 per year

The allotment per cleaner and replacement or reissue quantities in this section are based on one paper baler having continuous duty baling paper.

• Where the situation exists that other related duties are assigned to the paper baler to secure 100 per cent utilization of his time, consideration should be given to a reduction in the replacement quantities in accordance with the actual amount of paper baling assigned.

31B / PAPER AND TRASH COLLECTION

Job Description

The job of paper and trash collection includes picking up this material, both combustible and noncombustible, and transporting it to the trash room for baling and/or storage until picked up by the disposal contractor. The paper and other trash is collected from predetermined locations (usually outside the offices) where the room cleaners have placed it. Employees assigned to this job should be instructed to notice lights out or any other similar service defects and report them to their supervisor.

Equipment and Materials

A. *Equipment:*

1. For collecting paper and trash:

   a. Four-wheeled push truck
   b. Dustpan

      c. Dustpan broom

      d. Gloves

      e. Container for noncombustible material

   2. For equipment care.

      Wiping cloths

B. *Materials:*

   1. For collecting paper and trash.

      Twine

   2. For equipment care.

      Detergent

Job Methods

Operation Steps	*Explanation*
A. *Paper and Trash Collection:*	
Step 1	
1. Obtain equipment needed.	Equipment and materials needed for collecting paper and trash are listed under "Equipment and Materials."
2. Proceed to assigned work area.	Take equipment and materials to where the paper and trash have been placed by the room cleaners at conveniently accessible points for collection, starting usually at the top floor.
Step 2	
1. Collect wastepaper and combustible trash.	Pick up all material that has been set out in bags, tie all bags securely and place them in a neat manner on the four-wheel push truck. Where baler is available in the building, paper may be collected loose in an open type truck.
2. Collect noncombustible trash.	Pick up all noncombustible trash and place it in the container (on truck) for that purpose.
3. Collect unused wastepaper bags.	Pick up all unused wastepaper bags

Operation Steps	*Explanation*
	and place them in a suitable spot on the push truck.

PRECAUTIONS

1. Bags should be tied securely to avoid scattering paper and trash.

2. *Be careful* of broken glass and materials with sharp edges. *Always* wear gloves. *Do not* handle any trash with bare hands.

3. Use the dustpan and dustpan broom to pick up any small particles that fall on the floor during collection.

4. Do not take the push truck into offices.

B. *Removal and Disposal:*

Step 3

Removal and disposal of paper, trash, and unused bags.	a. Take all paper and trash from the areas of collection and deliver it directly to the trash room.
	b. Stack the bags so they will not fall.
	c. Empty the container for noncombusible material at the disposal point.
	d. Keep all unmixed paper and cardboard separate. Tie and tag the cardboard.
	e. Deposit all unused bags in the bag storage room.

PRECAUTIONS

1. Bags should be stacked at least two feet from lights, sprinkler heads, or steam pipes in the trash room.

2. *Do not* smoke while in the trash room.

3. *Be careful* of broken glass and materials with sharp edges. Always wear gloves. Do not handle any trash with bare hands.

C. *Returning Equipment and Materials to Storages*

Step 4

Return equipment and materials to gear storage room.	a. Gather all pieces of equipment and unused materials and proceed to the designated gear storage room.

Operation Steps   *Explanation*

b. Empty and sweep out with dustpan broom any trash remaining in the push truck.

c. Clean dustpan broom, dustpan, and container for collecting noncombustible material, and return to proper storage location.

d. Clean off all dust from gauntlet gloves and store.

e. Return the ball of unused twine to proper storage place.

PRECAUTION

Be sure the ball of twine is tied and placed so it will not fall or unwind.

Care of Equipment

A. *Four-Wheeled Push Truck:*

1. Do not allow truck to strike walls, radiators, or other objects.
2. Clean truck daily by sweeping it out with dustpan broom and wiping it off with a clean, dry wiping cloth.
3. All defects and damage to the truck should be reported immediately to the supervisor. Arrangements should be made to have the wheels of the truck oiled regularly.

B. *Dustpan:*

1. Empty the dustpan into the waste can provided in the vicinity of the trash room.
2. Wipe with a damp wiping cloth to remove all dirt and other matter.
3. Wipe with a dry wiping cloth to prevent rust.
4. Hang dustpan in storage closet so that it will not be bent or otherwise damaged.

C. *Dustpan Broom:*

1. Comb the broom bristles daily.
2. When the broom becomes soiled, wash it in a solution of synthetic detergent and rinse in clean water.
3. Store broom so that the bristles hang down, are free to straighten out, and do not support the weight of the broom. (This is especially necessary when bristles are wet.)

D. *Gloves, Gauntlet:*

1. After each use, shake out all dust.

2. Always maintain gloves in first-class condition. When holes develop the gloves should be repaired or discarded.

3. Be careful not to snag gloves on nails or other sharp objects.

4. Gloves should be stored in a clean dry place.

E. *Container for Collecting Noncombustible Material:*

1. Empty container after each use.

2. Using a damp wiping cloth, wipe interior and exterior of container to remove all remaining particles of dirt, dust, or other material.

3. Wipe with a dry wiping cloth.

4. Replace container on the push truck so it will not be damaged and will be ready for the next day.

F. *Wiping Cloths:*

1. Rinse all cloths used, or when dirty wash them in warm detergent solution, rinse and wring out.

2. Spread or hang cloths to dry in their proper storage places.

3. Avoid leaving wet, dirty cloths piled together in the storage place.

Performance Inspection Guide

A. *Paper and Trash Collection, Removal and Disposal:*

1. Bagged paper should be neatly stacked in the trash room two feet or more from lights, sprinkler heads and steam pipes.

2. All unused wastepaper collection bags should be in the proper storage location.

3. Cardboard boxes should be broken, tied together and stacked in the trash room.

4. All wastepaper, unmixed at the time of collection, should be separate from mixed wastepaper.

5. Any paper and trash spilled during the collection process shall have been cleared up.

B. *Equipment and Materials:*

1. All pieces of equipment should be clean and in their assigned storage location.

2. All detergent, twine and other similar materials left over from the job should be returned to their assigned storage location.

Frequency and Production

Frequency and Production Schedule:

	Paper and Trash Collection
Frequency	Daily
Production Rate per Man-day	600,000 sq. ft. gross area

The job of paper and trash collection includes collecting daily wastepaper and trash, including cardboard, cardboard boxes and noncombustible material and transporting this material to the trash room for storage pending disposal or baling.

Staffing

The necessary staffing can be figured by dividing the building by the daily production shown in the frequency and production schedule.

Assume an office building with a gross area of 544,000 square feet. From the production standard shown in chart, each collector in an eight hour period should cover 600,000 square feet. Using the formula as follows, calculate the required number of man-days to collect the paper and trash in this area.

Quantity = 544,000 sq. ft.
Production = 600,000 sq. ft.

$$\text{Man-days} = \frac{544,000}{600,000} = .907 \text{ jobs}$$

To this figure must be added a factor to compensate for leave in order to arrive at the total number of positions needed to accomplish the work.

Conversion factor including compensation for leave = 1.15
Conversion factor × jobs = total positions required
1.15 × .907 = 1.04 positions required

• Where the duty of collecting paper and trash does not require full time of an employee, other duties should be assigned to secure 100 per cent utilization of his time.

Equipment and Materials Allocation and Replacement

Items	Allotment per Cleaner	Replacement or Reissue
Broom, dustpan (hearth size corn broom)	1	1 per year
Ash can container (10 gallon container for collecting non-combustible material)	1	1 per 3 years
Detergent (soapless, synthetic, powdered)	1 pound	6 pounds per year
Dustpan	1	1 per 2 years
Gloves, gauntlet	1 pair	3 pair per year
Push truck (four-wheeled)	1	1 per 5 years
Wiping cloths (waffleweave)	1 ball	40 balls per year
Twine, jute	1	6 per year

The allotment per cleaner and replacement or reissue quantities are based on one collector of paper and trash having continuous duty collecting paper and trash.

• Where other duties are assigned to the paper and trash collector to secure 100 per cent utilization of his time, consideration should be given to a reduction in the replacement quantities in accordance with the actual amount of trash collection assigned.

32 / WINDOW WASHING

(*Sources:* General Services Administration, Washington, D. C.; "Ladder Code," National Safety Council, Chicago.)

Job Description

The job of window washing consists of washing and drying both the inside and outside glass surfaces of windows; washing the draft deflectors; washing window sills and frames on the inside; dusting the outside frame and sill with a counter brush; checking safety anchors, glass and frames, and reporting any defects or stuck windows to the supervisor; careful observance of all prescribed safety practices; making daily reports of work accomplished and proper care and storage of equipment and materials.

This includes windows in buildings where the use of safety belts, ladders, scaffolding and/or boatswains' chairs are required to accomplish the work satisfactorily; and windows which present unusual problems in cleaning because of their construction, or other reasons.

Schedules for window washing should be planned so that windows are washed on the shady side of the building to slow up drying time and eliminate streaks.

Squeegees should be cut to fit the size of windows to be washed.

General Requirements

1. Window cleaners and other window maintenance personnel should observe the requirements of all pertinent standards.

2. No one under eighteen or over fifty years of age should be permitted to do window cleaning at heights, except in cases where specific approval is given by the personnel department and safety engineer after a special review of the circumstances and conditions of the case. The physical condition of persons engaged in this work should be thoroughly checked before employment, and rechecked twice yearly thereafter. Persons physically unfit should not be permitted to engage in this type of work.

3. No one should be allowed to work at window cleaning without first being properly trained.

4. No outside window maintenance work should be performed when weather

conditions (such as high wind or ice) are such that they add to the hazards of the operation.

5. Smoking is prohibited while cleaning windows.

6. Acids should not be used for cleaning windows.

7. All persons engaged in window cleaning or window maintenance should be provided with and instructed to use safety belts, ladders, scaffolds, or boat-swains' chairs, as herein prescribed, *unless* such work can be safely performed from one of the following locations:

a. From the ground or floor level.

b. From a roof or similar area which has a slope of less than one inch to twelve inches, is at least six feet wide, and is capable of safely sustaining the weight of the worker and his equipment.

c. From a balcony or fire escape with a railing at least three feet high.

d. From some other safe walking surface which is similarly protected.

8. Where the window sill extends less than four inches out beyond the window frame, an approved portable auxiliary sill or other approved device should be used, together with the approved safety belt.

a. If portable sills are used, they should provide a working platform at least ten inches wide and at least twenty-six inches long.

b. Portable sills or other such devices should be so constructed that they can be readily put in position, will be safely held in place, and can be removed easily.

9. Windows should be cleaned by means of a safety belt only (1) safe foot-hold and a safe unobstructed passage through the window are obtainable, and (2) where it is possible to attach the belt terminal to the anchor by extending only one arm beyond the window sash opening.

10. The use of boxes or other makeshift arrangements for window cleaning should be prohibited.

11. Persons performing window cleaning or window maintenance from a sitting position, with their legs inside the room, should use an approved construction-type safety belt securely fastened to some substantial anchorage inside the room. Approval should be obtained from the safety engineer to do any such work without an approved safety belt.

12. Squeegees, brushes and other equipment used for cleaning windows should be attached to the window cleaner's person, either by a strong rope or a chain. This is to prevent the articles from falling and injuring someone, or damaging property. However, if there is no chance that such falling equipment could cause injury or damage, this requirement may be waived.

13. Pads for the protection of window sills should be of a nonskid or nonslip material to reduce the possibility of slipping and falling from a window. Sponge rubber or discarded carpeting with a nonslip backing are two of the materials which may be used.

14. Windows having frames or frame mullions that are not securely fastened in place, or that show evidence of decay, should not be cleaned from the outside

by means of a safety belt attached to anchors secured in such frames or mullions. Anchors should be *removed* from such windows.

15. Broken or defective sash chains, cords, windows or frames should be repaired before allowing anyone to work on the windows by means of anchors and safety belts. The windows should be maintained in good repair, and should be arranged to swing or slide freely.

16. On horizontally pivoted (reversible) sash windows, provision should be made so that the outside can be cleaned without the window cleaner leaning outside or putting his weight on or against the sash.

17. Casement windows, the entire surface of which cannot be cleaned while standing safely inside, should either have anchor fittings (installed as specified for other windows) to permit cleaning by use of a safety belt, or should be cleaned by some other method approved herein.

18. On reversible (pivot) windows which cannot be opened properly because of their design, or because of obstructions, work should be undertaken only in the manner and with the equipment described herein.

Equipment and Materials

A. *Equipment:*

For washing windows.

a. Belt, Safety
b. Brush, Counter Dust
c. Bucket
d. Chamois
e. Cloth, Wiping
f. Ladder, Safety
g. Pad, Floor
h. Pad, Sill
i. Scraper
j. Sponge
k. Squeegee

B. *Materials:*

For washing windows.

a. Ammonia (or other prescribed cleaning material)
b. Water

REQUIREMENTS GOVERNING PURCHASE OF SAFETY BELTS

Specifications to Be Used

Requirements Governing Installation of Window Anchors:

1. Specifications covering safety should be used for the installation of window anchors.

2. No anchors should be installed or repaired by building service employees, except where such work is undertaken with the approval of the safety engineer.

3. The use of lag screws or expansion bolts as a means of fastening anchors is prohibited.

4. Anchors should not be installed on windows where safe access to the anchor is not provided, or where a secure attachment cannot be made.

5. Where it is necessary to pass through a window to clean the outside, and where the only access to the window from either side is by means of a ladder, anchors should be provided on both the inside and outside of the window.

6. Installations where it is necessary to wear a window cleaner's belt while repairing or cleaning wide fixed-sash windows, and the only access to the window is from adjacent smaller windows, and the windows are separated by a substantial mullion, two anchors should be installed at each mullion.

Ladders:

All ladders used in connection with window cleaning and maintenance operations should conform to the requirements of the National Safety Council.

Boatswains' Chairs:

Boatswains' chairs should be used only upon approval of the safety engineer *after* it has been determined that other means of window cleaning or window maintenance as herein specified cannot be safely applied.

The absence of a requirement covering specific equipment, operations or hazards will not relieve the supervisor or the employee of the responsibility of taking further action to provide maximum safety in the performance of window cleaning and window maintenance. Each supervisor is expected to report to the safety engineer any unsafe conditions or work practices not covered by this standard or by related standards.

Use of Safety Belts:

1. The use of belts made entirely of leather, or in which any of the component impact-carrying parts are of leather, is prohibited.

2. Before fastening a belt terminal, the anchor installation should be checked for visible defects, such as loose, cracked or badly rusted anchors. Supervisors should spot check anchor installations frequently for such defects. If any are found, use of the anchors shall be prohibited until the defects are corrected. If there is any question as to the safety of anchor installations, the matter should be referred to the safety engineer, who will determine whether their use is to be continued. If investigation indicates that the window hangers are becoming defective due to corrosion or cracking, the safety engineer should arrange for testing of the hangers.

3. When using a safety belt, the user should attach one belt terminal to the anchor *before* stepping out onto the sill. Immediately after stepping out, he should attach the other belt terminal to the anchor. During the window cleaning or repair operation both belt terminals should be attached to the anchors. Just before re-entering the building the user should detach one belt terminal, leaving the other terminal attached until he is entirely inside.

4. Belt terminals should always be attached to both heads of a double-headed anchor.

5. Window cleaners using safety belts should not move from window to window by shifting their belt terminals from anchor to anchor, but should repeat the process described in paragraph three for each window.

6. In cleaning wide fixed-sash windows using window cleaners' belts and where access to the window is from adjacent smaller windows, the belt terminals shall be attached to the two anchors on the left mullion of the large window to permit work on the left half of the large window from the smaller adjoining window. The belt terminals likewise should be attached to the two anchors at the right mullion while working on the right half of the large window.

7. Extra holes should not be punched in a belt, nor should a belt be tied up with wire, or altered in any way.

8. On a two rope type safety belt the safety and service ropes should never be tied or bound together. The service rope should always be shorter than the safety rope to prevent wear on the safety rope.

Job Methods

Using Safety Belts

Operation Steps	Explanation
Step 1	
1. Obtain equipment and materials.	a. Obtain the necessary equipment and materials as listed under "Equipment and Materials".
	b. Inspect and put on safety belt.
2. Proceed to assigned work area.	Take equipment and materials to first window scheduled for washing.
Step 2—Prepare to Wash Windows	
1. Clear window area.	Raise blind or shade. Remove books, draft deflector glass, or any other items on window sill. Remove any furniture which is in the way.
2. Place sill pad.	Place nonslip pad on window sill inside room.
3. Place floor pad.	Place floor pad on floor and set bucket on pad, or where practical bucket may be set on sill pad.
4. Open window and place sponge.	Wet sponge and squeeze dry enough to prevent splashing and place on outside corner of window sill.

Operation Steps	*Explanation*
5. Hook safety belt.	Hook safety belt and test as stated for the type of belt used. Be sure squeegee and counter brush are attached.
6. Get positioned for outside washing.	If window washers belt is used attach belt as instructed, step outside and close window. If construction type belt is used sit on sill.

Step 3—Dust and Wash Windows

1. Dust outside window frames.	Dust frame and sill from top to bottom with counter brush.
2. Wash outside window.	Starting at the top, wash glass with sponge using straight overlapping strokes back and forth across the window.

NOTE

To remove spots, wet window thoroughly and starting at the top, use sharp scraper in up and down strokes. Never remove paint from other than glass surface as this breaks paint seal.

3. Dry outside window.	Wet the blade of the squeegee and pull across the window. Wipe the squeegee blade on the chamois after each pull.
4. Wipe corners.	Pick up water from the corners with the sponge braced with one finger. Wipe edges with chamois as necessary.
5. Unhook belt.	Unhook belt as stated for the particular type of belt used and close window.
6. Clean inside window.	Stand on sill pad and using same procedure as for outside of window, dust, wash and dry.
7. Clean sill.	Remove sill pad and wipe window sill with sponge and dry with cloth.
8. Clean draft deflector.	Replace draft deflector then dust, wash and dry it.

Operation Steps   *Explanation*

NOTE

Wash and rinse sponge frequently.

9. Replace items.   Replace all books, furniture, and other
                                     items that have been moved for ac-
                                     cess to the window.

Step 4—Proceed to Next Window

1. Inspect finished work.   Inspect window to make sure it has
                                     been washed properly. Be sure all
                                     moved items are replaced on the
                                     window sill.

2. Move to next window.   Move equipment and materials to
                                     next window location.

PRECAUTIONS

1. Make certain that safety equipment has been inspected.
2. Avoid any action which will break the glass.
3. Report all broken or loose glass and defective items to supervisor.
4. Avoid interference with activities carried on in the offices where
   windows are being cleaned.
5. Carefully observe all prescribed safety practices during performance
   of this work.

Step 5—Return Equipment and Material to Storage

1. Return to storeroom location.   Upon completion of day's assignment
                                     gather up equipment and unused ma-
                                     terials and return to the storage area.

2. Clean equipment.   Clean all equipment items as directed.

3. Store equipment and materials.   Store equipment and unused materials
                                     in their assigned storage locations.

Alternate Methods

There are so many types of windows, it will be necessary to give careful study
to each such job. For certain types of windows not equipped with safety hooks,
such study may result in a safe, more productive way to perform the job.

The *six step method* is one which has proved satisfactory for washing double
hung windows not equipped with safety hooks. This method is applicable to all
two sash double hung windows with not more than four foot height per sash.

Step 1: Raise bottom sash, lower top sash, wash and squeegee inside of top sash.

Step 2: Lower bottom sash halfway, wash and squeegee outside top half of bottom sash.

Step 3: Lower bottom sash, wash and squeegee outside top half of top sash.

Step 4: Raise both top and bottom sash halfway, wash and squeegee outside of bottom half of top sash.

Step 5: Raise top sash to top, wash and squeegee outside of bottom half of bottom sash.

Step 6: Close bottom sash and wash and squeegee inside of bottom sash. Relocate blind.

Care of Equipment

A. *Safety Belt:*

Safety specifications should establish the safety requirements for window cleaning and other window maintenance and should be closely adhered to in the performance of such work. The following excerpt should be included in the standard as it pertains to:

Care of Safety Belts:

1. Each safety belt should be identified and a record of the date it was put in service should be kept in the office of the building (or facility) superintendent and checked at frequent intervals to insure that proper inspection is made of belts which have been in service for a considerable period. A record should also be kept of any repairs or replacements made to the belt.

2. Belts should be maintained in good condition at all times.

3. All safety belts should be examined by the supervisor at the beginning of each work shift. The person using this equipment should also examine the belt at the beginning of each work shift and at other times during the day to make certain that no defects have developed.

4. All belts should be stored, transported and handled so as to prevent corrosion or injury. Belts which have been damaged by mildew by the action of an acid, by contact with sharp tools or equipment, or by any corrosive or deteriorating agent, should not be used.

5. Impact carrying parts of safety belts should be repaired *only by the manufacturer or his designated representative,* unless the regional safety engineer especially authorizes another individual to do so.

6. Persons responsible for supervising the use and care of window cleaning or window maintenance equipment should make certain that all unsafe belts and other unsafe pieces of equipment are disposed of at once, and not left lying about where they can be picked up and used.

7. When a belt has been exposed to rain or snow, wipe it off with a clean rag and allow it to dry at room temperature.

8. Never expose belts to heat in excess of ordinary room temperature.

9. Belts should never be dropped or thrown from one elevation to another, or otherwise mishandled.

10. For standard anchors and terminals a %₆-inch plug gauge can be used to check wear on the terminals. If the belt terminal slot is so wide at any point that it will accommodate a %₆-inch plug gauge, the terminal should be replaced.

B. *Counter Dust Brush:*

1. When brush becomes soiled, wash it in warm detergent solution.
2. Rinse brush in clear water and shake out excess water.
3. Hang up where air circulation is good so brush will dry.

C. *Bucket:*

1. Rinse and dry the bucket inside and outside after each day's use.
2. In storing, place the bucket so that it will not be damaged.

D. *Chamois:*

1. Rinse chamois in clean water and squeeze out excess water after each day's use.
2. Stretch chamois to original shape before placing to dry.
3. Avoid contaminating chamois with oils and polish.

E. *Wiping Cloths:*

1. Rinse the wiping cloths in clean water after each day's use.
2. Wash dirty cloths as required in a warm detergent solution, rinse thoroughly and wring or squeeze out excess water before spreading them to dry.

F. *Safety Ladder:*

The National Safety Council or your own local safety regulations should establish requirements essential to the safe use and care of ladders and should be adhered to closely in performing work in which these items are required.

   1. Ladder Maintenance and Inspection

   Ladder should be cleaned and inspected frequently and those which have developed defects should be withdrawn from service for proper repair or disposal. In either instance, such ladders should be tagged or marked "DANGEROUS! DO NOT USE!".

   2. During routine inspections all ladders should be carefully checked for the following defects:

      a. Loose steps or rungs (consider step or rung loose if it can be moved at all with the hand).
      b. Loose nails, screws or bolts.
      c. Loose or missing shoes or antislip bases.
      d. Cracked, split or broken uprights, braces, steps or rungs.
      e. Slivers on uprights, rungs or steps.

G. *Floor and Sill Pads:*

1. Shake pad free of dust and dirt.
2. Keep pad clean and in usable condition at all times.
3. Replace pad as necessary to avoid dirtying window sill or floor surfaces.

H. *Scraper:*

1. Wipe scraper free of moisture and dirt particles.

2. Use only for scraping glass surfaces.

3. Store scraper in a safe manner.

I. *Sponge:*

1. Rinse sponge in clean water after each day's use.

2. When dirty, wash sponge in warm detergent solution, rinse thoroughly and squeeze out excess water.

3. Store sponge in a clean dry location.

J. *Squeegee:*

1. Rinse the squeegee blade in clean water after each day's use.

2. Wipe squeegee dry and store in a clean dry location.

Performance Inspection Guide

A. *Preparation for Work:*

1. The safety equipment should have been properly inspected before using.

2. The window area should have been cleaned of obstructions before washing the window.

B. *Dusting Window Frame and Sash, and Washing the Glass:*

1. The window frame and sash should be free of dust and loose dirt.

2. The washed glass should be clean and free of dirt, grime, and streaks, and should be clear of all excess moisture.

3. The window sash, sill and other surroundings should be free of drippings and other watermarks.

4. Items moved during the washing operation should have been replaced to original position.

C. *Use of Safety Equipment and Methods:*

1. The safety equipment should have been used in accordance with prescribed instructions.

2. Safety practices should have been followed throughout the window washing work.

D. *Care of Equipment and Materials:*

1. All items of equipment used in the window washing work should be in a clean and well cared for condition.

2. Equipment should be stored in assigned storage location.

3. Unused materials should be properly contained and stored.

Frequency and Production

Frequency and Production schedule:

	Production Rate per 8-Hour Man-days	Normal Frequency in Work Days
Double hung, 2 pane, 4' × 7'	60	Every
Double hung, 4 pane, 4' × 6'	55	22
Double hung, 8 pane, 3.5' × 5.5'	45	Days

	Production Rate per 8-Hour Man-days	Normal Frequency in Work Days
Double hung, 12 pane, 2.5′ × 5.7′	40	
Double hung, 16 pane, 4′ × 6′	35	Every
Industrial, 20 pane, 4′ × 7′	30	22
Austral casement, 6 pane, 6′ X 7′	35	Days

Staffing

The man-day labor requirements for window washing can be determined by use of the following formula:

$$\text{Man-days (of labor)} = \frac{\text{Quantity}}{\text{Production}} \times \frac{1}{\text{Frequency}}$$

where quantity represents the number of windows (by type) to be washed; production is the number of windows, according to type washed per man-day; and frequency is the number of work days between washings.

If the frequency and/or production rate values as established have been modified to meet existing conditions at a particular location, then these modified values should be used in the formula shown for the labor requirement computation.

SAMPLE CALCULATION

Assume a building having 430 double hung type windows of 4 pane, 4′ x 6′ size plus 120 industrial type of 20 pane, 4′ x 7′, with existing conditions such that the established frequency and production rate values may be used.

Then the man-day requirement is computed as follows:

$$\text{Man-days} = \frac{430}{55} \times \frac{1}{22} + \frac{120}{30} \times \frac{1}{22}$$
$$= 0.355 \quad + 0.182$$
$$= 0.537$$

To this figure must be added a factor to compensate for leave in order to arrive at the total number of positions needed to accomplish the work.

Conversion factor including compensation for leave=1.15
Conversion factor × job = total positions required
1.15 × 0.537 = 0.62 positions required

• Where the window washing duty does not require the full time of the employee, other duties should be assigned to secure 100 per cent utilization of his time.

Equipment and Materials Allocation and Replacement

Items	Allotment per Cleaner	Replacement or Reissue
Ammonia, household *	As required	As required
Belt, safety	1	As required
Brush, counter dust	1	2 per year
Bucket (12 qt.)	2	2 per year
Chamois	1	4 per year
Cloth, wiping	2	24 per year
Ladder, safety	1	As required
Pad, floor ⎫ Pad, sill ⎬ Scraper ⎭	As required	As required
Sponge, cellulose	2	12 per year
Squeegee, window	As required	As required

* Or other prescribed cleaning material.

33 / AIR COMPRESSORS

(*Source:* Minneapolis-Honeywell Regulator Co., Minneapolis)

A definite schedule of maintenance should be provided for all compressors. Frequent inspections of the unit, along with cleaning and lubrication, will add considerably to the life of the compressor. While certain time intervals are indicated for maintenance procedures, in most instances they are considered minimum, and actual scheduling should depend upon usage.

A compressor inspection program should always include an examination of the air lines and thermostats near the compressor. Remove one or two thermostats and check for oil or water in the connections or in the controllers. Continued presence of oil is an indication that the compressor motor is worn. Moisture in the lines (assuming that the tank is drained regularly) is an indication that the source of air intake is improper or that moisture is condensing because piping runs through cold areas.

Frequently it is necessary to remove the air intake from the boiler room to the outside where the relative humidity will not be as great. Because of the prevailing high relative humidity in some sections, it may be necessary to install a moisture condenser downstream from the compressor.

Checklist for Servicing Compressor

Weekly

1. Check level of oil in crankcase.
2. Drain water from tank.
3. Drain combination filter-regulator.
4. Check regulator setting.

Every Three Months

1. Clean cooling fins on compressor.
2. Change crankcase oil.
3. Check oil in motor bearings.
4. Clean and reoil intake filter.
5. Clean filter element in combination filter regulator.
6. Check motor brushes and commutator.
7. Check pressure switch contacts.
8. Test belt tension.

Lubrication. Check the oil in the crankcase weekly. Drain the crankcase and change the oil after every 500 hours of operation or every three months, whichever comes first. Fill to level shown on sight gage. Do not overfill. A good grade of oil should be used, and temperature consideration should be given in choosing the correct oil.

Temperature above 50° SAE No. 30 High Detergent
Temperature below 50° SAE No. 10 High Detergent

Draining. Drain the tank periodically by opening the drain cock. Allow the compressed air to blow for a few seconds to insure a complete purging of oil and water from the storage tank. If increasing amounts of oil are present in the tank, the compressor pump should be checked for wear. The frequency of draining depends on climatic conditions. Weekly draining properly, is the minimum requirement, however in humid climates, daily draining may be necessary.

Drain the combination regulator, low pressure relief valve, and discharge air filter whenever the tank is drained. A drain cock is provided on the base of the assembly.

Filters. The air intake filter silencers have an oil wetted filter element. To maintain low resistance to air flow, this filter element must be thoroughly cleaned and reoiled whenever the compressor oil is changed. Unscrew the wing nut and remove the screen filter element. Wash it with trichlorethylene. *Do not* clean the filter element in gasoline. Gasoline fumes in the unit would be an explosion hazard. Allow the filter to dry, then dip it in SAE 30 to 50 weight engine oil. Drain off excess oil and reassemble the unit. The filter element may require periodic replacement.

The resin impregnated cellulose filter element of the discharge air filter should also be cleaned when the compressor oil is changed. Loosen the nut at the end of the regulator, remove the end plate, and slip the filter element off its shaft. Wash the filter element in trichlorethylene, allow to dry, and reassemble the unit.

Regulator setting. Check the setting of the combination regulator, low pressure relief valve, and discharge air filter weekly by reading the low pressure gage. If delivery pressure is incorrect, see that the tank pressure is adequate before changing the setting. To change the setting of the regulator, loosen the locknut on the adjusting screw, turn the screw clockwise to raise the outlet pressure, counterclockwise to lower the outlet pressure, and tighten the locknut.

V-belt adjustment. For maximum belt life, compressor efficiency and quiet operation, the V-belt must be kept at proper tension. Adjust the tension by sliding the motor in its slotted mounting holes. Correct tension will allow approximately one-half inch of belt depression.

Electric motor servicing. *Inspection.* The motor should bring its normal load up to speed within a few seconds. If it does not, check the connections between the motor and the power supply and the overload protector, if the motor is so equipped. Be sure the power supply is the correct voltage and frequency.

Cleaning. Clean the commutator when the copper is scored or blackened, by lightly touching No. 000 grade sandpaper against it while the motor is running. *Do not* use emery. If the motor has been exposed to dirt, grit, or moisture, clean and dry thoroughly. Remove dirt and oil from rotor and from stator windings.

Lubrication. Use only a light grade electric motor oil. Lubrication of bearings should be checked three or four times a year. When oiling be sure that the oil wells are completely filled. Wool yarn packing bearings, properly filled, will hold enough oil for a year's operation. As a safety precaution, be sure to disconnect the power supply before attempting to oil the motor.

Cleaning compressor. The cooling fins on the compressor cylinder should be cleaned periodically to prevent oil and dirt from collecting on them. When cleaning the compressor, check to see if any bolts have worked loose, or if the compressor, valves, gages, or connections have been loosened or damaged.

Low pressure relief valve. Every six months raise the outlet pressure of regulator (see "Regulator Setting"), watching the low pressure gage to be sure the pressure does not exceed the safe operating psi (pounds per square inch). Turn the adjusting screw of the regulator back to its original setting. The excess pressure should bleed off through the two orifices in the front of the regulator. This bleeding will clean the orifices and prevent clogging.

Tank relief valve. Manually open the tank relief valve and allow air to blow through for a few seconds once every six months to keep the valve from freezing in the closed position.

Trouble Shooting Chart

Trouble	*Possible Causes*	*Suggested Remedy*
Compressor inoperative— motor will not run.	Electrical:	
	blown fuse	Replace fuse.
	open switch	Close switch.
	overload protection cutout	Press reset button.
	broken connection	Call electrician to repair connection.
	pressure switch—open	Release air in tank, inspect switch. ("Repair, Pressure Switch.")
Compressor inoperative— but motor runs.	Check valve	Release all air in tank. (See "Compressor Failure and Repair Procedure.")
	Relief valve	"Repair Relief Valve."
	Broken compressor	Rotate fly wheel by hand. ("Repair Compressor.") Consult manufacturer's manual.
Compressor runs all the time or takes too long to build up.	Faulty devices or leaks in system	Check system.
	Undersized compressor	Replace, or add additional compressor.
	Carbon on compressor valves	Remove carbon. ("Repair Compressor.")

Trouble	*Possible Causes*	*Suggested Remedy*
	Combination regulator set wrong or not working	Reset. (Repair regulator).
	Plugged intake filter or line	Clean filter and/or line.
	Pressure switch set for wrong cut-in and cut-out pressure	Reset. (Repair switch).
	Compressor needs repair	"Repair Procedure." Consult manufacturer's manual.
Noisy compressor	Carbon buildup	Remove carbon. ("Repair Procedure.")
	Worn bearings or rods	Replace or repair. ("Repair Procedure.")
	Lubricating failure	"Lubrication Failure." Procedure—consult manufacturer's manual.
Compressor uses too much oil.	Worn parts or overfilling	"Excessive Oil Consumption." Procedu consult manufacturer's manual.

NOTE

See additional trouble shooting information following.

Treatment of Malfunctions

Compressor does not operate. *Motor failure.* If the compressor does not function, disconnect the belt and see if the motor will run. If the motor does not run, if it is possible that it is not receiving power. Check for blown fuse, faulty manual or magnetic starter (if installed), corroded contacts on pressure switch, or a short circuit in the wiring.

If the power supply checks out, the motor is probably burned out and will require replacement.

The cause of the motor burnout should be determined.

Examine relief valve and check valve. Check for a faulty pressure switch.

If proper maintenance has been ignored, the tank may be full of water, resulting in reduced tank capacity and placing unusual demands on the equipment. Water in the tank may also damage or corrode other equipment in the system.

Compressor failure. If the motor runs when the belt is disconnected, the compressor may be seized or "frozen." Rotate the flywheel by hand for evidence of binding which is proof of serious damage to the compressor pump.

Other points of inspection are the check valve, the relief valve, and the pressure switch. The failure of any of these may prevent the motor from starting the compressor.

Compressor operates but ineffectively. If the compressor operates, but the system does not function effectively, check out the system for bad connections, faulty devices, leaks, plugged intake or intake filter, or other difficulty. If the system is in good condition, check the operating time of the compressor and the rapidity of the pressure build-up in the tank.

Make a check with intake line and filter off. If the compressor is running most of the time, or if the compressor does not build up tank pressure from 50 to 65 psi within three minutes with the system disconnected or not calling for air, the compressor, or its auxiliary equipment is not operating correctly.

This may result from the installation of an undersized compressor, or from incorrect V-belt tension, leaky air lines and connections, carbon on the valves, or incorrect setting or action of the pressure reducing valve.

The combination regulator may be defective. Check performance with the following procedures:

1. Stop air flow through the regulator. Use the shutoff valve if one is provided or disconnect the line from the regulator and install a pipe plug in the regulator outlet. If the compressor is running, wait until it stops before performing the next step.

2. Manually open the tank relief valve and allow air to bleed off until the compressor starts. Close the relief valve and note the reading on the low pressure gage.

3. Take a second reading when the compressor shuts off. A difference in psi between the two readings may indicate that the regulator needs repair.

Cut-in and cutout pressures should be checked to determine the performance of the pressure switch.

There are two springs in the pressure switch. The larger one determines the cut-in pressure. Cut-in pressure is factory set for psi, but may be raised by compressing the spring. The amount which you are allowed to raise is generally described in the maintenance and repair bulletin accompanying the compressor when purchased.

The smaller spring determines the differential between cut-in and cutout pressure. This is factory set for psi.

If no other cause for inefficient operation can be located, the trouble probably lies in the compressor itself. Carbon build-up or dirt and foreign matter lodged under the compressor valves will cause the compressor to operate below normal efficiency.

Valve difficulty is particularly likely whenever the compressor loses efficiency although it continues to operate at normal speed and although the system has been checked and is in good condition. (See "Repair Procedure.")

Noisy compressor. Noise may be caused by carbon on the piston head and on the valves. Carbon on the piston head will cause rapping as the piston hits the valve plate. Valves must be inspected for carbon build-up which may limit port or lift area. (See "Repair Procedure.") Remove carbon carefully with putty knife. Carbon build-up may be decreased by improving ventilation for compressor room and by providing cooler intake air.

A noisy compressor may also be caused by worn main or connecting rod bearings, excessive end play, or excessive wear of other parts. (See manufacturer's manual.)

Lubrication failure. If the source of difficulty has been located in the compressor itself, or if it appears that the motor has had an insufficient power supply, or the motor or the sheaves have been replaced at any time, a lubrication failure may have damaged the compressor. The size of the sheaves, speed of the motor, and required power supply to the motor are indicated on the installation sheet. Actual speed of the compressor can be checked with a tachometer or by calculating rpm (revolutions per minute) as follows:

$$\text{Motor rpm (see motor nameplate)} \times \frac{\text{Dia., motor sheave}}{\text{Dia., comp. sheave}} = \text{Compressor rpm (approx.)}$$

Excessive oil consumption. A piston type air compressor must pass some oil for proper lubrication of the upper cylinder. It is difficult to determine just how much oil constitutes excessive oil consumption, because the rate of consumption is influenced by several types of operating conditions:

1. High ambient temperature increases oil consumption.
2. A high percentage of running time will cause greater oil consumption than a small percentage of running time.
3. Light weight oil will be consumed faster than heavy weight oil.

Consideration must be given to the preceding influences before a decision can be made that the rate of oil consumption is excessive.

Certain compressor conditions will cause oil consumption:

1. A plugged or dirty intake air line filter.
2. Worn cylinder walls, piston rings, etc. cause increased oil consumption.
3. Loose or worn gaskets or seals will cause oil loss.

Corrective action should be taken to eliminate the cause of oil consumption or loss when these conditions exist.

In general it is recommended that the crank case be drained and oil replaced once each year. If it is necessary to add oil (because the oil level in the sight gage has dropped from the full line to the add oil line) more than twice between changes it can be assumed that the compressor is using excessive oil and the above conditions should be checked to determine the cause, and the cause eliminated.

The discharge air filter should be checked and cleaned every three months.

34 / AIR CONDITIONING-VENTILATING SYSTEMS

(*Source:* The Trane Company, La Crosse, Wisconsin)

Failure to maintain proper conditions of cleanliness in air duct systems and carelessness in connection with repair operations have been important contributing causes of several fires which have involved air-conditioning systems. Systems operated only part of the year should be given a thorough general checkup before starting operation and again after shutting down.

The following recommendations apply, in general, to the period of operation of the system:

General Recommendations

Fresh air intakes. Conditions outside the fresh air intake should be examined at the time of inspection of the ducts. Items to be noted are: accumulations of combustible material near the intake; presence of buildings or structures which may present an exposure to the intake allowing smoke and fire to be drawn in; operating condition of any automatic damper designed to protect the opening against exposure fire.

If accumulations of combustible materials are noted, they should immediately be removed, and arrangements made to avoid such accumulations.

Inspection and cleaning of ducts. Inspections to determine the amount of dust and waste material in the ducts (both discharge and return) should be made semiannually, except that if after several inspections such frequent inspection is found unnecessary. The interval between inspections may be adjusted to suit the conditions.

Cleaning should be undertaken whenever inspection indicates the need.

Cooling and heating coils should be cleaned, if necessary, at the time of cleaning the ducts. Thorough cleaning of ducts may require scraping, brushing, or other positive means. Vacuum cleaning may not remove dust of an oily or sticky nature, or heavy accumulations in the elbows or seams. The amount and kind of dust and dirt will depend greatly on the occupancy and the arrangement of the duct system.

Inspection and cleaning of plenum chambers. Plenum chambers should be inspected quarterly, except that if after several inspections such frequent inspection is found unnecessary. The interval between inspections may be adjusted to suit conditions.

Cleaning should be undertaken whenever inspection indicates the need. Where plenum chambers are found used for storage, arrangements should be made to prevent this, such as keeping the doors locked.

Filters. All air filters shall be kept free of excess dust and combustible material. Unit filters shall be renewed or cleaned where the resistance to air flow has increased to five times the original resistance, or when it has reached a maximum of 0.5 inch water gage, whichever is higher. A suitable draft gage should be provided for this purpose. Draft gages, of a type, which will operate a warning light or produce an audible signal when excessive dust loads have accumulated, are recommended. If the filters are of the automatic liquid adhesive type, sludge shall be regularly removed from the liquid adhesive reservoir.

Disposable filters should never be cleaned and reused.

Care should be exercised in the use of liquid adhesives, use of an adhesive of a low flash point would create a serious hazard.

Electrical equipment of automatic filters should be inspected monthly, observing the operation cycle to see that the motor, relays and other controls function as intended. Drive motors and gear reductions should be inspected at least semiannually, and lubricated when necessary.

Fans, fan motors, and controls. Fans and fan motors should be inspected at least quarterly, and cleaned and lubricated when necessary. Care should be exercised in lubricating fans to avoid allowing oil to run onto the fan blades. Fans should also be checked for alignment, and to see that they are in operable condition.

Fire doors and fire dampers. Each fire door and fire damper should be examined once a year, giving attention to hinges and other moving parts, to see that it is in good operable condition.

Repair work. Great caution should be exercised in the use of open flames or spark emitting devices, inside of ducts or plenum chambers, or near air intakes.

Air Conditioning Equipment

Today's air conditioning plant deserves good periodic care the year round, not just the customary spring and fall checkups at the time the system is started up or shut down. If some sort of regular year-round maintenance program is followed, any air conditioning system will pay off with years of good dependable service, to the advantage both of the user and the contractor who installed it.

Certainly there are a great many items within an air conditioning system which require attention at the beginning of each cooling season. In far too many cases, however, inspection and maintenance are performed only when a breakdown occurs and perhaps at the time the system is started up at the beginning of each hot weather period.

There are no shortcuts to adequate year-round servicing of air conditioning equipment. By following recommendations, it is possible to increase the life and efficiency of any air conditioning installation.

Before considering the frequency of maintenance or inspection periods, let us first establish that there are two times in the year when a complete changeover must be made within the system. These changeovers occur in the spring and fall, when changing from heating to cooling and from cooling back to heating.

These two periods are important ones. A major portion of the maintenance work on the system can be accomplished at these times. The amount of work can be greatly reduced if good periodic attention is given throughout the year. In this connection there are three important points to be considered.

First, it is important that a program of periodic inspection be initiated. The work can be done either by the owner's maintenance personnel or by a reliable air conditioning service contractor.

Second, it is important to see that once such a program is started, it is kept up and each inspection made carefully and completely. After all, there would be no point in conducting such a program if the inspector did not carry out his work faithfully and consistently.

Third, it is important that records be kept of each inspection, indicating the work performed and the parts or material required. From these records the maintenance man and the owner will always know the frequency of replacements, and have a good running account of the cost required to maintain the air conditioning system.

To assist in the program, a complete check sheet should be made up to cover the system. The check sheet should be printed in such a way as to provide a record as well as a checkoff list for inspection points. This could be done on a single sheet to cover a one year period, or on a form that would be complete after each inspection.

A suggested form is shown in Figure 34-1 together with some notation on the work involved. You will note that in our check sheet we have divided the inspection periods into four columns: spring start-up, summer, winter and fall shutdown. In each column are dots indicating those points to be checked during the different inspection periods. Where no dot is shown, that particular point of inspection may be skipped during that period of the year. (See Figure 34-1.)

Figure 34-1

AIR CONDITIONING CHECK SHEET

CONTROLS	Start-up	Monthly S	W	Shutdown
Clean commutators D.C. motors.		•	•	•
Check thermostat setting and clean points.	•	•	•	•
Check contact points in starters. (Clean if necessary.)		•	•	•
Check steam valves and traps, dirt, leakage, etc.			•	•
Check damper motor.	•	•	•	•
Observe operation of system.	•	•	•	•
Check controls on water cooler.	•	•		
Air controls—drain water.		•	•	
Clean air compressor intake.		•	•	

CONTROLS	Start-up	Monthly S	W	Shutdown
Check air compressor oil.		•	•	
Oil air compressor motor.		•	•	

As mentioned previously, it is important that each spring and fall the entire air conditioning system be gone over carefully and checked for signs of wear and deterioration. Through the winter months, when cooling is not required, the maintenance man will have a good opportunity to make repairs of a preventive nature.

The Refrigeration System

There are three important things to protect in the refrigeration system: the compressor, the motor driving the compressor, and the refrigerant.

Cooling required in the colder months can usually be accomplished with outside air. It is not necessary for the refrigeration equipment to operate. In the fall, the refrigeration system should be pumped down and the refrigerant valved off in the condenser or receiver. By confining the refrigerant in the part of the system with the smallest number of joints, changes for loss will be minimized. Preventing loss of refrigerant is most important in shutting down a system.

After pumping down the refrigerant, the piping and vessels in which it is stored should be checked for leaks with a halide torch. If leaks are found in the condenser or receiver, the refrigerant should be removed from the system and leaks repaired. The system should be pressure tested after repairing. When pressure testing is completed, the system should be evacuated with an auxiliary vacuum pump. The refrigerant should then be charged back into the system and the system pumped down.

In the spring when the valves in the system are opened, the system should again be checked for refrigerant leaks. If, when opening the valves, only a small amount of refrigerant is allowed to pass from the condenser into the rest of the system and the system checked for leaks, a large amount of refrigerant can be saved should a leak exist.

The Compressor Unit

When the refrigeration system is pumped down, the compressor should be valved off thus preventing any loss of refrigerant through the compressor. If the compressor unit is belt driven, the motor should be shifted toward the compressor to lessen the tension on the belts. This prevents the drive belts from taking a "set". In the spring when the compressor unit is being readied for the summer months, the belt drive should be adjusted so as to give the proper belt tension. This is a good time to check the condition of the belts themselves and to order replacements if necessary.

If the compressor and motor are connected by a drive coupling, the coupling should be examined for tightness and wear.

The drive motor should undergo an annual cleaning. If the motor has ball bearings, the existing grease in them should be removed and replaced.

Once a year, the compressor should be checked thoroughly to determine if parts need repair or replacement. Overhaul services are best performed in the colder months when the unit is out of operation. If the crankcase oil is dirty or contains sludge, it should be drained from the compressor, the crankcase cleaned, and new oil put in.

Water Cooled Condensers

In the fall, water cooled condensers should be drained and checked for scale and mud. If the condenser tubes are fouled, they should be cleaned to deliver maximum efficiency. If the condenser is located in an area where subfreezing conditions are experienced, water should not be allowed to remain in the condenser, in the supply lines, or in the fittings.

Towers and Evaporative Condensers

Due to location and methods of evaporative condenser and cooling tower operation, complete and careful maintenance is required to preserve and prolong their operating efficiency. The fact that they are continuously handling warm moist air promotes fouling and deterioration. If not attended to periodically, their operating efficiency will be lost rapidly.

The unit should be drained at regular intervals and all sediment flushed from the sump. In the fall and spring, all metal surfaces should be examined for signs of rust. Where corrosion is taking place, the surface should be scraped or cleaned with a wire brush. The bare metal should then be covered with a protective coating of good chlorinated rubber base paint.

The following four suggestions have to do with care for working parts of the evaporative condenser:

1. Check the condition of the condensing coil. If scale has formed, it should be removed.
2. Spray nozzles should be checked and cleaned if necessary.
3. Pump strainer, water strainer, and air intake screen should be flushed and cleaned.
4. Float controls should be inspected and repaired as necessary.

In the fall, the entire unit should be prepared for winter inactivity. The fan motor and drive and the water pump and drive should be removed and taken indoors, or at least protected from the elements. If this is done each year, the work of placing the system into operation in the spring will be greatly reduced.

When shutting down the tower or the evaporative condenser in the fall, all points should be examined for low spots where water might collect and freeze, thus causing breakage. If such a condition should be discovered, it is suggested that the piping be disconnected at certain points to permit venting and complete drainage.

When water is drained from the pump, the pump should be protected against corrosion while it is allowed to remain idle. In many installations it may be possible to fill the pump with a light oil. When placing the pump in operation in the spring, the major portion of oil can be poured out of the pump and the remaining oil flushed from the pump by the water passing through it.

Wherever drives include belts, the belts should be removed and stored indoors during the winter months. Gear boxes should be cleaned, serviced and protected against cold weather conditions. All shafting should be inspected for rust, cleaned and protected. In the spring before placing the equipment in operation all of the belts should be inspected for wear and replaced as necessary. Gear boxes should be cleaned and filled with fresh lubricant.

In the spring, all bearings should be checked for wear, tightness, alignment and proper lubrication. Worn bearings should be serviced or replaced. Grease type motor bearings should be examined, flushed and regreased.

Air Units

At the beginning of the heating season and again at the beginning of the cooling season, the air handling units of the air conditioning system should be gone over completely as follows:

1. All motor, fan shaft and pump bearings should be inspected, serviced and lubricated.

2. All set screws and drives should be checked and tightened. If they have worked loose, the shafting should be checked for damage.

3. Drive belts should be checked for drive alignment and wear.

4. Air filters should be cleaned or replaced. Outside air intakes should also be examined and cleaned.

5. Other important components to be checked and cleaned are pumps, line strainers and spray nozzles.

6. In the fall, water cooling coils should be drained. The surfaces of direct expansion coils, water cooling coils and steam coils should be inspected for lint or dirt collection. Where necessary the coil surfaces should be cleared and cleaned. Since most of the foreign material deposited on coils is grease, a good detergent should be used to clean the surface.

Controls and Accessories

Last, but certainly not least, are the following check points on controls and accessories:

1. Clean thermostat points.
2. Clean or replace contact points of starters as necessary.
3. Check steam valves and traps for dirt or leakage.
4. Check damper motor and action of dampers.
5. Check the condition of dampers and damper linkage.
6. Check the seal around dampers and replace felting or other material as required.
7. Observe operation of the system.
8. Periodically, water should be drained from the pneumatic controls. The settings should be checked, adjusted or calibrated as required.
9. Clean intake on air compressor.
10. Check air compressor crankcase oil level.
11. Oil the air compressor motor.

It is important that this plan for systematically starting up, checking, and shutting down cooling systems be followed as closely as possible if they are to be expected to operate with maximum efficiency and minimum emergency service. Compared with the investment represented by the original cost of such systems, the small amount of time, effort or money which must be expended for periodic checkups is more than compensated for by the longtime trouble-free operation which will result. (See Figure 34-2.)

Figure 34-2

CHECKLIST FOR YEAR-ROUND AIR CONDITIONING SERVICE

COMPRESSORS	Start-up	Monthly S	W	Shutdown
Lubricate motor bearings.	•	•		•
Check drive.		•		
Check oil level.	•	•		
Check head pressure.	•	•		
Check suction pressure.	•	•		
Check pressure at settings.	•			

Figure 34-2 *(Continued)*

CHECKLIST FOR YEAR-ROUND AIR CONDITIONING SERVICE *(Continued)*

COMPRESSORS	Start-up	Monthly S	W	Shutdown
Check rotation.	•			
Slack off belts.				•
Align and adjust belts.	•			
Check head bolts.	•	•		•
Check cylinder bypass pressure at setting.	•			
Check oil pressure.	•	•		
WATER COOLED CONDENSERS				
Check water regulating valve for best economy.	•	•		
Drain condenser.				•
Reconnect condenser piping.	•			
EVAPORATIVE CONDENSERS				
Lubricate motor bearings.	•	•		•
Grease fan bearings.	•	•		•
Grease pump bearings.	•	•		•

Figure 34-2 *(Continued)*

CHECKLIST FOR YEAR-ROUND AIR CONDITIONING SERVICE *(Continued)*

EVAPORATIVE CONDENSERS	Start-up	Monthly S	W	Shutdown
Check drive belts.		•		
Slack off belts.				•
Check rotation of fans.	•			
Check condition of coils (scale).		•		•
Check spray nozzles. (Clean if necessary.)	•	•		•
Clean drip pan and drain.		•		•
Clean water strainer.		•		•
Clean pump strainer.		•		•
Clean air intake screen.		•		•
Check float control.	•	•		•
Drain water piping and drip pan.				•
Reconnect water piping.	•			
Align and adjust belts.	•			
Check fans for alignment.	•			

Figure 34-2 *(Continued)*

CHECKLIST FOR YEAR-ROUND AIR CONDITIONING SERVICE *(Continued)*

AIR UNIT	Start-up	Monthly S	Monthly W	Shutdown
Lubricate motor bearings.		•	•	
Grease fan bearings.		•	•	
Grease pump bearings.		•	•	
Check drive belts.				
Check fans for alignment.	•			
Check filters.		•	•	
Clean expansion valve, drip pan and drain.				•
Clean drip pan and drain.				•
Clean oil air intake screen.		•	•	
Check expansion valve adjustments.	•	•		
Clean pump strainer.		•	•	
Check spray nozzles. (Clean if necessary.)		•	•	
Drain water coils. (Water cooling.)				•
Reconnect lines to water coils.	•			

35 / **AUTOMATIC CONTROLS**

(*Source:* "Pneumatic Temperature Controls," *National Engineer*, illustrations—Johnson Service Co., Milwaukee)

Types of Systems

Automatic control systems may be classified into four general types, depending upon the basic operating principles. *Self-contained* control systems have their own motive power, while *pneumatic, electric* and *electronic* types are dependent upon an external source of energy. In addition to heating, ventilating and air conditioning, other building, mechanical and electrical systems may be supervised by central control.

Self-contained devices have many applications. They are not generally used where sequence operation is required. They cannot be interlocked to provide integrated control of valves, or motors and dampers. Installation is difficult where the thermostat must be remote from its controlled valve, and the self-contained system is difficult to repair. However, the construction is rugged and simple and there is little in the mechanism to cause failure.

Pneumatic temperature controls. Pneumatic temperature control systems have been in existence nearly as long as the temperature control industry itself. "Pneumatic" simply means air and a pneumatic control system is operated by air. All of the control equipment in a system is interconnected by small concealed copper or plastic tubing into which compressed air is released.

The thermostat controls the pressure of the air. This air pressure furnishes the power for the system. "Thermo" comes from the Greek word meaning heat, and "stat" is derived from the Greek word meaning apparatus to render something stationary. Put them together and you have an instrument that will keep the temperature always the same.

The thermostat automatically allows just the right amount of compressed air to pass to the dampers or valves which respond by opening or closing to let more or less heat into a room. Inside the thermostat is a small paper-thin bimetal strip or element composed of two different strips of metal bonded together. One is a metal that expands when it gets warm, while the other remains constant. Because of this, the strip bends as the surrounding temperature changes.

The bimetal element in the thermostat is mounted in front of a small hole or control port. Through the control port, varying amounts of compressed air is allowed to escape, regulating the output pressure of the thermostat. By its movement, the bimetal element determines how much or how little air escapes. As more air escapes, the output pressure is decreased and the valve gradually opens. As less air escapes, the output pressure increases, gradually closing the valve. All controllers and controlled devices in an air conditioning, heating, or ventilating control system generally operate on the same principle.

Electric and electronic controls. All electric controls, which may be interlocked and easily extended, are generally the choice for smaller simple mechanical systems. When the number of control points needed and the amount of synchronization and correction increases, pneumatic controls become more economical. When a system becomes larger and more complicated, it is far less costly to link the control elements with pneumatic tubing and pneumatic operators than with electric wire conduit and motors. Electric controls are known to offer economy, when packaged air conditioning does not exceed fifty tons. Pneumatic equipment is most effective for central systems of thirty tons or over, where there are several zones to be controlled.

It has become common practice to integrate electric and electronic devices with pneumatic control systems. The difference between electric and electronic systems lies in what is known as esoteric refinement in equipment. In an electric control system, the thermostat transmits its signal directly to the component which must be regulated (valve, motor, or dampers, etc.). In an electronic control system, the thermostat sends its signal to an electronic relay which measures and amplifies it. These command pulses are transmitted to the receiving relays by means of carrier current pulses superimposed on the light and power mains of a building or a group of buildings, and are produced by two basic types of equipment: tube transmitters and rotary generators. After receiving the pulsed signal and decoding it, special relays activate various types of motor starters, motorized valves, magnetic switches, etc. Electronic control equipment may be used to energize any type of electric or pneumatic device in a number of ways, or operation cycles.

A thorough semiannual inspection should be made of all operating equipment—relays, breakers, motors, valves, thermostats, and associated equipment. They should be cleaned and calibrated, oiled and greased if necessary, packing should be checked, and they should be tested for correct operation monthly. Compressors should be blown down weekly, motors and other operating parts should be checked monthly.

This work can be done by *trained* men from the maintenance department. It is imperative that they thoroughly understand the complete function of the system. The disconnecting of a relay-shutting of a valve may prove costly, and may be the cause of much discomfort. Maintenance agreements are generally available to cover periodic maintenance, emergency service, parts and replacements.

35A / FUNDAMENTALS OF AUTOMATIC COMBUSTION CONTROL

(*Source:* "Fundamentals of Automatic Combustion Control," by Crayton H. Schwestka, Hays Corporation, Michigan City, Indiana)

As a prerequisite to solving any control problem, whether it is the servicing of an existing control system or the working out of a new application, an understanding of basic control theory is helpful. This, at first glance would seem to be a bewildering undertaking.

Especially with automatic combustion control, much confusion lies in the fact that there are numerous mechanisms used in such control systems, different in application and design yet quite similar in operation. It is with the purpose of explaining and illustrating some of this basic theory that these articles are presented.

Controller Actions

Automatic controllers may be broadly divided into two groups: discontinuous and continuous controllers.

Typical of the discontinuous type are the on-off, two position, multiple position, and average position controllers. A good example of this type controller is a room thermostat which simply turns the heating equipment on and off to hold room temperature close to its set point.

Although the discontinuous controller is inexpensive and easy to adjust, it is useless in most process control because there is insufficient capacitance in most of the processes to allow its use.

Continuous-type controller action is subdivided into three classifications: proportional position, integral, and derivative action.

As an introduction to the formal discussion, refer to Figure 35-1. This is a representation of a control system involving a human being, a tank of water with variable demand, a water supply, and a control valve. All three controller actions are involved as follows:

Proportional position. The lower the water level, the wider will the valve be opened and vice versa.

Integral. As the level stabilizes at a new control point, valve is readjusted to return the level to its correct set point.

Derivative. If level changes suddenly, valve will be opened or closed beyond that position required for normal control in order to bring the level back to set point more quickly.

Each of the above controller actions will be discussed individually in this section.

Proportional Position Action

Proportional position action is that action in which a continuous linear relation exists between the value of the controlled variable and the position of the final control element.

This is illustrated by Figure 35-2. Here, water level is to be controlled at a predetermined point. There is an exit pipe through which a demand flow of water leaves the tank. Inlet water is varied by means of the gate valve so as to maintain desired water level in the tank. In this diagram, the automatic controller consists of the float, balance beam with its pivot or fulcrum, and the gate valve. The float controls one end of the beam and the opposite end of

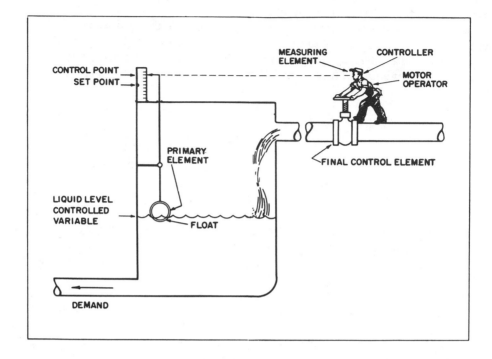

Figure 35-1 / **Three basic controller actions are involved when human operator controls water level in tank. Automatic controllers can do the same job—better, if properly applied.**

the beam determines position of the valve in the inlet water line. The float is the primary element of the measuring means. It obtains information as to the value of the controlled variable, the water level. The end of the beam "Y" is the operator. The gate valve is the final control element which varies water flow, the manipulated variable.

Assume that the inlet water line is 10 inches in diameter and that the amount of water flow is directly proportional to the position of the gate valve. For example, if the gate valve is fully open, water flow is 100 gallons per minute; if it is 50 per cent open, the flow is 50 gpm. Assume that the fulcrum is at the exact center of this beam balance so that the arm X equals the arm Y and that, when the outlet valve is completely closed, the water level is 40 inches in the tank. It can be seen that when there is a 50-gpm demand for water in the outlet pipe, water level in the tank will drop and position of the valve will be directly proportional to that water level until the water level has dropped to the 35-inch level.

Assume the demand for water at a 50-gpm rate and level of water in the tank to be maintained at 35 inches when the inlet flow equals the demand. The 35-inch level is termed the "set point," in this case at 50 per cent of the range of the controlled variable. As the demand for water increases or decreases, the level will deviate from this set point and actuate the automatic controller in order to allow more or less water to flow through the inlet pipe.

Departure of the controlled variable from the set point is termed deviation. Sustained deviation, which always occurs when the demand exceeds or falls below the 50-gpm rate, is termed offset. The new value at which the controlled variable is maintained for this new demand is termed the control point.

Proportional band. Proportional band is the range of values of the controlled variable which corresponds to the full operating range of the final control element. If the position of the final control element (or operator) is represented by V and the deviation from the set point of the controlled variable is represented by θ, it follows that:

$$\frac{\theta}{V} = -k$$

where k is the proportional band factor. The value of k is negative to show that the corrective action is opposite to the deviation.

As shown in Figure 35-2, full operating range of the final control element is 100, and the range of values of the controlled variable is 100, then:

$$-k = \frac{\theta}{V} = \frac{100}{100} = 1.0$$

or, as is more common, expressed in percent of controller scale:

$$\% \text{ proportional band} = 100 \frac{\theta}{V}$$

In Figure 35-2 proportional band is adjusted by changing the ratio of the beam lengths x and y. With the fulcrum as shown, arm x equals arm y, so that in this case:

$$\% \text{ k} = \frac{x}{y} \text{ } 100 = 100\%$$

This may also be represented by the graph (Figure 35-3).

Suppose the fulcrum is moved to the left, so that the beam length y is 10 times that of the length of arm x. In this case:

$$\frac{x}{y} = \frac{1}{10} \text{ } 100 = 10\% \text{ proportional band factor.}$$

By further analysis, it will be seen with this ratio if the 50 per cent open position of the valve corresponds to a 35-inch level of water in the tank; then, as the demand for water increases to the maximum flow capabilities of the inlet pipe, or 100 gpm, deviation of the level will be only ½ inch below the set point. Conversely, if the demand decreases to zero, upward deviation of the controlled variable will also be only ½ inch to enable full closing of the valve.

Thus, in an automatic controller having proportional position action, as the proportional band factor is decreased the offset is also proportionally decreased. The reverse is also true. In addition, the set point can be located at any point in the normal operating range of the controlled variable. In Figure 35-2, the set point can be changed by adjusting the position of the float sus-

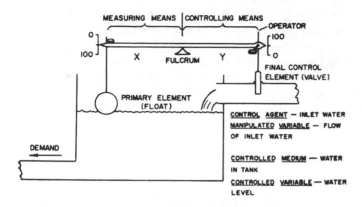

Figure 35-2 / **Simple proportional position controller will do adequate job on applications such as this. The automatic controller consists merely of the horizontal beam.**

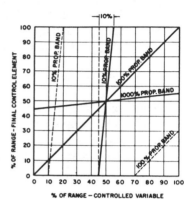

Figure 35-3 / **Dotted line indicates proportional band can be shifted in set point without affecting its slope.**

pension rod in the x end of the beam to the desired level for equal inlet to outlet flow conditions:

Stability. Referring to Figure 35-2, if the proportional band were made extremely narrow, for example, by moving the fulcrum so that the ratio $\frac{x}{y} = 1\%$, deviations of the level will be only 5 per cent above or below the set point for full motion of the gate valve. This action closely parallels action of the two-position or on-off control. Here, any disturbance will cause oscillation or hunting of the automatic controller. It has been shown, conversely, when the proportional band is high, the sustained deviation or offset will be proportionally high. Thus it can be seen that in an automatic controller having proportional position action, a compromise must be made between decreasing the proportional band to decrease offset and keeping it large enough to prevent instability. In a process in which large load changes are apt to occur, proportional position type action is apt to be inadequate unless a narrow proportional band can be used. Danger of oscillation or hunting then becomes a major factor. The above problems are expressed graphically in Figure 35-4, which shows the response and tendency toward hunting or the lack of it for various proportional bands.

Combustion control. The application of a proportional position master controller in a combustion control system to fairly high capacity, low-pressure boilers is quite acceptable. Steam pressure, as the controlled variable, reflects boiler load almost exactly. On such installations some variation in the con-

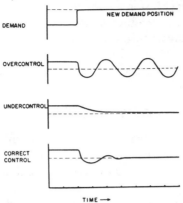

Figure 35-4 / **Characteristic of proportional position action allows stabilization by increasing the proportional band.**

trolled pressure is generally permissible and the proportional position action is sufficient for control purposes. The master controller sets up a loading impulse for the remaining controllers, proportional to steam pressure (boiler demand), in order to maintain a uniform steam pressure.

In the Electric Master Controller, illustrated in Figure 35-5, steam header

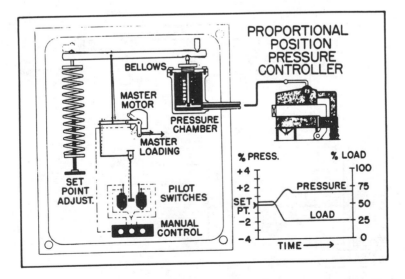

Figure 35-5 / **Hays Electric Master Controller for Combustion Control** uses mercury switch pilot device to operate master motor. Approximate 1% pressure variation results from 25% load swing.

pressure is applied to the metallic bellows on the right and is balanced by effect of the calibrating spring on the left. A drop in load (increase in pressure) is shown. Through linkage and the pilot device, the loading motor, located in the rear of the Master Controller, resets center horizontal lever to return pilot device operating magnet to its neutral position. In so doing, however, the loading motor travels through an angular distance in direct proportion to the increase in boiler pressure. Since the fuel and air controllers are loaded by the motor, firing rate is also reduced proportionally.

Integral Action

Most common forms of integral action are usually expressed in terms of floating action or as reset. Reset, however, is a term applied only to a particular form of floating action when used with proportional position action. This will be discussed later.

Floating action may be subdivided into two common types: (1) single speed floating, and (2) proportional speed floating.

To conveniently illustrate single speed floating action, the mechanical level controller previously described is converted to an electrical controller. Initially, its proportional position action is retained.

This is shown in Figure 35-6. Assume that the transmitting slidewire position is exactly balanced by the rebalancing slidewire position. When the condition holds, there is electrical balance. If water level in the tank drops, the transmitting slidewire moves downward.

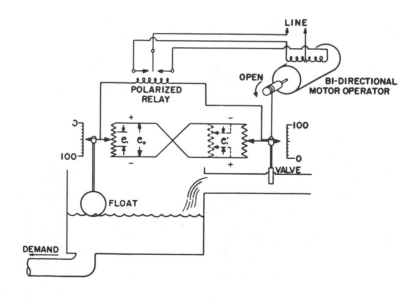

Figure 35-6 / **Proportional position controller, using electric components, has advantage of permitting remote operation of valve from float. Removal of link between valve and wiper provides single speed floating action.**

This puts the transmitting slidewire at a potential more negative than that of the rebalancing slidewire. Current flows from negative to positive through the coil of the polarized relay. When current flows in this direction, the armature is moved to the right to make contact with the counterclockwise terminal of the motor operator or power unit.

The motor shaft revolves in a counterclockwise direction, and the wiper of the rebalancing slidewire is moved upward to a more negative position. At the time it reaches the equivalent negative position to which the transmitting slidewire has moved, no current flows in the polarized relay and its armature returns to the neutral position. Thus, rebalancing has occurred and, in so doing, the motor operator has also lifted the gate valve or final control element to allow more water to enter the tank. The converse is also true.

Proportional band control is added to this circuit by controlling voltage to either or both of the slidewires. The dotted lines in Figure 35-6 show a proportional band adjustment potentiometer added to adjust voltage across the rebalancing slidewire.

Suppose that the proportional band adjustment potentiometer wipers are moved closer together as shown so that the potential e_1 is 50 per cent of the potential e_0 applied to the transmitting slidewire. When this is done, it can be seen that full range movement of the rebalancing slidewire wiper must be obtained for a movement of the transmitting slidewire wiper equal to the voltage range shown by e_1. Since the full operating range of the final control element is 100 per cent and the range of values of the controlled variable is only 50 per cent, the proportional band can be expressed as:

$$100 \ \frac{e}{V} = 100 \ \frac{0.5e_0}{e_0} = 50\%$$

Single speed floating action. To illustrate this action, the link between the final control element and the wiper of the rebalancing slidewire shown in Figure 35-6 is removed. The wiper is then moved manually to a position representing the desired set point, termed the set point adjustment.

Assume that equilibrium conditions are obtained, i.e., the position of the float or transmitting slidewire wiper is electrically equal to the set point. When demand increases, tank level decreases, thus dropping the float and its slidewire to a more negative position. This unbalances the circuit, and the gate valve is opened wider. It continues to be opened more and more until the level of water in the tank increases and the float and its wiper are again raised to an electrical position equal to the set point. At this time, the electrical circuit again becomes balanced and the motor operator is deenergized.

However, the gate valve may have been raised or opened to a position greater than that needed to maintain the level at the set point. If this is true, the level will continue to rise and the circuit will again become unbalanced but in the opposite direction. This will allow the gate valve to drop and decrease the flow of inlet water. However this can become a cyclic recurrence known as "hunting" or "oscillation." Tendency of the single speed floating action controller toward hunting or continuous oscillation is one of its great disadvantages. One of its great advantages, however, is that of not suffering from any offset plus the fact that it is adapted to control of nonlinear final control elements.

Proportional speed floating action. It can be seen in the following example that, when a small increase in demand flow occurs, the rate of change of tank level will be small. Motor speed is required proportional to the deviation. This type of action can be obtained and is known as proportional speed floating. (See Figure 35-6A.)

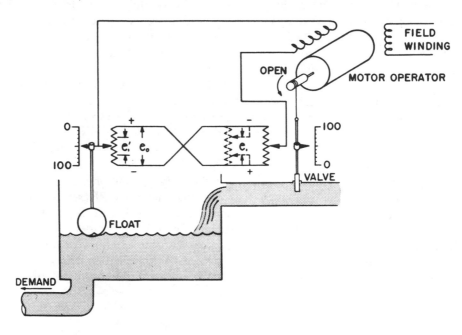

Figure 35-6A / **Proportional speed floating action controller has great advantage of no offset control. This is extremely important in control of furnace draft, for example.**

Here the polarized relay is replaced directly by the armature winding of a DC motor having a separately excited field. Rotation speed of this motor is directly proportional to the voltage across the armature circuit. Direction of rotation is determined by direction of the current through the armature winding. With this arrangement, if the change in demand flow is small, the level change is small, and vice versa.

Therefore, if a large sudden drop in level occurs, the motor speed will at first be great and raise the valve at a fast rate. As the inlet flow rapidly increases and raises the level of the float in the tank, the voltage difference is reduced, producing a gradual reduction in motor rpm proportional to the increase in level until the set point is reached. The proportional band potentiometer in this example controls the voltage appearing across the controlled variable slidewire which affects, in turn, the amount of voltage appearing across the motor for a given deviation from set point. This determines the motor speed for that deviation.

Combustion control. Proportional speed floating controllers are the "work horses" in most phases of boiler control. Typical of these applications are boiler feedwater control, superheat and reheat controls, and fuel air ratio controls.

Proportional speed floating action is obtained in the majority of diaphragm unit controllers by adding a piston type pneumatic pump, driven by the motor operator or power unit. This pump is referred to as the feedback or stabilizer piston. Its output is connected into the diaphragm unit impulse line to partially oppose any change.

Electric control. A typical controller of this type is illustrated in Figure 35-7, which shows a Constant Suction Controller (CSE) used to operate an induced draft fan damper. Furnace draft applied to the diaphragm is balanced by the calibrating spring. Unbalance produced by an increase or decrease (deviation) of furnace draft from the set point moves the magnet arm to close either the decrease or increase mercury switch, respectively, causing the power unit to so move.

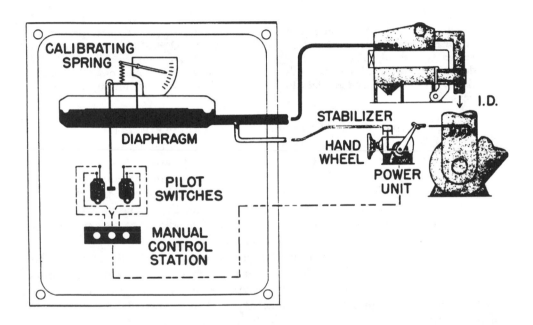

Figure 35-7 / **Hays Type CSE Constant Suction Controller as applied to furnace draft control.**

Initial motion of the power unit is at a constant speed. This motion, while changing the damper position in a direction to correct the draft deviation also operates the feedback piston, which develops air pressure in the diaphragm in a direction opposing the change in furnace draft. This causes the mercury switch to open and the power unit to stop.

Within a short period, pressure developed by the piston is absorbed in the furnace draft line. If a deviation from the set point still exists, the mercury switch is again closed and the above action is repeated. Thus, power unit motion is a series of on-off steps which, when integrated with respect to time, exhibit speeds proportional to the deviation. This is illustrated graphically in Figure 35-8.

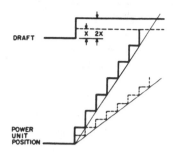

Figure 35-8a / **On-off motion of motor operator (power unit) plotted versus two steady state values of deviation. Integrated motion with time (speed) is proportional to deviation.**

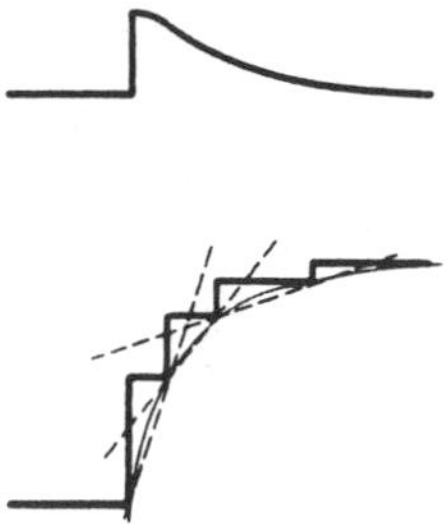

Figure 35-8b / **Typical proportional speed floating action showing deviation, corresponding motor operator motion, and deviation reduction occurring during a load change.**

Electronic control. The Electronic Ratio Controller is applied whenever greater threshold and resolution sensitivity that can be obtained with the mercury switch controller are desired. Its dead band and hysteresis characteristics are correspondingly less. Action of the EARE can be either proportional position or proportional speed floating. Figure 35-9 shows its application as a fuel-air ratio controller, for instance. An electronic amplifier raises the deviation signal level, detects it, and controls a pair of relays, for controlling the increase and decrease windings of the power unit or motor operator.

While the rotational speed of the power unit is at a fixed rpm when energized, action of the electronic amplifier is to energize the power unit intermittently. The on-time to off-time ratio of applied motor energy is proportional, within limits, to the deviation and can be adjusted by means of the dwell control of the EARE. This, as in the CSE, provides a series of on-off steps of the power unit shaft rotation which when integrated with respect to time effectively provides proportional speed floating action.

Desirable Features of Proportional Position Action Combined in a Single Unit

As previously mentioned, proportional position control action will always exhibit offset or droop when load changes occur. On the other hand, integral control action will not immediately compensate for sudden load changes and can exhibit considerable oscillation when measuring lags are high.

To overcome these limitations the two actions are combined to provide proportional plus integral control action. Either single speed floating action or proportional speed floating action in the classification of integral action is used, the latter being more common. When proportional position action is combined

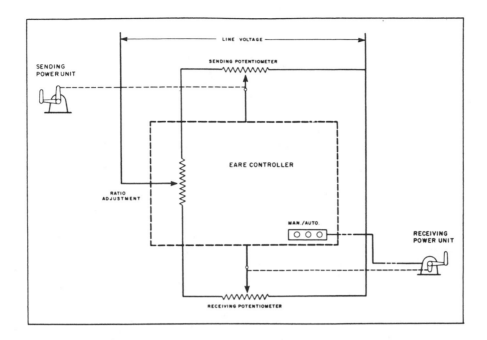

Figure 35-9 / Hays **EARE** electronic proportional position controller provides extreme sensitivity through use of thyratron relay tubes. Ratio between sending and receiving potentiometer position is adjustable.

with proportional speed floating action, the result is given the term "proportional plus reset action."

Proportional position action has been defined as that in which a continuous linear relation exists between the value of the controlled variable and the position of the final control element.

Proportional speed floating action has been defined as that in which the final control element is moved at a speed proportional to the deviation.

Proportional plus reset action is the result of combining proportional position action and proportional speed floating action. The reset action which then results is defined as that in which the final control element is moved at a speed proportional to the extent of proportional position action.

Figure 35-10 again shows the proportional position action controller applied to regulate water level. For example, with a 50 per cent valve position, tank level is maintained at 50 inches. If water demand increases so that the tank level drops and a stabilized condition is reached at 40 inches, the 10-inch difference in level from the 50 inches set point is a sustained deviation or offset.

How can this offset or droop be corrected? One method might be for the operator to loosen the set screw holding the float to the transmitting slidewire wiper of the controller, and readjust the float upward to the desired set point. When the float and wiper assembly is released, the proportional action of the controller will open the valve allowing more water to enter until the level in

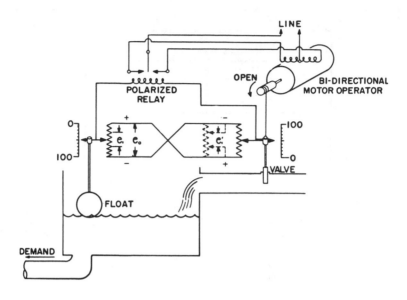

Figure 35-10 / **Proportional position controller illustrated suffers from offset whenever control point is different from set point.**

the tank again reaches the set point. At this point equilibrium conditions are again obtained. The same effect could also be accomplished by the operator loosening the set screw attaching the gate valve to the output of the automatic controller and adjusting the gate valve to a more open position. Again, the water level would rise to the desired level or set point. This action of the operator to "reset" the automatic controller to the set point describes "reset action" in proportional plus reset.

Either of these methods would be time consuming and would be required for each load change. The automatic controller with proportional plus reset action does this resetting action automatically.

Proportional Plus Reset Action

Proportional position action. Figure 35-11 illustrates one way in which this can be accomplished. To the right of the dotted line is the familiar proportional position section. For simplicity, float motion is shown transmitted by a cable passing over a pulley actuating the transmitting slidewire wiper. Unbalance, due to water level fluctuation between it and the rebalancing slidewire wiper connected to the valve actuates the motor operator until balance is effected.

Proportional speed floating action. To the left of the dotted line is the proportional speed floating section of the controller. The float level here also determines the position of the transmitting slidewire wiper of this section. At equilibrium, this is electrically equivalent to the position of the set point

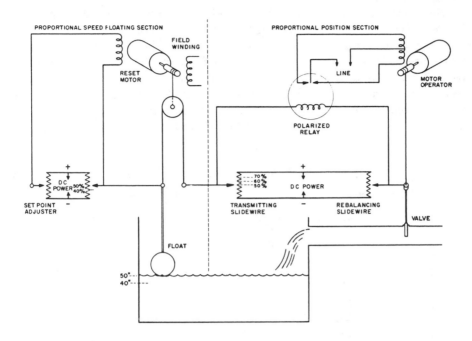

Figure 35-11 / **Proportional plus reset controller includes means for overcoming offset. For simplicity, method for setting proportional band and reset rate is deleted.**

potentiometer wiper. However, when a deviation exists between the float wiper and the set point, the error voltage causes the reset motor to rotate at a speed proportional to the deviation. The pulley over which the cable to the proportional band transmitting wiper travels is thus moved to reduce the proportional speed floating section error voltage.

Assume the control is in equilibrium at a water level set point of 50 inches. The proportional position section wiper is at 50 per cent and the valve is also at 50 per cent position. The proportional band is 100 per cent. A step change disturbance lowers the water level to 40 inches, a drop equal to 10 per cent of range. Since the transmitting wiper is then moved to the 60 per cent position, the proportional position section will immediately (neglecting operating time) open the valve to the 60 per cent position. Equilibrium would then be attained for this section, if used alone.

However, the error voltage, existing in the proportional speed floating section between the level slidewire wiper and the set point slidewire wiper, acts to rotate the reset motor clockwise. This raises the pulley to attempt to decrease the error voltage. Initially, however, the level has not changed, resulting in the wiper of the proportional position transmitting slidewire being raised. The proportional position section is thus again unbalanced and the motor operator is actuated to raise the valve. The resulting increased water flow eventually raises water level to the set point. Thus the proportional position section is stabilized, and a re-setting action has occurred to overcome offset.

Reset rate. The amount of integral action in a controller with proportional plus reset action is usually expressed in "repeats per minute" i.e., the number of times per minute that the integral action repeats the motion of the proportional position action after a step deviation from the set point. In the example given, if one minute is required for the reset motor to raise the wiper of the proportional position transmitting slidewire an additional 10 per cent the reset rate is 1.0 repeats per minute. This action is plotted graphically in Figure 35-12 for a condition (not normally encountered) where the deviation remains constant.

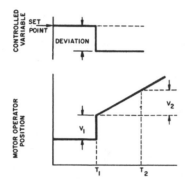 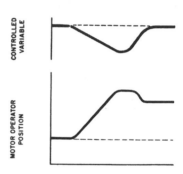

Figure 35-12 / **Proportional plus reset action for a step deviation which then remains constant. If $v_2 = v_1$ and $t_2 - t_1 =$ one minute, the reset rate is 1.0 repeats per minute. Reset rate is usually adjustable and may be calibrated in steps from 0.1 to 10 repeats per minute.**

Figure 35-13 / **A typical proportional plus reset action restoring set point to overcome offset.**

Reset rate is, therefore:

$$\frac{V_2}{(t_2 - t_1)\ v_1}$$

A typical plot is shown in Figure 35-13 of the action obtained when the deviation is corrected by proportional plus reset controller.

Combustion and process control. The type LPME master steam pressure controller was shown to have proportional position action. When its inherent offset characteristics are intolerable, a similar master controller may be used in which integral action is combined with the proportional position action. This is known as the type HPMEC Master Controller.

As we mentioned previously describing integral action, the majority of mercury switch controllers equipped with feedback pistons normally provide proportional speed floating action. However, when the combustion or other process to be controlled exhibits considerable lag or sudden load changes, proportional position action is added to provide greater speed of response. This action is then proportional plus reset.

Typical of the diaphragm-actuated controllers utilizing proportional plus reset action is the Adjustable Differential Pressure controller type ADE-A, illustrated in Figure 35-14. This shows its use to control the speed of a forced draft fan turbine. By comparison with Figure 35-7 the similarity is apparent. However, a diaphragm unit with three chambers, termed a trifferential unit, is employed together with a "proportional band tank" and a bleed valve or floating rate adjuster. The top diaphragm of the trifferential unit is simply a sealing diaphragm. The large center diaphragm is the differential pressure measuring diaphragm while the small bottom diaphragm measures the feed back pressure.

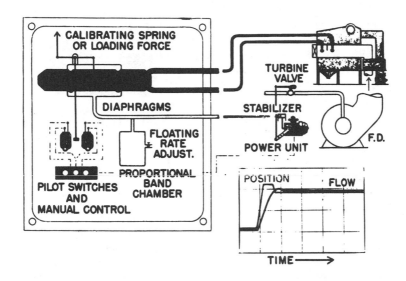

Figure 35-14 / **ADE-A, Adjustable, Low Differential Pressure Controller, applied to air flow control. This is a typical Hays design diaphragm-actuated controller including feedback or stabilizer piston and proportional band chamber.**

Differential air pressure equivalent to air flow is balanced by the calibration spring, or external loading force. When an increase in differential air pressure occurs, the magnet is moved by the diaphragm to close the decrease mercury switch to operate the power unit. While the power unit is rotating to decrease steam to the turbine, the feedback piston acts to build up air pressure in the proportional band tank and the small "stabilizing" section of the trifferential diaphragm unit. Since the capacity of the proportional band tank is large compared to the stabilizing section of the trifferential unit, considerable time is initially required to build up pressure in the system. When sufficient pressure is built up in the tank and differential unit to balance the change in draft, the magnet is returned to center position and the power unit stops.

Meanwhile, the open bleed or floating rate adjusting valve allows air to slowly escape from the proportional band tank. If the deviation persists, pressure loss causes the mercury switch again to operate the power unit. Now, however, the proportional band tank is almost at full air pressure so that a relatively smaller power unit motion is required to balance the diaphragm and stop the power unit motion. This action is repeated with successively smaller steps until equilibrium is obtained.

This action, then, is the same illustrated graphically in Figure 35-12 with true proportional plus reset action resulting.

Derivative Action

Derivative action is that in which there is a predetermined relation between a time derivative of the controlled variable and the position of the final control element. The first derivative, or rate action, is most commonly used. Nonstandard terms such as anticipatory action, preact, damping, or stabilizing action are used occasionally, and may refer to some type of rate action. Rate action is similar to the action applied by a manual operator when he "anticipates" the amount of corrective action required by observing the rate at which a deviation from the set point is occurring. Hence the term "anticipatory control."

Rate action. Rate action is that in which there is a continuous linear relationship between the rate of change of the controlled variable and the position of the final control element, or motor operator acting upon the final control element. Because rate action is not affected by the error signal itself, but only by the rate of change of the error signal, it is not used as a single control mode.

Rate action may be added in limited amounts to a controller having proportional position action. When so used, it will decrease or narrow the effective proportional band. Otherwise, decreasing the proportional band of a proportional position controller usually increases its instability. Hence the term "stabilizing action" which is sometimes used. Some processes have inherently excessive lag. Rate action is added to automatic controllers to overcome this lag in such processes.

Figure 35-15 illustrates one means of adding rate action to an automatic controller having proportional position action. The proportional position section shown has 100 per cent proportional band and will open or close the valve in proportion to a respective decrease or increase in water level. Here the motor operator is equipped with a direct armature winding so that its motion is proportional to the voltage across that winding.

Action of the proportional position section is conventional. If a difference in voltage between the float wiper and the motor operator or valve wiper is created due to a change in water level, the motor operator will be energized. Its movement will position the valve and change water flow until the rebalancing wiper is moved to a position at which no potential difference exists. However, the rate action bridge, here shown connected to a separate source of direct current, will also have a difference of potential existing between its float wiper and the valve wiper. This voltage is impressed across the differentiating network, consisting of resistor R_1 and condenser C_1. If the voltage was to remain constant, in a short time the condenser would be charged up to full value of this voltage, and no current would flow through R_1. However, when the float wiper is moved with respect to the valve or rebalancing wiper position, a changing voltage is produced

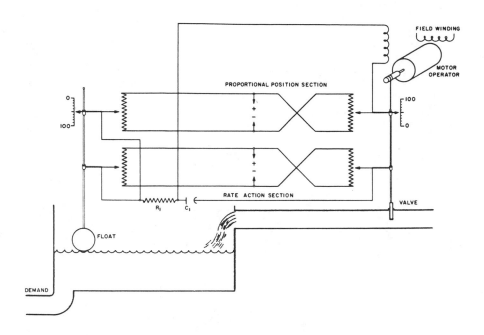

Figure 35-15 / **Proportional position plus rate action controller will help
overcome inherent process lags. Reset may be added to eliminate offset.**

which is impressed across R_1C_1. When this occurs, there will be a current, I_1,
flowing through R_1C_1 which is proportional to the rate at which this voltage is
changing.

The product of this current times resistance (I_1R_1) is a voltage proportional
to the rate of change of the error between water level and valve position. This
voltage is added algebraically (and vectorially) to the proportional position
section error voltage. This increases or decreases the normal motor speed as a
function of the rate at which the error is increasing or decreasing.

For example, from a stabilized position, if the water level and therefore the
float start to drop at a constant rate with respect to the valve position, a voltage
will be produced across R_1. If, by the addition of this voltage, the voltage appearing across the proportional position bridge alone is doubled, the motor will tend
to rotate at twice its previous speed. The resulting action is graphically illustrated
in Figure 35-16. The solid line indicates the speed with which the valve position
would be changed with proportional position action alone. Dotted line indicates
the faster response time when rate action is added.

At this point, it may appear that considerable overshoot might result from
this combination. However, in practical application, the rebalancing action will
reduce the rate of change and the resulting rate action signal. Also, after a sudden
discreet change in level, the motor operator is still rebalancing that portion of
the bridge, and a signal will be produced proportional to the rate of change of
motor operator position. This signal will have a polarity opposite that due to a
change in float level and will therefore add to the stabilizing action.

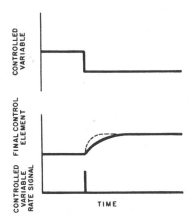

Figure 35-16 / **Response time of final control element due to proportional position action alone (solid line) is made faster (dotted line) by addition of rate signal.**

Note that this controller does not have reset action and therefore suffers from offset. When desired, reset action can be added to a controller having proportional plus rate action.

In practice, the amount of rate action added to proportional position action is made variable to meet varying application requirements. This may be done by changing the voltage applied to the rate action bridge circuit, or by changing the value of R_1 or C_1. Various combinations of these three also may be used.

Rate time. Rate time is the interval by which rate action increases the motor operator motion due to proportional position action alone. It is determined for a constant given rate of change of controlled variable. The time required for a given motion of the motor operator due to proportional plus rate action is subtracted from the time required for the same motion due to proportional position action alone. The difference is rate time and is usually expressed in minutes. It is desirable to have the rate time adjustable with respect to the proportional band adjustment in order to meet varying process characteristics. Representative rate times vary from .01 minute to 10 minutes. Rate narrows proportional band.

As we stated earlier, the proportional band is the range of values of the controlled variable which corresponds to the full operating range of the motor operator. In Figure 35-17, if 100 volts are applied across both the float slidewire and the value or rebalancing slidewire, the proportional band would be 100 per cent. If, however, the float slidewire has a voltage across it of twice the value of that across the rebalancing slidewire, the proportional band is only 50 per cent. This is because the float slidewire need move only through 50 per cent of its range to equal the voltage difference appearing across 100 per cent of the value or rebalancing slidewire.

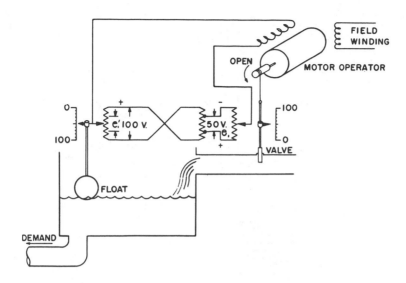

Figure 35-17 / Increase of voltage across transmitting (float) slidewire with respect to receiving (valve) slidewire narrows proportional band. Rate action accomplishes this automatically during change of float position.

We have shown that the voltage produced across R_1 of the rate action section during a change in float level is added to that developed by the proportional position action section. This, in effect, momentarily increases the voltage at the float slidewire with respect to the rebalancing slidewire. Thus the effective proportional band is decreased, or narrowed, during this change. The magnitude of this decrease is proportional to the rate of change of the error and the relative setting of the proportional band and rate adjustments on the controller.

Combustion control. The electronic totalizer Model EART is an extremely versatile controller. It is frequently used whenever it is desired to totalize varying amounts and types of fuels in multiple burner installations for the purpose of proportioning and controlling the necessary air input. This controller, by suitable arrangement and connection of input auxiliary devices, can be applied to exhibit proportional position action and proportional speed floating action. By using other suitable auxiliary equipment, reset action can be added.

Ability of this controller to function similarly to the proportional plus rate action controller is illustrated in Figure 35-18. This shows a Model EART applied to control gas flow as a function of boiler steam pressure. The resulting gas flow is metered, and is connected into the totalizer circuit in such a way as to provide 'stabilizing action."

Three equal independent voltages are supplied from separate windings of the power transformer. Each voltage is individually variable by means of its respective slidewire winding as shown. The instantaneous polarity shown by the

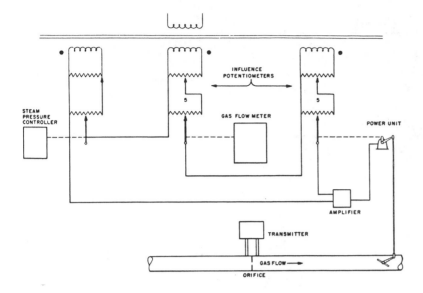

Figure 35-18   /   **Hays Model EART controller applied to control gas valve to maintain steam pressure. Gas flow influence as used here permits narrowing proportional band during steam pressure changes, a distinguishing feature of rate action.**

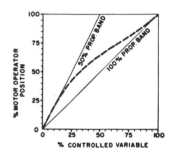

Figure 35-19   /   **Rate action narrows proportional band during a disturbance. Dotted line typified initial narrow proportional band obtained at start of upset and shift to wider band at completion of corrective action.**

dots adjacent to the secondary windings indicates that the steam pressure signal may be balanced by either the gas flow or gas valve slidewires, or both.

For example, assume the steam pressure influence potentiometer at 100 per cent, the gas flow influence potentiometer at 0 per cent, and the gas valve influence potentiometer at 100 per cent. With this arrangement, the gas flow slidewire is effectively out of the circuit. Thus a 100 per cent proportional band exists between the steam pressure and the gas valve position. This is illustrated in Figure 35-19. If the gas valve influence potentiometer now is set to 50 per cent, the 50 per cent proportional band characteristic plotted in this graph results. Now if the gas flow influence potentiometer is brought from 0 per cent to 50 per cent, a proportional band of 100 per cent can be seen to exist, providing the gas valve and gas flow slidewires move in unison.

However, if initially there is no motion of the gas flow slidewire due to lag, while the gas valve slidewire does move, a proportional band of 50 per cent will be obtained. As the gas flow slidewire then eventually moves to a position equivalent to the gas valve slidewire, the 100 per cent proportional band is again

ACTION	APPLICATIONS TO SYSTEMS HAVING:	ADVANTAGES	DISADVANTAGES
Proportional Position	Need for direct proportionality Good self regulation Little or no lag Little or no dead time Slow response time Tolerance at offset	Precise proportionality Fast response time No overshoot No phase shift	Set point error or offset oscillation at narrow proportional bands
Single Speed Floating	Good self regulation Little or no lag Little or no dead time Fast response time	No offset	Overshoot Oscillation Requires dead band Negative phase shift Slow response time
Proportional Speed Floating	Good self regulation Little or no lag Little or no dead time Low disturbance rate Fast response time	No offset Requires no dead band Can be applied to nonlinear system	Tendency to overshoot Tendency to oscillate Negative phase shift Slow response time Floating rate must be matched to system response
Proportional + Single Speed Floating	Little self regulation Some lag Little or no dead time Fixed response time Medium disturbance rates	No offset Fast response time Can be applied to nonlinear system	Oscillation on systems having dead time Some negative phase shift Single speed Reset not applicable to wide load changes
Proportional + Reset	Little self regulation Some lag Little or no dead time Variable disturbance rates Variable response time	No offset Fast response time Compensates for wide load variations Good stability Can be applied to nonlinear systems	Oscillation on Systems having dead time Some negative phase shift
Proportional + Rate	Need for proportionality Good self regulation Large lags Large dead time Variable response time Variable disturbance rate Tolerance of offset	Positive phase shift Compensates for systems lags	Offset Rate increases gain Rate amount limited
Proportional + Reset + Rate	Poor self regulation Large lags Large dead time Variable response time Variable disturbance rates	Ultimate in controller action Rate phase shift compensates for reset phase shift	Some degree of complexity in adjustment of made controls for system condition

Figure 35-20 / **Three basic control modes (proportional, reset, and rate) and their combinations are listed in the left-hand column. Note that rate action can never be used by itself in control work.**

obtained. The resultant dynamic variation in proportional band characteristic is illustrated by the dotted line in Figure 35-19. This is equivalent to the dynamic proportional band characteristic for the controller illustrated in Figure 35-15.

This describes only one possible application of the Model EART controller. This unit and the Electronic Feedwater Controller (Model EFW) are typical examples of the flexibility of controllers in combining various modes of automatic controller action to meet a wide range of practical applications.

Proportional position action and integral action modes may be used singly; however the derivative mode can only be used in combination with one or more of the other two. There are advantages and disadvantages peculiar to each mode. Combinations of these modes are made in a single controller to gain the advantages of each.

These basic modes as used in practical control applications and the majority of their combinations possible are listed in Figure 35-20. This chart shows, in general, the process or system characteristics to which each is applicable, and in brief, the respective advantages and disadvantages.

Automatic controllers or their combinations are available to meet the application requirements of each of the seven modes and combinations listed. A typical installation of automatic controllers applied to a multiple-fuel-fired boiler is illustrated in Figure 35-21. This system of control is designed for boilers equipped with gas burners, oil burners, pulverizers, and constant speed fans.

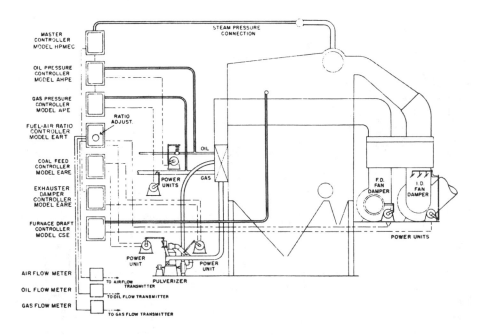

Figure 35-21 / **System schematic of 3-fuel-fired boiler illustrates use of automatic controllers to do the work of the human operator.**

All of the advantages of electrical transmission, direct-connected power units, fuel pressure control, convenient load distribution and fuel-air ratio adjustments, boiler base loading control and simple remote manual control are provided. Three fuels can be controlled automatically at substantially equal ratings, one fuel may be controlled automatically with the others base loaded, or a single fuel may be controlled.

Master controller. This is a model with proportional plus floating action. It measures steam pressure and electrically loads the Fuel Feed Controllers to maintain the correct firing rate to maintain set point steam pressure.

Fuel feed controllers. An Adjustable Pressure Controller with its power unit controls the gas valve. A High Pressure Controller controls the oil valve. Where control of atomizing steam is required, a constant differential valve is installed in the steam line or a lever-operated valve is controlled in parallel with the oil valve. Each controller measures fuel pressure at the burner and balances that value against Master Controller loading. Burner fuel pressures are, therefore, independent of valve position, supply pressure, number of burners in service, and the condition of the burners. Exact minimum pressures are established to safeguard ignition at low loads.

The Model EARE Controller and its electric power unit control the exhauster fan damper to regulate the flow of pulverized coal to the burner. The transmitting potentiometer for the EARE Controller is mounted on the Master Controller. The receiving potentiometer is adjusted by the exhauster fan damper power unit. A manually adjusted biasing potentiometer in the EARE Controller changes the ratio of coal flow to Master Controller loading, thereby changing load distribution.

A Model EARE Controller adjusts the lever on the pulverizer feeder to proportion coal input to primary air. The transmitting potentiometer is adjusted by the exhauster damper power unit and the receiving potentiometer is operated by the fuel feed power unit. A biasing potentiometer in the EARE Controller permits changing the ratio of coal feed to primary air flow.

The system is designed for automatic firing of the three fuels in equal amounts, or with one at a higher value than others. It is also possible to fire only one fuel or to fire them in any combination of automatic and manual control.

Fuel-air ratio controller. A Model EART Totalizing Controller with its power unit controls the forced draft damper, fan turbine valve, hydraulic coupling lever, or magnetic coupling rheostat. The EART Controller has three transmitting potentiometers. One of them is adjusted by the gas flow meter, one by the oil flow meter, and one by the pulverizer feeder. The voltages established on these potentiometers are added and balanced against a single voltage on a receiving potentiometer adjusted by the air flow meter. The EART Controller proportions air flow to the total fuel input, regardless of whether the fuels are controlled automatically or manually. The biasing potentiometer in the EART Controller makes it possible, either manually or automatically, to vary the air flow with respect to total fuel input.

Furnace draft controller. A Model CSE Constant Suction Controller measures furnace draft and, with its electric power unit, controls either the boiler outlet or induced draft fan damper. If control of induced draft fan speed, instead of the damper, is desired, a Model CSE-A Proportional Plus Controller is used. Both types of controllers have calibrated manual adjustments for draft set point.

Safety and interlocking controllers. In addition to the basic controllers shown, additional equipment can be provided to limit fuel and air input to the available draft, to limit fuel input to the available air, and to shut off the fuel if the limiting action is ineffective after a definite time interval.

Although this example concerns itself with combustion control, all principles used are applicable to any control problem.

36 / **BOILER MAINTENANCE**

(*Source:* International Correspondence Schools, Scranton, Pa.)

When performing any type of repairs, it is wise to make an estimate of the length of time, the number of people required, the kind of tools to be used, and the probable cost. This is especially true when undertaking *boiler repairs*, because the work must frequently be done rapidly and on an overtime basis to prevent more than a minimum of outage or down time (the period of inactivity of the boiler during normal working hours). Every item necessary for repair should be available prior to starting the work. A stock of main repair parts should be maintained. The tools needed to make various kinds of repairs should be available to enable the repair work to be made promptly. In determining the outage or down time of a boiler, sufficient leeway should be allowed, depending on the work to be done, to insure a measure of protection, because the work does not always progress as rapidly as planned, especially if workmen are required to be on the job for many successive hours. No repairs of any kind should be attempted on a boiler while it is under pressure. Repairs attempted under these circumstances are not only exceedingly dangerous to the workmen, but also endanger the continuity of operation of the equipment. This should be taken into account in requesting a scheduled outage, for it will take some time for the pressure to be worked off the boiler after it has been taken off line, and additional time will be required for it to cool sufficiently for repair work to proceed.

Repair of Cracks in Steam Boilers

Fusion welding. Progress in the art of fusion welding, particularly electric arc welding, has progressed to the point where repairs made by welding do not affect the strength of the original metal for all practical purposes. Fusion welding is a well recognized and effective method of making many types of repairs to steam boilers, provided the work is done carefully and proper procedures, materials and qualified welders are employed.

The American Society of Mechanical Engineers' Boiler Code Committee has issued detailed regulations covering all aspects of permissible fusion welding, including materials, procedures and qualifications of welding operators. The Code is continually revised as progress is made in the art of fusion welding, and as various questions arise. Most regulations by state or municipal authorities recognize the A.S.M.E. Code and do not permit fusion welding on pressure vessels unless the work is done in accordance with the Code requirements.

Repairing cracks in boiler plates. The repair of cracks in boiler plates can be accomplished most effectively by the fusion welding process. This work is generally subject to inspection by a certified inspector. He will require that all traces of the cracked metal be chipped, ground, or otherwise machined out so that only solid parent metal exists over the area to be repaired. The edges of the groove are to be properly beveled; the welding rod to be suitable for use with the parent metal; each layer of weld metal to be properly cleaned and possibly peened before the next layer is placed; preheating to be done; and the completion of work be followed by stress relieving by heating the area around the weld to at least 1100° F. or higher if this can be done without distorting the plate. If the crack extends entirely through the boiler plate, a backing up plate is generally required under the narrow portion of the V-groove.

Repairing cracks in tubes. Cracks in tubes with the presence of small blisters accompanied by a rupture at the surface of the blister frequently can be repaired by heating the tube to a red hot temperature and driving the metal back into position. This is followed by cleaning and lace welding over the surface to restore the tube strength.

Caulking cracks. When welding tools are not available, small cracks may be temporarily repaired by caulking. It must be remembered that this method merely closes up the crack, but does not restore the original strength of the parent metal. The crack should be repaired by welding.

Plugging cracks. Plate cracks less than 1¾" long are sometimes repaired by plugging. The plate is center punched on the crack midway of its length, and a hole drilled through the plate of sufficient size to remove all cracked metal and to permit the hole to be threaded with a standard pipe tap. The hole is carefully tapped and a steel or wrought iron pipe plug is screwed in tightly. After a plug is installed as tightly as possible, it is cut off flush with the fire side of the plate and the head is peened over to further expand the plug in the hole. When the repaired plate is not directly exposed to the fire, the plug need not be cut off flush.

Stop rivet. When a crack extends inward from the edge of a plate on an outer seam, it will generally go farther unless stopped. Repair with a stop rivet can be made when the crack does not go beyond the seam. A hole of suitable size is drilled at the end of the crack and a properly heated and tightly fitted rivet is driven in the hole. The crack should then be closed by caulking.

Rivets. All loose rivets should be replaced as soon as possible because they cannot be made permanently steam tight. However, they may be tightened up temporarily by caulking. The effectiveness can be tested by the use of a hammer determining the sound of the rivet.

Removing rivets. The heads of rivets ordinarily used in boilers can be cut off by using a cold chisel and hammer. After the rivet is cut off flush, the rivet can usually be backed out of the hole by using a driftpin and hammer.

Replacing rivets. When loose or defective rivets have been removed from a boiler plate, the rivet holes are usually found to be in bad condition and it is advisable to ream the holes for rivets one size larger than the old ones.

Seal welding rivet heads. When rivet heads have been improperly caulked so they still leak, and when they cannot be replaced, seal welding may be employed under the supervision of an authorized inspector. Care must be taken to use the proper welding material. The welding should be done preferably from the pressure side of the plate to prevent entrance of chemicals which could cause embrittlement of the metal. It must be remembered that seal welding merely stops off a leak. It will not strengthen the holding power of the rivet.

Soft patches. When a patch plate is applied to a damaged boiler sheet that is not exposed to the fire, the patch may be installed without removing the damaged portion of the sheet. This type of patch plate is called a soft patch to distinguish it from other types of patches. The patch plate should be of the same material as the boiler plate and of equal thickness. It should never be applied to a boiler sheet exposed to incandescent gases. The double thickness would restrict heat transfer, causing the patch plate to be damaged by overheating.

Hard patches. A hard patch differs from a soft patch in that the damaged metal is cut out of the boiler plate before the patch is applied. It is riveted in place and then securely caulked. A hard patch constitutes a permanent repair when authorized by the inspector having jurisdiction. A hard patch is usually applied on the water side of the damaged plate so that internal pressure on the boiler will press the plate into position and so that the rivets will not be put under tensile stress. Furthermore, placing the hard patch on the water side of the shell plate prevents the formation of recessed pockets into which sludge or loose scale flakes might settle, thus leading to overheating owing to the insulating qualities of such foreign materials.

Fusion welding of patches. The common practice today is to weld a patch into a boiler to take the place of riveting. Such fusion welded patches can be depended on to develop almost the full strength of the original boiler drum plate, provided the work is done properly and in accordance with A.S.M.E. Code requirements. It must of course, be done with the permission of, and under the jurisdiction of a qualified inspector.

Repair of Furnaces

Replacement of corrugated furnaces. A corrugated furnace in an internally fired boiler may collapse because of overheating, as shown in Figure 36-1, in which case it will probably have to be entirely replaced.

The replacement procedure depends on the make of boiler and the type of furnace lining used. In some cases it may be necessary to remove the boiler tubes and take out the front head before the furnace can be replaced. In other boiler units, the furnace is of the removable type as shown in Figure 36-2.

Repair of corrugated furnaces. When the furnace is bulged inward to such a small extent that it does not require entire replacement, it can be repaired by heating the metal to a cherry-red temperature and then forcing the bulged part back to its original position.

The preparatory work consists of removing all fittings from the interior

Figure 36-2 / **Inserting new furnace in scotch boiler.**

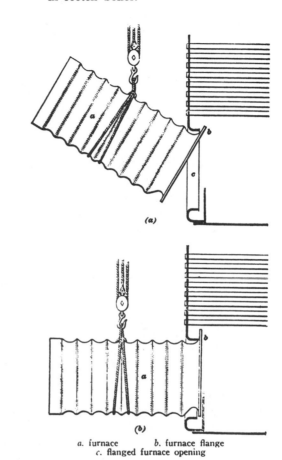

(a)

(b)

a. furnace   b. furnace flange
c. flanged furnace opening

Figure 36-1 / **Collapsed scotch boiler furnace.**

of the furnace, and providing the necessary tools and a source of heat, generally gas torches. Care must be taken to have all water drained from the boiler and all vents open. In sledging the heated metal, care must be taken not to dent the plate or nick it with sharp edges of a hammer. It is much better to drive with a maul or against hardwood blocks, if there is enough space to use such equipment. It is still better to use blocks corresponding to the corrugations and to employ a hydraulic jack to force the heated plate back into position. The opposite end of the jack is supported against the cold shell with suitable planking, to distribute the force over a large area.

Repair of cracks in corrugated furnaces. Cracks sometimes occur in corrugated furnaces as shown in Figure 36-3.

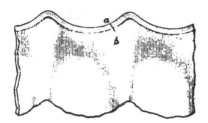

Figure 36-3 / **Crack in corrugated furnace.**

A crack like the one shown in the illustration presents a sizable repair problem, because considerable stress exists at a point *a,* and the crack may run in a ragged line as shown by *b.* The shell plate expands and contracts; hence, there is considerable bending force and nothing but fusion welding under the direct supervision of an authorized inspector and in accordance with the A.S.M.E. Code can be considered a satisfactory repair.

Refractory Repairs

Refractories are used in furnaces largely as a convenient means for containing the fire and forcing the hot gases to travel over boiler heating surfaces without damaging other portions of the structure or the setting. It is evident, therefore, that the refractories must be kept in good repair; otherwise the equipment may be extensively damaged, or the efficiency may be reduced. The frequency or type of refractory repairs needed depends largely on the severity of the service, the type of materials originally used, their design and construction. When refractories must be replaced, it is usually advisable to use the same or better grade, because a large portion of the expense involved is in the labor required for repair, not in the cost of the material. Repair work must be done carefully by skilled workmen, otherwise the repair cannot be expected to stand up to the severe requirements of boiler service.

Plastic refractories. Plastic refractories consist largely of materials normally used for making a good grade of firebrick. Plastic refractory should be rammed down solidly by repeated blows from a pneumatic rammer or wood mallet. The ramming should be done in a downward direction, for gravity assists in keeping the fill solid. Proper ramming closes out all voids in the material and works the mass together, thus avoiding cracks in the finished job.

Castable refractories. Castable refractories ordinarily do not have as good temperature resisting characteristics as plastic refractories, because various materials such as water glass or lumnite cement are used in order to give them air setting properties. Castable refractories will develop considerable strength without the application of heat. They are valuable for installation as furnace linings in close contact with water cooled surfaces, because the water cooled surfaces prevent the lining from reaching a temperature at which the refractory would develop strength by vitrifying.

Chrome-ore plastic refractories. Chrome-ore plastic refractory is used extensively where temperature and slagging conditions are severe. It is a dense, hard material, having good heat-conducting properties and considerable resistance to slag penetration. The spaces between water wall tubes are frequently filled with this type of refractory to make a practically indestructible furnace lining.

Super refractories. A furnace wall will generally last longer if super refractories are used in the fuel bed area. An 85 per cent silicon carbide brick is known to have good resistance, very dense and not subject to corrosion by coal ash, and resists clinker adhesion and erosion far better than fire clay. Silicon carbide works best when placed to fuel bed depth because it resists abuse and wear. Hand fired furnace walls last longer when side and back walls have three rows of silicon carbide bricks at fuel bed area. For the hotter area above, use four rows of mullite bricks on side walls, five rows on back wall. Generally the front wall will not need any. This refractory is ideal because it forms inert crystals of mullite that prevent fluxing to ash. Bonded mullite and bonded silicon carbide brick both stand temperatures above 3000° F. For single retort stoker jobs, silicon carbide bricks need only be laid to protect the side and bridge walls. Follow the fuel bed pattern with only the trouble spots being laid. For multiple retort stoker furnace walls, silicon carbide refractory should be placed on all four walls. For spreader stoker walls, silicon carbide should be installed at grate line, and higher on the back wall to resist abrasion of coal chucks during heavy feeding rates. Mullite can be used above fuel bed. They should be laid in cements of their own composition with thin joints, and bonded to the fire brick backing. Use standard brick laying methods. When chosen and installed correctly, their life is generally several times that of common refractories.

Mortars used with firebrick. Air setting mortar has the property of bonding with brick to produce a solid structure before the application of heat. It is usually supplied in a prepared state, requiring only a thorough mixing with the liquid in the container to make it ready for use. The joint should not exceed approximately ⅟₁₆ inch in thickness. When walls are laid with air setting mortar, they will air dry and afford a good bond between adjacent bricks.

Heat setting mortar is usually premixed, requiring only the addition of fresh water to make it ready for use. No other substance should be added, because the fusion point of the mortar is likely to be decreased, causing the formation of a weak wall structure.

Laying firebrick. There are a number of ways of laying firebrick to produce a satisfactory wall, but the following method is one most generally used. One end and one side of the brick are dipped into the mortar. It is good practice to use a sliding, edgewise motion to eliminate air pockets or bubbles and completely cover the desired surfaces of the brick with mortar. The brick is then lifted and any excess mortar is allowed to drip off before the brick is placed in position. No additional mortar should be used, the coating adhering to the drip brick is sufficient. The brick is then placed in position and tapped into place with a wooden mallet to work the mortar out of the joints and produce a mortar joint thickness of approximately ⅟₁₆ inch. The bricks should be carefully placed. Any excess mortar should be removed from the wall. The completed wall must be true, without projecting edges of brickwork or bulges.

Drying out refractory lining. Much damage can be done to a new refractory lining in a furnace, or to one that has been extensively repaired, by warming it up too rapidly. The brickwork should be allowed to air dry for at least twelve hours before a fire is placed in the furnace. Circulation of air should be provided during this time for the purpose of removing as much moisture as possible. The actual drying out period under fire will depend on such factors as the extent of the furnace repair job and the size of the setting. If there is any doubt about time, it is better to prolong the drying out period and keep the firing rate low, then to try to hurry the job by firing too rapidly.

Tube Repairs

Tube repairs are required when blistering, rupturing, cracking, corrosion, leaky joints or other types of failure occur. Repairs may also become necessary when inspection reveals progressive deterioration and possibility of tube failure before the next outage can be scheduled. The method of repair to be used depends on the type of weakness found. It may vary from the rerolling of a tube end to the complete replacement of the tube.

Fusion welding. Small ruptured blisters and certain types of cracks in tubes may be temporarily repaired by fusion welding, using a qualified welder to do the work in accordance with A.S.M.E. Code. This type of repair should never be undertaken without a qualified inspector's permission and supervision. To do so may mean cancellation of insurance or infraction of legal, local or state regulations.

Rerolling tubes to stop end leaks. Boilers that are fired up too rapidly or cooled down too rapidly may develop leaks at the rolled joints. The reason for this is that the tubes are affected by the temperature changes more rapidly than the thick drum plates into which they are rolled; hence, the compression stress in the original rolled joint is released. The same effect may occur when a boiler is overheated by being operated with low water and when other damage has not resulted. A similar condition may be found in new boilers when the tubes have been underrolled at the time of installation. Rerolling by use of suitable expanders may constitute a thoroughly satisfactory and permanent repair in such cases.

Caution: The tube expander is a powerful tool and it is readily possible to overexpand a tube to such an extent that the plate is distorted or the end of the tube thinned out or crystallized, making more extensive repairs immediately necessary.

Temporary fire-tube repair. When a fire-tube boiler cannot be taken out of service for any great length of time to enable permanent tube repairs to be made, boiler inspectors may permit plugging of damaged tubes by the use of a tube stopper. Figure 36-4 shows a steel ferrule used for stopping leaks at the ends of tubes in Scotch marine and other types of fire tube boilers.

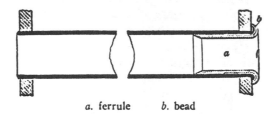

a. ferrule   b. bead

Figure 36-4 / **Steel ferrule for stopping tube leak.**

Figure 36-5 shows the use of plugs and a tie rod for the purpose of closing off the ends of a tube having a crack or other type of failure that cannot be covered with a ferrule as shown in Figure 36-4.

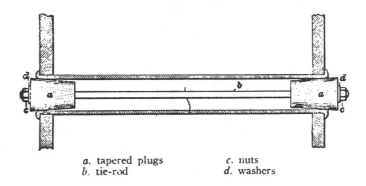

a. tapered plugs   c. nuts
b. tie-rod   d. washers

Figure 36-5 / **Tube stopper for interior tube leaks.**

There are a number of forms of the tapered plug arrangement. A fault common to all of them is that the plugs are subjected to boiler pressure tending to force them out of the tubes, also they are vulnerable to temperature because they are exposed at the tube sheets. Their use should be of a most temporary nature only.

Temporary water tube repair. Water tubes are subjected to internal pressure and also to exterior application of heat from hot gases. When they are plugged off, water and steam cannot circulate through them, hence, they will quickly overheat to a dangerous degree. Boiler inspectors occasionally may permit leaving a ruptured tube in place to permit temporary plugging at tube sheets, if the tube is located in a cool zone where the metal is not apt to be destroyed by high temperature and when there is no possibility of building up internal pressure. It is considered good practice to remove the tube by cutting it off at the tube sheets, removing the expanded portions from the tube sheets and rolling in key caps or stub ends. The tube may be left in the boiler setting if there is not sufficient repair time available to effect its removal, for it can do little damage when it is cut loose from the tube sheets and when the holes through the sheets are plugged off properly.

Removing tubes. When the portion of a tube on the side of the sheet opposite the flared end can be reached externally, an oxyacetylene blowpipe or torch can be employed to cut the tube off at a distance of about ¼ inch from the tube sheet. A blunt chisel is then applied on one side of the cut tube end remaining in the sheet and is driven in forcibly to collapse and loosen this end. Once it has been forced inward, and rolled joint is loosened, hence a hammer can be used to drive the tube end out of the hole.

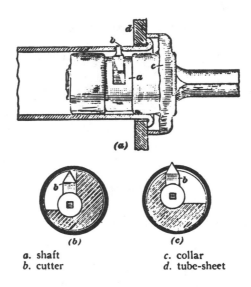

(a)

(b)   (c)

a. shaft   c. collar
b. cutter   d. tube-sheet

Figure 36-6   /   **Tube cutter for fire tubes.**

Various types of tube cutters, such as those shown in Figures 36-6 and 36-7 are available for cutting off tubes that cannot be reached readily to allow the use of a torch. After the tube has been cut off by use of these or similar tools, a cape or ripping chisel may be used to make one or more longitudinal grooves in the circumference of the tube end. Care must be taken not to make this groove so deep as to damage the tube sheet.

Installing new boiler tubes. The basic principles are employed for installing tubes in both fire tube and water tube boilers. The ends of fire tubes are usually beaded to prevent overheating owing to the intense heat of the gases of combustion at the hot side of the tube sheet, however the ends of the water tubes are usually belled because they are cooled by contact with the boiler water.

Before installing new tubes, the tube holes must be thoroughly cleaned and freed of burrs or irregularities. If the hole has been damaged by the cutting action of escaping steam or water, the hole should be repaired by welding in a suitable bead of metal. This should be followed by filing and grinding to make a smooth surface. In case the seat has been grooved to insure tightness of rolled tubes in high pressure boilers, the grooves should be cleaned out and formed properly

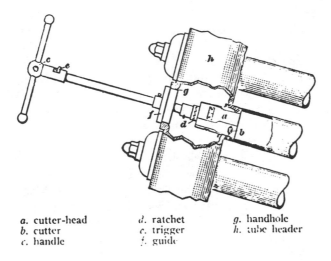

a. cutter-head   d. ratchet   g. handhole
b. cutter   e. trigger   h. tube header
c. handle   f. guide

Figure 36-7 / **Tube cutter for water tubes.**

in a circumferential direction. The tube ends are prepared by removing any paint or grease and by polishing with a wire brush, emery paper, or other effective means.

Some codes require tubes be belled or beaded under existing codes, but some boiler manufacturers recommended that the belling operation be dispensed with in case of water tube boilers.

Tube expander. The general form of a tube expander used for rolling and belling a tube is shown in Figure 36-8.

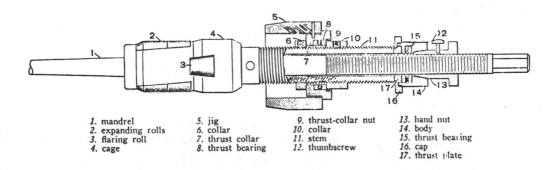

1. mandrel   5. jig   9. thrust-collar nut   13. hand nut
2. expanding rolls   6. collar   10. collar   14. body
3. flaring roll   7. thrust collar   11. stem   15. thrust bearing
4. cage   8. thrust bearing   12. thumbscrew   16. cap
    17. thrust plate

Figure 36-8 / **Tube expander.**

By use of this tool, the boiler tube is expanded by the pressure exerted when a tapered mandrel is rotated inside the rolls. The mandrel, when turned by hand or by motor, feeds into the rolls because they are set at a small incline in the cage. When the direction of rotation of the mandrel is reversed, it feeds outward because of these inclined rolls.

The expander shown has two sets of rolls, one set of expanding rolls and one set of flaring rolls. Both the expanding and flaring rolls are contained in the same cage.

Method of expanding tubes. The improper use of a tube expander can severely damage both the tubes and the tube holes. Underrolling will result in leaky joints that may cut the metal before being discovered. Excessive rolling may crystallize and crack the tubes and may also enlarge the tube holes. Overrolling has been known to stress the tube sheet so greatly as to loosen otherwise perfect joints in adjacent tubes.

An experienced tube roller develops a "feel" of the expander which tells him the precise point at which to stop rolling. The success in rolling operation, regardless of the method used, or of the size, gage, or composition of the tubes, is tightness of the joint and freedom from overrolling as evidenced by a damaged tube sheet. The completion of the rolling process is always followed by a hydrostatic test one and one-half times the operating pressure of the boiler.

General reminders. Water tube boilers generally have $\frac{1}{32}$ inch clearance between the tube and tube sheet hole at both ends. Due to the fact that fire-tube boiler tubes become scaled, their clearance is generally $\frac{1}{32}$ inch at the fire end and $\frac{1}{16}$ inch at the opposite end, which aids in their removal. The A.S.M.E. Code requires end tubes in water tube boilers, extended $\frac{1}{4}$ to $\frac{1}{2}$ inch and to be flared $\frac{1}{8}$ inch in diameter larger than the tube. Never overroll tubes. It is good practice to roll slowly and to stop when tiny flakes start leaving the tube sheet around the tube. Overrolling will enlarge and deform the tube sheet, thins the tube and weakens the tube sheet, hardens the inside of the tube and tube sheet causing early failure and can cause surface tears.

Casing Repairs

There are various types of boiler casings which serve the general purpose of providing airtightness with structural support, improvement in appearance, aid to cleanliness, and reduced cost of maintenance. The type of casing used, the severity of service, and the will to maintain a generating unit in first-class operating condition at all times, will largely determine the kind of repairs, and frequency.

Various types of casings:

Steel Clad. This casing is entirely enclosed by steel plates fastened together.
Plaster Covered Insulation. Insulated blocks or blanket insulation wired to structure, covered with metal or chicken wire to form a base for plastering.
Canvas Covered Insulation. Hard block insulation covered with canvas.
Plastic Air Seal Coatings. Several types are available. Check with the manufacturer to ascertain the maximum temperature they will withstand.

Casings of all types must be kept tight if the generating unit is to operate efficiently. All leaks, cracks, and voids must be closed immediately. Their appearance is the signal for investigation and sometimes repair.

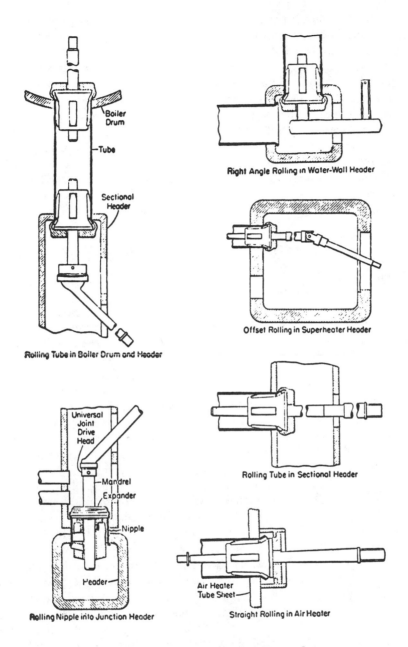

Figure 36-9 / **Methods of rolling tubes.**

Dampers

The boiler damper is of great importance in the successful operation of a steam generator or hot water boiler. When the damper is in the open position, it should provide as little obstruction in the damper frame as possible, and when it is in the closed position, it must seat as tightly as possible to prevent leakage. At intermediate positions, it must work freely, so that its effectiveness as a control means will not be impaired.

Lost motion should be taken out of the actuating mechanism as lost motion in automatically controlled damper installation cannot develop all the sensitivity of which it is capable. Damper bearings should be maintained in first-class repair to prevent sagging of the damper shafts, binding and sticking, and excessive air infiltration. Damper shafts and blades are subject to high temperatures and considerable stress which sometimes causes deformation which may prevent them from opening widely or seating tightly. The shafts and damper blades should be straightened or replaced if they are damaged beyond repair. When the boiler is out of service, a complete inspection of the damper assembly should be made and all repairs and replacements made at this time.

Cooling Down and Emptying Boilers

When taking a boiler out of service, it is desirable to reduce the load gradually and to maintain a low load for a time before actually killing the fires. This procedure will insure that any suspended sludges and scale flakes have a maximum opportunity to settle out of circulation in proximity to the blow off line so that they can be readily removed. When fires are being killed, the water level of the boiler should be raised to a high point, because this can be done safely under a low load condition and there will be ample water reserve to blow down the water under pressure after the fires have been killed. When the boiler is at full pressure, the blow off valves should be given a quick blow (drain) to remove sludges and loose scale flakes as far as possible. This will reduce the water level, but should not be continued to a point where low water level is obtained, or to the point where feed water has to be admitted to restore the water level. Cold feed water should not be put into a hot boiler as it may contribute to loosening of rolled tube joints. After being blown down under pressure, the boiler should be cooled slowly, at a recommended rate of approximately 100°F. per hour. A slower rate of cooling will do no harm, but a more rapid rate may be harmful by causing unequal contraction strains. The water level will settle as the boiler cools down; so long as it remains in sight at the bottom of the gage glass, feed water should not be admitted to the cooling boiler.

When the furnace has cooled to a temperature at which a person can enter and remain in the furnace, the boiler may be entirely emptied without danger. It should be emptied at as rapid a rate as possible by opening the main blowoff valves wide because this will assist in removing sludges that settle out in the vicinity of the internal blow off lines. The steam drum vents or safety valves should be wide open, thus permitting the entrance of air to the drum, otherwise, the water will not drain from the boiler entirely.

Boilers containing superheaters of a nondrainable type should be cooled down in such a manner as to prevent admission of air to the superheater tubes. The superheater tubes will condense full of pure water, having a greater affinity for oxygen, hence oxygen corrosion on the interior of the superheater tubes will occur if air is permitted to enter. The recommended procedure is to open the superheater vent valves while cooling down the boiler to make sure that the steam flows through the superheater while pressure is being worked off. These vent valves should then be closed when there is still a steam pressure of 10 psi on the boiler so that the superheaters will condense full of water. They should remain closed as the boiler is emptied to make sure that air cannot enter the superheater to cause corrosion during the outage period.

Parts, Tools, and Equipment

Boilers and their auxiliaries are used to produce steam, or hot water as it is needed by an office building, plant, or factory; failure of some minor part can often cause a forced outage with a resultant loss of time and production. It is important that all equipment be kept in first-class condition at all times. Items which might cause plant or building shutdown are:

1. Parts that move and are consequently subject to wear.
2. Parts subject to extreme temperature.
3. Parts damaged by foreign material.
4. Equipment damaged by improper handling.

Inspection, adjustment and repair are the general rules for successful boiler room operation. These operations require the use of certain tools and replacement parts. A card filing system is generally used containing such information as, manufacturer's name and address, style, serial number, speed of operation, and purchase date. Space for inspection dates, conditions found, and adjustments and repairs should be provided for.

An inventory of boiler tools, and spare parts should be maintained to insure immediate information when required. Overhaul schedule should be established during low load periods and can also be noted on the card.

Inspection, Preparation, and Operation

Inspection schedules. Inspection of boilers and pressure vessels is required at periodic intervals by the state or municipal inspectors having jurisdiction, or by your insurance company. A convenient date for both the chief engineer and the inspector should be arranged. This is generally scheduled during the low load periods, to reduce the number of outages. The chief engineer will overhaul all his equipment prior to inspection, and discuss the results of the inspection with the inspector to insure that any discrepancies and recommendations are thoroughly understood. Supplemental inspections should also be scheduled to insure proper operation.

Blowing tubes. Prior to taking a boiler out of service, the soot should be cleaned from the fireside of the boiler. This is generally done by blowing air or steam against the tubes, with automatic soot blowers or a hand lance. Tubes should not be blown with wet steam as moisture cakes the soot, and can plug the area between tubes. Gas passages are stopped and the tube heat absorption surface reduced, and sulphuric acid formed, which will corrode the tubes.

This can be avoided if superheated steam is used, or in any case all condensate should be drained by a drain valve in the header prior to using the soot blower.

Boiler cleaning. After the boiler has been drained, the manhole covers on any upper drum should be removed and the drum and tubes entering it should be washed down with a hose, using warm water, just as soon as the drum is cool enough to enter. Other covers on lower drums or water wall headers may be removed to provide air circulation to accelerate the cooling process. A heavy stream of water should be used to wash down the drums and tubes before the sludge has had time to dry out. The access doors to the furnace and setting should be opened next and all refuse removed as thoroughly as possible by working through these openings with available cleaning tools. Opening the boiler damper will create a draft into the openings, pulling any dust up the chimney, and reduce to a considerable extent, the physical hardship of removing refuse. It is good practice to clean from the furnace front toward the rear of the unit so that any dislodge dust will float out ahead of the workmen.

Certain precautions should be observed by workmen engaged in the cleaning operation:

1. No workman should be permitted to enter any drum until it has been thoroughly ventilated, to insure sufficient air is available.

2. Some person should stand outside the drum to offer assistance if needed.

3. All drum openings such as blowoff lines, control valves, feed pumps, fans, fuel valves, and others should be shut off and tagged to be kept inoperative.

4. Extension cords and electric lamps used for illumination *must be* ruggedly constructed, and kept in good repair. A breakdown in insulation, or breakage of an unguarded bulb can result in electrocution if space is damp.

Fusible plug. The cross tee plugs in the water column should be removed so the piping can be examined, and a brush run through every time the boiler is cleaned internally. The fusible plugs in the steam drum should be renewed each year.

Stack. The breeching should be examined for indications of structural weakness in the plates or supports. Air leaks into joints and connections, through corroded places or through leaking dampers on idle units will seriously interfere with the stack capacity and reduce the available capacity of steaming units. Soot and fly ash accumulations inside the breeching will reduce the area, overload the supports and cause corrosion in the presence of dampness. All dampers and damper operating mechanisms should be operated and inspected to insure that they are in a serviceable condition.

Valves. Valves should be kept packed and in operating condition because steam leaks are costly, as an inoperative valve may constitute a hazard.

Piping. Water piping, pressure connections to gages, automatic control lines and all other piping should be inspected to insure that they are in good condition. Safety valve vent piping should be inspected to insure that there are no obstructions at the discharge end, and that it is securely anchored to withstand the force of the steam released by the valve, and that the piping is self-supporting and does not put any strain on the safety valve body. The piping between the drum and the water column should be checked for tightness to insure that the correct water level will be shown in the gage glass. Column connections should be checked to insure that they are unobstructed. If there are any valves in the water column lines, they should be left open. The water piping between the drum and the bottom of the column should be level, or pitched toward the drum. The water column supports should be checked to make sure that the column is being maintained at the proper height in relation to the drum. The high and low water alarms should also be checked for proper functioning.

Feedwater regulating equipment, check valves, control valves, and all piping and supports should be inspected, because a failure at any point in the feed system will create a low water hazard. Blowoff valves and their piping should be thoroughly inspected, as the failure of any part of the equipment may have serious consequences. The soot blower piping and the soot blower heads should be thoroughly examined to ascertain whether the soot blowing system can operate efficiently. No water pockets should be allowed in the piping because slugs of water driven into the soot blower heads or elements may cause serious trouble.

Corrosion. Rusting, or corrosion, on the fire side of boilers is not likely to be caused by oxidation, unless it occurs in a natural way. *For example,* when the boiler is washed down and then allowed to stand in a humid condition. Fireside corrosion more generally results from acid attack on the metal caused by moist accumulations of soot or fly ash within the boiler or setting. Acid accumulations of this nature will readily absorb moisture from even a moderately dry atmosphere and will rapidly destroy the boiler metal. This destructive action can be detected by removing the soot deposit and then scraping the metal surface. The black, pockmarked surface and the thin layer of corroded iron that flakes away with the soot are typical indications of destructive corrosion.

Abrasion. The wearing away of the metal by rubbing or sandblasting action is called abrasion. It may occur whenever there is relative movement between parts in contact; where fly ash particles concentrate and impinge on the metal; where fuel rubs against headers or water wall tubes by stoker action, or where cleaning tools are used frequently. The detection of abraded metal and the general appearance of the defect usually enable you to determine its cause and remedy.

Leakage at joints. Leakage at rolled joints inside boilers may be revealed by the appearance of moisture or a flowing stream of water when the boiler is subject to a hydrostatic test; by accumulations of chemicals or cemented soot deposits when the boiler is dry, or by a hissing sound when the boiler is under steam pressure. Leaks may develop when boiler tubes are not rolled hard enough to create the required stress in the tube seat; when the tubes do not enter the tube hole squarely; when the tube is belled without being supported or rerolled in the straight portion of the seat; or when the tube and seat have not been properly cleaned. Leaks at rolled joints may also result from overrolling to the point where the ductibility of the metal is lost; from overheating caused by scale or sludge accumulations; by some types of foaming action; and by low water. Rapid temperature changes such as those that occur when cold water is fed to a hot boiler, or when a boiler is fired up too rapidly or cooled down too fast, can loosen tubes that have been rolled perfectly.

The inspector. The inspector must bear in mind all the various types of metal deterioration, and should carefully examine all portions of the pressure parts of the boiler for indications of such deterioration. A thorough inspection should also be made of the refractories, supports, baffles, soot blower supports, tube clamps, and all other items which may fail and lead to damage of the boiler metal.

Brickwork, refractories and insulation should be closely examined to determine if there is any dangerous cracking, loosening or deformation that might lead to failure. Air spaces provided for cooling support members should be checked to insure that they are not obstructed and especially that they are not filled with combustible refuse. Mud drums should be externally inspected for any indications of overheating or corrosion. All connections to drums should be examined for indications of leakage, as indicated by damp spots on the refractory or insulation cover, damp soot accumulation, and scouring of the tube wall, or deposit of chemicals from escaping boiler water.

Plates exposed to the fire or to the hot gases, such as the lower shells of horizontal return tubular boilers, upper drums of some types of water tube boilers, crown sheets, water legs, and other parts must be given attention to reveal any evidences of overheating or cracking metal. Moderate overheating will produce discoloration accompanied by slight bulging, whereas severe overheating will produce pronounced bulging and scaling. Repeated heating and cooling will eventually produce cracks. Any such evidence must be explored fully to determine the cause. Immediate action should be taken on the cause. Feed water entrance nipples should be given special attention because cold water entering the hot boiler creates severe localized stress in the plate at the point of entrance.

Boiler tube inspection. Straight tubes in either fire-tube or water tube boiler should be examined for straightness because any sagging or warping is usually an indication of overheating, even when blistering or bulging is not present. Upper tubes in water tube boilers are particularly subject to such deformation as a result of alternate heating and cooling. Temperature variations of great magnitude can occur rapidly when the boiler is operated at high rating with the boiler water in a foaming condition. Foam bubbles move sluggishly through the tubes and slugs of water follow them. The tube metal heats rapidly when the tube is filled with foam and chills quickly when the foam is displaced by the boiler water. Tubes may be inspected by using any of the following methods, flashlight mirror, cupping of the hand, or measuring the correct diameter against area of suspected bulging.

In boilers of bent tube type, remove the insulation from the outside surfaces of both the mud drum heads and between the brick setting and the heads. They should be checked for overheating and cracking, particularly if they are subjected to high temperature gases or to repeated bending stresses by boiled drum or header movements.

Soot blowers. Soot blower elements are subjected to extreme and rapid temperature changes, especially when they are close to the furnace. Their failure may result in damage to boiler metal or baffles. They should be examined for straightness, length, and fire cracks. Their supports should be checked for soundness and for distance from adjacent boiler tubes, to prevent rubbing. Nozzle spacing in tube lanes should be checked. When doing so it should be remembered that nozzles are often set closest to tubes nearest the soot blower head so that the nozzles will be in mid position of lanes when operating temperature is reached. Nozzles should never be closest to tubes farthest away from the blower head because expansion of the element would then probably produce tube impingement.

Baffles. Boiler baffles should also be examined for indications of overheating, misplacement, or shifting of supports. As a general rule, gas leakage through damaged baffles will increase flue gas temperature and reduce efficiency. Leakage of high temperature gases may be harmful to the circulation characteristics of some types of boilers, and leakage of gases carrying abrasive particles may lead to erosion of tubes. Steel baffles are often attached to boiler tubes by hook bolts or they may rest on tube clamps. The baffle plates may vibrate or the tubes may move relative to the supports. All such action may possibly cause rubbing and erosion of boiler metal.

Internal inspection of boiler heating surfaces. The principal items to be looked for when checking the interior of the boiler heating surfaces are scale deposits, sludge accumulations, corrosion and pitting, the mechanical condition of internal fittings, and evidence of improper circulation.

Scale deposits are formed on boiler heating surfaces when impurities in the boiler water are deposited on the boiler metal as the water is heated or as steam is generated. Scale of any thickness whatsoever, acts as an insulator and reduces the rate of heat transfer through the boiler metal. As a result, the boiler metal becomes hotter for any given steaming rate and the flue gases reach a correspondingly higher temperature. Hotter boiler metal implies the possibility of metal damage and hotter flue gases reduce the boiler efficiency.

Sludges are also the result of impurities in the boiler water. They consist of small particles of matter which do not readily attach themselves to the boiler metal but generally remain suspended in the water while it is moving at a high velocity through the boiler tubes. These sludges may settle out in the drums or headers in which water movement is slowed down. Blowoff valves have been provided for their removal. The general purpose of the chemical treatment of boiler water is to produce such sludges from unavoidable feed water impurities and thereby prevent the impurities from depositing as scale. Settled out sludges may block circulation through drums, headers, or supply tubes, and may pile up in drums to cause overheating. If such sludge deposits are found, the adequacy and operating frequency of the blowoff equipment should be questioned.

Corrosion and *pitting* are both destructive to boiler metal, although the physical appearance of their destructive action differs materially. General corrosion will likely produce a black magnetic iron oxide powder on the surface of a dried boiler metal and a typically rough etched surface when this coating is wire brushed. This type of metal deterioration is not likely to be present when the surface is protected by scale deposit. The best preventative measure against corrosion consists of proper feed water treatment of boiler water conditioning. Pitting is a form of localized attack, usually by oxygen, that corrodes the metal in small areas at first but which continues rapidly if not checked. The pits will be filled with hard, crusted iron oxide and will have pimples or small bumps of iron oxide all over them. Upon finding such pimples or bumps on the surface of boiler metal, it should be determined what lies under them. If pits are found, a number of them should be thoroughly cleaned out to their full depth so that the amount of damage can be appraised and appropriate repairs or preventive action taken.

Improper circulation is seldom found in well-designed standard boilers. Failure of tubes is often attributed to faulty circulation or to flame impingement when no reason other than scale deposit or a foaming condition of the boiler water exists. If scale has flaked off or if sludges have settled out in sufficient quantity to fill lower headers or supply tubes, there will naturally be inadequate circulation until the obstruction is removed. If improper water treatment deposits sticky sludges on supply tubes, their effective areas are reduced. This also causes inadequate circulation. If oil finds its way into the boiler water, or if other foam producing elements are present, some of the tubes may be filled alternately with foam and water. Overheating followed by sudden quenching, then occurs. The tubes will warp and bulge. Fatigue cracks may be expected at points where hot gases come in contact with the tubes. Clean tubes that have adequate water supply and an adequate steam releasing area, will take very large absorbing rates for years without any signs of distress.

Water level. The boiler manufacturer or builder determines the correct normal working water level for the boiler unit and specifies this in his construction drawings that the center of the gage glass must correspond to this level.

As the steam flow increases, more water must be fed. The object is to maintain the water level in the boiler at a point as close to the design level as is consistent with other operating features. The boiler operator should become thoroughly familiar with the feed water system, as he may be required to take quick action in case of improper performance of any equipment. Auxiliary feed pumps are normally installed for use in the event of failure of the main feed pumps. Auxiliary feed lines are often installed to permit delivery of water to the boiler.

Boilers are generally equipped with a stop valve in the feed line system immediately adjacent to the boiler. The valve is usually mounted on a nipple extending from the boiler drum. Next to the stop valve is a check valve so positioned as to prevent the discharge of water from the boiler in the event of failure of any part of the feed system. Both valves should be kept in first-class condition. Regulation of feed water is generally accomplished by valves exterior to the check valve, and may be of the automatic or hand type. This principle is applicable in both a steam or hot water boiler. However, in low pressure hot water systems, normal main water pressure is sufficient.

In most installations, the center tri-cock of the gage column corresponds to this mid-point on the gage glass. Water columns and gage glasses should be blown down at least once each watch to prevent sludges from settling in the connections and interfering with the correct registering of the water level. The column should normally be blown first because it is larger and will accumulate a greater quantity of sludges. Blowing the column first will also insure removal of cooler water and return of hotter water to the column and gage assembly, so that when the gage is later blown there will be less temperature change on the gage glass. The blowoff of water and steam through the lines will rapidly remove any sediment that may have collected. If the lines are entirely clear, the water will return when the valves are closed. If the water does not return promptly or the action seems sluggish, it should be reported to the chief engineer for slowness may be an indication of clogged lines, representing a dangerous condition.

Gage glasses. Gage glasses are subjected to very quick temperature changes, and difficulty is sometimes experienced in maintaining them. Leakage at gage glass connections will accelerate deterioration and breakage and will cause a false water level to be indicated, therefore, it is doubly important that leakage at these connections or at any other valves or fittings in the gage or water column assembly be avoided at all times.

Tubular gage glass is generally used up to 300 psi. The glass should be handled carefully prior to installation. A scratch or chip may cause failure. A tubular gage glass may be cut to the proper length with a glass tube cutter.

When a gage glass breaks proceed as follows:

1. Shut off top and bottom cocks.
2. Open drain valve that is used for blowing down gage glass and/or bottom pet cock.
3. Remove broken glass, then open top and bottom cocks to blow out any broken pieces.
4. Place nuts then washers on new gage glass, add a small amount of graphite to the washer so the nut does not twist the glass, close pet cock.
5. Install, open top cock slowly to heat glass gradually, open bottom cock slowly, close drain.

If valves are used instead of cocks, they should be closed and opened daily, so that cleaner extension or stem can remove any possible scale.

Flat gage glass. High pressure boilers are normally equipped with flat gage glasses of which there are a number of designs. Instead of glass, strips of transparent mica are used in some high pressure gages. When flat glasses are used, they are normally protected by mica gaskets to prevent etching of the glass and to provide a certain amount of resiliency in order to prevent breakage of the glass when it is tightened up in assembly. The flat type gage glass usually must be disassembled from time to time for cleaning purposes. Particular attention must be given to the mica gaskets because mica has a laminated structure and therefore may cause difficulty by flaking or by wearing through.

When a gage glass breaks proceed as follows:

1. Close upper and lower gage cocks. Remove the clamps, glass gaskets, mica, graphite, the threads, and run them down.
2. Remove the old gasket.
3. Polish surface, perfectly smooth.
4. Clean both ends of gage.
5. Clean surface thoroughly and oil with molybdenum disulfide.
6. Install new gasket, mica, new glass, another new gasket and clamps. Let glass heat by conduction.

7. The bolts should be pulled up hand tight, then those at the center should be pulled up lightly, then the remaining bolts should be tightened lightly.

8. All bolts should be again tightened progressively until the pressure is uniform. Crack top cock first to heat glass gradually.

9. Open both gage cocks slowly. If additional tightening is required close gage cocks so they are not being tightened under pressure.

10. Keep safe distance away the first time the gage glass is blown down.

Discoloration of gage glasses can occur from a variety of causes. Glasses should be discarded if they are etched or scratched in any way. Glasses can sometimes be cleaned by using a solution containing six ounces of fresh water, one ounce of common salt, one ounce of acetic acid to which a small amount of baking soda is added.

Causes of Low Water Level in Boilers:
1. Failure of the feed pump.
2. Leaks in the discharge relief line of the feed pumps.
3. Improper regulation or defective valves.
4. Low water in feed tank.
5. Water at pump inlet too hot.
6. Leak in feedwater heater.
7. Leak in boiler.
8. Priming.

The first two causes are the result of faulty feed pump operation and may be overcome by using the auxiliary feed pumps until the main pump has been repaired. Improper operation of feed water regulators or defective valves can be compensated for by using the auxiliary feed line or auxiliary feed line valves. Low water in the feed tank is the result of lack of make up or supply water. If the water temperature is too high, vapor will form, causing the pump to become vapor bound. This trouble results when the reduced pressure in the feed suction pipe caused the hot water supply to flash into steam. It may be corrected only by increasing the pressure at the pump inlet, or by using cooler water. When leaks occur in the feed water heater system or in the boiler, the equipment must normally be taken out of service immediately, pending repairs or adjustments. Priming of the boiler can have serious consequences; hence, the load on the boiler should be reduced or the boiler water concentration should be lowered until the cause of the priming is ascertained and corrective measures can be applied.

Low water level. When the water level drops in a boiler so that it is out of sight in the gage glass, its exact location is unknown. Blowing down the glass for a couple of seconds and then closing the valve may cause the water to momentarily reappear in the glass and give some indication of just how low it may be. It is not safe to continue operation of a boiler for any period of time after the water has dropped so low as to become invisible in the glass. The fires should be put out immediately when this condition develops.

The fire should be secured, and removed from the fire box as soon as possible. Every effort should be made to prevent doing anything which would increase the furnace temperature and cause overheating of the boiler. The boiler should be allowed to cool slowly. Make up water should not be added under any circumstances while the dangerous overheated condition prevails. No attempt should be made to reduce the pressure by venting the safety valves, as a sudden rush of water lifting in the boiler may cause an explosion. It is considered good practice to just let the boiler cool gradually, preventing further danger or damage.

The unit should then be entered for inspection to determine the extent of the damage. If no damage is found, a hydrostatic test should be placed on the unit. If tube failure resulted from low water, the water level should be maintained with additional water until the pressure is worked off. The boiler dampers should be left open to carry any escaping steam up the stack. The fires should be killed immediately.

Filling the boiler with water. It is customary to open the drum vents to provide a means for the air to escape as it is displaced by the water. It is normally proper to fill a boiler to the level of the lower tri-cock, remembering that the water will expand when heated. In some cases of boilers equipped with water walls, some engineers prefer to fill the boiler to a high level and to drain off excess water through water blowoff valves as steam pressure is raised on the unit. This will assist in starting circulation and avoid sudden temperature changes in the water wall system.

Foaming. When steam boiler water containing chemicals that will produce a foaming condition is heated, it will produce a foaming condition of the water of foam bubbles that will pass out of the boiler with the steam and will carry considerable quantities of water and chemicals with them. Foaming can sometimes be detected through abnormal agitation of the water level and the condition will be shown in the gage glass. When a boiler is foaming badly, a sudden increase of load will cause a larger than usual use in water level and enormous quantities of foam to be carried into the steam liner. A sudden reduction of load will cause the water level to drop rapidly, and possibly disappear from the glass entirely. The remedy for a foaming condition is to remove the impurity by blowing down the boiler. This may be accomplished by blowing down through the lower blowoff valve and replacing the water by increased flow of feed water, or by increasing the quantity of surface or continuous blowdown when the boilers are equipped with a surface blowoff system.

Priming. Priming in a boiler is the entrainment or carry-over of water with the steam in the absence of a foaming condition of the boiler water. Priming in a boiler may be caused by forcing the boiler beyond its rated capacity or increasing the load too quickly; carrying the water level too high; sudden opening of nonreturn valves when the boiler is cut into the line with excess pressure, lifting of the safety valve when the boiler water level is carried too high, and faulty design of the boiler or faulty adjustment of steam purifying and steam washing apparatus in the boiler drum.

The most effective remedy is to reduce the load on the boiler, or reduce the water level by blowdown. If the load cannot be reduced, the only remedy is to carry the water at a low point. When cutting a boiler on line, the water level should be reduced.

Blowing down boilers. The primary reason for blowing a boiler down is to reduce concentration, remove scale and sludges as much as possible. Blow-down to reduce concentration can be either from the lower main blowoff valves or by use of a continuous blowdown system. Blowdown to remove sludges and scale flakes is always from the lower main blowoff valves. Another reason for blowing a boiler down is to reduce the water level to prevent carry-over. The frequency of blowing down a boiler is determined by the type of treatment, makeup of feed water and type of boiler. Blowdown valves must work freely, be kept properly packed, and all external piping and fittings should be thoroughly inspected for corrosion or other damage when the boiler is out of service.

Safety valves. If the pressure in a boiler were allowed to increase without restriction, it would become so great that with even the strongest boilers, an explosion would occur. To prevent this from happening, safety valves are set to open at a pressure far below the bursting pressure of the boiler. There are three types of high pressure steam drum safety valves which are generally used: (1) jet flow, (2) huddling chamber, and (3) nozzle reaction. The type of safety valves on your boiler should be determined along with the name of the manufacturer. Written adjusting instructions and overhaul procedures should be secured from manufacturer. If it is set by the inspector, it should not be tampered with, as it is the only insurance against excess boiler pressure.

Internal cleaning. After a boiler drum has been taken out of service for cleaning, the interior of the drum will normally be found to contain various types of deposits. The nature of the deposits determines the cleaning process to be used. Mud and sludges are easiest to remove when they are still damp, therefore the boiler drums should be entered for cleaning purposes just as quickly as their temperature permits. The temperature should be brought down to 150° or less before sluicing with cold water. A fire hose will loosen much of the mud and sludge, and will be an aid in saving on labor. Sludges remaining on the shell plates after the sluicing operation can sometimes be removed by scrubbing the plates with brushes. Wire brushes curved to the drum radius, and of a thickness that will permit them to slide between adjacent tube ends will be found helpful. Wire brushes mounted on a hand drill or pneumatic tool will save much manual labor.

When mud and sludges have been allowed to dry out, they can often be removed by flaking off the larger particles with a putty knife or a thin chisel. After the large flakes are removed, smaller flakes and areas not accessible with the scraper or other tools can be cleaned by using wire brushes and in some cases may be mounted to portable drills or other pneumatic tools. When adherent scale is present, it can sometimes be removed by tapping the surface with a light hammer. Some types of scale will separate readily from the drum metal and can be removed in rather large flakes by this method. Other types of scale adhere so tightly that they must be removed by the chiseling action of a hammer and a cold chisel or similar tool. In the latter event, much labor can be saved by providing the workmen with pneumatic or electric tools equipped with chisel points of shape and size to meet the requirements best. In using chisels, care must be taken not to nick the shell plate or to drive against the expanded tube ends.

After all sludges or scale have been cleaned down to the bare metal, the surface of the shell plates should be examined for pits that might have been caused by corrosion. These pits can normally be detected by the presence of pimples or by irregular spotted appearance of the surface before cleaning. A knife blade or chisel may be used to remove the pimples and thus expose a black core with an underlying depression corroded into the boiler metal. All of these pits in the metal should be cleaned out thoroughly to expose bare boiler metal. To retain the corrosion will invite further attack at the same point when the boiler is returned to service. Progressive corrosion of the drum shell will eventually destroy it and require its replacement, or necessitate a reduction of the operating pressure on the boiler. The surest preventive measure is to treat or condition the boiler feed water to remove or inhibit the causes of corrosion. Repair of the pitted surfaces may consist of painting them with a protective coating to prevent further attack, or to weld them up under the supervision of a qualified inspector, grind down the weld to bring the metal flush with the general surface of the shell, and follow that by the application of a protective coating.

Cleaning fire side of tubes. Before taking a boiler out of service, the soot blowers should be operated thoroughly to remove as much dust and soot accumulations as can be removed by these means. When the unit is cool enough to permit the access doors to be opened, further cleaning can be done by blowing accessible areas, using steam or air lances. When the gas passages can be entered, air lances can be used for further cleaning. This procedure can be followed by brushing with any available cleaning equipment. Care must be taken to remove thoroughly all refuse, especially refuse lodging in constricted spaces, for this material will absorb moisture and corrode the boiler metal if the boiler is laid up for any length of time.

Cleaning water side of tubes. Mud, sludges, and loose scale particles can usually be removed in large quantities from boiler tubes by washing them with a fire hose before they have dried. In the case of a water tube boiler, the drum must be sufficiently cooled to allow entrance in order to proceed with the washing operation. In the case of a horizontal return tubular boiler, or similar type of boiler, considerable areas of tubes can be reached by using a fire hose through the manhole opening, while further cleaning can be done after the boiler is sufficiently cooled to allow entry.

Tube scrapers and brushes. The fire side of fire-tube boiler flues sometimes acquire a tenacious, hard coating of soot requiring more than usual effort for removal.

The flexible steel blades (a) are of a spiral form to give a slicing action when they are pushed through the tube, and they are springy to give them good contact with the tube wall. They are firmly attached to the heads (b) and (c) and a pipe handle of proper length is screwed into one of the heads so that the scraper can be readily pushed through the entire length of the tubes. When the deposits are too tenacious to be removed readily with the scraper shown in Figure 36-10, a combination scraper and brush as shown in Figure 36-11, will be found more effective.

The cutting blades (a) are formed in a steep spiral so that they will have a powerful scraping action when the tool is pushed through the tubes. They

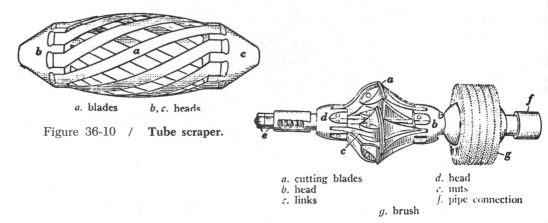

a. blades   *b, c.* heads

Figure 36-10 / **Tube scraper.**

a. cutting blades   *d.* head
b. head   *e.* nuts
c. links   *f.* pipe connection
                g. brush

Figure 36-11 / **Combination scraper and brush.**

are mounted on the cleaner by being flexibly connected to the head (b), and to the head (d) through the links (c). The head (d) is mounted loosely on the shaft so that it may be forced toward the head (b) by applying spring tension when the nuts (e) are tightened. This action forces the cutting blades outward, hence the cleaner may be adjusted to fit tubes of different diameters and to apply internal pressures. The brush (g) is normally made of stiff wire to facilitate brushing the loosened material from the tubes. A pipe handle of suitable length is screwed onto the cleaner at (f) for convenience in handling.

Power driven cleaner for fire-tube boilers. A tube cleaning motor driven by compressed air or water can be equipped with a variety of attachments for specific purposes. One such attachment is a rotary brush arrangement for cleaning the interior surfaces of fire-tube boilers or air heaters. Another attachment for a power motor is illustrated in Figure 36-12.

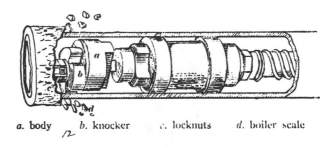

a. **body**   *b.* knocker   *c.* locknuts   *d.* boiler scale

Figure 36-12 / **Turbine cleaner for fire-tube boilers.**

This tool is essentially a vibrator consisting of a body (a) on which the knocker (b) is eccentrically mounted and secured by the locknuts (c). The construction of this attachment is such that when the head is rotated rapidly, the three lobes of the knocker deliver repeated light blows to the inside of the tube, causing scale to be loosened from the exterior of the tube as shown. Power driven cleaners of this kind are known as turbine cleaners.

Power driven cleaner for water tube boiler. Power driven cleaners are manufactured in a variety of sizes and degrees of flexibility to fit tubes of various diameters and to allow working around bent tubes of normal curvature. One form of turbine cleaner used for removing scale from the inside of water tubes is shown in Figure 36-13(a).

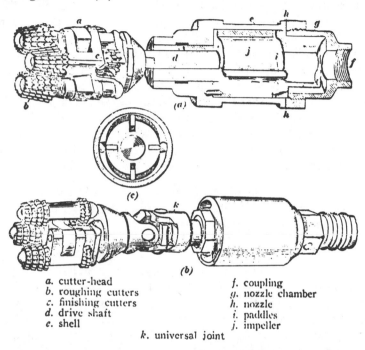

a. cutter-head
b. roughing cutters
c. finishing cutters
d. drive shaft
e. shell
f. coupling
g. nozzle chamber
h. nozzle
i. paddles
j. impeller
k. universal joint

Figure 36-13 / **Turbine cleaners for water tube boilers.**

Figure 36-14 / **Drill-type tube cleaner.**

It consists of a cutterhead containing three sets of roughing cutters and three finishing cutters. The head is keyed to a shaft which is, in turn, attached to the motor. The motor or turbine may be driven by compressed air or water admitted by a flexible hose which is also used for moving the cleaner through the length of the tube. The operating pressures required are from 80 to 100 psi. The exhaust from the turbine is forward toward the cutterhead; hence, loosened material is driven out the far end of the tube and is not blown back to inconvenience the operators. View (b) shows the cutterhead connected to the motor by a flexible joint to enable the cleaner to be moved readily around the curvature of bent boiler tubes.

Scale on the interior of boiler tubes sometimes may close a tube almost entirely or be very hard and difficult to remove. In this event, the type of cutter-head illustrated in Figure 36-14 is used.

This type of drill will enlarge the opening through the tube sufficiently to permit the use of the larger diameter, flexible cutterheads. The action of the turbine cleaner is quite fast and the cutters used are hardened so that they will not wear rapidly in breaking up and cutting the various types of hard scales. The cutterheads must therefore be used with judgement to prevent damaging tubes. If the cutter should strike hard material which may impede the forward movement of the tool, the operator should back the cutter up and immediately try again, thereby saving the tube wall from damage.

Laying up a boiler. Moisture is one of the primary causes of boiler deterioration. This is likely to occur during the nonheating season. The external surfaces, and the fireside should be kept dry. If gas and oil fired boilers are not used in the summer, it is recommended that the boiler be kept warm (approximately 100°F.). The boiler should be filled and the temperature raised to 200°F. to drive off gases. If feed water is normally treated, add the compound at this time. Leave the boiler filled until the beginning of the next heating season.

The following precautions should be taken if firing is to be discontinued:

1. All heating surfaces should be cleaned of soot, ash, or other residue.
2. The heating surfaces should be given a coating of oil or grease.
3. Stack connections should be cleaned.
4. The water side should be cleaned thoroughly with a high pressure hose.
5. The boiler should then be dried out with a small kerosene burner or small wood fire. Care should be taken to see that the boiler is not overheated.
6. The grates and ash pits should be thoroughly cleaned.
7. Remove any rust or other foreign matter from external surfaces, and apply a coat of preservative paint where required, if the parts are generally painted.
8. All firing equipment, operating and safety controls should be checked. If repairs are required, they should be made immediately.

Alkalinity of Feed Water

Hydrogen-ion concentration. The corrosion-forming characteristics of boiler feed water depend to a large extent on the degree of acidity of the water. Whether the water will be acid or alkaline is determined largely by the manner of ionization of the water molecules. Although water is a very weak electrolyte, it does dissociate to a very slight extent into positive hydrogen (H^+ ions) and negative hydroxyl (OH^-) ions, as indicated by the equation:

$$H_2O = H^+ + OH^-$$

Equal numbers of both H^+ and OH^- ions are formed in the dissociation of pure distilled water. Since neither H^+ nor OH^- predominates, the solution is termed neutral. A solution that contains an amount of H^+ ions in excess of the OH^- ions is spoken of as an acidic because it has an acid reaction. Conversely, an excess of OH^- ions over H^+ ions will produce a basic or alkaline condition. The H^+ ions liberated by the dissociation of the solution, attack the metal of the boiler, causing it to dissolve. The greater the hydrogen-ion concentration, the more rapid will be this attack and vice versa.

It has been established that the product of the hydrogen-ion concentration and the hydroxyl-ion concentration in water is a constant. By electrical measurements, this constant has been found to be .00000000000001 or 10^{-14}, at 21°C. Therefore since, in pure water the H^+ ions exactly equal the OH^- ions, each must have a concentration of .0000001 or 10^{-7}. It also follows that any increase in the concentration of one group of ions must result in a decrease in the concentration of the other group. Hence, a measure of the hydrogen-ion concentration is actually also a measure of the hydroxyl-ion concentration. By measuring the hydrogen-ion concentration in a sample of the feed water, it can be determined what corrective measures should be taken to counteract the corrosive characteristics of the water.

*p*H **scale of H-ion concentration.** The hydrogen-ion concentration in water is very small. Pure water contains only 0.0000001 grams of hydrogen ions per liter. When such water contains more hydrogen ions, it has an acid reaction, and with fewer hydrogen ions, it has an alkaline reaction. Because of the extremely small hydrogen concentration, it is inconvenient to express the percentages of such concentrations in decimal form. Instead a scale has been evolved based on numbers from 0 to 14 and known as the hydrogen ion, or *p*H scale. Each number represents a definite hydrogen ion concentration.

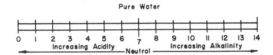

Figure 36-15 / **pH scale.**

The *p*H number of water of any solution indicates whether the liquid has an acid or an alkaline reaction. A neutral liquid, such as pure water, is neither acid nor alkaline and has a *p*H value of 7. Liquids having *p*H values between 0 and 7 indicate acidity, the degree of acidity being less for a high *p*H value than for a lower value. Thus, a *p*H of 4.5 indicates a greater degree of acidity than a value of 7.2. Liquids having *p*H numbers between 7 and 14 have an alkaline reaction, the degree of alkalinity increasing as the *p*H value of the liquid increases.

The *p*H scale is illustrated in Figure 36-15. Each successive number on the scale represents 1/10 the concentration of the preceding number. For example, a *p*H reading of 5 indicates a hydrogen-ion concentration 1/10 as great as that for a *p*H reading of 4, and 10 times the concentration for a *p*H reading of 6.

Boiler Tube Failures and Water Treatment

Boiler generating tubes fail for one of three reasons: (1) overheating, (2) attacked chemically on water side or fire side, and (3) thinned down from the fire side.

Overheating may be caused by deposits on the water side which prevent adequate cooling of the tube metal, or by unusually hot zones in the furnace. Two of the common deposits found in boiler tubes are hard scale, and sludge.

Hard scale treatment: Lime and soda ash or zeolite ahead of the boiler, and phosphate internally.

Sludge treatment: Concentrates of phosphate silica and alkali. Frequent wash outs and blow downs.

Chemical attack is in some cases related to overheating, as high temperatures greatly speed up certain chemical reactions involving iron. Most chemical attacks can be prevented by proper boiler water treatment.

Chemical Attacks	*Treatment*
Steam	Increase circulation, change baffles.
Caustic	Use sodium nitrate.
Acid	Frequent alkalinity checks.
Concentrated boiler water	Amine treatment to reduce deposits of corrosion products.
Stress corrosion	Baffle repair, and other repairs, if necessary.
Oxygen	Aeratoring with sodium sulphite.

External thinning may be caused by the improper placement of soot blowers and, in rare cases, by abrasive particles in the fuel. Many of the difficulties arising on the water side of the tube are the result of overheating. This condition may be the result of difficulties on the fire side, such as defective baffling or poor furnace design. Sometimes the arch is not properly designed in forced draft chain grates, and also in underfeed stokers there is a blow pipe effect near the lower ends of the tube, due to the constriction of gases in this region, extreme temperatures are obtained. Corrosion on the fire side of tubes is sometimes a problem in coal fired boilers equipped with an economizer section. Most bituminous coal contains some sulfur which decomposes to acid gases on burning. In some economizers, temperatures are sufficiently low to allow moisture containing these gases to condense on the outside of the tubes, thus producing sulfuric acid which is extremely aggressive to the tube metal. Tubes may also fail from the fire side by the abrasive action of fly ash and dirt particles in the fuel.

On closed hot water systems, chromate is generally used, the amount dependent on the characteristics of the raw water and cycles of concentration.

A complete analysis of your boiler water should be made to determine the proper chemicals to be used. This should be done by an expert, and personnel should be trained in making the required analysis and the frequency established by the need.

37 / DATA PROCESSING ROOMS

Basic elements are required for *data processing rooms*. Some of these are ceiling height, floor strength, lighting levels, electrical power, temperature and humidity controls, sound levels, air flow, space for movement between auxiliary equipment, and future expansion requirements.

Acoustical ceilings, and sometimes walls acoustically built down to shoulder height are recommended. Noise levels for this room and also that of adjacent areas must be considered, along with the problem of noise traveling through ducts from this room. Proper insulation and use of acoustical materials will correct these conditions.

A *ground floor* site for the computer will reduce the problem of vibration. Floor and wall construction should be capable of preventing transmission of vibrations to other areas of the building. The weight point load of the equipment should be secured from the manufacturer, and the floor designed accordingly, including possible future increases. A raised or "floating" floor is an accepted type floor for computed installations. This is a secondary floor over the original floor of the building. This type of construction allows power cables, receptacles, temperature controls, air-conditioning ducts and other facilities to be concealed yet readily accessible. Access to the underfloor area is achieved by lifting any of the panels. This type of floor is equipped with leveling devices. With a raised floor, a ramp rather than steps leading into the data processing room will make it much easier to get equipment in and out of the area. When new floors are poured in a new building which will contain a data processing room, raceways are generally provided in the floor, or large segments of floor duct installed. Sometimes channel raceways are provided in side walls approximately six inches off the floor.

The equipment manufacturers generally agree that a 40-foot candle, measured thirty inches above the floor is sufficient. Direct sunlight is not desirable, as easy observation of the console and the signal lights requires a *low level* of *illumination.* It is recommended that lighting be zoned, so that sections can be turned on or off as desired.

Electrical power requirements are critical, as the computer requires a constant rate of power in order to perform accurately. Separate feeders should be considered and separate transformers or installation of a motor alternator may be required. Separate breakers should be installed in the data processing room, and should be turned off when the equipment is used for only one or two shifts per day. When considering the power requirements, air conditioning and lighting must also be provided for.

Careful *temperature* and *humidity control* must be maintained. The temperature must generally be maintained between 65 to 90 degrees, and the humidity between 20 and 80 per cent. A midpoint between the two figures is considered good for operating, and the temperature and humidity should be kept constant with as little variation as possible. Independent air conditioning units are recommended for data processing rooms. Heat dissipation is an important part of any computer installation, and must be figured when sizing the air-conditioning unit along with walls (inside or out), glass, windows, insulation, lights and outside temperatures. In some cases, exhaust hoods may be desirable, with a canvas flap between the hood and the machine, making a tight seal. If windows are provided, special glass is recommended to prevent condensation during low outside temperatures, and transmission of heat during high temperatures. It is also recommended that the glass be tinted slightly to prevent excessive sunlight which can effect signal lights on the computer.

Physical maintenance is of utmost importance. Changing of filters in the air handling units should be done periodically upon the basis of inspection needs. Computers also maintain filters and should be changed by the customer service engineer when required. If the data processing room has carpeting, it should be vacuumed nightly. If it is equipped with tile, it should be dry mopped nightly with a treated mop. Dusting of equipment should be done nightly with a treated duster or wool duster. Doors should be equipped with automatic door closers to keep dust to a minimum.

Supplies or other items should be dusted and cleaned thoroughly before being brought into the room. Dust is inclined to gather under raised floors and must be removed periodically, before it can circulate into the machine area. Perforated acoustical tile or perforated walls should be vacuumed periodically to prevent dust accumulations. Walls should be washed a minimum of twice a year. Tile floors should be scrubbed, waxed, and buffed periodically as required. Ventilation should be sufficient along with temperature control to keep windows closed at all times. Windows should be caulked thoroughly to prevent the entrance of dust and dirt. If drapes have been installed, they should be vacuumed twice per year.

The best way to insure savings in time, cost, and future maintenance, is to consider all areas mentioned and to provide for them in the original design or conversion.

37A / PROTECTION OF ELECTRONIC COMPUTER SYSTEMS

Reproduced, with editing, from "Standard for the Protection of Electronic Computer Systems" (NFPA No. 75), 1962 edition, copyrighted by the National Fire Protection Association. Copies of the complete Standard in its current edition are available from the Association, 60 Batterymarch St., Boston, Mass.

Purpose. The purpose of this standard is to set forth the minimum requirements for the protection of electronic computer systems from damage by fire or its associated effects.

Scope. This standard covers the fire protection requirements for installations of electronic computer systems where special building construction, rooms, areas, or operating environment are required for protection of the system.

This standard presently does not cover systems which can be installed without this special construction.

This standard, however, may be used as a management guide for the protection of electromechanical processing equipment, small table top or desk type units and electronic computer systems that do not require specifically constructed rooms or areas.

Building construction. The computer area shall be housed in a fire-resistive, noncombustible or sprinklered building, or when the portion of a nonfire-resistive structure housing a computer area is a separate fire division, only that portion of the structure housing the computer area is required housed in construction as noted above.

Location of computer area. The electronic computer area shall be located to minimize fire, water, and smoke exposure from adjoining areas and activities. The computer room shall not be located above, below or adjacent to areas or other structures where hazardous processes are located unless adequate protective features are provided.

Computer room construction. The computer system shall be housed in a room of noncombustible construction. All materials including walls, floors, partitions, finish, acoustical treatment, raised floors, raised floor supports, suspended ceilings, and other construction involved in the computer room, shall have a flamespread rating of 25 or less. (See NFPA Standard Method of Test of Surface Burning Characteristics of Building Materials, No. 255.)

1. Floor covering materials, such as asphalt, rubber or vinyl floor tiles, linoleum or carpeting may be used to cover any exposed floors.

2. All metal floors should be grounded.

In multistoried buildings, the floor above the computer shall be made reasonably watertight to prevent unnecessary water damage to equipment. Any openings at beams, pipes, etc. shall be sealed to watertightness.

Computer room fire cutoffs. Where exposure to the building housing the computer is unfavorable, good protection against exposure shall be provided. This protection should consist of blank masonry walls, or other suitable exposure protection, depending upon local conditions. (See NFPA Suggested Practice for Protection Against Fire Exposure of Openings in Fire Resistive Walls, No. 80A.)

The computer room shall be cut off from other occupancies within the building by noncombustible, fire-resistance-rated walls, floor and ceiling. The fire-resistance rating shall be commensurate with the exposure, but not less than one hour.

The fire-resistant walls or partitions enclosing the computer room shall extend from the structural floor to the structural floor above, or the roof.

Raised floors (where required). Raised floors including the structural supporting members shall be made of concrete, steel, aluminum or other noncombustible material. Pressure impregnated, fire-retardant treated lumber having a flame-spread rating of 25 or less may be used. (See NFPA Method of Test of Surface Burning Characteristics of Building Materials, No. 255.) Other types of wood construction shall not be used for raised floors. Existing combustible, structural floors shall be covered with an insulating noncombustible material before installing a raised floor. Access sections or panels shall be provided in raised floors so that all the space beneath is readily accessible. Openings in raised floors for electric cables or other uses shall be protected to minimize the entrance of debris or other combustibles beneath the floor. This may be accomplished by covers, grilles, screens, or by locating equipment directly over the openings.

Cable openings. Electric cable openings in floors shall be made smooth or shall be otherwise protected to preclude the possibility of damage to the cables.

General Computer Room Requirements

Materials and equipment permitted in the computer room. Except as noted below, only the actual electronic computer equipment and such input-output or other auxiliary electronic equipment electronically interconnected with the computer, or which must be located in close proximity to the electronic computer equipment, shall be permitted within the computer room itself.

All office furniture in the computer room shall be metal.

Small supervisory offices and similar light hazard occupancies directly related to the electronic equipment operations may be located within the computer room if all furnishings are metal and adequate facilities are provided for containing the necessary combustible material. Supplies of paper or other combustible material shall be strictly limited to a minimum needed for efficient operations.

Limited records may be kept in the computer room.

The following shall not be permitted within the computer room:

1. Any activity or occupancy not directly associated with the electronic computer system(s) involved.

2. Supplies of paper or other combustible material in excess of that necessary for efficient operation.

3. Service and repair shops and operations except for that servicing and repairing performed directly on machines which are impractical to remove from the computer room.

4. Bulk storage of records.

5. Any other combustible material, equipment or operation which constitutes a hazard and which can be removed.

Combination of systems. Separate electronic computer systems should not be combined in a single computer room unless the systems are interconnected electronically, use the same input-output equipment or must be located in the same room for other operational reasons. Computers may be located in adjacent rooms with properly protected communicating openings in separating walls.

General storage. The operation of an electronic computer system frequently requires considerable quantities of stationery supplies and other combustible support materials. This material can present a serious fire exposure within the computer room capable of causing serious damage to vital equipment or records.

Paper stock, unused recording media, and other combustibles within the computer room shall be restricted to the absolute minimum necessary for efficient operation. Any such materials in the computer room shall be kept in totally enclosed metal file cases or cabinets.

One or more storage rooms outside of the computer room shall be provided for reserve stocks of paper, unused recording media and other combustibles.

Construction of Computer Equipment

Types of computer equipment:

Type I. So designed that, when any component or part is ignited, the fire will be confined to the immediate area where the source of ignition is located, to the extent that the affected parts can be readily replaced; or so designed, by the use of special construction and material, to be inherently free from the possibility of ignition of any component or material. The equipment shall include automatic means to de-energize the circuits before components or units are caused to operate at, or to be subjected to, hazardous temperature conditions.

Type II. So designed that when de-energized fire is not likely to spread beyond the external housing of the unit in which the source of ignition is located.

Type III. Includes all equipment not defined in Types I and II above.

Classification of all equipment into the three types described above is being developed by the testing laboratories. Until this information is available, Underwriters' Laboratories listed equipment may be considered as meeting at least the Type II rating.

Use of approved or listed equipment. Wherever possible, each installed electronic computer system or individual computer unit shall be a recognized Type I or Type II construction.

"Recognized" equipment is that equipment which has been accepted by the authority having jurisdiction as meeting the requirements for Type I and Type II. Approval or listing as Type I and Type II by Underwriters' Laboratories, Inc., Factory Mutual Engineering Division or other nationally recognized independent fire testing laboratories shall be considered as proof that the equipment has met such standards.

Any equipment not of a recognized construction shall be considered to be of a Type III construction in determining the applicable installation requirements pertaining to that particular equipment.

Design features. Approved flexible cord and plug assemblies, not to exceed 15 feet in length, may be used for connecting the computer to building wiring to facilitate interchange.

Interconnecting cables and wiring between units should be of a type approved for the purpose by a nationally recognized testing laboratory. Such cables shall be considered as a part of the computer system and suitable for installation on the floor or under a raised floor. If cables or other interconnecting wiring is of any other type, the equipment shall be so designed that the cables or wiring can be installed in accordance with the NFPA National Electrical Code (No. 70).

Individual units of a system should be housed in metal or noncombustible enclosures with suitable subdivisions to minimize the likelihood of fire spreading from one section to another within a single unit structure. Enclosures shall be designed to permit easy access to all interior sections in the event of an emergency.

Air filters for use in individual units of a computer system shall be of approved types that will not burn freely or emit a large volume of smoke or other objectionable products of combustion when attacked by flames, so arranged that they can be readily removed, inspected, cleaned or replaced when necessary.

Each electronic computer system shall be so designed that, in the event of an emergency, the system can be de-energized by the operation of a suitably marked control at one location.

Except as noted below, oil shall not be used as a component of a unit of an electronic computer system. If the design of the unit is such that oil or equivalent fluid is required for cooling or other purposes, it shall have a flash point of 300°F. or higher, and the container shall be of a sealed construction, incorporating automatic pressure relief devices.

All sound-deadening material used inside of computer equipment shall be noncombustible.

Protection of Computer Rooms and Equipment

Protection of computer rooms. If the construction of the computer room contains any combustible material other than that permitted, or if the computer housing or structure is built all or in part of combustible material, then the computer room shall be protected by an automatic sprinkler system.

If the operation in the computer room involves a significant quantity of combustible materials, the computer room shall be protected by an automatic sprinkler system.

Automatic sprinkler systems protecting computer rooms or computer areas shall conform to NFPA Standard for the Installation of Sprinkler Systems (No. 13). Sprinkler systems protecting computer rooms should preferably be valved separately from other sprinkler systems.

To minimize water damage to the electronic computer equipment located in sprinkler protected areas, it is important that power be off prior to the application of water on the fire. In facilities which are under the supervision of an operator or other person familiar with the equipment (during all periods that equipment is energized), the normal delay between the initial outbreak of a fire and the operation of a sprinkler system will provide adequate time for operators to shut down the power by use of the emergency shutdown switches. In other instances where a fire may operate sprinkler heads before discovery by personnel, a method of automatic detection should be provided to automatically de-energize the electronic equipment as quickly as possible.

Smoke or fire detectors shall be provided in the air space below existing combustible raised floors to sound an audible as well as visual alarm and to shut down all electric power passing through the air space.

Air spaces below existing combustible raised floors shall be subdivided by tight, noncombustible bulkheads into areas not exceeding that required for one system or, in any case, not more than 10,000 square feet.

The air space below a raised floor or above a suspended ceiling may be used as a plenum chamber for air conditioning if construction is noncombustible and all wiring is of an approved type. Interconnecting cables and wiring between units should be of a type approved for the purpose by a nationally recognized testing laboratory. Such cables shall be considered as a part of the computer system. If cables or other interconnecting wiring is of any other type, the equipment shall be so designed that the cables or wiring can be installed in accordance with the NFPA National Electrical Code (No. 70).

Portable fire extinguishers. Approved portable carbon dioxide extinguishers shall be provided and maintained for electrical fires. See NFPA Standard for the Installation, Maintenance and Use of Portable Fire Extinguishers (No. 10).

Approved Class A type extinguishers shall be provided and maintained for ordinary combustible materials such as paper.

If it is desired to provide other types of extinguishers, advice should be obtained from the computer equipment manufacturer and the authority having jurisdiction as to their acceptability.

In installations where conditions may require the provision of inside hose, it shall be 1½-inch rubber lined with shutoff combination solid stream, water-spray nozzles.

Training. Designated persons working in the computer area shall be thoroughly trained in how to use each of the available types of manually operated fire fighting equipment. This training should show the capabilities and the limitations of the extinguishing equipment.

All hand-type extinguishing equipment shall be plainly marked to indicate the type of fire for which it is intended, and installed and maintained in accordance with NFPA Standard for Portable Fire Extinguishers (No. 10).

Protection requirements for equipment. In addition to the protection required elsewhere in this standard, each unit of an electronic computer system shall be provided with the following special protection:

Type I Equipment. Type I equipment requires no special protection.

Type II Equipment. There shall be available to each unit of Type II equipment an adequate means of extinguishing the maximum fire which may occur as follows:

1. Carbon dioxide fire extinguishers or carbon dioxide hand hose systems installed in accordance with NFPA Standard for Carbon Dioxide Extinguishing Systems (No. 12), shall be considered as providing adequate extinguishing protection provided all of the following conditions are met:

a. The equipment, during all periods that it is energized, is under supervision of an operator or other person familiar with the equipment and trained in the operation of the types of extinguishers or hand hose systems involved.

b. Adequate controls are readily accessible to shut down power and air conditioning to the involved equipment.

c. All interior sections are readily accessible to manual application of the extinguishing agents.

d. There is located within the computer room and not more than 50 feet from the equipment under consideration either a carbon dioxide fire extinguisher or carbon dioxide hand hose system having a capacity of at least one pound of carbon dioxide for each cubic foot of volume of the unit under consideration if the equipment is on open racks; one-half pound for each cubic foot of volume if the unit under consideration is enclosed in a cabinet.

2. Type II equipment not meeting the requirements of (1) above shall be protected by a fixed carbon dioxide extinguishing system conforming to the requirements of NFPA Standard for Carbon Dioxide Extinguishing Systems (No. 12).

Type III Equipment. Because of the hazard presented by the possibility of communication of fire to other equipment outside of Type III units, hand applied extinguishing agents shall be considered inadequate except in the case of small (table top or desk size) units. Except for the previously mentioned small units, all Type III equipment shall be protected with a fixed carbon dioxide extinguishing system.

Fixed carbon dioxide extinguishing systems installed may be manually actuated when the equipment is, during all periods when it is energized, under the supervision of an operator or other persons familiar with the equipment. In all other instances, the extinguishing systems shall be provided with both manual and automatic actuation means. The automatic actuation should be by an approved method of detection meeting the requirements of NFPA Standard for Proprietary Protective Signaling Systems (No. 72). Particular attention shall be given in the choice of actuation means, to insure detection, considering the air flows usually involved in such systems, and the small heat released under fire conditions.

When called upon to operate, each fixed carbon dioxide extinguishing or carbon dioxide hand hose installation shall be arranged to automatically sound an alarm, and shut down power and air conditioning supplied to the equipment involved.

Protection of Records

General. The operation of most electronic computer systems involves obtaining, using, creating and storing large amounts of records. In many operations these records are as important to the continuity of the operation and its mission as the computer itself.

Record media. Records may be commonly encountered paper records, punch cards, plastic or metal base electronic tapes (on metal or plastic reels and in metal, plastic or cardboard containers), paper, control panels, magnetic discs, memory drums, memory cores or various other means of maintaining for future use information in plain or machine language, inside or outside of electronic equipment. Some of these records such as magnetic discs, memory drums and memory cores are usually found as an integral portion of electronic equipment and as such the protection of these records is covered in a previous section.

It is extremely important to note that the degree of resistance of magnetic tape to fire exposure is not completely known. It is known, however, that fire exposures (heat and/or steam) that would not damage records on paper media may damage records on magnetic tape. The protection of records or magnetic tape by storage methods presently available must be considered limited.

Types of records. Records involved in computer operations fall into five basic types which must be safeguarded according to their importance and the difficulty involved in their replacement as follows:

Input data. Raw or partially refined information to be entered into the computer system, either as memory for later use or for immediate use in the solution of a problem, development of a statistic or production of some other product.

Memory. Information previously converted to language or symbols immediately recognizable to the computer equipment and held for future use. Memory may be on any media which can be directly read by the computer system.

Program. Data, which may be on paper, punch cards, photographic, magnetic or electronic media, used to direct the computer as to which input or memory data to use, how to use it and the type of results to obtain. Also to be considered are any diagrams or other records which can be used to reproduce programs.

Output data. The final product of the computer system. This may consist of printed material or electronic data.

Engineering records. Those plans, specifications, and other records which provide the engineering record of the construction, wiring, and arrangement of the computer system and its housing area. Of particular importance are records of modification made following the original installation.

Value of records. The evaluation of records should be a joint effort of all parties concerned with the safeguarding of computer operations. The amount of protection provided for records shall be directly related to the importance of the records as measured by evaluation of what the loss of a particular record would mean in terms of the mission of the computer system and the re-establishment of operations after a fire. In order to maintain a reasonable sense of consistency, it is assumed that computer equipment capable of properly using the records will be available. The following classifications of records are based on the recommendations of NFPA Standard on the Protection of Records (No. 232). All records shall be evaluated and assigned to one of these categories to ensure that adequate protection is provided where necessary and that unimportant records are not overprotected.

Class I (Vital) Records. Records that are essential to the mission of the equipment, are irreplaceable, or would be needed immediately after the fire and could not be quickly reproduced. Examples might include key programs, master records, equipment wiring diagrams, and certain input-output and memory data.

Class II (Important) Records. Records that are essential or important but which, with difficulty or extra expense, could be reproduced without a critical delay of any essential missions. Some programs, wiring diagrams, memory and input-output data have this level of importance.

Class III (Useful) Records. Records whose loss might occasion much inconvenience but which could readily be replaced and which would not be an insurmountable obstacle to prompt restoration of operations. Programs and procedures saved as examples of special problems are typical of records in this category.

Class IV (Nonessential) Records. Those records which on examination are found to be no longer necessary.

Protection required. *Records kept within the computer room.* 1. The amount of records kept within the computer room shall be kept to the absolute minimum required for efficient operation. Nonessential records shall not be kept in the computer room.

2. Any records regularly kept or stored in the computer room shall be provided with the following protection:

a. Class I (Vital) or Class II (Important) records shall be stored in Class C or better records protection equipment.

b. Class III (Useful) records on paper based or plastic materials shall be stored in metal files or cabinets.

c. Class III (Useful) records on metal based material require no special protection.

Records stored outside of the computer room. 1. To the maximum extent consistent with efficient operation, all record storage shall be outside of the computer room.

2. Record storage room.

a. Class I (Vital) and Class II (Important) records shall be stored in fire-resistive rooms. The degree of fire resistance shall be commensurate with the fire exposure to the records, but not less than two hours.

b. Unless the records are contained in metal files, cabinets or other non-combustible containers, records storage rooms shall also be provided with an automatic sprinkler system.

c. Class III (Useful) and Class IV (Nonessential) records do not require any special fire protection unless these records are stored with vital or important records. In such case the requirements for the most valuable records apply to all records.

d. The records storage room shall be used only for the storage of records. Spare tapes, however, may be stored in this room if they are unpacked and stored in the same manner as the tapes containing records. All other operations including splicing, repairing, reproducing, etc. shall be prohibited in this room.

When records are kept in cases, boxes or other containers, protection shall be that required for the highest level of damageable media in the total assembly of records and containers.

It is recommended that the following be considered as limitations in the design of record storage rooms:

1. Rooms containing only paper records shall not exceed 50,000 cubic feet.

2. Rooms containing plastic based records in noncombustible containers shall not exceed 10,000 cubic feet.

3. Rooms containing plastic based records in combustible containers shall not exceed 5,000 cubic feet.

Duplication of records. The best protection for records consists of storing duplicate records in separate areas not subject to the same fire. In some electronic computer operations the duplication of records on the same or different media is a common practice. The keeping of duplicate records is particularly important when records on magnetic tape are involved.

1. All Class I (Vital) records shall be duplicated on the same or different

media and the duplicates stored in an area which is not subject to a fire that may involve the originals, preferably in a separate building.

2. Whenever practical, Class II (Important) records shall be similarly duplicated and stored.

3. Class I (Vital) records not duplicated shall be protected in accordance with NFPA Standard on the Protection of Records (No. 232).

Protection against building collapse. In any building where a fire may result in a building collapse which could either drop the records storage equipment or drop structural members on the records storage equipment, the records storage equipment shall be of types designed to protect the records against damage from the impact involved.

Utilities

Air conditioning systems. Air conditioning equipment shall conform to the requirements of NFPA Standard for the Installation of Air Conditioning and Ventilating Systems of other than Residence Type (No. 90A), and to the additional requirements set forth below.

A separate air conditioning system should be provided for the computer area.

Air ducts serving other areas should not pass through the electronic equipment area. When it is impractical to reroute such ducts, they shall be encased in a fire-resistive duct, equivalent to the fire resistance of the enclosure for the electronic equipment area.

Air ducts serving other areas shall not pass through any computer records storage room.

All duct insulation and linings shall be noncombustible, including vapor barriers and coatings.

Air filters for use in air conditioning systems shall be of approved types that will not burn freely or emit a large volume of smoke or other objectionable products of combustion when attacked by flames and shall be so arranged that they can be readily inspected, cleaned and/or replaced when necessary.

If the computer area is within an area which is air conditioned and additional air conditioning capacity is not required, the ducts serving the computer area should have suitable fire dampers.

Electrical service. The requirements in this section apply to all power and service wiring supplying the electronic computer equipment. They do not apply to wiring and components within the actual equipment or to wiring connecting various units of equipment.

Service equipment supplying the main power requirements of the computer room area should be of a type arranged for remote control or located to fulfill the requirements as covered previously.

All wiring shall conform to the NFPA National Electrical Code (No. 70).

Service transformers should not be permitted in the electronic computer area. However, if such a transformer must be installed in this area, it shall be of the dry type or the type filled with a nonflammable dielectric medium. Such transformers shall be installed in accordance with the requirements of the NFPA National Electrical Code (No. 70).

Protection against lightning surges shall be provided where needed in accordance with the requirements of the NFPA National Electrical Code (No. 70).

The number of junction boxes in underfloor areas should be kept to a minimum. If they must be used, they shall be metal, completely enclosed, readily accessible, properly grounded and in compliance with the NFPA National Electrical Code (No. 70) requirements as to construction. They shall be securely fastened to the floor. No splices or connections shall be made in the underfloor area except within junction boxes or approved type receptacles or connectors.

Emergency power controls. In addition to any emergency shut-down switches for individual components or other units of equipment, controls for the disconnecting means provided as a part of the main service wiring supplying the electronic computer equipment shall be located near the operator's console and next to the main exit door to readily disconnect power to all electronic equipment in the electronic computer area and to the air conditioning system.

Provision should be made for emergency lighting.

Fire Emergency Procedures

Preplanning for continued operation in a fire emergency. The continued operation of an electronic computer system is dependent on information stored on cards, tape, discs, drums, etc. Therefore, the preplanning for continued operation should include:

1. A program to protect records in accordance with their importance as set forth by the section on protection of records.

2. Arrangements for emergency use of other installed computer equipment to cover:

a. Plans for transportation of personnel, data and supplies to emergency computer locations.

b. Agreements and procedures for the emergency use of the computer equipment.

3. Programs designed with adequate number of checkpoints and restarts to ensure rapid recovery to normal operations.

Personnel should receive continuing instructions in:

1. Method required for turning off all electrical power to the computer.
2. Turning off the air conditioning to the area.
3. Alerting the fire department or company fire brigade.
4. Evacuation of personnel.
5. The location of and proper operation and application of all available fire extinguishing and damage control equipment. Because of the noise and of the need for skillful operation of carbon dioxide extinguishing equipment, computer room personnel should be fully trained in carbon dioxide usage through actual operating of the equipment on a practice fire.
6. The importance of records and their storage requirements.

Emergency fire procedure. A written emergency fire plan should be prepared for each installation which assigns specific responsibilities to designated personnel. The following major items are suggested as minimum features of this plan:

1. *Remove all power to the computer system.*
   a. MEANS
   Main line circuit breaker or equivalent for turning off all power.
   b. LOCATION OF CONTROL FOR DISCONNECTING MEANS
   Remote controls for operating the disconnect located near the operator's console and next to the main exit door.

2. *Shut down air conditioning system.*
   a. IN CASES OF COMPLETELY SEPARATE SYSTEMS ONLY
   Emergency means as previously described provided to turn off the computer room air conditioning. They should also be located near the emergency power shutoff device.
   b. IN CASES OF REGULAR BUILDING SYSTEMS ONLY
   Emergency means as previously described provided to close off all duct dampers leading to and from the computer room. They should be located near the emergency power shut-off device.
   c. IN CASES OF COMBINING THE REGULAR BUILDING SYSTEM WITH A SUPPLEMENTARY SYSTEM
   Emergency means provided to simultaneously accomplish the similar action as described in the preceding (a) and (b) items.

3. *Notification of proper authority.*
   a. BUILDING FIRE BRIGADE
   Fire brigades shall be called immediately.
   b. OUTSIDE FIRE-FIGHTING COMPANY
   Outside fire departments shall be called in accordance with the emergency fire plan.

Damage control. Means should be provided to prevent water damage to electronic equipment. The proper method of doing this will vary according to individual equipment design. Consideration should be given to the provision of waterproof covers.

Whenever electronic equipment or any type of record is wet down, smoked up or otherwise affected by the results of a fire or other emergency, it is vital that immediate action be taken to clean and dry the electronic equipment. If the water, smoke or other contaminations are permitted to remain in the equipment longer than absolutely necessary, the damage may be grossly increased.

Glossary

Associated effects. Smoke, wind, heat, water.

Business interruption. Loss of use of the equipment and restoration to the former level of operation.

Check point (restarts). A predetermined point in the programming of information where the operator can return to continue that portion which was accidentally interrupted. These points are spaced to minimize lengthy reruns.

Console. Unit containing main operative controls of the system.

Control panel (plug board). A removable wiring panel for manually changing the operation of a component of an electronic computer system.

Electro-mechanical processing equipment. Individual units which are not electronic in nature and do not constitute a system or complete a total function.

Electronic computer system. Any electronic digital or analog computer, along with all peripheral, support, memory, programming or other directly associated equipment, records, storage and activities. The most common types of electronic computer systems are of the digital computer type and are usually classed as Electronic Data Processing Machines (EDPM), Automatic Data Processing Machines (ADPM), and/or Integrated Data Processing systems.

Electronically interconnected. Units that must be connected by a signal wire to complete a system or perform an operation.

Input-output (I-0). That equipment electronically associated with the computer which feeds information into the computer for computation or receives information which has been computed and displayed in a form familiar to the operator.

Interconnecting cables. Signal and power cables for operation and control of system (usually supplied by computer manufacturer).

Master record. A record of information on a medium which can be referred to whenever there is a need to rebuild a program.

Program. Instructions to direct systems operation.

Raised floor. Platform on which machines are installed for housing interconnecting cables, and at times as a means of supplying conditioned air to various units. Sometimes referred to as a false floor or secondary floor.

Readily accessible. When the covers, panels, doors or other enclosure for the electronic components within the equipment or the flooring can be removed by quick, simple operations to expose any area which might be involved in fire and permit the application of an extinguishing media. Preferably no special tool or other removable device should be required to perform this operation. Where safety from electrical shock requires the extra security for electronic components, a simple tool may be required provided the tool is kept in a convenient tamperproof location on or near the machine. Quick removal of the machine enclosure means that any component can be exposed by the average male employee in not over one minute. Where a special tool or device is required to assist in facilitating accessibility to underfloor areas, the number and arrangements of the devices shall be such that any space beneath the floor can be exposed to application of the extinguishing media by the average male employee in not over one and one-half minutes.

Separate fire division. A portion of a building cut off from all other portions of the building by fire walls, fire doors and other approved means adequate to prevent damage from any fire which may occur in the building from involving more than one such separate fire area. In nonfire-resistive buildings, this includes protection against building collapse as a result of a fire outside the separate fire area.

38 / THE "INS" AND "OUTS" OF DOOR CLOSERS

(Reprinted with permission from Norton Door Closer
Service Manual. Copyright 1968 by Norton Door Closer
Division, Eaton Yale & Towne, Inc.)

Proper door control is essential to the efficient operation of any building and to the comfort and safety of those who use it. More than any other building product, door controls may cause irritation and difficulty if they do not perform properly.

Door closers require minimal periodic maintenance to insure efficient operation and long life. An improperly adjusted or installed door closer shortens its useful life and places severe strain on the door, frame, and other door hardware. All too often, a door closer is called upon to perform a function for which it was not intended, or is installed in such a manner that prevents it from operating as it was designed.

It should be kept in mind that a door closer is not designed to stop or limit the travel of a door. This function should be provided for through the use of a unitized door control, a separate door holder, or a doorstop mounted on the floor or wall.

Most companies manufacturing door closers offer several models, each designed for a specific function and application. Your local dealer will be happy to assist you in selecting the correct model for your needs.

Each closer is packed with a detailed instruction sheet for installation and adjustment. In the event that no instructions are found, be sure to get them either from the dealer or direct from the manufacturer before you attempt installation.

All closers are stamped with the model number and the size on the shell. Be sure, in any correspondence or conversation with either the dealer or the manufacturer, to mention the specific model involved.

Major manufacturers offer maintenance and service manuals, available upon written request. It is to your advantage to maintain a file of instruction sheets covering the specific types of closers under your care.

Maintenance Inspection

Periodic maintenance inspections, every six months, will prolong the useful life of any closer. During a maintenance inspection, all screws, bolts, and nuts should be carefully examined and tightened. This should include all bracket screws, arm bracket screws, hold-open nuts and screws, and adjustable arm screws. If a closer is not operating efficiently, an examination of the door, frame, and other door hardware should be made. Many times a door may be jammed or a closer and/or arm may be improperly mounted.

Here are some of the most common causes of closer inefficiency, and their solutions:

PROBLEM: The door closes too fast or too slow.

SOLUTION: Adjust the speed valve. Use the regulating screw as shown in the instruction sheet for the particular closer. The first closing speed (or "sweep speed") controls the speed of the door to within 5 inches of a closed position (see Figure 38-1). On regular surface closers, the first closing speed is controlled by turning the adjusting screw IN or OUT by full 180° turns. The fastest closing speed is obtained by turning the screw out toward its extended position. Turning it inward produces a slower closing speed. On other model closers (see Figure 38-2), the regulating screw may only need to be turned 90°–180° for necessary adjustment. It may be necessary for the custodian to experiment. The second closing speed (or LATCH SPEED) controls approximately the last 5 inches of closing travel. The second closing speed is generally controlled by turning the adjusting valve screw a partial turn. This may vary from 0°–180.° Again, experimentation may be necessary. On closers with a back-check feature, where the closer controls the door opening approximately the last 20° of its opening, the regulating screw is adjusted in the same manner as the latch speed. The regulating screw is generally marked "BC."

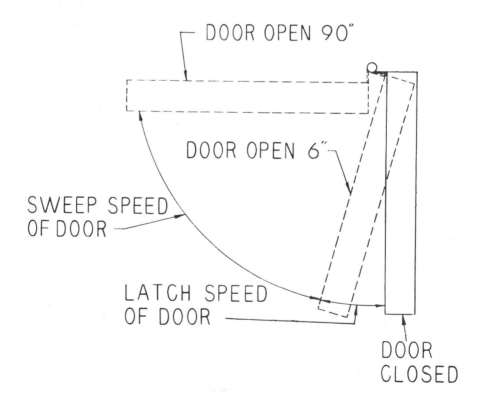

Figure 38-1

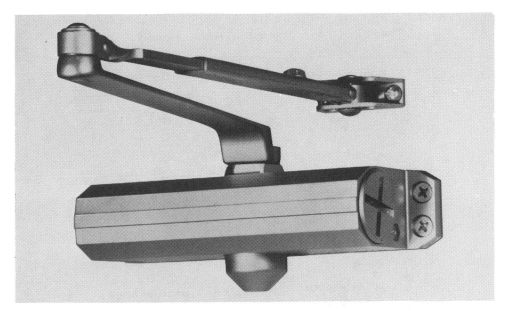

Figure 38-2

PROBLEM: Door doesn't close smoothly.

SOLUTION: Remove closer from door. The probable cause is a lack of fluid in the closer. A door closer dealer or repair station will check the sealing gasket and fill with the manufacturer's specified fluid.

PROBLEM: Door doesn't close at all.

SOLUTION: Most likely an internal part is broken. Remove closer from door and return to door closer dealer or repair service station for repair.

PROBLEM: Door bangs against doorjamb.

SOLUTION: The closer is out of adjustment. See adjustment instructions above. If adjustment fails to correct the checking action, the oil level is low. Remove closer and return to door closer dealer or service station for inspection and refilling.

PROBLEM: Reduced power. Door opens wider than necessary.

SOLUTION: Closer is applied too close to door hinge. The closer should be reinstalled per manufacturer's instruction sheet.

PROBLEM: Door with holder arm closer does not stay open or does not open wide enough.

SOLUTION: Adjust the door to less than desired hold-open position and tighten holder nut securely. If existing hold-open position is less than desired, loosen holder nut first, position door a little less than desired hold-open position, and tighten nut securely.

PROBLEM: Door slams on short opening or doesn't close properly.

SOLUTION: Closer arm is set at wrong angle. For regular arm closers, open door slightly and loosen set screw. Adjust main arm so that forearm is perpendicular to face of the door. (See Figure 38-3a and 38-3b.) Tighten set screw securely. For parallel arm closers, loosen set screw and adjust forearm until main arm is approximately 2° from a parallel plane of the door face when closed. (See Figures 38-4a and 38-4b.) Tighten set screw securely.

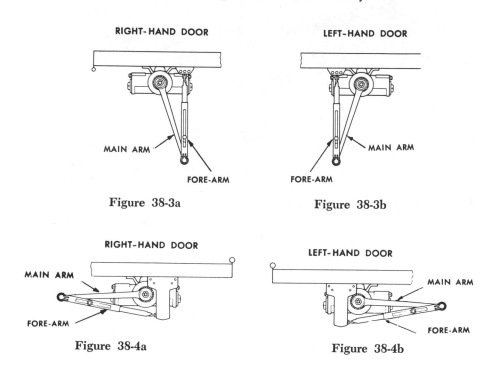

Figure 38-3a

Figure 38-3b

Figure 38-4a

Figure 38-4b

PROBLEM: Door does not open wide enough.

SOLUTION: Closer is applied too far from door hinge. Reinstall per manufacturer's instruction sheet.

PROBLEM: Door opens with resistance.

SOLUTION: Closer is oversized for door. Replace with smaller size.

PROBLEM: Sluggish closing.

SOLUTION: Closer is too small for door and conditions or set too close to hinge. Replace with larger-size closer, after trying alternate arm pivot pin setting to increase power (the latter is available on later closer models).

PROBLEM: Arm strikes top of door.

SOLUTION: Closer is mounted too low or the arm is loose on shaft. Tighten main arm, reset closer, or hold the main arm at the shaft and bend arms up slightly at the elbow.

As mentioned in the beginning of this article, closers that require removal can be sent to a qualified door closer repair service. If a well-equipped repair shop is available on the premises, a manufacturer's service manual should be used for detailed instructions in the servicing of these products.

Because closers are subjected to continuous and sometimes abusive use, servicing may be required after several years of operation. Major service can usually be avoided easily by making sure that the closers are properly installed and that they are not being subjected to unnecessary abuse.

Before disassembly is considered, a check for proper closer installation should be made. Commonplace causes of unsatisfactory operation are improper arm installation and improper adjustment.

If improper arm installation or adjustment is the cause of the closer malfunction, the manufacturer's maintenance manual should be used for detailed instructions in the servicing of these products.

If overhauling is indicated, the closer should be taken apart for thorough cleaning, broken or badly worn parts replaced and tightly reassembled per instructions.

NOTE: Servicing is not performed in the same way on all model closers. Proper closer identification is therefore important and can be made by consulting the illustrations in the manufacturer's manuals.

Norton, for example, recommends that only special Norton tools, designed for its own closers, be used. Otherwise, it is advisable for the user to simply remove the closer and send it to the nearest service station where special tools are available.

The disassembly and reassembly instructions which accompany this section, therefore, are intended as a guide to regular surface closers only.

Regular Surface Closers

DISASSEMBLY

1. Because closer seals are necessarily tight, a brief immersion of the closer in boiling water will facilitate removal of sealed parts. Carefully place closer in vise. Do not tighten more than necessary. Relieve spring tension. Loosen and remove top nut and arm from closer. (See Figure 38-5.)

2. Remove ratchet by prying with screwdriver. (See Figure 38-6.) Remove cover in same way.

3. Note whether the spring hooks are on the right or left side of the shell as indicated by the letters on the shell. (This determines the hand of the closer.) Remove spring with screwdriver. (See Figure 38-7.)

4. Reposition closer in vise and remove one end plug using special end plug wrench. (See Figure 38-8.) Remove regulating screw assembly. (See Figure 38-9.)

5. Remove closer from vise and drain oil. Carefully replace closer in vise and remove remaining end plug and regulating valve assembly.

6. Place socket wrench over packing nut. (See Figure 38-10.) Place master handle over socket wrench; loosen and remove packing nut and shaft assembly. (See Figures 38-11 and 38-12.) Note: A steady pull on the master wrench will loosen the packing nut more effectively than a sudden jerk.

7. Remove shaft from packing nut. (See Figure 38-13.) Carefully remove "O" ring from packing nut with a small steel pick. Avoid nicking or scratching "O" ring groove and packing nut bore. (See Figure 38-14.)

8. Remove the piston from either end of closer. (See Figure 38-15.)

9. Immerse parts in a noncaustic solvent solution; flush with clear running water to remove oil residue and foreign matter. Dry all parts with a lint-free cloth or clean air jet if available. *Caution—do not allow parts to remain in solvent solution.*

Figure 38-5

Figure 38-6

Figure 38-7

Figure 38-8

Figure 38-9

Figure 38-10

Figure 38-11

Figure 38-12

Figure 38-13

Figure 38-14

Figure 38-15

REASSEMBLY

1. Replace any worn or broken parts.

2. Place new "O" ring in packing nut. Use care to avoid damaging "O" ring, ring groove, and packing nut bore. (See Figure 38-16.)

3. Replace piston in closer. Replace pinion as per Figure 38-17.

4. Apply oilproof sealing compound to threads of packing nut assembly and slip over pinion shaft. Use care to avoid cutting "O" ring. Tighten packing nut securely with socket wrench and master handle. (See Figure 38-18.)

5. Apply oilproof sealing compound to threads of one end plug; insert and tighten securely with end plug wrench. Insert and tighten regulating screw assembly, with new packing.

6. Reposition closer in vise with open end of closer up. Lift ball valve in piston with wire. Fill closer with Nortol liquid and pump piston up and down with arm on shaft until liquid level is midway between the portholes on inside of shell. (See Figure 38-19.) Turn regulating valve to open position to allow oil to flow throughout the closer.

7. Apply oilproof sealing compound to threads of remaining end plug and tighten securely with end plug wrench. (See Figure 38-20.) Insert and tighten remaining regulating valve assembly with new packing, into closer body. (See Figure 38-21.)

8. Replace spring in original position. Replace cover; tap lightly with hammer to insure proper fit. (See Figure 38-22.) Replace ratchet, arm, and nut. Clean closer thoroughly and refinish with Norton Spray Pak quick drying lacquer.

9. If spring was assembled for left hand, open regulating valve on right hand of closer. The other regulating valve should be turned in until you can feel checking power when pulling on arm. For right hand spring assembly, reverse the procedure.

The closer is now ready for installation. Mount and adjust in accordance with installation instructions.

Figure 38-16

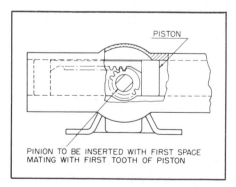

Figure 38-17

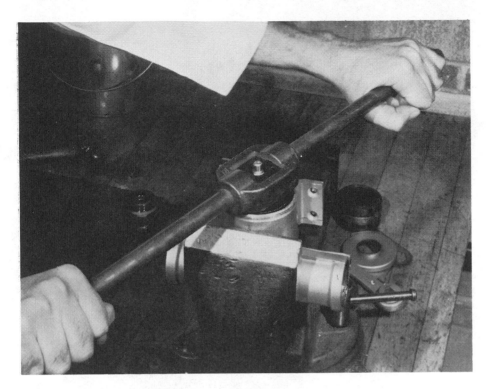

Figure 38-18

Figure 38-19

Figure 38-20

Figure 38-21

Figure 38-22

The battery compartment must be ventilated in a manner to keep out water, oil, dirt, and other foreign matter. It is essential that the battery be properly blocked in the compartment to prevent excessive movement or shifting.

Any surfaces which are to be bolted together should first be made clean and then coated with a film of "NO-OX-ID" grease or pure vaseline which also should be applied to the bolt studs before making the connection. A soft wire brush should be used for cleaning the lead plate connections. After all connections have been securely tightened, they should be gone over a second time. All connections should then be wiped off with a clean cloth to remove the surplus grease. Do not leave excess grease on any surface. Only a thin film is needed to accomplish the desired purpose.

Adding Water

Approved or distilled water should be added to all cells of the battery at sufficiently frequent intervals to keep the electrolyte level above the top of the splash cover. Do not fill above the level recommended by the manufacturer, which is generally indicated on the battery or in literature accompanying the battery when purchased. The battery manufacturer will generally test the local water for you, determining if it is usable in the battery.

During freezing weather, water should be added just before the battery is placed on charge, so that the water will be thoroughly mixed with the electrolyte during the charge and thus avoid any danger of the water freezing. All cells of the battery should require the same amount of water. If one cell takes more water than the others, examine it for leakage. A written record should be kept of the date water is added and the amount.

Cleanliness

Keep the battery, its connections and surrounding parts clean and dry, but do not remove the grease from the seal nuts. The vent plugs should be kept in place and tight, including when the battery is on charge.

If any electrolyte is accidentally spilled or if any parts are damp with acid, apply a solution of bicarbonate of soda (baking soda), using one pound to one gallon of water, then rinse with water, and dry. An old paint brush is effective for getting the soda solution under the connectors. Do not allow solution to get into the cells. If this treatment is given three times a year and the battery kept clean between times by regular washings with water or blowing off with an air jet, the life and service of the battery in general and the trays in particular will be increased considerably. Before washing a battery without removing it from the compartment, the equipment manufacturer should be consulted for advisability.

If the terminals or connections show any tendency to corrode, clean the corroded surface with a soft wire brush, wash it with a soda solution, rinse with water, wipe dry, and coat it thinly with Vaseline or "NO-OX-ID" grease. No corrosion will be experienced unless electrolyte is spilled and allowed to remain. Soda solution or ammonia applied promptly, will neutralize the effect of acid on clothing, cement, or other material.

Battery Storage

If the use of a battery is to be temporarily discontinued, give it an equalizing charge, and add water to the cells during this charge so that the gasing will insure thorough mixing and prevent its freezing in cold weather. Add enough water to meet the level specified by the manufacturer. After the charge is complete, remove all fuses to prevent the use of the battery during the idle period. *Make sure* all vent plugs are in place. The battery should be reconnected, water added and charged every two months in climates averaging 70° to 80°F., and every six months in climates averaging 40°F.

Placing Batteries in Service Again

Add water if needed, and give a charge until the gravity of the electrolyte has ceased rising over a period of three hours.

40 / ELECTRIC MOTORS

(*Sources: Plant* magazine; F. C. Osterland, Allis-Chalmers Mfg. Co.)

Each building and plant has its own special operating conditions due to geographic location and plant or building environmental conditions.

General Recommendations

General recommendations to aid in the understanding of motor care are as follows:

Dust. Dust is probably the most natural enemy of a motor. Dust is contantly settling on motors, housings, windings, slip rings, and commutators, trying to work its way into the bearings. This danger is naturally at a minimum when totally enclosed, fan cooled motors are used. On windings, dust acts as a layer of insulation, confining heat until it may reach dangerous temperatures. Dust also plugs ventilation spaces which further interferes with proper cooling. It will act as an abrasive, and insulator, on slip rings and commutators, multiplying the wear and blocking full passage of current. Once inside bearings, dust can be harmful as sandpaper to the highly polished surfaces. If it is allowed to fill the open spaces in a winding, it turns the entire wound section into a sponge for soaking up harmful oil, moisture, and acid fumes.

Methods of cleaning. The time to catch dust in motors is before it has a chance to unite with water or oil to form a gummy mass. This means wiping off the motors on regular inspections, their housings, slip rings, commutators, and occasionally blowing dust out of the wound section with fairly low pressure (not over 40 psi) compressed air. The compressed air must not contain grit, metal, or moisture. If there is any danger that blowing may shoot abrasive or conductive material into the air gap or windings, suction may be the preferable method of cleaning. Oil filter caps must be kept closed at all times, and dust seals and gaskets must be kept in good condition. Worn or rotted seals should be replaced, their replacement is much easier than a burned out bearing.

Oil. Inside bearings, oil can be the "lifeblood" of motors; but outside the bearings, oil is strictly poison to them. As mentioned, dust can soak up stray oil. This trouble-making teamwork operates the other way, too. Sticky oil catches dust. When both oil and dirt cover a commutator, good commutation is impossible. The faces of the brushes become glazed and packed with dirt. Harmful sparking is the result. Oil also harms commutators by deteriorating the mica insulating segments between bars. Oil is even more harmful to taped and varnished insulation on windings. Once a winding is thoroughly oil soaked, the motor is in immediate danger of a burn-out or breakdown. Oil is most dangerous to a motor when it has had a chance to unite with dust in the windings to produce a greasy gum. Then ventilation is smothered, windings are under continuous attack from oil, and metallic dust present is caught and held in this "gum," a constant threat of shorting or grounding.

Build-ups of oil and dust on varnished windings should be removed with the aid of a solvent, preferably a non-inflammable one. If brushes or scrapers are used, care must be taken not to scrape the insulation.

Solvents must be handled with caution. Insulation should not be soaked as the solvent is apt to have a softening effect. When all the oil and dirt have been removed, windings should be dried and insulating varnish applied.

If a motor is insulated with silicone, rubber, or epoxy resin, time can be saved in cleanups by washing off oily deposits with detergent and warm water. The dry out period can be eliminated since these materials are highly resistant to moisture. In addition such insulation need not be revarnished as a part of maintenance procedure.

Sometimes a certain amount of cutting oil and other oils reach a motor in the form of a mist. That which lands on the outside should be wiped off before it can travel inside. Much can be accomplished in fighting stray oil by keeping a motor's own lubricating oil where it belongs. For example, sleeve bearings equipped with oil rings should never be lubricated when the motor is running. The oil level then does not indicate full supply. A certain amount of oil is riding the ring above the level. Overfilling and escape of oil can result.

Moisture. Motors are available for high humidity applications with sealed insulations being employed to keep out moisture. Conventional insulations however, are actually hygroscopic and really absorb moisture. One general aim of effective maintenance is to keep conventionally insulated motors completely dry.

If your motor does acquire moisture, it should be disposed of as quickly as possible for two reasons. First, it takes time for moisture to soak and soften insulation. Often, it is possible to get it out before the damage is done. Also evaporated moisture is pure water when it first condenses in a motor. It is then at its least dangerous stage. But every extra hour water stays in a motor, it gives it just that much more time, to absorb harmful compounds. Water in liquid form should be prevented from dripping on, splashing or flooding a motor. When resistance has dropped to a dangerous point, the motor should be dried by one of the following methods:

1. Heating in an oven, as this is the best method of drying.
2. Circulating current through the windings to produce heat inside the insulation to drive moisture outward. The armatures should be locked to prevent rotation and a rheostat employed to apply low voltage current.
3. Forcing air through hot elements (resistors, steam pipes, and other elements) and into the windings with a fan.
4. Enclosing the motor with nonflammable materials and conveying heat to it through ducts.
5. Covering the motor with a tarpaulin, leaving openings at top and bottom for air circulation, and providing heat with one or more electric light bulbs.

Care should be taken that the drying is accomplished without damage to the insulation. It is better to interrupt the baking for a resistance test than to bake insulation to the point of brittleness or burning. (The baking temperature should not greatly exceed the boiling point of water.) When dry, the motor's resistance to moisture can be renewed by applying a good grade of insulating varnish.

Lubrication. No wear can take place within a bearing if the surfaces are smooth and properly lubricated. A film of oil keeps them apart and prevents metal to metal contact. But in actual practice, bearings often receive an insufficient supply of oil, or the wrong type of oil. Sometimes grit gets into the oil and causes scraping, or so much load is put on the bearing, that the film of oil breaks down. Manufacturer's lubrication instructions should be followed closely. Oil is the lowest item of operating costs, yet no one thing is more certain to cause trouble and expense than incorrect lubrication. Only best quality oil and correct grade suited to the requirements of each individual motor should be used. In case of doubt, the manufacturer will be pleased to send you proper lubrication instructions.

Sleeve bearings require the oil ring be free and turning with the shaft. The oil reservoir should be flushed out, refilled, and the dust seals checked at regular intervals.

With the antifriction type of bearings (ball or roller) it is well to keep in mind that the main purpose of the grease is to guard the steel rolling elements and races from corrosion, not friction. Too much grease promotes friction and heat. In any type of bearing, the effective film lubricant is often microscopically thin. Excess load can break it down and cause metal to metal scraping and bearing failure all in a matter of seconds.

Misalignment. Examples of motor damage commonly caused by misalignment are: sprung or broken shafts, burned out bearings, and overload failure. The damage can occur in drive or driven machinery too. When these elements are lined up incorrectly, bending, breaking, or excessive wear must result.

Misalignment is not always the result of improper installation, settlement of foundation, heavy floor loading, excessive bearing wear are other factors to be considered. Mounting motor and machine on a common, rigid base is not proof against misalignment. The base itself may have been mounted off level.

Often spur gears, when employed in the drive, tend to push apart and in time disturb alignment. Sometimes misalignment gives warning. Excessive temperature may warn that misalignment is causing bearing overload. Increased vibration may be the warning signal, or a rapid knocking, a sign that a shoulder of a shaft is being driven against the end of a bearing.

It may be possible, if the installation is not too heavy, to loosen the mounting bolts of the motor or machine and experiment with the variations in position while the set is in operation. Changes in degree of noise or vibration will guide workers making the adjustments until the set is in line and the symptom disappears. If a motor is found to be off level, shims can be placed under the motor legs until the units are in line.

Vibration. One important cause of motor vibration is misalignment, but other causes should be considered. Vibration in the driven machinery may be transmitted to the motor. Also careless servicing can put rotors out of balance, motor mounting bolts can work loose, bearings can wear to the point where shaft oscillation of considerable amplitude is permitted. Because the vibration resulting from such conditions can shake loose motor parts and electrical connections, crystallize metals, and multiply friction wear, excessive vibration must be located by a process of elimination; a suggested order of checking points is as follows:

1. Tighten all mounting bolts and check motor for loose parts.
2. For solidity, compare foundation with that of non-vibrating motors. Further improvement of conditions leading to vibration may be unnecessary.
3. Check bearings for looseness. Generally, allow two thousandths of an inch clearance plus one thousandth of an inch for each inch or fraction of an inch of journal diameter.
4. Disengage vibrating motor from driven machine. If motor operates far more smoothly when disconnected (proper alignment is assumed), then the machine should be examined for sources of vibration.
5. If vibration trouble seems to date back to its repair, check rotor for balance (dynamic balance), if possible, otherwise static. Cap screw the needed weight to the point of unbalance—hooking weight around part of motor to relieve pull on screw.

Uneven wear. Wear of slip rings and commutators of electric motors is accepted as inevitable, because none of these rotating parts exposed to constant friction can be lubricated with petroleum products. However, the wear can be minimized by preventing conditions leading to "grooving" (concentration of wear in narrow rings or ruts). Dust is an abrasive agent and should be wiped off the surfaces of slip rings and commutators at regular intervals. Be sure that a thin film of oil does not collect on their surfaces, as oil catches dust and also prevents proper contact with the brushes. When the surfaces of rings or com-mutators are rough or otherwise uneven, the first opportunity before grooving, pitting, sparking, and increasingly accelerated wear can occur. Running the rotor, or slip rings do not need to be removed when a commstone in a suitable holder is employed. After commutators are turned down, the mica separating bars should be relieved to a depth of about one thirty-second of an inch. Mica should always be undercut when high, or even level, to forestall excessive brush wear and sparking. Brushes themselves, can contribute to excessive wear. When worn unevenly, they should be true or replaced with a new set before they can damage slip rings or commutator. To assure maximum, smooth contact, new brushes should be "sanded in" to the proper arc. Should a motor with sleeve bearings and belt drive not oscillate (the shaft floats back and forth axially), a change in alignment may permit this movement. The advantage is that wear is spread over a wider area.

Overload. Overloads will cause motors to heat up and eventually break down or burn out. To prevent this, manufacturers commonly rate the capacity of their motors below their particular breaking point. The margin of safety thus provided should never be narrowed by deliberately making a motor carry too great a mechanical load. Overloading is not always intentional. It can be pro-duced by error in application, obstruction in the drive, or driven machine, ex-cessive friction within the motor itself, or by efforts to obtain greater output from the driven machine than the motor is capable of carrying. If any of these conditions cause the current in a motor to exceed its nameplate current rating, heating may increase as much as the square of the current increase. The danger is that insulation will be "fried", soldered connections melted, and bearings burned out.

High temperature insulation should not be depended upon as a guarantee of motor capacity. Nameplate ratings should never be exceeded, except in emergency, regardless of the "class" of insulation used on the motor windings. The insulation might be able to take temperatures that will cause trouble with mechanical fits or bearing life.

Motors are given various forms of "overload protection". In most cases, a thermal element is connected in the power circuit to the motor. Heat from the element operates an overload relay which opens the circuit to the motor. But whether the protective unit be thermal element or a fuse, it should be of the proper capacity as listed in the national code. A unit of more than this recom-mended value should never be employed. Continuous production should be pro-tected by guarding against the conditions that trip or blow the overload protec-tion devices. Fuses and heater coils are cheaper than motors.

Underload. Underload penalties are less severe, but underloading an induc-tion motor is just as clearly improper as overloading it. When induction motors are working only partly loaded, "power factor" (ratio of power to total power) is lowered. A low power factor, produced by induction motors working underloaded, results in the following disadvantages:

1. Higher line losses, higher voltage drop and motors lose torque.
2. Increased load on a utility line and equipment. In many cases this causes breakdowns and interrupted service which slow down production.

3. Frequently a higher energy cost to the user.

4. Motors are apt to overheat as a result of excessive currents set up by the high proportion of wattless current.

5. To correct a condition of underload causing low power factor, the rated capacity of each induction motor and its actual load should be determined and the data listed on a large sheet. Data may indicate that some motors, except the largest, can be advanced to heavier jobs. Small motors may be replaced with even smaller capacity. Such handling of equipment almost suggests itself, once the facts are gathered and organized.

Minimizing underload is the preventive way of improving power factor. Corrective measures include use of synchronous condensers, synchronous motors, and capacitors.

Checklist of Symptoms, Causes and Cures of Electric Motor Troubles

The trouble shooting chart is divided into symptoms that can be seen at a glance, those that can be heard in passing, those that can be felt, and some that are worth searching for. The type of machine covered by each notation is coded as follows:

Critical equipment. Critical equipment should be so identified on the card. It is generally considered vital if failure directly or indirectly interrupts or delays production. Preventive maintenance on this type of equipment is a must.

Planning. Downtime can be minimized in two ways:

1. Carry a stock of critical renewal parts.
2. Schedule preventive maintenance and overhaul well in advance.

Close coordination is required to insure that men and materials are available for such programs, to prevent delays, and overstaffing. Normal maintenance and overhauls should be scheduled during normal downtime, weekends, or during scheduled vacations, whenever possible.

Pillow tanks (electric). Many man-hours in the maintenance of oil bearing substation transformers and circuit breakers can be saved by using rubberized fabric pillow tanks for storing oil during periods of routine overhaul or repair. The pillow tank is placed near the transformer or circuit breaker with the oil being pumped into it, then returned after the repairs or maintenance have been performed. This eliminates the time-consuming job of setting up a system using 50-gallon drums. This type of equipment is especially valuable during times of emergency.

Trouble Shooting Chart

KEY FOR CODE

M	Motor
G	Generator
DC	Direct Current
AC	Alternating Current

KEY FOR CODE (*Continued*)

C   Cage (motor)
WR   Wound Rotor (motor)
S   Synchronous (motor
          or generator)

SYMPTOMS THAT CAN BE SEEN

Symptom	Cause	Code	Cure
1. Motor will not start (with or without humming sound).	No power. Faulty control.	M	Check power supply and connections to control and motor.
	Wrong rotation.	M	Replace fuses and check clip contacts, or allow overload trip to cool. Investigate cause.
	Low voltage.	M	Check control rating and connections.
		M	Reverse any two main leads.
		M	Check nameplate against power supply.
		M	Use adequate wire size. Check for excessive voltage drop during starting.
	Short or open-circuited or grounded winding.	M	Starting tap too low.
		M	Check stator and rotor windings (including cage or damper bars) and repair.
	Single-phasing.	AC M	Replace fuse. Check for external or internal open circuits.
	Field excited.	S M	Check operation of control.
	Wrong field connections.	DC M	Check for reversed field connections or opposing shunt and series field.
	Blocked or locked rotor.	M	Remove foreign matter from between stator and rotor. Make sure rotor is not touching stator.
		M	Free or replace frozen bearings. Check pressure lubrication system. Use special lubricant if conditions are unusual.
	Excessive load.	M	Check unloading device.
		M	Start pressure lubricating system.
		M	Reduce belt tension.
		M	Reduce load or change motor.

SYMPTOMS THAT CAN BE SEEN (*Continued*)

Symptom	Cause	Code	Cure
2. Motor starts but: Stops.	Power stoppage.	M	See Item 1, this section.
		DC M	Check rheostat and setting; starting with weak field may actuate overload relays.
Runs too slow.	Excessive load.	M	See Item 1, this section.
	Low voltage.	M	See Item 1, this section.
	Brushes ahead of neutral.	DC M	Correct brush setting.
	Short circuit in armature or commutator.	DC M	Locate and repair.
	Open circuit in rotor.	WR M	Locate and repair.
	Field excited too soon.	S M	Check control.
	Abnormal bearing friction.	M	Check bearings and lubrication.
Runs too fast.	Weak field.	DC M	Check for poor field connection or ground.
	High voltage.	DC M	Correct voltage.
	Brushes back of neutral.	DC M	Correct brush setting.
Fails to "pull in."	Excitation troubles.	S M	Make sure field-applying contactor is operating, but not too soon.
		S M	Make sure exciter and rheostat setting provide sufficient excitation.
		S M	If starting voltage drop is large, greater discharge resistor may be helpful. Consult motor manufacturer.
	Excessive load flywheel effect.	S M	Consult motor manufacturer about reducing discharge resistor.
Pulls out of step.	Excitation troubles.	S M	Check exciter output and circuits for reversed field coil.
	Low voltage.	S M	Correct, or raise excitation if safe.
	Excessive torque peaks.	S M	Check driven machine adjustment.
3. Excessive sparking, or flashing of brushes and rigging and/or blackened commutator.	Overload.	M G	Reduce load or change motor.

SYMPTOMS THAT CAN BE SEEN (*Continued*)

Symptom	*Cause*	*Code*	*Cure*
	Dirty, rough or eccentric commutator or rings.	M G	Sand, grind or turn commutator or rings.
		DC	Undercut mica.
	Brush troubles.	M G	Adjust springs for sufficient, equal brush pressure.
		M G	Clean, sand or replace sticking, incorrectly fitted, burned, wornout, or unsuitable brushes.
		M G	Check connections between brushes and holders.
		DC	Check brush position and holder and bracket alignment.
	Shorted or open armature winding.	DC	Check and repair.
4. Smoke, charred insulation. Solder "whiskers" on armature.	Overload.	M G	(See Item 2, "Symptoms That Can Be Felt.")
5. Commutator grooving.	Brushes not correctly staggered.	DC	Adjust staggering.
6. Fine dust under coupling with rubber buffers or pins.	Misalignment.	M G	Realign unit and repair coupling.
7. Excessive brush wear.	See Item 3, this section.	M G	See Item 3, this section.
8. Generator fails to develop normal voltage.	Short-circuited or partially reversed shut field.	DC G	Check and correct.
	Loss of residual magnetism (self-excited machine).	DC G	Separately excite fields for few minutes.
	Incorrect brush position.	DC G	Check and correct.
	Poor fit or riding of brushes or commutator.	DC G	Clean and sand brushes. Check commutator surface.
	Excessive resistance in field circuit.	G	Check and correct.

SYMPTOMS THAT CAN BE SEEN (*Continued*)

Symptom	Cause	Code	Cure
	Below normal speed.	G	Check prime mover.
	Faulty instruments.	G	Check and repair or replace.
9. Generator requires excessive overhauling.	Partially short-circuited field coils.	G	Check and correct.
10. Motor draws excessive current at normal speed.	Overload.	M	Reduce load. See Item 1, this section.
	Wrong brush setting.	DC M	Check and correct.
	Armature short circuit.	DC M	Check and repair.
	Abnormal bearing friction.	M	(See Item 3, "Symptoms That Can Be Felt.")

SYMPTOMS THAT CAN BE HEARD

Symptom	Cause	Code	Cure
1. Excessive hum.	Uneven air gap.	M G	Check air gap. Center rotor. Replace bearings if necessary.
	Unbalanced rotor.	M G	Rebalance, especially after repairs.
	Loose punchings.	M G	Tighten holding bolts.
	Single-phasing.	AC M	Stop motor and try to restart. If it will not start, check power supply, fuses, and for internal or external open circuits.
2. Regular clicking, scraping or grinding sound.	Foreign matter in air gap or between other rotating and stationary parts.	M G	Remove and check for damage.
	Rotor rubbing stator.	M G	Replace worn (sleeve) bearings, check air gap and alignment.
	Noisy ball bearings.	M G	Check lubrication, and replace bearings if necessary.
3. Rapid knocking.	Shaft shoulder hitting bearing.	M G	Realign unit.
4. Brush chatter or hissing.	Extreme vibration.	M G	(See Item 1, "Symptoms That Can Be Felt.")
	Brush troubles.	M G	Check brushes for correct fit in holders.

SYMPTOMS THAT CAN BE HEARD (*Continued*)

Symptom	Cause	Code	Cure
		M G	Check holders for excessive clearance or incorrect angle.
		M G	Use brushes suited to service.
		M G	Check brush spring pressure.
	High mica.	DC	Undercut.
5. Miscellaneous noises.	Loose coupling, foundation bolts, and other items.	M G	Tighten bolts, recheck alignment.

SYMPTOMS THAT CAN BE FELT

Symptom	Cause	Code	Cure
1. Vibration.	Misalignment.	M G	Realign unit.
	Vibration in drive or driven machine.	M G	Eliminate source if possible. Consider belt drive, and other parts.
	Unbalanced rotor.	M G	Rebalance.
	Loose bearing.	M G	Correct or replace.
	Loose parts or mounting bolts.	M G	Tighten and recheck alignment.
2. Overheating.	Overload.	M	Check for excessive friction in motor, drive, driven machine.
		M G	Reduce load or starting requirements, or replace unit.
	Dirt.	M G	Clean windings, vents, air filter, or wherever dirt may be seen. Remove external obstructions.
	Incorrect voltage and/or frequency.	M	Check and correct.
With vibration, noise and low starting torque.	Defective rotor bars or joints.	AC M	Install new bars or rods. Clean and rebraze or resolder joints.
Part of winding hot, and part cool. Field coils hot.	Uneven air gap, with or without rotor rubbing stator.	M G	Replace bearings, check shaft.
	Shorted or grounded winding. Single-phasing.	M G	Check fuses. Locate and repair breaks, loose connections or grounds.
	Excessive excitation.	M G	Reduce excitation. (Avoid operating dc motor below normal speed.)

SYMPTOMS THAT CAN BE FELT (*Continued*)

Symptom	Cause	Code	Cure
3. Bearing overheated. (*Warning:* Keep shaft turning until bearing cools—to prevent "freezing.")	Misalignment.	M G	Realign unit. Check anti-friction bearing fit, end-shield or pedestal alignment, and other parts.
	Faulty lubrication.	M G	Use correct amount of clean, high-grade, suitable oil or grease.
		M G	Check oil ring operation, oil flow, and other functions.
	Excessive chain or belt tension.	M G	Reduce tension.
	Excessive end thrust.	M G	Reduce thrust. See that unit is level, not tilted.
	Improper radial clearance.	M G	Check bearings and seals for excessive wear or misadjustment.
	Damaged bearing parts or shaft journal.	M G	Check and correct or replace.
4. Commutator or rings overheated.	Excessive load.	M G	Reduce load.
	Brush troubles.	M G	Check particularly for correct grade of brush and absence of excessive brush pressure. (See Item 3, "Symptoms That Can Be Seen.")
	Rough commutator or rings.	M G	Sand, grind or turn.

SYMPTOMS IN HIDING

Symptom	Cause	Code	Cure
1. Dropping insulation resistance.	Usually moisture and dirt, sometimes damage.	M G	Check and record at least once a year. Clean, dry and repair before trouble develops.
2. Faltering air gap.	Sleeve bearing wear.	M G	Periodic checking with feeler gages, at least once a year, can prevent work stoppages.
3. Production process change.	Increased load peaks, higher ambient temperature, and other causes.	M G	Check motor frequently to make sure it is suitable for revised application.

Symptom	Cause	Code	Cure
4. Aging or vanishing lubricant.	Mostly just the nature of things.	M G	Periodic inspection and appropriate action.

Electric Shock

The amount of current does not always determine the severity of the injury, which may be a shock, a burn, or death. However the higher the voltage, the greater the danger. Lower voltages are also known to be dangerous. The path which the current takes in passing through the body generally determines the extent of damage. The heart is the most vulnerable organ, however in the average heart beat, there is a time when it is less susceptible to shock. For this reason the instant of the shock can determine the amount of injury.

The majority of electrical shocks in industry are in the low voltage range. Low voltage current is extremely hazardous when the body is wet, which can be caused from perspiration, rain, damp floors, and other moisture. The body would have to be in contact with the ground and the live conductor.

High voltages across the body will cause a non-breathing shock condition. Generally over half of the cases of this type of accident can be revived if artificial respiration is promptly applied. Artificial respiration should be started immediately, in some cases, you may not be able to hear the heart beat, however it should be continued for several hours (at least two) or until a physician arrives.

Electric shock does not always show immediately. The victim's temperature will generally rise fast, then he may become nauseous, then unconscious. Persons that have received severe shock should be kept under observation for several hours, as the shock sometimes causes damage to organs, tissues, or muscular system which becomes uncontrollable, causing a fall. Persons that survive electric shock completely recover in a short time without any ill effects or permanent disability.

41 / CURBING TRANSFORMER FAILURES

(Reprinted with permission from General Electric. Copyright 1962 by General Electric Company.)

The more we see of transformer failures and study their causes, the more convinced we become that many power interruptions can be avoided if adequate transformer maintenance is provided.

Of all the electrical equipment a plant engineer has under his care, transformers are perhaps the easiest to neglect. This section describes what can be done to avoid transformer failures through inspection, testing, and evaluation.

In many plants, disconnecting the transformer can be tolerated only during plant shutdowns for important holidays or vacations. However, important maintenance steps can be taken by trained personnel while voltage is on the transformer. These include visual inspections and transformer oil sampling and testing.

Visual Inspection

At least once a month have someone on the maintenance crew visually inspect the transformer and its accessories. This includes reading the liquid-level gage, checking ambient temperature, and reading liquid and winding temperature indicators. See that these items are within safe limits and record them. Accurate records give the best indication of operating performance.

Keep informed about the load on the transformer. Check the switchboard instruments for load current in amperes on the bank. Kilowatt records are not entirely satisfactory, inasmuch as heating is determined by the load current and voltage and not by the kilowatt output, unless the load is at the unity power factor.

Check visually for any indications of leaking oil, bad gaskets, or restricted breathers. Observe the bushings and note if there are any deposits of dirt, which, under moist conditions, may cause flashovers. Check for any excessive peeling of paint on the tank, or rusting of exposed metal parts.

Look for refuse on the lid of the transformer. A wire mill, producing copper magnet wire, suffered the loss of one 2,000-kva transformer, which cascaded into a shutdown of their entire electrical system. When the smoke had blown away, the failed transformer was inspected and a bird's nest was found on the lid. In it were discovered several pieces of copper wire. It was concluded that the transformer flashover had occurred when the bird, flying near the nest, dropped a piece of wire across two bushing studs.

Transformer Oils

First, a word about the characteristics and functions of transformer oils. There are two basic types: one is a specially refined mineral oil, a flammable material; the other, an insulating liquid, is a nonflammable chemical, askarel.

Transformer oil serves as an insulating and cooling medium. The insulation and coil spacings of the windings are based on use of an oil having an average dielectric (by ASTM test) of not less than 22,000 volts. New oil has a dielectric of 30,000 volts or better. When it falls below 22,000 volts it should be filtered or replaced. It is frequently not recognized that this dielectric strength of oil must be the same for all transformers, whether rated 220 volts or 220,000 volts.

The cooling qualities of the oil are no less important than its insulating properties. Circulating between the coil spacings and through the ventilating ducts, the oil carries heat from the core and coils to the tank walls, where the temperature is lowered by the cooling effect of outside air. Transformer capacities are often increased by supplementing this natural cooling process with fans installed on the outside of the tank.

Effects of Oil Sludging

Oil exposed to air oxidizes and forms sludge. The amount of oxidation increases with the temperature and surface area of the oil that is in contact with the air. As sludge is formed, the insoluble portion settles on the core and coils and sometimes builds up on the tank's walls. This blanket of sludge is a barrier to the cooling function of the oil, and therefore causes overheating of the windings. Failure can result.

Acidity Tests of Oil

The products of sludge are acidic in nature, although the organic acids may be present without any visible indication of sludge. Acidity can readily be determined by test.

The degree of acidity is expressed by a "neutralization number" which represents the weight in milligrams of potassium hydroxide required to neutralize the acid in 1 gram of oil. New oil has a "neut" number of less than 0.08 mg. When the neutralization number rises significantly beyond 0.1, filtering may be desirable.

Sludge has been observed to accumulate rapidly as the neutralization number approaches 0.5. At 0.6 and higher it may be necessary to replace rather than recondition transformer oil.

The use of a filter press can remove moisture, insoluble sludge, and other foreign particles, but it will not remove the soluble acids. These can be removed only through a purification process involving the use of fuller's earth or activated alumina. Such purification is generally uneconomical unless the oil quantities involved are substantial.

The nonflammable insulating liquid, known as askarel, has characteristics entirely different from transformer oil. Askarel is more stable and is not subject to oxidation, hence does not form sludge. Since it is heavier than water, any moisture which forms in the transformer will settle on top of askarel. Therefore, samples of it drawn for testing must be taken from a sampling device at the top of the tank. Its dielectric is higher than oil and voltage breakdown tests often run to 35 kv or higher. The minimum acceptable dielectric is 25,000 volts.

Taking an Oil Sample

When you prepare to take oil samples, supply yourself with a clean, dry, quart mason jar for each transformer. A cork, rather than a rubber ring, should be used under the cap. Draw off 1 quart of oil from the sampling valve at the bottom of the tank and discard this, since it may not represent a true sample. Draw a second quart and cap the jar immediately, labeling it with the kva and serial number of the transformer. Have it delivered for testing without delay. There are certain advantages to having the sample drawn and tested right on the spot. A number of service shops offer this type of service.

Testing Oil Samples

The present ASTM test for measuring dielectric strength of transformer oil has limitations: It is not truly representative of the dielectric stresses occurring in a transformer. It is not sensitive enough to detect minor contamination deterioration which may be significant. Solid contaminants in test samples tend to settle, falsely improving test readings.

A test which gives a truer reading of transformer oil condition is presently being used by General Electric Service Shops. Developed by the G-E Power Transformer Department, it replaces the old 1-inch vertical flat electrodes (0.1 inch apart) with spherically capped electrodes 0.081 inch apart. An electrically driven impeller stirs the oil, simulating movements caused by convection or forced flow in the transformer.

The stress distribution on the sphere shape is similar to that occurring in an operating transformer. Close gap improves sensitivity. Gentle stirring assures that the entire sample will be tested. Dielectric strength criteria with the new gap differ somewhat from those appropriately used with the conventional gap. This improved test gap is now accepted as an alternate ASTM standard.

Reconditioning Procedure

Where acidity and sludge are encountered, the usual procedure for reconditioning is as follows: The transformer is de-energized and the lid or hand-hole cover is removed. The old oil is pumped out of the tank into empty drums and all accessible sludge is removed by brushes or wedges. Fresh, clean oil, under pressure supplied by a filter press, is then used to flush down the core and coils and the tank interior. This cleaning oil is in turn pumped out of the tank. New oil, as recommended by the transformer manufacturer, is allowed to fill the tank to the proper level as indicated either by markings inside the tank or in the transformer instruction book.

Occasionally, the sludge will be found to be heavy and tenacious, or inaccessible for hand cleaning. Then it will be necessary to remove the transformer to a service shop where it can be untanked and the exposed windings and core given a thorough cleaning.

In one Pennsylvania plant, a bank of three single-phase transformers developed so much sludge that it was necessary to send them to a service shop for cleaning and for new oil. However, the three transformers could not be removed at the same time, since some amount of light and power had to be maintained in the plant. This was solved by removing one transformer each weekend, until all three were reconditioned. While one transformer was away from the plant, the other two were connected open-delta, to keep the plant in operation at a somewhat reduced rate of production.

How Often Should You Take Samples?

Most transformer manufacturers will recommend that a sample be drawn once a year for dielectric and acidity tests. The extent of maintenance on any transformer will be governed by its size, importance of service continuity, its location on the system, and operating conditions, such as ambient temperatures, unusually dirty atmospheres, heavy fogs, etc. In these cases, a greater degree of attention is justified. A regular sampling program can help to discover cracked bushings, leaky gaskets, faulty breathers, and other defects which could cause serious failures.

A New Jersey plant, operating a 2,500 kva, three-phase transformer as its principal source of power, turned down a suggestion that an oil sample be taken once a year. "Our headquarters engineering department has instructed that samples will be tested every two years," said the Maintenance Superintendent. Six months later, when the trouble developed in the transformer, he put in a frantic call for assistance from the manufacturer. After it was untanked at a service shop, maintenance engineers discovered that the failure had been caused by a considerable amount of water in the transformer. This had entered through a cracked pressure-relief diaphragm. It is highly probable that an oil test, if it had been made at the recommended time, would have disclosed the presence of excessive moisture in the oil in time to filter the oil and replace the defective relief diaphragm before a failure occurred. It took a drastic curtailment of production and a $3,500 repair bill to convince plant management that oil should be tested at least once a year.

Under-the-Cover Inspections

As a general rule, any transformer that has seen ten years or more of service should have an under-the-cover inspection. Such visual inspections with oil removed to expose the top of core and coils may indicate the presence of water not disclosed by dielectric tests.

Recently, an inactive government plant, having 16 2,000-kva transformers, suffered the failure of two on which there was only a small lighting load. Previous dielectric tests of the oil showed acceptable values. After an under-cover inspection was suggested and authorized, the oil was dropped below the core level. In one of the operational transformers, the inspectors found, sitting on a terminal board, a pool of water 3 inches across. This had leaked down through a defective bushing gasket and was just sitting there waiting for enough vibration to start it rolling off into the winding and cause a failure.

It is sometimes difficult to schedule an outage to make these under-cover inspections. Often it will be necessary to schedule them for a Sunday, holiday, or vacation-shutdown period. On the other hand, just imagine how inconvenient it would be if a transformer failure should cut off your entire plant's electric power.

42 / STATIC ELECTRICITY

(U. S. Department of Commerce)
(Bureau of Mines—Bulletin 520)

To assist the general reader in interpreting the data presented, a few simple facts concerning static electricity and its nomenclature will be given. The material universe, as nearly everyone knows today, is composed of atoms, and the outermost parts of atoms are composed of electrons. There are many ways in which an outer electron may be detached from an atom. Detached electrons often manifest themselves in electrical phenomena, of which perhaps the simplest are those of static electricity. Static electricity is extremely commonplace, as electrons can be detached and transferred from surface atoms through mere contact and separation of material bodies. Fortunately, however, in the majority of cases, contact is made only at a few points, and as there is often little resistance to the return of electrons at the moment the contacting bodies are separated, the external effects are feeble and transitory, and pass unnoticed.

In certain materials, notably metals, electrons can move from atom to atom with considerable natural freedom. This makes these materials good conductors of electricity. Other materials, such as glass, stone, rubber, plastics, and textiles, have molecular structures that offer great resistance to the flow of electricity, and consequently, these are called nonconductors or insulators. Materials that are neither good conductors nor good insulators are sometimes called semiconductors. Moderate opposition to the flow of electricity is expressed in units called ohms. A larger unit, the megohm (1,000,000 ohms), is generally employed in high-resistance measurements.

When dissimilar materials are pressed together, free electrons from the surface structure of one material may shift across the contact or interface to the other. If the materials are separated, the new distribution of electrical entities probably will persist if one or both of the materials are nonconductors. The extent and direction of the electrical shift between contacting surfaces depends largely upon the nature of the materials and is usually in accord with their position in tabulations known as triboelectric series. The object that acquires extra electrons is said to have a negative (−) charge and the one that loses electrons, a positive (+) charge. Considered minutely, a normal atom or molecule that gains an electron is a negative ion, and one that loses an electron is a positive ion. The transfer of electrons or ions by contact and separation is often facilitated by rubbing and friction, whence the name "frictional electricity." Positive and negative charges (quantities) of electricity produced and kept apart by nonconductors are virtually at rest; that is, they are *static*. However, owing to their difference of potential, they are acted upon by a force that tends to unite them. This reunion will take place instantly if a low-resistance circuit is provided. It will occur in any event because there are no perfect insulators. Charges acquire voltage or potential in proportion to the amount of work or energy required to separate them. Stress in the electric field or region around charged bodies is manifested not only in the attraction of opposite charges but also in the repulsion of charges of like kind. As this effect is elemental, static charges in the aggregate are self-repellent and, thus, normally reside only on the external surfaces of electrified objects. Although static charges produced in operating rooms often have high potential, they are always extremely small in quantity compared with the most meager demands of power electricity. This smallness of quantity may be appreciated by considering the following facts. A current of ½ ampere or a quantity of ½ coulomb of electricity per second is needed to light an ordinary

4.5-volt flash-lamp bulb to full incandescence. (1 coulomb of electricity is the aggregate charge of 6.289 x 10^{18} electrons.) Certainly the amount of electricity involved in such illumination for an interval of 1 second might be considered insignificant; however, it would require 500,000 persons, or perhaps five to ten times as many operating tables as there are in the United States, charged to 5,000 volts each to hold a total quantity of ½ coulomb of electricity. The charge carried on each person or table at this voltage would, nevertheless, be that of more than 6,000 billion electrons. The energy comparison is quite another matter, as may be seen in the following formulas:

Charge quantity in coulombs, $Q = CV$.

Charge energy in joules or watt-seconds, $W = \frac{1}{2} CV^2$ or $\frac{1}{2} QV$, where C is the electrical size or capacitance of the charged body in farads and V is the charge potential in volts.

The quantities of electricity dealt with in electrostatics are generally so small that it is convenient to employ the terms micro-microfarad (10^{-12} farad), micro-microcoulomb (10^{-12} coulomb), and millijoule (10^{-3} joule). Static-electric potentials are conveniently expressed either in volts or kilovolts (10^3 volts). In the above illustration, the charged persons or tables were assumed to have a capacitance of 200 micro-microfarads each.

Although static electricity usually develops on nonconductors, a charge can be readily transferred to and stored on insulated conductors. Sometimes this is accomplished by a subsequent intimate contact of large surface areas or by slow leakage from one body to the other. More often, however, the transfer of electrons is apparent, and the charge exhibited by the conductor is the result of an action known as induction or influence. In considering separation and movement of charges and the effects of induction, it is important to know that charges are not produced or lost singly. A positive charge implies an equal negative charge. If the complementary charge seems to be lost or unnecessary, the circuit is simply longer or the phenomenon more involved than suspected. Positive elemental charges are virtually immobile, except in gases and liquids; and if a positive charge on a body is to be moved, the body upon which it resides will have to be moved with it. In general, therefore, it is well to think of the transfer of electrons or negative charges rather than of positive charges, to regard the appearance of a positive charge as evidence that a negative charge has moved away, and to understand that the removal of positive charge is due to the invasion of an equal negative charge. Thus, when charges are neutralized or grounded, they are not annihilated; they are permitted to assume merely a close atomic association, such as they had before initial separation. However, in this process charges lose energy, and their potentials are reduced virtually to zero. Static electricity of high tension on an insulated conducting body, such as a metal stand with ordinary rubber casters or a person wearing nonconductive rubber shoes, can discharge through the air if another conductor of considerably different potential is brought near. Such a discharge or spark, even though sometimes too small to be seen or felt, may contain more than enough thermal energy to ignite sensitive explosive mixtures of anesthetic gas and oxygen.

One obvious way of preventing the development of static electricity would be to eliminate nonconductors. This method should certainly be employed wherever possible by finding substitutes for easily electrified materials. Cotton sheets, cotton clothing, and other definitely essential nonconductors can be regarded as safe if all conducting bodies they may touch or be near have suitable means for the rapid escape of acquired charge. Sparks will not occur if there are better means for neutralization of charges. Casters, wheel tires, stool-leg tips, anesthesia-machine bags, breathing tubes, face masks, operating-table mattresses, and the shoes of personnel, therefore, should be of conductive rubber or other conductive material or be bridged by suitable conductors. To insure the neutralization of all charges, conductive paths should be continuous or merge with others. Inasmuch as operating-room personnel and almost all furniture make some contact with the floor, it is a matter of convenience and importance that the latter should be conductive. A suitable conductive floor is one that always has a resistance low enough to remove or bring about the safe neutralization of charges at a rate considerably greater than the rate at which they can develop on operating-room furnishings or personnel, yet high enough to prevent sparks or shocks from electric-power lines or equipment that might become defective. In other words, the floor resistance should lie somewhere between an upper and a lower limit of safety. An ideal value long recommended by the Bureau of Mines is 1 megohm to ground. However, tests have shown that, for practical considerations, the upper limit of floor resistance may be placed as high as 2 megohms, and the lower limit at about 0.2 megohm to ground. A floor maintained within these limits may be regarded as satisfactory for spark prevention and for protection against electric shock, provided that shoes, casters, stool-leg tips, etc. introduce no more than a similar amount of resistance. The overall effective resistance of the charge-neutralizing circuit should not be allowed to exceed 5 megohms.

43 / **LUBRICATION**

(*Source:* "How to Select Grease for Bearings," *Power Transmission Design*)

Lubrication greases are solid to semifluid products composed of oil and thickening agents. The amount of thickening agent determines, to a great extent, consistency of the mixture. The type of agent also affects consistency, but gives other qualities too. Consistency is the most important difference between grease and oil. Grease is semifluid or plastic, it does not run like oil and, therefore, tends to remain where it is put. It does, however, flow in the direction of motion in a bearing and lubricates in the same way oil does. Grease has much higher adhesive properties than oil. This means that more pressure is needed to squeeze a grease film into and out of two bearing surfaces.

Grease is good on applications not readily accessible to lubricant replenishment. The sealed for life bearing is probably the best example of this. An automobile wheel bearing is another example where grease needs replenishing only at long intervals, of many thousand miles. Grease works well in oscillating bearings where there is no continuous rotary motion of the journal in the bearing. Under these conditions, hydrodynamic lubrication cannot exist and grease will do a better job than oil because of its thicker film and higher adhesion.

Grease forms an effective seal that prevents entrance of dirt into a bearing. It does not necessarily have to be sealed in the bearing to stay in. Under normal operation, it stays put and only the simplest seals are needed.

How Grease Lubricates

Grease becomes fluid when sheared, or worked in a bearing. Therefore, the bearing is actually lubricated by an oily film at points of contact. But, where the grease is not being sheared, as at the edges of the bearing, it remains a semisolid, and therefore, continues to do its sealing job. In some bearings it is believed that lubrication is performed by oil which bleeds from the main body of the grease to the moving surfaces. There are also indications that the grease itself may flow to the bearing due to heating, vibration and temperature.

Dropping point. Temperature at which grease melts or changes to liquid from semisolid state is the dropping point. This point is important because it is a measure of the highest temperature at which the grease is expected to perform satisfactorily. Some oil will separate or bleed from some greases before the whole mass melts.

Penetration. Consistency of grease is important because it is a measure of how easily grease may be squeezed out from between the parts being lubricated. Also, it is a measure of the resistance to motion the grease offers in an antifriction bearing. Grease with too high consistency in an antifriction bearing may offer more resistance to the motion of the bearing than the friction it is supposed to reduce.

The National Lubricating Grease Institute (N.L.G.I.) has set up a system of numbers for expressing grease consistencies. Each number corresponds to a particular range of penetration numbers.

N.L.G.I. Number	Worked Penetration Number
0	385-355
1	340-310
2	295-265
3	250-220
4	205-175
5	160-130
6	115-85

Thickening agents. Thickening agents used to make grease from oils largely determine the characteristics and applications. Most thickening agents are metallic soaps, such as: calcium, sodium, aluminum, barium, and lithium soaps. Some

Figure 43-1

TABLE OF THICKENING AGENTS

Property	*Thickener*							
	Calcium Soap	*Calcium-Resinate Soap*	*High-Melting Point Calcium Soap*	*Sodium Soap*	*Sodium-Calcium Soap*	*Aluminum Soap*	*Lithium Soap*	*Solid Thickener*
Texture	Buttery	Buttery	Buttery	Fibrous or smooth	Buttery to fibrous	Buttery to stringy	Buttery to stringy	Buttery to stringy
Highest continuous usable temperature	175°F.	200°F.	225°F.	300° to 400°F.	250°F.	150°F.	300°F.	Depends on oil
Dropping point	200° to 225°F.	275°F.	280°F.	300° to 450°F.	315°F.	200°F.	360°F.	500°F.

Reaction when heated above melting point and then cooled	Separates unless chemically stabilized	Separates	Separates	None if worked	None if worked	Texture changes, does not separate	None if worked	Retains texture, but may harden
Water resistance	Good	Good	Good	Poor to fair	Poor to fair	Good	Good	Good
Resistance to work softening	Poor to fair	Poor to fair	Excellent	Poor to excellent	Poor to excellent	Poor to fair	Poor to excellent	Poor to excellent
Application	Plain bearings	Heavy bearings at low speeds	Ball, roller and plain bearings automotive and industrial	Ball and roller bearings at low to medium speeds, vehicle bearings and chassis	All types of ball and roller bearings at high temperatures, depends on Ca-Na percentages	Special applications where high adhesion, resistance to centrifugal throw-out are needed	Aircraft, 100° to 300°F, also automotive and industrial	For high temperatures

503

greases use solids such as silica, carbon black, finely divided clay and clay derivatives. Figure 43-1 shows how the characteristics of greases vary with thickener, and the applications or uses for the different types.

Maintenance

HOT BEARINGS

Causes	Remedies
No oil.	Check lubricators and bearings on scheduled periods to insure safe operation.
Dirty oil (causes undue friction).	Keep oil supply clean and stored properly.
Out of line (load on small area).	Bearings should be lined up properly.
Wrong grade oil (will not support shaft).	Use correct grade of oil, as specified by the manufacturer.
Uneven bearing.	Reduced bearing area cause heating, replace the bearing.
Tightly keyed (no space for oil between bearing and shaft).	Loosen bearing bolts until shutdown.

STEPS TO COOLING

1. *Flood with oil*, remove cup and pour oil into the bearing.
2. *Remove wicks*, pour oil into box to flood bearing.
3. *Feel shaft near bearing*, if temperature goes up after flooding, investigate further.
4. *Examine oil*, if gritty and dirty keep flooding bearing to wash out grit.
5. *Increase water flow*, if water cooled bearing, check feed and discharge line to insure water is flowing.
6. *Pour oil on shaft*, if temperature continues to increase.

EMERGENCY MEASURES

1. Remove the load. (Do not stop engine.)
2. Throttle engine to slow speed, but do not stop engine, as the bearing may seize to the shaft or pin.
3. Remove oils and flood with oil while engine is turning slow.
4. Slacken setscrews on bearing cap nuts and flood bearing with heavy oil, as it stands the heat much better.
5. When bearing cools, increase load or speed slowly keeping constant check on the temperature.

44 / LUBRICATION GUIDE FOR PACKINGS AND SEALS

(*Source:* "Lubrication Guide for Packings and Seals," *Plant Engineering*)

The prevention of fluid leakage between relatively moving surfaces calls for a close conforming device described as a packing or seal which may be defined as follows:

1. A *dynamic packing* is a nonautomatic or adjustable item installed in a stuffing box and adjusted by a gland to the desired shaft fit. Such items are often called jamb packings.

2. A *dynamic seal* is an automatic or nonadjustable item installed in a predetermined space.

Packings, whether adjustable or nonadjustable, usually function under fluid pressure. This fact demands close conformity with the moving surfaces and the ability to carry this pressure which is the equivalent of a bearing load.

Therefore, packings are in a sense, bearings—because every effort is made to eliminate an oil wedge between the surfaces, and to maintain so-called "contact" with the shaft around its entire periphery.

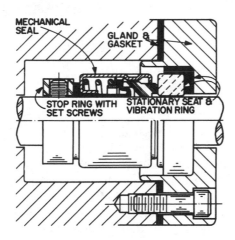

Figure 44-1 / **An efficient packing seal.**

No true contact is established, of course, since a lubricating film is necessary for the operation of any dynamic packing.

Without the film, excessive heat would be produced, and the result would be wear, scoring, and destruction of both the packing and its journal, the shaft. On the other hand, the film must be exceedingly thin to prevent it from flowing through the clearance to produce excessive leakage.

The first consideration of packings discussed is limited to dynamic applica-

Packings and Seals / 505

tions of the broad range of nonautomatic devices—that is, to jamb packings as they are employed between two relatively moving surfaces. Included among these packings are braids, duck and rubber coils and rings, jacketed packings, and certain of the plastics—bulk mixtures of various fibers; solid lubricants such as graphite, mica, or molysulfide; and bits of soft metals commonly bonded together with grease or an elastomeric binder.

In *braided packings,* the type of lubricant employed varies with the specific function of the braid. For cold water service, a waterpoof grease must be supplied; for work against solvents, castor oil, soap or hydrogenated fish oil may be used.

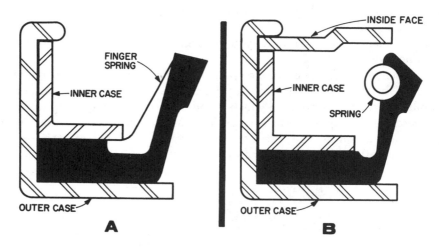

Figure 44-2 / **Oil seal sections show how springs maintain lip contact with shaft.**

Duck and rubber rings are normally impregnated with a mineral oil, but occasionally must depend upon water which they attract in service. For plastic packings the oil or grease in the binder usually provides the initial lubrication, and there is also a high percentage of graphite or similar material which functions as a dry lubricant. Some of the plastic packings contain bits of soft metal, such as babbitt, resiliently supported in the mix. The metals are in themselves excellent bearings; combined with the plastic, they provide superior service.

In *automatic packings,* such as molded V-packings, U and O-rings, flanges, and cups, the fluids being handled act as lubricants. Since these packings are automatic, they squeeze the shaft more tightly as the pressure increases. The film is thus thinned and the probability of leakage is decreased.

The lubricant is drawn by the shaft under the packing by the reciprocating action of the rod or piston. (While automatic packings are not generally used with rotating shafts, partly because there is no feasible method of drawing the fluid under, they are sometimes employed on slow speed, low pressure jobs.)

Because of this condition, lubrication engineers have attempted to determine the ideal shaft surface. Under normal conditions, a surface finish of between

16 rms (root-mean-square) and 32 rms has been found most practical. This degree of roughness produces excellent pockets to carry the lubricating liquid under the packing with a minimum cutting or wear.

With high pressures, a finer finish (approximately nine rms) is considered necessary, and in extreme conditions a finish of four rms is used. These recommended surface finishes may vary slightly with the lubricity of the liquid, its ability to wet the shaft and the type of packing.

Metal packings, in the form of floating segmental rings held in initial sealing relationship with the rod or cylinder by a spring, are also in this category. They require that the lubricant be drawn under the rings by the reciprocating rod. Where such lubricant is not present in the gas being handled (this packing is normally for high pressure gases or steam) lubrication must be provided from external sources.

"Dry" Bearing Materials

Some good results have been achieved on dry gases, where the temperatures are low without external lubrication, by using "dry" bearing materials such as special mixtures of "Teflon". Such mixtures are usually referred to as "filled" and consist of minute glass fibers or carbon graphite filling in a matrix of "Teflon".

The oil seal, a lip type packing, is designed for nonpressure service; however it does respond to act automatically under pressure. Since it operates on a very narrow wear band, any fluid loading must be kept very low to avoid excessive unit loading on the wear band which would burn up the seal. (See Figure 44-2.)

Under seal design conditions of no pressure, it cannot be automatic. However, such action is achieved by building in a spring loading device. A finger or coil spring is provided around the lip thus forcing it inwardly against the shaft. The loads used are low, normally ranging from one-quarter pound to three-quarter pound per linear inch of circumference. High pressure velocity ratios are not satisfactory in oil seals.

Another method of introducing the lubricant employs the oil cup, and depends upon the lantern, an annular ring having a spool or H-shaped cross section. The lantern ring which is associated with a lubrication connection through the stuffing box wall, may be machined from a short metal cylinder with clearances adequate to fit within the stuffing box and around the shaft.

The mid-section of the cylinder is provided with annular grooves—one each on the i.d. (inside diameter) and o.d. (outside diameter)—which are interconnected by a series of holes through the remaining web. A lubricant, introduced through the side wall of the stuffing box, fills the voids in the lantern, thus reaching the shaft and packing.

The stuffing box lantern ring serves not only as a means of introducing the lubricant, but also as a reservoir for it, and may be fed by a grease cup, by a drip oiler, or by oil under pressure. A positive means, such as the latter, is much preferred to the grease cup or grease gun fitting, which supplies only intermittent servicing.

An oil bottle may also be used to good effect in keeping the oil under pressure. The bottle is energized by pressure from the discharge side of the pump; and since the shaft is on the inlet side, the discharge pressure creates an adequate differential.

Another applicable device is the differential cylinder, used chiefly when it is desirable to maintain the pressure in the lantern at a slightly higher level than in the medium being packed. The differential pressures involved are not great, generally ranging from five to ten pounds.

Finally, it is possible to employ a completely independent lubricating system, which may operate either continuously or intermittently.

Lantern Rings

Besides serving as a means of introducing the lubricant and as a reservoir for it, the lantern ring also performs another function—cooling. By locating an outlet equipped with a bypass valve on the opposite side of the stuffing box, it is possible to adjust a constant flow of the lubricating agent to the packing and around the journal, thereby removing a high percentage of the heat generated inside.

The thin film which lubricates a packing is also used as a sealing agent. Since it is generally introduced at a pressure several pounds higher than the fluid being handled, co-mingling is expected. For this reason the two substances must be compatible.

In packing mayonnaise, *for example*, pure food regulations must be observed: vegetable or cotton seed oil is often used as the lubricating agent. Other examples are more obvious: one would not use glycerine in the lantern ring or seal ring of a pump that was pumping nitric acid—nitroglycerine would be formed. The object, in all cases, is to assure compatibility while maintaining the greatest possible lubricity. (See Figure 44-3.)

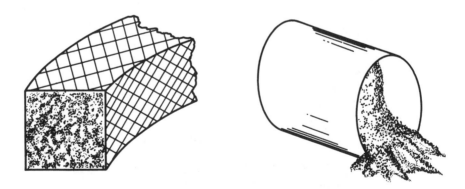

Figure 44-3 / **In plastic packings, oil or grease in the binder provides initial lubrication.**

Adjusting a set of soft (nonautomatic) packings, should be accomplished by tightening the gland stud by not more than one flat of a hex nut at a time, after which the packing should be allowed to run uninterrupted for ten or fifteen minutes. Any attempt to eliminate leakage completely is almost certain to result in a burned-up set of packing, often in a few minutes' time.

In cases where the set is in good condition and still contains adequate lubricant, the lubricant acts as a safety valve. However, if the packing is over-tightened, friction becomes excessive, heat is produced and the set tries to expand, as a result the lubricant becomes thinner and more mobile and is squeezed out, thus the volume of the set decreases and equilibrium is reached. However, should there be insufficient lubricant remaining, the result may be a cloud of smoke, destroyed packing and a scored journal.

Automatic packings should not require adjustment. Cups, U's, O-rings, V-packings and flanges, have a predetermined depth of the stuffing box or axial space between the retaining surfaces.

In replacing a packing, cleanliness is the major factor to be considered. Rings should be carefully slipped into place with joints of soft packings staggered. They should be positioned true and square and free of all foreign matter. A split wooden block or cylinder that one can slide into the box can be used to tamp the packing in place uniformly. After the stuffing box has been filled, it should be tightened up snugly, then backed off to allow the packing to expand.

In this manner, provision is made for a certain "looseness", so that the pump can be started up without danger of overheating or damaging the packing or journal in any way. The shaft and especially the piston and the rings supporting it must also be checked to insure that it is true and properly supported.

High-Pressure Work

In high-pressure work, precautions concerning clearance must be observed, since failures in the operation of automatic packings, cups, O-rings, and V's are most often caused by extrusion into the clearances. Elastomeric materials, whether used alone or as a binder in a fibrous packing, are deformable. Even with recommended fine clearances, the O-ring requires special back-up or anti-extrusion rings above 1500 psi.

Under several thousand pounds of pressure, very fine clearances, on the order of one or two thousandths, are required. If such close tolerances are not prac-ticable (as is usually the case), a strong backup ring is considered necessary. These rings range from the familiar adapter rings used with V-rings to metal-backed cups.

The choice of a packing is not an easy one to make, since thousands are available. A brief guide, however, might be useful:

Braided packings are manufactured from a variety of fibers, including cotton, flax, jute, ramie, asbestos (both blue and white), "Teflon", rayon, nylon, and dacron. Some of these packings are equipped with metal jackets.

Metal makes a good bearing. When used around a core to make a foil jacket, shaft movement beyond that which a solid piece of metal can handle is normally expected. The core provides the resilient springback to the metal surface. Most such packings are used in rotary service or mild reciprocating work.

Coil and *ring packings* are fabricated from duck and rubber, and are

Figure 44-4B / **Sections show various packing shapes to fit machine contours.**

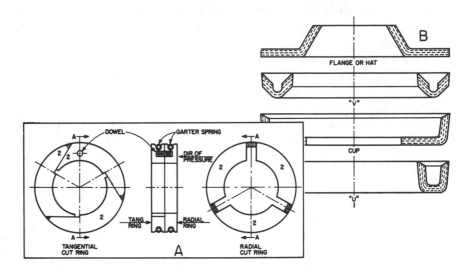

Figure 44-4A / **Metal packing rings are used almost exclusively for high-pressure reciprocating work involving gases.**

generally used in reciprocating work. Occasionally, however, they come into service as end rings for rotary service as well.

Molded and *automatic* types of *packings*, some of which are reinforced with certain of the fibers mentioned above or they may be a homogeneous rubber compound, are generally recommended for use in reciprocating work.

Floating metal packings are segmental rings of packing generally held in initial contact with the shaft by means of a garter spring circling the surface of the ring, and are manufactured of cast iron, bronze, phenolic, carbon, and "Teflon". This type of packing is used almost exclusively for high-pressure reciprocating work involving gases.

The standard oil seal employs a nitrile rubber element, although other elastomers and leathers are sometimes used. It is used more often for rotating service than for reciprocating service and is designed for "nonpressure" applications.

Mechanical seals, though expensive, are probably the most efficient packings. They are used exclusively for rotary service, and are manufactured from many materials and combinations such as metals, ceramics, plastic, elastomers, carbon, and many others.

Figure 44-5A / **Braided packing. Lattice weave of the material increases its durability.**

Figure 44-5B / **Spiral packings are generally recommended for reciprocating service.**

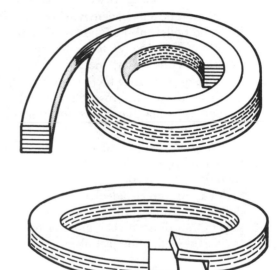

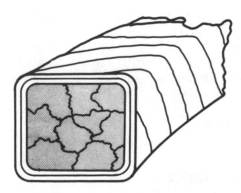

Figure 44-5C / **Asbestos core with metal foil is used in rotary or mild reciprocating work.**

Regardless of how colorful plumbing fixtures are, they can perform their function only if the piping behind the walls and in the basement is made of quality materials, correctly designed and installed, and periodically inspected.

Periodic inspections of piping should be made where conditions are particularly severe, such as in elevator shafts and boiler rooms, and where stresses of high temperatures prevail. In some types of buildings, the life of piping is shortened by vibration, chemicals, condensation, electrolysis, and by alternation of temperatures over short periods.

Insulation of piping will minimize piping hazards resulting from condensation and excessive heat or cold. It will add life to piping and effect economies in heating and cooling. It will also keep the water at a higher temperature.

Water heavily laden with oxygen is particularly destructive to hot water lines and boiler tubing. The installation of deaerators or deactivators will lengthen the life of the pipe. Water softening equipment is sometimes required due to the hardness of the water, and aids in keeping the pipe free of scale.

Leaky pipes may be the result of chemical decomposition of the pipe caused by electric current. Inspection should be made periodically to determine if anyone is using the pipes for ground wires. If so this practice should be discontinued.

A good quality valve or fitting should always be used. Valves should be turned open and closed at least once every three months. Valves which stick should be replaced.

The use of an adequate number of valves both on industrial lines and on individual fixtures will prevent inconvenience to tenants should it become necessary to shut down certain areas of the building for repair. A union should always be placed in the line just ahead of the valve to aid in its replacement.

45A / **REPAIRING WATER FAUCETS AND VALVES**

(A publication of the U. S. Department of Agriculture)

Water faucets and globe valves serve the same purpose, in that they control the flow of water. The essential difference is that faucets are used at discharge points over fixtures such as sinks, lavatories, and tubs, while valves are used to close off portions of the plumbing system.

Other types of valves, such as check valves, gate valves, and pressure-reducing valves, are required in plumbing.

Faucets and globe valves are very similar in construction (Figure 45-1) and repair instructions apply to both. (Your faucets or valves may differ somewhat in general design from the one shown in Figure 45-1, because both faucets and valves come in a wide variety of styles.)

Mixing faucets, which are found on sinks, laundry trays, and bathtubs, are actually two separate units with a common spout. Each unit is independently repaired.

If a faucet drips when closed or vibrates ("sings" or "flutters") when opened, the trouble is usually a worn washer at the lower end of the spindle. If it leaks around the spindle when opened, new packing is needed. To replace the washer:

- Shut off the water at the shutoff valve nearest the particular faucet.
- Disassemble the faucet—the handle, packing nut, packing, and spindle, in that order. You may have to set the handle back on the spindle and use it to unscrew and remove the spindle.

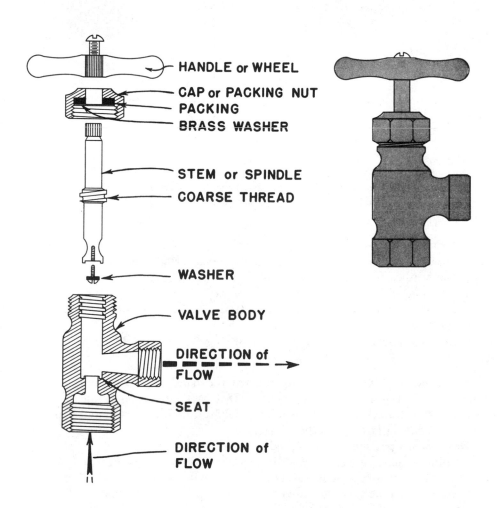

HANDLE or WHEEL
CAP or PACKING NUT
PACKING
BRASS WASHER
STEM or SPINDLE
COARSE THREAD
WASHER
VALVE BODY
DIRECTION of FLOW
SEAT
DIRECTION of FLOW

Figure 45-1 / Globe-type angle valve.

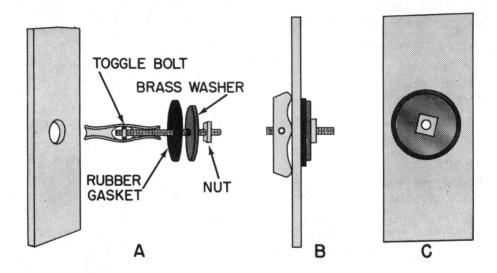

Figure 45-2 / Closing a hole in a tank: *A*, The link of the toggle bolt is passed through the hole in the tank (hole is enlarged if necessary). *B*, Side view of tank edge (nut is drawn up tightly to compress washer and gasket against tank). *C*, Outside view of completed repair.

• Remove the screw and worn washer from the spindle. Scrape all the worn washer parts from the cup and install a new washer. If you do not have the proper size washer, file down a larger one; do not use one that is too small.

• Examine the seat on the faucet body. If it is nicked or rough, reface it. Hardware or plumbing-supply stores carry the necessary seat-dressing tool. Hold the tool vertically when refacing the seat.

• Reassemble the faucet. Handles of mixing faucets should be in matched positions.

To replace the packing, simply remove the handle, packing nut, and old packing, and install a new packing washer. If a packing washer is not available, you can wrap stranded graphite-asbestos wicking around the spindle. Turn the packing nut down tight against the wicking.

Other faucet parts may be replaced as necessary.

Complete faucet inserts in which the washer does not turn on the seat are available. This feature prolongs washer life indefinitely.

Several new faucet designs aimed at easier operation, eliminating drip, and promoting long service life, are on the market. Instructions for repair may be obtained from dealers.

If a shower head drips, the supply valve has not been fully closed, or the valve needs repair.

Precautions

Polluted water or sewage may carry such diseases as typhoid fever and amoebic dysentery. If you do your own plumbing work, be sure that:

• There are no leaks in drainpipes through which sewage or sewage gases can escape.

• There are no cross-connections between piping carrying water from different sources, unless there can be reasonable certainty that all sources are safe and will remain safe.

• There can be no back siphonage of water from plumbing fixtures or other containers into the water-supply system.

Once a pipe has become polluted, it may be difficult to free it of the pollution. For this reason, building codes do not permit the use of secondhand pipe. All initial piping and parts and subsequent replacements should be new.

Since a plumbing system will require service from time to time, shutoff valves should be installed at strategic locations so that an affected portion can be isolated (water flow to it cut off) with minimum disturbance to service in the rest of the system. Shutoff valves are usually provided on the water closet supply line, on the hot- and cold-water supply line to each sink, tub, and lavatory, and on the water-heater supply line. Drain valves are usually installed for water-supply piping systems and for hot-water storage tanks.

A pressure-relief valve should be installed for the water-heater storage tank to relieve pressure buildup in case of overheating.

45B / REPAIRING LEAKS IN PIPES AND TANKS

(A publication of the U. S. Department of Agriculture)

Pipes

Leaks in pipes usually result from corrosion or from damage to the pipe. Pipes may be damaged by freezing, by vibration caused by machinery operating nearby, by water hammer, or by bumping into the pipe.

Corrosion. Occasionally waters are encountered that corrode metal pipe and tubing. (Some acid soils also corrode metal pipe and tubing.)

The corrosion usually occurs, in varying degrees, along the entire length of pipe rather than at some particular point. An exception would be where dissimilar metals, such as copper and steel, are joined.

Treatment of the water may solve the problem of corrosion. Otherwise, you may have to replace the piping with a type made of material that will be less subject to the corrosive action of the water.

It is good practice to get a chemical analysis of the water before selecting materials for a plumbing system. Your State college or university may be equipped to make an analysis; if not, you can have it done by a private laboratory.

Repairing Leaks. Pipes that are split by hard freezing must be replaced.

A leak at a threaded connection can often be stopped by unscrewing the fitting and applying a pipe joint compound that will seal the joint when the fitting is screwed back together.

Small leaks in a pipe can often be repaired with a rubber patch and metal clamp or sleeve. This must be considered as an emergency repair job and should be followed by permanent repair as soon as practicable.

Large leaks in a pipe may require cutting out the damaged section and installing a new piece of pipe. At least one union will be required unless the leak is near the end of the pipe.

Vibration sometimes breaks solder joints in copper tubing, causing leaks. If the joint is accessible, clean and resolder it. The tubing must be dry before it can be heated to soldering temperature. Leaks in places not readily accessible usually require the services of a plumber and sometimes of both a plumber and a carpenter.

Tanks

Leaks in tanks are usually caused by corrosion. Sometimes, a safety valve may fail to open and the pressure developed will spring a leak.

While a leak may occur at only one place in the tank wall, the wall may also be corroded thin in other places. Therefore, any repair should be considered as temporary, and the tank should be replaced as soon as possible.

A leak can be temporarily repaired with a toggle bolt, rubber gasket, and brass washer as shown in Figure 45-2. You may have to drill or ream the hole larger to insert the toggle bolt. Draw the bolt up tight to compress the rubber gasket against the tank wall.

Identification

Valve identification. Valves should be identified so that their function can be easily identified. This can be accomplished by the use of a tag on the valve explaining its function, or with a number or letter system that corresponds to a keyed chart.

Pipe identification. Pipes or their covering can be painted for identification. Color codes generally used are:

Cold Water	Blue	Waste Lines	Yellow
Hot Water	Red	Vent Pipes	Black

Water faucets should be inspected for leaks once every three months. Filter screens should be cleaned every other month. Toilet bowls and urinals should be inspected for leaks, cracks, and correct seating, once a year.

45C / **WASHROOMS**

Important factors in washroom planning are the layout and the equipment and materials to be used which will effect the overall maintenance costs. Non-porous materials are best for washrooms because they keep maintenance to a minimum. Floor materials in this category are concrete, terrazzo, and ceramic tile. Wall surfaces should be impervious to moisture. Non-porous surfaces are the easiest to be kept clean.

Wall hung fixtures provide an unobstructed washroom floor for rapid hosing or mopping. It pays to buy good quality faucets and fittings made for durability. Combination faucets are preferred to separate hot and cold water faucets for lavatories. This permits the user to temper the water to suit his wishes. Do not economize in buying a toilet. The better toilets have a larger water seal which will reduce stoppages. Better toilets are equipped with quieter and more positive flushing action. Elongated bowls are recommended, and are sometimes required by the plumbing code.

The best way to provide for ease in servicing is to have an access chamber in back of each washroom. If men's and women's washrooms are back to back, one chamber will do for both. All supply, waste piping, and the flush valve mechanism will be easily accessible, and it will be rarely that a maintenance man will have to enter the women's rest room for the purpose of making adjustments.

Washroom facilities per person. State or local laws normally will specify minimum washroom facilities to be provided in industrial buildings. Office occupants should have facilities above this minimum. A desirable standard to cover these facilities provides for:

WOMEN

No. of Employees	Toilet Booths	Washbasins
1-5	1	1
6-15	2	2
16-25	3	2
26-40	4	2
41-70	5	3
71-100	6	4
101-140	7	5

Note: One toilet booth is allowed for every 45 employees over 140, and three basins are allowed for every four toilet booths.

MEN

No. of Employees	Toilet Booths	Urinals	Washbasins
1-5	1	—	1
6-25	2	1	2
26-50	3	2	2
51-80	4	2	3
81-125	5	3	4

Note: One toilet booth is allowed for every 45 employees over 125, one urinal for every 60 persons and three basins are allowed for every toilet booth.

Stoppages of toilets. Fill the fixture approximately two-thirds with water. Then use a rubber suction and force type plunger to try to dislodge the stoppage. The overflow pipe should be covered to help concentrate the pressure produced on the drain. If this does not clear the stoppage, a flexible spiral wire (plumber's snake) should be inserted into the clogged pipe either through toilet bowl or a cleanout. If a pipe is only partially blocked and water will still flow, drain pipe solvents can sometimes be effective. Caution should be noted in using lye or other strong alkali, as soap may be formed with grease, thus increasing the problem of removing the stoppage.

Elements of Maintenance

Cleanout covers. These should be removed at least twice per year to determine the condition of the pipe, and to prevent the covers from sticking. If a cover cannot be turned with an ordinary pull on a wrench, sometimes tapping the cover at the same time force is applied to the wrench will aid in loosening the cover. A cleanout cover can also be loosened by applying heat to the outside of the pipe. *Be sure* to keep the cover cool.

Traps. A plumbing trap is a fitting placed in a drain pipe for the purpose of holding water or other fluid to form a seal that will prevent the passage of gases and odors from the drain pipes into the immediate area. The seal may be destroyed when water is blown or sucked out of the trap as a result of variations in pressure in drain pipes, when the water discharging through a trap at a high rate of velocity does not fall sufficiently to maintain or restore the seal, or when water evaporates from the trap. Proper venting will control these difficulties, *except* evaporation.

Grease traps. Grease traps are required in waste pipes from sinks and fixtures in which greasy foods are prepared, and where dishes are washed which contained greasy foods. Traps of this type are generally installed in restaurants, hotels, schools, laboratories, hospitals, and all institutional type buildings.

These are installed in the waste pipe for the purpose of separating grease from the liquid retaining the grease. The construction of the trap permits cooling of the liquid sufficiently to precipitate some grease from solution and to permit suspended grease to rise to the top of the trap where the grease is held.

Leaks in water closet flush valves. When water runs continuously from a water closet flush tank, either one of two conditions may exist.

1. The float on the float valve is submerged more than it should be, and water is running through the overflow. The float should be inspected to see if it is waterlogged and needs to be replaced. If the inlet valve is leaking, a new seat or repacking may be required. If both the float and valve are in satisfactory condition, the bar holding the float should be bent so that the float will rise sooner and close the valve tightly.

2. The flush valve is not seated and water runs into the toilet bowl. The rubber ball, the guide, or the flush valve seat may need adjustment or replacement.

45D / CARE OF WATER PIPES

Cleaning. Pipes can be cleaned by pushing or pulling through them, a cutting tool attached to a flexible wire twisted by hand or power. Pneumatic tools are also available which deliver a slug of compressed air hydraulically.

Thawing frozen water pipes. Pipes burst due to force of the expansion of water as it turns into ice. If ice is allowed to expand as it forms, the pipe will not burst. The moment the movement is obstructed, such as at a valve, a bend, or a rough spot, it will burst at that obstruction.

A frozen pipe may be thawed by:

1. Applying cloths soaked in hot water.
2. Playing steam on the pipe.
3. Injecting steam into the pipe.
4. Pouring hot water into the pipe.
5. Unslaked lime wrapped in a waterproof material—water poured over the lime.
6. Or other methods of applying heat to the pipe that *are not considered dangerous.*

Air in water pipes. Air may enter pipes from draining them, or from a period when the pressure was low enough to siphon air in through an open faucet or flush tank valve. To correct this condition:

1. Put an aerating attachment on the faucet from which air is escaping.
2. Insert an air relief valve at the high point of the piping system.
3. Place a frequently used faucet near the point of supply concentrating the trouble there.

Water hammer. Water hammer may be caused by the sudden closing of valves or faucets, or any other method of quickly stopping the flow of water in a pipe. Other causes are:

1. Displacing air from a closed tank or pipe from the top.
2. Condensation of steam in water in a closed pipe.
3. By pumps, sudden stopping, damaged or worn gaskets.

Water hammer can be prevented by:

1. When filling a closed tank or pipe, it should be filled from the bottom up to allow the air to escape at the top.
2. Steam and water should not be allowed to come into contact in a closed pipe.
3. Downward dips in steam pipes should be avoided or drains should be provided.
4. Installation of air chambers may control water hammer. Air chambers should be installed near the valve that is causing the water hammer in a vertical position over the top of a riser pipe.
5. Use of pressure reducing valves.

Combating Destructive Return-Line Corrosion

(*Source:* "7 Ways to Combat Destructive Return-Line Corrosion," by J. G. Roussos, *Power*, February 1963)

Corrosion of return lines and steam lines mean expensive piping and valve replacement, curtailed production because of forced shutdowns and inefficient operation of heat-transfer equipment. Virtually all corrosion in steam and condensate lines is caused by oxygen or carbon dioxide in the presence of moisture. And when both gases are present the rate of corrosion is appreciably greater than either could produce alone.

Oxygen corrosion is characterized by deep pit marks that ultimately penetrate the pipe wall. CO_2 attacks grooves, the pipe proper and produces marked thinning at threaded joints.

A classic example of return-line corrosion induced by CO_2 is shown in Figure 45-3. Note deep grooving of piping on the bottom surface and actual metal failure at the threaded end. Corrosion has removed about 50 per cent of the pipe-wall thickness. This specimen was taken from the condensate system in a large industrial plant where steam produced at 600 psi generates electricity and meets both process and general heating needs.

Fouling with corrosion products is marked in Figure 45-4. Deposits of this type are frequently found in traps, valves and other areas where flow is restricted. Section of piping removed from a large textile mill (Figure 45-5) is also fouled by corrosion products in its upper section and the lower pipe wall is deeply gouged.

Seven proven methods. Many factors contribute to steam and return-line corrosion. One plant's solution to the problem is not necessarily the best answer for another. But here are seven proven ways plant engineers can combat corrosion:

1. *Good insulation practice is a must.* Well-insulated, dry steam lines show very little corrosion, if any, but poorly insulated mains increase amount of steam condensation and permit absorption of oxygen and CO_2. Carbonic acid forms as moisture combines with the CO_2 ($H_2O + CO_2 = H_2CO$). *Result:* accelerated oxygen and acid corrosion attack. When repairing or replacing steam or return-line piping make sure insulation is replaced.

2. *Provide enough return-line pitch.* Improperly pitched return lines decrease flow rate of the condensate, giving it more time to cool. And as temperature drops, solubility of oxygen and carbon dioxide increases. So if flow is low or stagnant in a line, condensate can absorb more oxygen and CO_2. *Result:* low pH and severe corrosion. Vertical lines where condensate velocity is high almost always show much less corrosion.

3. *Reduce steam and condensate losses.* Keeping required flow of makeup water at a minimum means less dissolved gas enters the feedwater system. Dividends include lower cost of makeup, fuel and water treatment.

In many cases, steam losses can be drastically reduced by switching from direct to indirect steam heating and recovering condensate for boiler feed. Cost of installing necessary heat-exchange equipment is then charged against saving in operating cost.

Figure 45-3

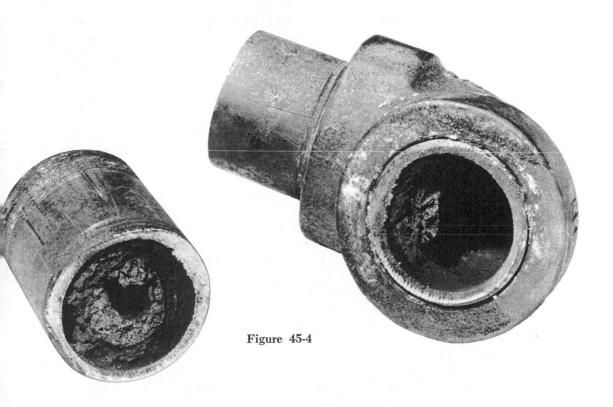

Figure 45-4

Figure 45-5

Leaks in steam and condensate lines are obvious losses that cannot be tolerated. Compare chlorides in the feedwater with chlorides in the makeup to obtain actual percentage of makeup and condensate returns. These figures should check closely with separate estimates of condensate-return flow. If not, chances are good that steam and condensate leaks account for the difference. Some plants bury their return lines underground, making leak detection very difficult. But even then it pays to locate and repair system leaks.

Process condensate is frequently discarded because of possible contamination that could harm the boiler and steam system. But how much high-quality condensate is wasted to avoid this contamination? As a more economical arrangement, return condensate to the boiler but set up automatic controls to detect contamination. Then discard condensate only if quality is unacceptable.

4. *Minimize air leakage.* Temperature drop in condensate-storage tanks can produce a partial vacuum which sucks outside air in through vents or the overflow connection. One way to minimize this fairly common trouble is the addition of a heating coil in the tank to keep temperature at 180°F. or higher at all times. Another calls for installing a vertical unloaded check valve in the vent or overflow pipe.

5. *Proper system lay-up is vital.* It's good practice to drain all idle lines in steam and condensate systems completely. If condensate is left in return lines when the system isn't operating, it picks up CO_2, which corrodes the piping in the presence of air. This type of attack is often called sectional return-line corrosion. It takes place in parts of the system not in continuous use, such as circuits used only during the heating season.

6. *Provide efficient deaeration.* Most economical approach is to remove oxygen from the feedwater as completely as possible. This can be done by mechanical deaerating, by chemical oxygen scavengers such as catalyzed sulfite, or a combination of both methods. In any case feed enough chemical to maintain an adequate sulfite reserve in the boiler water. Any oxygen not removed will go over with steam from the boiler and corrode return-line piping. That's why it's extremely important to continuously check deaerator performance.

Temperature of water in the deaerating section should correspond with temperature of saturated steam at heater's temperature above 216°F. (corresponding 1.3 psi). When heater is working properly, temperature of water in the deaerating section is normally one or two degrees higher than water in the deaerated storage compartment.

Watch sulfite dosage as a possible clue to faulty operation. When heater is not removing oxygen, sulfite requirement goes up—often to many times its normal level.

7. *Use the proper feedwater treatment.* Even though oxygen and CO_2 are removed from the feedwater by deaeration, CO_2 enters the steam cycle and reaches condensate-return lines in another way. Bicarbonate and carbonate alkalinity in the feedwater makeup decompose to liberate free CO_2 at boiler operating pressure and temperature. This CO_2 leaves the boiler as a gas and is absorbed by condensate to form carbonic acid. The higher the bicarbonate or carbonate alkalinity of the makeup the more CO_2 is liberated in the steam.

Basic aim of feedwater treatment is to prevent damage from scale or corrosion. This is generally handled by external treatment followed by application of internal-treatment chemicals or, in many cases, internal treatment may be all that is actually required.

Many factors have to be considered and carefully weighed before you know which type treatment will be best for a particular plant. Chemical analysis of raw-water makeup required for the boiler are paramount. Boiler pressure, type of boiler, amount of makeup, space available, initial investment required, operating cost are all part of this picture.

Treatment. External treatment, from a return-line corrosion point of view, should provide for reducing alkalinity of the makeup before it enters the feedwater cycle. Common treatment systems include: (1) hot and cold process softening (2) sodium-zeolite softening followed by acid treatment and degasification (3) split-stream zeolite (4) dealkalization by anion exchange (5) demineralization and (6) evaporation. The split-stream technique involves blending hydrogen-zeolite-softened water with sodium-zeolite-softened water, then passing the water through a degasifier.

Internal treatment for corrosion prevention involves feeding some chemical as a corrosion inhibitor. Ammonia was one of the first. Added to the boiler, this alkaline chemical evaporated with the steam, redissolved in the condensate and raised its pH. But ammonia is not in common use today because it dissolves copper in the presence of oxygen, attacks brass or bronze fittings and valves.

A class of chemical compounds known as amines is now employed. There are two basic types: a volatile neutralizing amine and a high-molecular film-forming type.

Selecting an amine for an individual plant depends on: (1) steam pressure (2) amount of steam and condensate piping in the system (3) distance to farthest point in the system (4) per cent of blowdown (5) temperature of feedwater and (6) makeup-water analysis.

Neutralizing amines in common use have low boiling points. They are alkaline in nature and very soluble in water. They behave much like ammonia, raising pH of the condensate, with one vital difference—they will not attack copper. Special feeding equipment is not required and dosage is easily controlled by making simple checks of pH on samples of condensate.

In many plants, operating conditions vary greatly at different points in the same system. Conditions like this are best handled by a mixture of these amines with different boiling and condensing characteristics.

High-molecular film-forming amines, unlike the neutralizing type, depend on the film to protect internal surfaces of steam and condensate piping. In theory this film acts as a barrier to keep acid condensate or oxygen from touching iron piping. Satisfactory results require film formation on all metal surfaces—a thin, uniform film that will not affect heat-transfer efficiency.

Because of their adhering and insoluble characteristics, film-forming treatments have been known to produce deposits in steam traps and boilers. Plants with steam turbines should add these amines to the steam after it leaves the unit. Filming amines call for their own tank and pump setup; dosage is determined by daily evaporation.

45E / **FROSTPROOF HYDRANTS**
(A publication of the U. S. Department of Agriculture)

Frostproof hydrants are basically faucets, although they may differ somewhat in design from ordinary faucets.

Two important features of a frostproof hydrant are: (1) the valve is installed underground—below the frost line—to prevent freezing; and (2) the valve is designed to drain the water from the hydrant when the valve is closed. Figure 45-6 shows one type of frostproof hydrant. It works as follows: When the handle is raised, the piston rises, opening the valve. Water flows from the supply pipe into the cylinder, up through the riser, and out the spout. When the handle is pushed down, the piston goes down, closing the valve and stopping the flow of water. Water left in the hydrant flows out the drain tube into a small gravel-filled dry well or drain pit.

As with ordinary faucets, leakage will probably be the most common trouble encountered with frostproof hydrants. Worn packing, gaskets, and washers can cause leakage. Disassemble the hydrant as necessary to replace or repair these and other parts.

46 / **SEPTIC TANK CARE**
(A publication of the U. S. Department of Health, Education, and Welfare)

Most of the water used in our buildings only carries off wastes. Drinking, cooking, running the garbage grinder, and washing the company cars use less water than doing the dishes, bathing, and flushing the toilet. Wastes carried away by water from kitchens, bathrooms, and laundry rooms should be collected in sewers and carried away to a community sewage treatment plant or central disposal point, operated and maintained by trained operators to insure proper control.

Individual septic-tank–soil-absorption systems are most frequently used in rural areas and in some unsewered suburban areas. A septic tank system will serve a building satisfactorily only if it is properly designed, installed, and adequately maintained. Even a good system which does not have proper care and attention may become a nuisance, and a burdensome expense.

Remember, a septic-tank–soil-absorption system which does not function properly frequently becomes a health hazard. To obtain satisfactory service, the building engineer must know something about the design, operation, and maintenance of his own septic tank system.

Where it is impossible to connect to a community sewer, the building buyer should satisfy himself that his septic-tank–soil-absorption system is properly designed and installed to serve the anticipated number of occupants of the build-

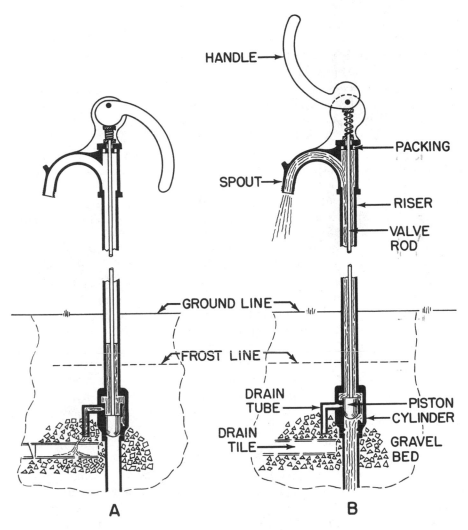

Figure 45-6 / Frostproof hydrant: *A,* Closed; *B,* Opened. As soon as the hydrant is closed, water left in the riser drains out the drain tube as shown in A. This prevents water from freezing in the hydrant in cold weather.

ing. He should also be sure that the system is located where it is not likely to endanger water-supply systems, and that the absorption system is capable of disposing of liquid wastes under year-round weather conditions. This information is usually available at your local health department.

When septic tank systems are improperly designed or maintained, liquid wastes may overflow to the ground surface or the plumbing in the building may often be stopped up. These overflows not only create offensive odors but are also a health hazard. Sewage may contain dysentery, infectious hepatitis, typhoid and paratyphoid, or other infectious disease organisms. Ponded sewage creates breeding places for some kinds of mosquitoes and other insects.

The purpose of a septic tank is primarily to condition building wastes, including water from the laundry and the bath, discarded food scraps, and body wastes so that they may be more readily percolated into the subsoil of the ground. The normal use of bleaches, detergents, soaps, and drain cleaners does not harm or interfere with the operation of the system.

A septic tank is a watertight structure in which organic solids are decomposed by natural bacterial processes. The flow of sewage is slowed in its passage through the tank so that larger solids settle to the bottom and accumulate as sludge. Grease and lighter particles rise to the surface and form scum.

The partially treated sewage, or effluent, flowing from the tank still contains large numbers of harmful bacteria and organic matter in a finely divided state or in solution. Foul odors, unsightly conditions, and health hazards will develop if this effluent is ponded on the surface of the ground or carried away in open ditches. Final disposal of the effluent in a subsurface soil absorption system is necessary to avoid these problems.

The bacteria present in a tank are able to thrive in the absence of oxygen. Such decomposition in the absence of air is called "septic," which led to the naming of the tank. Solids and scum are digested and reduced to a smaller volume by the bacteria in the tank. However, a residue of inert solid material remains which must be stored during the interval between tank cleanings.

The frequency of cleaning depends on the size of the septic tank and the number of people it serves. When a garbage grinder is used, more frequent cleaning will be required. With ordinary use and care, a septic tank usually requires cleaning every two or three years. However, in many cases septic tanks can be satisfactorily operated even longer. The building owner can make measurements and decide for himself when his tank needs cleaning. (See Figure 46-1.) When the bottom of the scum is within 3 inches of the bottom of the outlet device or the top of the sludge is within the limits of the following table, the tank should be cleaned. The accumulated solids are ordinarily pumped out by companies that make a business of cleaning septic tanks. Your neighbors or your local health department usually know which local companies do this work satisfactorily. The solids removed should be disposed of in a manner approved by your local health department to avoid obnoxious odors and health hazards.

Allowable Sludge Accumulation

Liquid Capacity of Tank, Gallons (a)	Liquid Depth			
	2½ feet	3 feet	4 feet	5 feet
	Distance from bottom of outlet device to top of sludge, inches			
750	5	6	10	13
900	4	4	7	10
1,000	4	4	6	8

(a) Tanks smaller than listed will require more frequent cleaning.
(b) In large commercial septic tank systems, two compartments are normally used, with the first section serving as a primary sludge settling basin, being 2–3 times the size of the second section.

There are no known chemicals, yeasts, bacteria, enzymes, or other substances

DEVICES FOR MEASURING SLUDGE AND SCUM

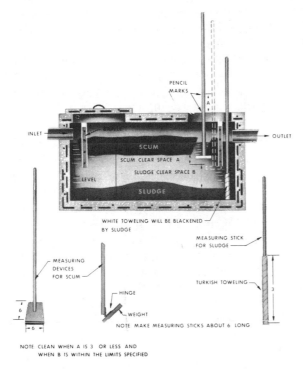

Figure 46-1

capable of eliminating or reducing the solids and scum in a septic tank, so that periodic cleaning is unnecessary. The addition of such products is not necessary for the proper functioning of a septic-tank–soil-absorption system.

To facilitate cleaning and maintenance, the building owner should have a diagram of his septic tank system, showing the location of the building, the septic tank manholes, the piping, and the soil absorption system. Figure 46-2 is a suggested chart for keeping a tank cleaning record. This information should be kept on the premises, regardless of a change in occupancy.

Tank Inspection Record

Date	Cost	Contractor	Description of Work

Figure 46-2

Septic tanks and soil absorption systems frequently are damaged when heavy trucks or other equipment drive over them. An accurate diagram of the system enables the building owner to keep heavy vehicles away from the critical area. A line of cast iron pipe covered with concrete instead of tile should be installed under any necessary crossings for heavy vehicles.

Neglect of the septic tank, however, is the most frequent cause of damage to soil absorption systems. When the tank is not cleaned, solids build up until they are carried into the underground soil absorption system, where they block the flow of the liquid into the soil. When this happens, the soil absorption system must be rebuilt—a costly undertaking. The precautions of periodic inspection and cleaning of the tank prevent this needless expense and work.

Clogging of the absorption field is the most common trouble with septic tank systems. This trouble may be due to improper design or construction, improper use, or neglect of necessary servicing. A tank that is too small, overloaded, improperly proportioned, or that agitates or short-circuits the sewage flow is liable to allow excessive amounts of small sewage particles to carry over to the absorption area, where they clog the pores of the soil. Neglect of cleaning produces the same effect. If the absorption area is in an unsuitable soil, or is too small, overloaded, or poorly constructed, the small amount of sewage particles normally in the effluent may lead to early clogging of the soil pores.

Clogged fields can sometimes be cleaned by opening up and flushing the lines with a hose. However, this does not open up clogged soil pores. More often it will be necessary to dig up, clean, and re-lay the absorption lines in new locations to get the benefit of unclogged soil. When doing this, the system should be checked to determine and correct the cause of the trouble.

House sewers also clog—usually from entry of roots and less frequently from paper, rags, sticks, or other trash and foreign materials that get in through the water closet or a floor drain. If the slope of the sewer is too flat to give the sewage a cleansing velocity, greases and solids may deposit in the pipe and cause trouble. Obstructions due to congealed grease can sometimes be cleared by adding hot water or drain solvents that generate heat. Other obstructions may yield to rodding or mechanical root cutters inserted at clean-outs. Caution should be exercised in using these devices in fiber-type sewers as they may damage the pipe or joints. In some cases, it may be necessary to dig up a line to reach an obstruction.

If the trouble is due to root entry, the only permanent remedies are to make the sewer line rootproof or move either it or the vegetation so that the roots cannot reach the line. Merely removing roots presently inside a sewer will not prevent future re-entry. Willow, poplar, and Chinese elm roots are especially troublesome.

For more detailed information concerning special conditions in your area, consult your local or State health department. Public Health Service Publication No. 526, "Manual of Septic Tank Practice" may also be helpful in providing additional information, including cleaning procedures for septic tank systems designed in accordance with this manual. It can be secured from the Superintendent of Documents, U. S. Government Printing Office, Washington, D.C., 20402.

(*Source:* The material on pump maintenance is taken
largely from an article entitled "Preventive Maintenance
of Pumping Units"by Otto M. Kristy, *Journal American
Water Works Association.*)

Preventive Maintenance

Preventive maintenance of a pump begins the day the pump is selected.
Important considerations in choosing a pump in order to keep maintenance to a
minimum are:

1. The materials selected should be capable of handling the fluid to be
pumped. The materials of which the pump is made is an important factor in the
initial cost of the equipment. A high initial cost for better materials to withstand
corrosion, abrasions, temperature, dust, and dirt, may be much cheaper in the
long run than continuous overhaul and repair of a pump made of less expensive
but inferior materials. When costs of labor, materials, and production time lost are
added up, it frequently is found that better materials actually begin to save dollars
after a few years of operation.

2. The bearings and lubrication should be adequate with respect to ambient
temperature, dust, and moisture conditions.

3. In order to handle the fluid in a correct manner, a decision should be
made whether to use a mechanical seal or packing in a stuffing box. Consideration
should be given as to the use of mechanical seals if no leakage from the pump is
desired, if the fluid being pumped is not abrasive, and if the temperature and
pressure are not excessive.

If a packing stuffing box is used, a shaft-sleeve material should be chosen
that will give maximum service and will not be attacked chemically. The packing
should be chosen with the sleeve material in mind, and should be capable of
withstanding the chemical attack of the fluid as well as the effects of temperature
and pressure.

One of the important factors in pump maintenance is the coupling alignment.
Coupling misalignment can damage couplings, bearings, shaft sleeves, packing or
mechanical seals, wearing rings or shafts. The vibration and noise level of the
machinery will also be affected. Periodic checks should be made to determine that
the alignment is being maintained.

The bearings should be carefully checked for signs of overheating. It is wise
to inspect, drain, flush, and relubricate bearings during the first few months of
operation in order to establish an individual timetable for bearing lubrication. The
ambient temperature, the dust and dirt condition, and the humidity will make the
timetable for one installation differ greatly from that of another. There may even
be considerable differences in timetables between one pump and another pump
in the same plant. The pump manufacturer generally furnishes recommendations

covering grease or lubricating oil. Lubrication suppliers, or supply houses, can normally give advice as to what type of lubricant should be used. It may be possible to use the same lubricant for several applications, thus requiring the storage of fewer types of lubricants.

Each plant or building should set its own lubrication timetable, oil-lubricated bearings usually requiring at least daily attention and grease lubricated bearings usually requiring attention at infrequent intervals—usually every two to three months. Care should be taken with oil lubricated bearings to prevent the various fittings from leaking and allowing the oil to drop below the safety level or fill above the full line, causing too high an oil level, which will make the bearing overheat. Grease lubricated bearings must not be too full or under pressure. Excessive fullness or pressure will pack the grease in and around the balls and will prevent them from rolling. This will result in the balls sliding, which in turn will cause scoring, galling of the bearing races, and rapid failure of the bearing.

The stuffing box should be checked daily to determine that the proper amount of leakage is taking place. If a mechanical seal is being used many units will require either flushing or cooling and lubrication flows to the sealing faces. This pressure and capacity must be regulated and maintained during operation. When a stuffing box is repacked, it should be checked to make certain that the box is clean down to the base metal, and that the shaft sleeve is not cut worn. The packing should be cut to an exact length so that the ends just butt firmly. Each ring should be pushed firmly to the bottom of the box with a packing tamping tool. The joints of each ring should be staggered 90 degrees. Special attention should be paid to the water seal ring, if one is used, to make sure it is in position directly under the flushing line connection to the stuffing box. This requires the correct number of rings placed both in front of and behind the seal ring. When the packing is all installed and there is no pressure in the pump, the gland should be securely tightened to squeeze the packing out against the stuffing box wall and around the sleeve. Then the gland should be loosened to permit the packing to expand. Gland bolt nuts should be retightened only finger tight. The packing should then be ready for service.

If further adjustment is required, it is preferable to stop the pumps, drain the casing and repeat the entire process. Often, however, this cannot be done. If adjustment must be made on a running pump, the gland should be tightened only one flat of the gland bolt nut, no oftener, then once every twenty minutes. This will enable the packing to expand slowly so that it will not score the sleeve or burn the packing.

The shutdown procedure for the pump is also important in preventive maintenance. It is recommended that a gate valve be located in the discharge line. Prior to complete stopping of the pump, this valve should be closed to keep the casing filled with fluid and to prevent metal to metal contact of the wearing surfaces. A centrifugal pump should never be throttled on the suction side, nor should it be stopped by closing the valve on the suction side. These actions will prevent the fluid being pumped from entering the casing and lubricating the close rotating clearances.

If the pump is to be shut down for a long period of time, the fluid should be drained from the casing to prevent corrosion products from filling the close run clearances. Additional protective steps should be taken to prevent corrosion of bearings, stuffing box parts, and couplings. The shut down pump should be turned over slowly by hand at least once per week.

Wearing rings should be replaced when the clearance has increased to a point at which either the driving motor is overloaded or the pump is no longer capable of producing rated capacity. In general, when the clearance has increased from two to three times its original size, the rings should be replaced.

In most cases it is preferable to use factory replacement parts for a pump because this will make it possible to keep the pump equipped with parts of correct dimension, size, and material. The factory supplying replacements may also be able to recommend a change of materials or design which will make possible longer and better service.

Annual Inspection

If it is possible without jeopardizing operations, the pump and drive should be subjected to a thorough annual inspection. The inspection should cover the following points:

1. The bearings should be removed, cleaned, inspected, and replaced, if necessary.

2. The entire casing and the bearing body should be cleaned.

3. The old packing should be removed and the shaft sleeve should be inspected. Both should be replaced if worn.

4. All of the rotor parts should be inspected and replaced if necessary.

5. If a new rotor is required, it should be checked for straightness and assembled in the pump, using new gaskets where necessary.

6. The casing should be checked for pipe strain and realigned and doweled if necessary.

7. The drain, vent, recirculation, and lubrication lines should be disconnected and cleaned and flushed.

8. All gages and instruments should be tested and recalibrated.

If the service requirements will not allow taking the equipment down for an annual inspection, then the preventive maintenance program increases in importance. This program should be designed to periodically check the various items that prevent a major breakdown of the pump.

48 / ROOF COOLING

A large part of the heat load in a building is produced by the sunshine on the roof. Surface temperatures of walls are approximately equal to those of the air. If the wall temperature were 100° and the inside temperature 75°, there would be 25° F. pushing inside. Since roof temperatures sometimes are around 150° F., this would result in three times as much heat pushed in per square foot of roof area. This multiplied by the length and by the width of the plant, can often be more than half the load.

Roof cooling can make your plant or building more comfortable in which to work. Roof cooling is also recommended to cut down on air-conditioning requirements. The rate of heat flow through each square foot of roof is obtained by multiplying the overall heat transmission coefficient by the equivalent temperature differential across the roof. Values of equivalent temperature differential for sprinkled roofs are known to be only one-third the maximum of that for roofs exposed to the sun.

Keeping the roof temperature down preserves the life of the roof. If a pond is used, the roof must be drained in order to make repairs, or for painting equipment. Ponds are sometimes breeding places for mosquitoes, algae, fungus and bad odors. The pond has a heat storage effect also. An even spray is considered a much better method of cooling a roof.

Water is the most efficient refrigerant, as one pint (pound) will handle over 1000 units of heat. Melting ice handles 144 units, and Freon 12, 72 units. Therefore less than 0.04 gallons of water are needed per hour for a square foot of roof surface. Effectiveness of any system of roof spraying depends upon the quality of the installation to wet the roof. Controls should be installed for turning the water on automatically for short intervals at a time. A heavy spray or thick drops avoid loss of water into the air.

49 / WATER TREATMENT FOR COOLING TOWERS

(*Source:* Bulletin No. 47A, Allis-Chalmers Mfg. Co., Milwaukee)

As a result of receding water tables and general water shortage, industry is faced with the increasing necessity for adopting conservation measures.

Water used for cooling constitutes a large portion of industrial demands. One effective means of reducing this requirement is the use of a cooling tower. The following covers one aspect of this general field—water treatment for cooling towers.

Types of Cooling Systems

In general, there are three types of cooling systems: the *once-through* system, the *closed* system, and *recirculating* or *recycling* system.

The *once-through cooling system* is not common and finds application only where there is an abundant, inexpensive source of cool water. This system employs no cooling tower, hence no evaporation takes place, the water merely being increased in temperature and discharged to waste. It offers relatively no scale or corrosion problems and treatment, except for chlorination, is rarely used.

The *closed system* is illustrated by the "inner-cooler" or "jacket cooler" in a diesel engine installation. Water circulates through the engine jacket, absorbs heat and enters the shell and tube heat exchanger where it is cooled by the water from the recycling system, and then the cycle is repeated. In a closed system no evaporation occurs and therefore there is no concentration. Neither is the water subjected to aeration. Makeup requirements are very low and are due primarily to leakage. Little difficulty is encountered with scale unless extremely hard water is used for makeup. Since the makeup requirements are so low, zeolite softened water is economically justified. In a system of this sort, corrosion is controlled by using inhibitors which render the metal surfaces passive.

The recycling system utilizes a cooling tower to lower the temperature of the water which is then recirculated. Evaporation of the water in the tower concentrates its impurities and obviously presents a wide range of problems.

In general, difficulties arising from the use of cooling water are those resulting from the reactions promoted during the transfer of heat from a metallic surface to water in contact with this surface.

Cooling Systems Problems

Cooling system problems usually encountered are: (1) corrosion of metal surfaces; (2) inorganic scaling; (3) organic growths such as algae and bacterial slime; and (4) delignification of the wood in the tower.

Corrosion of a surface condenser to the point of leakage leads to condensate contamination with its attending difficulties. A similar situation in a diesel plant results in contamination of the water in the closed system and damage to the unit because of insufficient cooling. Failure of heat exchangers through corrosion results in forced outages and production loss.

In the case of *scaling*, a deposit of water origin adheres to metal surfaces and, because of its insulating qualities, greatly reduces the efficiency of heat exchange. In a turbine condenser, the obvious result is loss of vacuum and drop in turbine efficiency. Scaling of heat exchange surfaces in diesel engine operation causes overheating. Under these conditions, maintenance costs increase.

In addition to forming insulating layers in themselves and aggravating fouling by combination with silt or calcium carbonate, *organic slimes* may promote corrosion by forming membranes for concentration cells and furnishing a source of corrosive gases as ammonia, hydrogen sulfide and sulfur dioxide during the process of bacterial metabolism.

The action generally termed *delignification* results in a weakening of the wooden sections of the cooling tower. If allowed to progress to the point where the wood fibers separate, extensive replacement of the wood is required.

Corrosion

The large volume of water used in cooling systems magnifies the problem of completely controlling corrosion by chemical treatment.

Aqueous corrosion of iron and steel is an oxidation reaction involving the formation of iron oxides; hence, the problem is accentuated in recirculating systems which permit the cooling water to become saturated with oxygen. Further the rate of attack is greater with waters of low pH. If cooling towers are located in highly industrialized areas, the water may become even more aggressive due to the absorption of corrosive gases.

Effective protection by chemical treatment requires care in selection of chemicals and a rigid control procedure. Further, the chemical balance to be maintained should be determined not only from the limitations presented by the water supply but also by the type of metals in the cycle.

In general, corrosion may be minimized by using chemical inhibitors such as polyphosphates, silicates or chromates which form protective films on the metal surfaces.

Inorganic Scaling

Calcium Carbonate. Calcium carbonate is by far the most troublesome deposit in heat exchange systems. It normally constitutes some 90 per cent of the total deposit.

The factors controlling the deposition or solution of calcium carbonate are very complex, but depend mainly upon (1) the total calcium hardness, (2) the total alkalinity, and (3) the pH value.

The curves in Figure 49-1 represent equilibrium conditions for calcium carbonate and indicate the interdependence of these three relations at a temperature of 25°C(77°F.). In the use of the figure the two conditions of calcium hardness and alkalinity locate the point either to the right of the proper pH curve (scale forming area) or to the left of the curve (scale dissolving area). *For example,* suppose a condition indicated by point A in which the total alkalinity of the water is 100, the calcium hardness is 700 and the pH value is 7.0. In this instance it occupies a point exactly on the curve for pH 7.0. Calcium carbonate will not be precipitated from solution under those conditions.

Consider the case of another water in which the pH and the calcium hardness are the same as the first, namely 7.0 and 700, but the alkalinity is 130 as at point B. Point B is above and to the right of the curve and, therefore, calcium carbonate would be expected to precipitate. This water could be brought back to equilibrium in two different ways, either by reducing the hardness or by reducing the alkalinity. This is the reasoning used in correcting a scale forming circulating water to one which is not scale forming.

Figure 49-1 must be used with considerable reservation, however, as it represents equilibrium conditions for a temperature of 25°C (77°F.) only. At higher temperatures lower values of calcium hardness, alkalinity, and pH are necessary to avoid calcium carbonate deposits.

Calcium sulphate. Calcium sulphate deposits are formed when the calcium sulphate concentration of the cooling water has been allowed to reach too high a value. Solubility of calcium sulphate is close to 1500 ppm (parts per million), and this is usually set as a maximum permissible concentration in cooling systems. It is considerably more difficult to remove chemically than calcium carbonate for it is only slightly soluble in acid.

Magnesium compounds. Magnesium deposits are rarely found since the

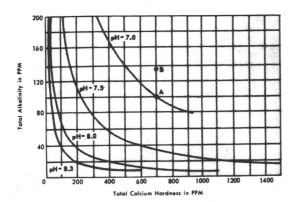

Figure 49-1 / Equilibrium conditions for calcium carbonate at pH values (77°F.).

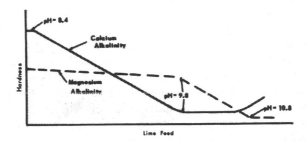

Figure 49-2 / Softening of water by lime treatment.

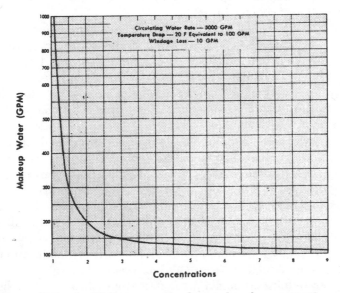

Figure 49-3 / Makeup compared to cycles of concentration.

necessary factors of time, temperature and concentration seldom exist in a cooling cycle.

Iron oxides. Iron oxide is occasionally found as a major constituent of a deposit. Its presence in quantity in a scale indicates deficiencies in corrosion control measures.

Scale Prevention

Scale preventive procedures fall into the following classes: (1) softening; (2) alkalinity reduction; (3) surface active treatment; and (4) deconcentration.

Softening. Softening by lime is an effective way of removing calcium bicarbonate from water. The calcium in the water and that added as lime are both precipitated by lime but it is not necessary to remove the magnesium because it does not ordinarily form exchanger deposits. As lime is added to water, the pH value increases and the removal of carbon dioxide, calcium bicarbonate and magnesium bicarbonate takes place in the order mentioned. Up to pH 8.4, carbon dioxide is neutralized and precipitated. As more lime is added, the pH value rises between 8.4 and 9.8, during which time calcium bicarbonate precipitates to the limit of its solubility. As more lime is added, magnesium begins to precipitate until a value of 10.8 is reached. After this, any further addition in lime increases the hardness due to excess calcium hydroxide remaining in solution. The water from the lime softener is supersaturated, and since the temperature will be increased in passing through the system, the supersaturation is broken by adding acid. Sufficient acid is usually added to bring the pH value below 8.2. (See Figure 49-2.)

Zeolite softening reduces calcium hardness to zero eliminating the danger of calcium carbonate deposits. The resulting high alkalinity, however, may cause delignification of the cooling tower.

Alkalinity reduction. Frequently a heat exchanger scaling problem can be overcome by the use of acid alone without any softening. The use of acid accomplishes a reduction in total alkalinity, by neutralizing part of the bicarbonate as well as reducing the pH value. This provides a more stable condition by increasing the solubility of calcium bicarbonate. In passing through the tower, CO_2 is released and the pH value increases, requiring additional acid. In practice, a continuous acid feed is employed.

Surface active treatment. One of the most effective methods of preventing precipitation of calcium carbonate is the use of surface active agents. These include certain phosphorous compounds such as metaphosphate and pyrophosphate. Several of the organic dispersives have also proved to be effective. Application can be as simple as dripping a solution into the suction of the circulating pump as continuously as possible. The hardness, alkalinity and pH values are not affected by the small amounts of treatment which are required. The usual control is to feed that amount required to maintain an effective concentration of the treatment in the circulating water.

Deconcentration. Salts will not precipitate from solution unless their solubility is exceeded. Therefore, another method of scale prevention is control of the concentration of the critical salts. This is accomplished by blowdown from the cooling tower basin.

The interdependence of calcium hardness, alkalinity, pH and temperature is shown in Figure 49-1. For any particular cooling system, if some of these values are unknown, they must be estimated. This method can be used to estimate the concentration of calcium hardness permitted before its maximum solubility is reached.

The maximum solubility limits of 600 ppm calcium hardness and 1000 ppm sulfates with a pH value of about 6.3, have been found to be applicable over a great percentage of the cooling systems observed. However, if the makeup water is zeolite softened, the sulfates can be allowed to increase above 1000 ppm since there will be very little calcium to precipitate the sulfate as calcium sulfate.

The number of cycles of concentration to be maintained in the system will be limited by either the calcium hardness or the sulfates. If the calcium hardness of the makeup water allows only three concentrations before reaching 600 ppm, but the sulfates could be concentrated four times before reaching 1000 ppm, the limit of the system will be three cycles.

The curve shown in Figure 49-3 illustrates the savings in makeup water that are gained from recycling water in cooling systems.

The greatest amount of water is saved in the first few cycles. Thereafter, increased cycling results in a smaller percentage of water saving, but greater economy in the use of treatment chemicals.

Part III / **MISCELLANEOUS**

50 / **CHARTS AND GRAPHS**

Statistics enable a building superintendent, plant engineer, or owner of a business to secure information by which he can plan better, compare costs, number of personnel, and other operating statistics by the month, year, or any particular period. Statistics is concerned with the collection, presentation, and analysis of numerical data. Techniques have been developed which help the reader understand the significance of large groups of figures by reducing them to smaller amounts or by portraying them on charts or diagrams.

Charts

As a means of simplifying the interpretation of figure facts, graphic presentation is often used. Figure 50-1 shows a comparison of five columns representing the shareholders' equity from 1953 to 1957.

Column charts are frequently used because they are easy to construct and understand. They show magnitudes by means of upright bars or columns, sometimes expressed in terms of money, physical units, percentages, or absolute amounts.

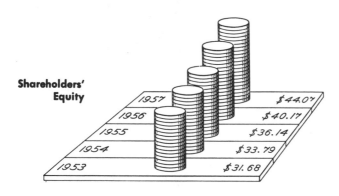

Figure 50-1

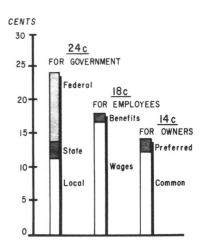

Figure 50-2

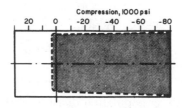

Figure 50-3

Bar charts differ from column charts only in the fact that the bars are laid off on a horizontal scale from left to right, rather than on a vertical scale from bottom to top. (See Figure 50-3.)

Pie charts are most logically used to indicate totals and the groups into which they are divided. A protractor is used to divide the circle. Pie charts are seldom used when results for more than one year are to be shown. (See Figure 50-4.)

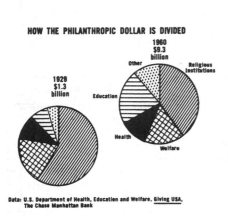

Figure 50-4

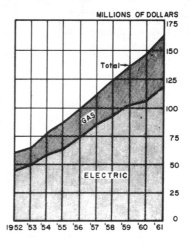

Figure 50-5

Graphs

The most popular method of displaying magnitudes over a period of years, however, is by means of graphs in which trends are shown by curves. A graph is a diagram that shows in graphic form the relation between two quantities. It may be a straight line, a smooth curve, a broken line, or an irregular curved line. Graphs are useful in engineering work, as data can be recorded in the form of diagrams. (See Figure 50-5.)

If two variable quantities are related so that a change in the value of one produces a change of the same proportion in the value of the other, the graph showing this relation will be a straight line. (See Figure 50-6.)

The graph below contains three curves showing the relation of speed, efficiency, and current to horsepower for an electric motor. (See Figure 50-7.)

1. As the horsepower increases, the speed steadily decreases.
2. The efficiency rises rapidly at first, then falls off.
3. The current increases at a rate that gradually grows more rapid.

Geographical charts can be used to show population densities, sales areas, or spot specific locations. (See Figure 50-8.)

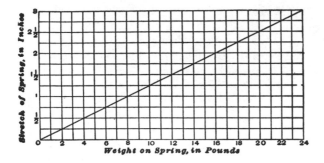

Figure 50-6

Figure 50-7

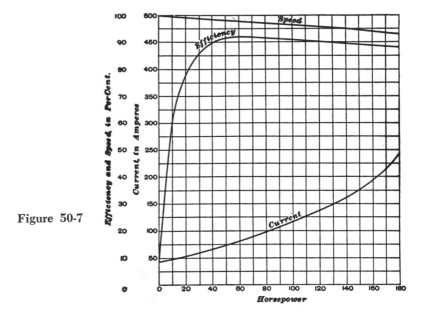

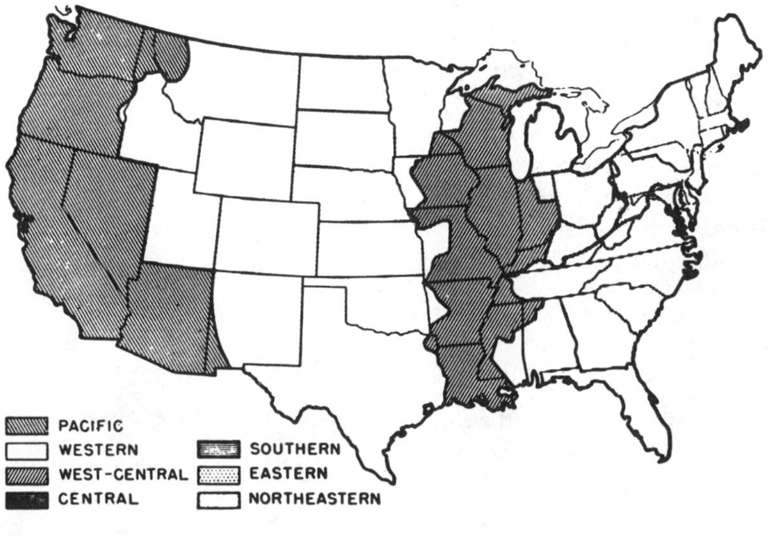

Figure 50-8

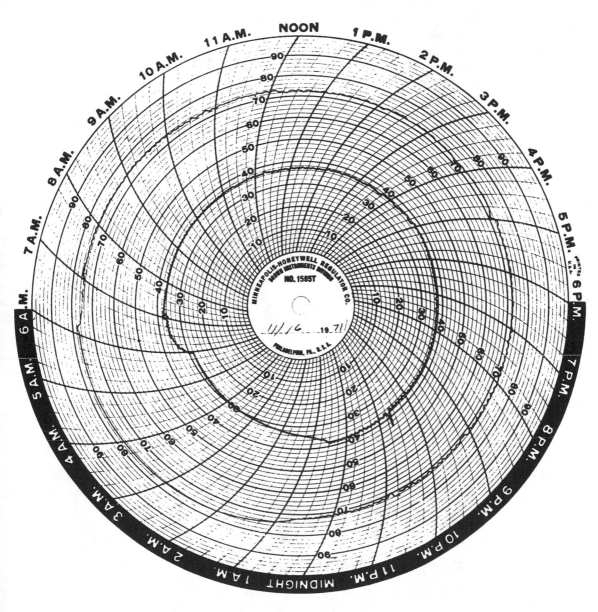

Figure 50-9

Temperature and humidity charts can be of great assistance in recording the temperature and humidity from a recording unit for all hours of the day (Figure 50-9) or week (Figure 50-10).

Many types of graph papers are available in standard stock or they can be printed to specifications.

Some useful charts and graphs to the building superintendent or plant engineer are:

Income   Number of employees
Expenses   Truck costs
Rent   Production records
Fuel   Repair costs
Electric   Attendance records
Wages   Safety records

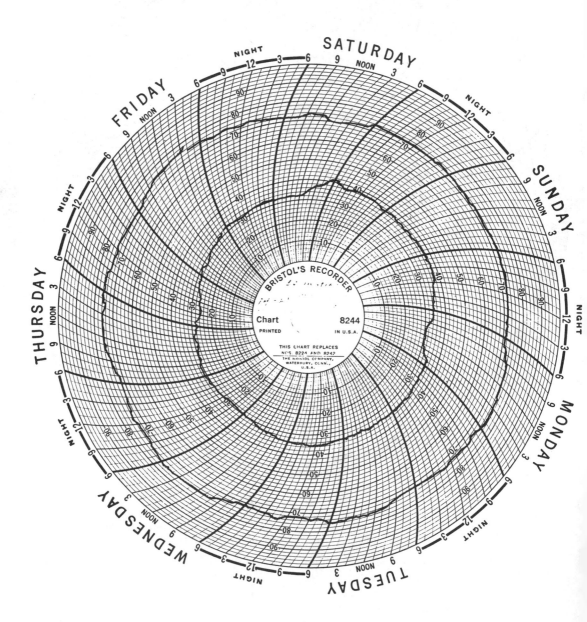

Figure 50-10

51 / EMERGENCIES AND DISASTERS

(*Sources:* Office of Civil and Defense Mobilization, Washington, D. C.; "Pre-Planning for Industrial Emergencies and Disasters" by John E. Lockhart and James A. Davis, *Plant Engineering*)

Pre-planning for emergencies and disasters is a necessity. In planning to be prepared for such occurrences, you should:

1. Insure the safety of employees.
2. Protect shareholders' investment.
3. Have direction to insure continued operation in the shortest expenditure of time.

Such occurrences may be caused by fire, power failure, explosions, collapse, flood, hurricane, tornado, freezeup, earthquake, contamination, or enemy attack. In planning for your particular situation, you should first list the emergencies which strike your building. This list may be influenced by your type of construction and geographic location. A review of such emergencies will reveal if your own or local resources are fully available and reliable. A review should be made annually. Persons responsible for drafting the plans should be appointed by top management, and a permanent committee formed. A good emergency plan will:

1. Anticipate probably hazardous situations.
2. Provide for speedy restoration of operation.
3. Insure protection for the employees.

An inspection of facilities should be made by the Emergency and Disaster committee to determine if personnel are obeying all underwriters' recommendations, if prevention methods are being adhered to, if personnel understand how to evacuate the building, isolate and secure equipment, and know their function in time of an emergency. When an emergency occurs with gas, electric, or water, all employees should know what to do immediately. Who should be notified, and how. Someone should be designated to assume immediate responsibility of setting the program in motion.

A thorough investigation should be made of insurance coverage, to determine if it is adequate, and that top management understands exactly what is covered. Recommendations should be made in writing if coverage is inadequate.

Committee Planning Considerations

1. Proper assessment of damage.
2. Resources on hand:
   a. personnel
   b. tools
   c. parts and materials

3. Outside Resources:
   a. utilities
   b. municipal
   c. contractors
   d. hospitals
   e. civil defense
   f. military

4. Plant security during emergency.

5. Information. A central source of information should handle all releases and interviews.

6. Welfare. A person should be designated to be responsible for coordinating food, shelter, first aid, and for securing aid from civic and governmental organizations, if required.

7. A financial advisor should be available for consultation of the emergency chief, or others.

8. Legal advisor.

9. A mutual aid organization can be formed with local companies, or local governments. An inventory of resources and facilities should be made.

10. Planning by top management for extreme situations—desirability of rebuilding, expanding, abandoning, financing.

11. Training of employees.

12. After the plan has been completed, have it typed for future reference and revision.

Emergency Lighting

Emergency lighting should be provided in key areas in order to permit orderly continued operation, or shutdown, if the main source of light power is put out of service. Such emergency lighting should be provided in control rooms, turbine rooms, boiler rooms, exit stairs, and passageways. This power supply should be completely independent of the normal source. This can be done through standby generators or batteries. The emergency luminaires themselves are the normal light units transferred to the new source of electricity as the occasion demands. The luminaires can be of low wattage, which will aid in giving greater area coverage (at low brightness level) with minimum power.

Boiler and Air Conditioning Emergencies

Consideration should be given to restoring a boiler plant or central air conditioning plant in emergencies. Plans should be made to determine if:

1. Any coverage can be secured from neighboring plants, or buildings, and the time required to adapt the system.

2. There are any units available locally, which can be used temporarily.

3. The system must be supplied with the same unit, and if it must come from the manufacturer, what is the availability of the equipment, and shipping time?

The use of mobile steam generating equipment in emergencies has proved very successful. Boilers are mounted on trucks, and can be quickly adapted to meeting an emergency. The mobile steam units are complete steam plants. It is generally only necessary to connect power and water lines to start producing steam. They carry their own water softening equipment, and usually a full day's supply of fuel oil. Sizes of the steam generating units vary from 15 to 100 hp mounted on trucks. Some trucks also carry motor generator sets. Investigation should be made to determine the nearest mobile unit location, to service your plant or building in times of emergency.

Fallout Shelter

Fallout shelter is needed everywhere. One thing is certain, if this country is attacked with nuclear weapons, our air and missile bases will be primary targets. The enemy would try to knock out our retaliatory power. He may try to destroy our cities. No one is sure how far he would go. However, it must be emphasized, that even if an enemy confines his attack to our retaliatory bases, the radioactive fallout from his nuclear bombs would threaten life in the entire country. An atomic burst on the ground sends up a mushroom cloud from which radioactive dust will fall hundreds of miles away. Fallout from one test explosion spread over 7,000 square miles of the Pacific Ocean.

FIVE STEPS TO SAFETY

Learn

1. Warning signals and what they mean.
2. Your community plan for emergency action.
3. Protection from radioactive fallout.
4. First aid and emergency preparedness.
5. Use of CONELRAD—640 or 1240 for official directions.

The safety of employees should be planned, in the event of an emergency attack. If safe below ground shelter can be provided, the route of evacuation should be properly posted

Shelters. You can protect yourself from fallout radiation. Any mass of material between you and the fallout will cut down the amount of radiation that reaches you. Sufficient mass will make you safe. Concrete or bricks, earth or sand, are some of the materials heavy enough to afford protection by absorbing radiation. There is about the same amount of shielding in eight inches of concrete for instance, as in twelve inches of earth, sixteen inches of books, or thirty inches of wood. In most of the country, except in areas hit by exceptionally heavy fallout, these thicknesses would give ample protection.

Buildings (plants, offices, schools, apartments) of masonry or concrete, provide better natural shelter than the usual family dwellings. In general, such buildings afford more protection than smaller buildings because their walls are thick and there is more space. The central area of the ground floor of a heavily constructed apartment building, with concrete floors, should provide more fallout protection than the ordinary basement of a family dwelling. The basement of such a building may provide as much natural protection as a specially constructed concrete block shelter recommended for the basement of a family dwelling.

The Federal Government will aid local governments to survey residential, commercial, and industrial buildings, to determine what fallout protection they would provide, and for how many people.

The problem for multi-story plants, office, or apartment buildings, is primarily to plan the use of existing space. Such planning will require the cooperation of other occupants and of the plant engineer, building superintendent, or apartment manager. The space available should be identified and assigned to those who are to use it. The plan will work more smoothly if it is rehearsed. The owner of the building may find it necessary to modify the basement ventilation, water supply, and sanitation system. Basic supplies should be stored in the basement to insure ample preparedness.

Basement shelters are one of the most common, least expensive, and most substantial. The best protection is an underground shelter with at least three feet of earth or sand above it. Two feet of concrete will give the same protection. If the shelter has an adequate door and air filter, it will give almost complete protection. Large buildings offer good protection, as their masonry or concrete construction generally makes it harder for radiation to get through. Basements, inside rooms, or corridors are safest.

The radioactivity of fallout decays rapidly at first. Forty-nine hours after an atomic burst, the radiation intensity is only about one per cent of what it was an hour after the explosion. However, the radiation may be so intense at the start, that one per cent may be extremely dangerous.

Radio reception is cut down by the shielding necessary to keep out radiation. A radio check should be made from the shelter. It probably will be necessary to install an outside antenna, particularly to receive CONELRAD broadcasts.

Lighting is an important consideration. Continuous low level lighting may be provided in the shelter by means of a four cell hot shot battery to which is wired a 150-milliampere flashlight type bulb. Tests have shown that such a device, with a fresh battery, will furnish light continually for at least ten days. With a spare battery, a source of light for two weeks or more would be assured. A flashlight or electric lantern should also be available for those periods, when a brighter light is needed. There should be a regular electrical outlet in the shelter, as power may continue in some areas.

Windows. If there are outside windows, they should be shielded or blocked during an emergency.

Sanitation supplies:

    Cans for garbage.

    Covered pail for toilet purposes.

    Can for human waste.

    Toilet tissue, paper towels, sanitary napkins, ordinary and waterless soap.

If there are children in the building, disposable diapers may be necessary.

Grocery bags, newspapers, and plastic bags.

Household chlorine.

Waterproof gloves.

Water and Food:

You should know:

Where to find safe water.

How to turn off water service valve.

How to purify water.

What foods to store and how to prepare them.

How to dispose of garbage.

How to dispose of human waste.

You should have:

Stored water or other liquid.

Supply of canned food.

Cooking and eating utensils.

Special foods for babies.

Wrench, screw driver, shovel, tool kit, radiation meter, fire fighting equipment.

First aid kit, and other medical supplies.

A complete analysis of supplying temporary water supply should be made. This should include a check with the local water company of other mains available, neighboring plants aiding in supply, or other sources of supply which may be located on your property or neighboring properties. (See Figure 51-1.)

Figure 51-1

SUPPLIES FOR SHELTERS

Assemble supplies, wrap them in a moisture-proof covering, and place in an easily carried box. Paste this sheet to the box cover and place the box in your shelter area.

For These Purposes	Use These	Or These	Suggested Quantity
For open wounds, scratches, and cuts. Not for burns.	1. Antiseptic solution: Benzalkonium Chloride Solution, U.S. P., 1 to 1,000 parts of water.	Quaternary ammonium compounds in water. Sold under trade names as Zephiran, Phemerol, Ceepryn, and Bactine.	3–6 oz. bottles.
For faintness, adult dose ½ teaspoon in cup of water; children 5 to 10 drops in ½ glass of water. As smelling salts, remove stopper, hold bottle under nose.	2. Aromatic spirits of ammonia.		4–2-oz. bottles.
For shock—dissolve 1 teaspoonful salt and ½ teaspoonful baking soda in 1 quart water. Have patient drink as much as he will. Don't give to unconscious person or semiconscious person. If using substitutes dissolve six 10-gr. sodium chloride tablets and six 5-gr. so-	3. Table salt.	Sodium chloride tablets, 10 gr., 50 tablets in bottle.	1 box.
	4. Baking soda.	Sodium bicarbonate or sodium citrate tablets, 5 gr., 50 tablets in bottle.	3–10 oz. boxes.

SUPPLIES FOR SHELTERS (*Continued*)

For These Purposes	Use These	Or These	Suggested Quantity
dium bicarbonate (or sodium citrate) tablets in 1 qt. water.			
For a sling; as a cover; for a dressing.	5. Triangular bandage, folded 37 by 37 by 52 in., with 8 safety pins.	Muslin or other strong material. Cut to exact dimensions. Fold and wrap each bandage and 2 safety pins separately in paper.	16 Bandages.
For open wounds or for dry dressings for burns. These are packaged sterile.	6. Eight medium first aid dressings, folded sterile with gauze enclosed cotton pads, 8 in. by 7½ in. Packaged with muslin bandage and 4 safety pins.	a) Eight emergency dressings 8 in. by 7½ in. in glassine bags, sterilized. One roller bandage, 2 in. by 10 yds. b) Four large sanitary napkins, wrapped separately and sterilized. One roller bandage, 2 in. by 10 yds.	As indicated.
For open wounds or for dry dressings for burns. These are packaged sterile.	7. Eight small first aid dressings, folded, sterile with gauze enclosed cotton pads and gauze bandage, 4 in. by 7 in.	2 doz. sterile gauze pads in individual packages, 3 in. by 3 in. One roller bandage, 1 in. by 10 yds.	As indicated.
For eyes irritated by dust, smoke, or fumes. Use 2 drops in each eye. Apply cold compresses every 20 minutes if possible.	8. Eye drops.	Bland eye drops sold by druggists under various trade names.	4–10 oz. bottles with dropper.
For splinting broken fingers or other small bones and for stirring solutions.	9. 2 doz. tongue blades, wooden.	Shingles, pieces of orange crate, or other light wood, cut to approximately 1½ in. by 6 in.	As indicated.
For purifying water when it cannot be boiled. (Radioactive contamination cannot be neutralized or removed by boiling or by disinfectants.)	10. Water purification tablets, Iodine (trade names—Globaline, Bursoline, Potable Aqua) Chlorine (trade name—Halazone).	Tincture of iodine or iodine solution (3 drops per quart of water) Household bleach (approx. 5% available chlorine) 3 drops per quart of water.	Tablets—Bottle of 200 or 500. Liquid—Three small bottles.
For bandages or dressings: Old soft towels and sheets are best. Cut in sizes necessary to cover wounds. Towels are burn dressings. Place over burns and fasten with triangular bandage or strips of sheet.	11. Large bath towels. 12. Small bath towels. 13. Bed sheet.		8 8 4

For These Purposes	Use These	Or These	Suggested Quantity
Towels and sheets should be laundered, ironed, and packaged in heavy paper. Relaunder every 3 months.			
For administering stimulants and liquids.	14. Paper drinking cups.		100-200
Electric lights may go out. Wrap batteries separately in moisture-proof covering. Don't keep in flashlight.	15. Flashlight.		4
	16. Flashlight batteries.		12
For holding bandages in place.	17. Safety pins, 1½ in. long.		4 Doz.
For cutting bandages and dressings, or for removing clothing from injured body surface.	18. Razor blades, single edge.	Sharp knife or scissors.	12
For cleansing skin.	19. Toilet soap.	Any mild soap.	1 Doz.
For measuring or stirring solutions.	20. Measuring spoons.	Inexpensive plastic or metal.	3 sets
For splinting broken arms or legs.	21. 3 doz. splints, plastic or wooden, ⅛ to 1¼ in. thick, 3½ in. wide by 12 to 15 in. long.	A 40-page newspaper folded to dimensions, pieces of orange crate sidings, or shingles cut to size.	As indicated.

First Aid

First aid training should be given to occupants of the building or plant. Six persons in every hundred should be formally trained. The American National Red Cross has a Standard First Aid Course in which they will instruct your personnel. Emergency portable first aid kits should be kept, and their responsibility assigned to persons for resupply and delivery to the scene of an emergency or disaster. Six emergency actions to save lives until medical help arrives are:

1. Keep the injured person lying down, with his head level with the rest of his body unless he has a head injury. In that case, raise his head slightly. Cover him and keep him warm.

2. Do not move the injured person except to remove him from fire, flood, smoke, or anything that would further endanger his life.

3. Examine the injured person to determine whether emergency action is necessary. If he is not in danger of bleeding to death, or is not in shock, it is better for the untrained person to leave him alone.

4. Do not give an unconscious or semiconscious person anything to drink.

5. Do not let an injured person see his wounds.

6. Reassure him and keep him comfortable.

For bleeding. Take this emergency action. Apply pressure directly over the wound. Use a first aid dressing, clean cloth, or even the bare hand. When bleeding has been controlled, add extra layers of cloth and bandage firmly. Do not remove the dressing. If the wound is in an arm or leg, elevate it with pillows or substitutes. Do not use a tourniquet except as a last resort.

For burns. Remove clothing covering the burn unless it sticks. Cover the burned area with a clean dry dressing or several layers of cloth folded into a pad. Apply a bandage over the pad, tightly enough to keep out the air. Do not remove the pad. *Do not use grease, oil, or any ointment* except on a doctor's order. On *chemical burns,* such as caused by acid or lye, wash the burn thoroughly with water before covering with a dry dressing.

For broken bones. Unless it is absolutely necessary to move a person with a broken bone, do not do anything except apply an ice bag to the injured area to relieve pain. If you must move him, splint the broken bone first so the broken bone ends cannot move. Use a board, thick bundle of newspapers, even a pillow. Tie the splint firmly in place above and below the break, but not tightly enough to cut off circulation. Use layers of cloth or newspapers to pad a hard splint.

Broken bones in the hand, arm, or shoulder should be supported by a sling after splinting. Use a triangular bandage or a substitute such as a scarf, towel, or torn width of sheet and tie the ends around the casualty's neck. Or place his forearm across his chest and pin his sleeve to his coat. In this way the lower sleeve will take the weight of the injured arm.

If you suspect a broken neck or back do not move the casualty except to remove him from further danger that may take his life. If you must move the casualty, slide him gently onto a litter or a wide, rigid board. Then leave him alone until trained help arrives.

If a bone has punctured the skin, cover the wound with a first aid dressing or clean cloth and control bleeding by hand pressure.

For shock. Shock may result from severe burns, broken bones, or other wounds, or from some acute emotional disturbance. Usually the person going into shock becomes pale. His skin may be cold and moist. His pulse may be rapid. He may become wet with sweat. He may become unconscious.

Keep the casualty lying down. His head should be level or lower than his body unless he has a head injury. In the latter case his head should be raised slightly. Wrap the casualty warmly but do not permit him to become overheated. Try to avoid letting him see his injury. If he is able to swallow, give him plenty of water to drink, with salt and baking soda added. Mix one teaspoonful of salt and one-half teaspoonful of baking soda to one quart of water. This will help to prevent severe shock.

Do not give anything by mouth to a person who is vomiting, is unconscious, or semiconscious, or has an abdominal wound.

For suffocation. Suffocation can result from pressure on the neck or chest, contact with a live electric wire, drowning, or breathing-in foreign substances such as liquids, smoke, or gas. The usual signs of suffocation are coughing and sputtering or other difficulty in breathing. As breathing becomes difficult or stops, the face may turn purple and lips and fingernails become blue. Unconsciousness will follow quickly unless you act at once.

First, remove the person from the cause of suffocation. If he is in contact with a live wire, do not touch him. Shut off the current if you can. If not, stand on a piece of dry wood or on paper, and remove the wire from the person with a long dry stick or other nonmetallic object.

If the person is in a room filled with gas, smoke, or water, get him out quickly. Remove any objects from his mouth or throat that may obstruct breathing. Then apply artificial respiration immediately, as follows:

Artificial respiration. *Mouth-to-mouth or mouth-to-nose method.* Tilt the head back so the chin is pointing upward, and pull or push the jaw into a jutting-out position. (These maneuvers should relieve obstruction of the airway by moving the base of the tongue away from the back of the throat.)

Open your mouth wide and place it tightly over the casualty's mouth. At the same time pinch the casualty's nostrils shut or close the nostrils with your cheek. Or close the casualty's mouth and place your mouth over the nose. Blow into his mouth or nose. (Air may be blown through the casualty's teeth, even though they may be clenched.) The first blowing efforts should determine whether or not obstruction exists.

Remove your mouth, turn your head to the side, and listen for the return rush of air that indicates air-exchange. Repeat the blowing effort. For an adult, blow vigorously at the rate of twelve breaths per minute. For a child, take relatively shallow breaths appropriate for the child's size, at the rate of about twenty per minute.

If you are not getting air-exchange, recheck the head and jaw position. If you still do not get air-exchange, quickly turn the casualty on his side and administer several sharp blows between the shoulder blades in the hope of dislodging foreign matter. Again sweep your fingers through the casualty's mouth to remove any foreign matter.

Those who do not wish to come in contact with the person may hold a cloth over the casualty's mouth or nose and breathe through it. The cloth does not greatly affect the exchange of air.

Mouth-to-mouth technique for infants and small children. If foreign matter is visible in the mouth, wipe it out quickly with your fingers or a cloth wrapped around your fingers.

Place the child on his back and use the fingers of both hands to lift the lower jaw from beneath and behind, so that it juts out.

Place your mouth over the child's mouth and nose, making a relatively leakproof seal and breathe into the child, using shallow puffs of air. The breathing rate should be about twenty per minute.

If you meet resistance in your blowing efforts, recheck the position of the jaw. If the air passages are still blocked, the child should be suspended momentarily by the ankles or inverted over one arm and given two or three sharp pats between the shoulder blades, in the hope of dislodging obstructing matter.

Other manual methods of artificial respiration. Persons who cannot, or will not, use the mouth-to-mouth or mouth-to-nose method of artificial respiration should use another manual method. The nature of the injury in any given case may prevent the use of one method, while favoring another. Other methods suggested for use by the American Red Cross are: "The Chest Pressure-Arm Lift Method" (Holger-Nielsen).

When performing any method of artificial respiration, remember to time your efforts to coincide with the casualty's first attempt to breathe for himself.

Be sure that the air passages are clear of all obstructions, that the casualty is positioned in a manner that will keep the air passages clear, and that air is forced into the lungs as soon as possible.

If vomiting occurs, quickly turn the casualty on his side, wipe out his mouth, and reposition him.

When the casualty is revived, keep him as quiet as possible until he is breathing regularly. Loosen his clothing, cover him to keep him warm, then treat him for shock.

Whatever method of artificial respiration you use, it should be continued until the casualty begins to breathe for himself, or until there is no doubt that the person is dead.

To move injured person. Do not move an injured person except to prevent further injury or possible death. If you must move him, keep him lying down flat. Move him on a wide board, such as an ironing board or door, and tie him to it so he will not roll off.

If you have nothing to carry him on, get two other persons to help carry the casualty. You must kneel together on the same side of the casualty and slide your hands under him gently; then lift him carefully, keeping his body level. Walk in step to prevent jarring, and carry him only far enough to remove him from danger.

Reference Material

Medical publications. Any one of the following publications from a list provided by the Bureau of Health Education of the American Medical Association would be helpful when a physician is not available.

What to Do Until the Doctor Comes
> Wilham Bolton, M.D., 145 pp., (Reilly & Lee Co., 325 West Huron St., Chicago, Illinois)

Book of Health "A Medical Encyclopedia for Everyone"
> Randolph Lee Clark, M.D., 768 pp., 1400 illustrations, Elsevier Press Inc., 402 Lovett Blvd., Houston 6, Texas)

Ship's Medicine Chest and First Aid at Sea
> U.S. Public Health Service and War Shipping Administration, 498 pp., illustrated, (U.S. Government Printing Office, Superintendent of Documents, Washington 25, D.C.)

Office of Civil and Defense Mobilization (OCDM) publications. The following OCDM publications can be requested through your local civil defense director, or purchased from the Superintendent of Documents, U.S. Government Printing Office, Washington 25, D.C. at nominal cost:

Civil Defense Preparedness, G.P.O. 825110

Fire Fighting for Householders, PA-B-4, revised May 1958, reprinted June 1959

First Aid: Emergency Kit, Emergency Action, L-2-12, revised April 1958, reprinted June 1959

Handbook for Emergencies, 1958

What do Do Now About Emergency Sanitation at Home, H-11-1, revised August 1958, reprinted December 1958

What You Should Know About Radioactive Fallout, PA-B-7, revised May 1958

Facts About Fallout Protection, L-18

OCDM films. The following films are available through state civil defense offices or regional headquarters of the Office of Civil and Defense Mobilization:

Bombproof
Crisis
Day Called "X", The
House in the Middle, The
New Family in Town
Operation Ivy
Rehearsal for Disaster
Time of Disaster
To Live Tomorrow

Red Cross Films. The following first aid films can be obtained through your local Red Cross chapter:

Easy Does It
First Aid
First Aid for Burns in Civil Defense

Educational material can be secured and passed out to all employees, especially a small wallet sized card, "Civil Defense Preparedness" G.P.O.* 825110, and training programs should be conducted to test your program, and coordination of employees and building occupants.

* Government Printing Office.

52 / FIRE EXTINGUISHING EQUIPMENT

(Publication of the Post Office Department, Washington, D.C. 20260 January, 1967)

What You Should Know About Fires and Fire Extinguishers

A. Kinds of Fires

To know the type of fire extinguishing equipment to use for a fire you must understand something about the three main kinds of fires. Fires are usually divided into three classes as follows:

1. *Class A Fires*—These are fires that start in ordinary material that burns easily such as paper, wood, cloth, and rubbish. They can usually be stopped with water or with mixtures containing large amounts of water.
2. *Class B Fires*—These are fires involving grease or flammable liquids such as gasoline, oil, or kerosene, which burn rapidly. These fires may start in garages, paint shops, spray paint booths, lubrication racks, gasoline pumps, or building maintenance shops. A smothering or blanketing effect is needed to stop them.
3. *Class C Fires*—These are fires that start in electrical equipment, such as battery chargers, air-conditioning compressors, motors, transformers, generators, and switchboards. Use of water or a chemical which carries an electrical charge may result in death or injury to persons trying to stop the fire. Therefore, it is of great importance that the extinguishing material used will not carry an electrical charge from the equipment when fighting Class C fires.

B. **Using the Different Types of Extinguishers**

1. *Why You Need to Know About Different Extinguishers*—There are many types of fire extinguishers, using different kinds of extinguishing agents. It is imperative for your personal safety, as well as for quick extinguishment of the fire, that you know which type extinguisher to use on each class of fire.
2. *Water-Type Extinguishers*
   a. When Used–Water-type extinguishers (soda acid, stored pressure, pressure cartridge) are used only for Class A (wood, paper, etc.) fires. Water does not give the smothering action needed for Class B (oil) fires; in some cases water will spread rather than stop oil fires. Do not use water for Class C (electrical) fires unless all current in the area is cut off, since use of water on electrical fires may result in serious injury to persons trying to stop the fire.
   b. How Used–Water-type extinguishers depend on pressure to force a stream of water on the burning area. The pressure comes from hand pumping, stored pressure, the mixing of soda and acid, or by piercing a pressure cartridge. For best results, direct the stream of water at the base of the fire.
3. *Foam-Type Extinguishers*—Foam-type extinguishers release a mass of carbon dioxide bubbles which smother a fire. These are very good for Class B (oil) fires and fires involving flammable liquids other than alcohol, acetone, ether, and carbon disulphide. In an emergency they can be also used for Class A fires (wood, paper, etc.). Do not use for Class C (electrical) fires unless all current is shut off in the area since the liquid in the foam solution conducts electricity easily which could result in serious injury to user.
4. *Conventional Dry Chemical Type Extinguishers*—The conventional dry

chemical extinguisher releases a stream of bicarbonate of soda which smothers a fire. Dry chemical extinguishers work very well on Class B (oil) and C (electrical) fires. They are not recommended for Class A (wood, paper) fires. The pressure needed comes from piercing a pressure cartridge in the unit.

5. *Nonconducting Liquid (Vaporizing) Type Extinguishers*—These extinguishers use chemicals (usually carbon tetrachloride) which do not conduct electricity. They place a smothering vapor blanket over the burning area. Use of this type of extinguisher is limited to Class C (electrical) fires. *The vapor created is toxic and great care should be taken to avoid extensive breathing of the fumes.* The unit ordinarily depends on a hand pumping action to create the vapor stream.

6. *Carbon Dioxide Type Extinguishers*—These extinguishers release carbon dioxide in the form of a very cold snow which blankets the fire. These extinguishers are very good for Class C (electrical) fires and can also be used for Class B (oil) fires. They are not effective for Class A (wood, paper) fires.

7. *Multipurpose Dry Chemical Type Extinguishers*—These extinguishers release a stream of ammonium phosphate which smothers a fire. This type extinguisher may be used on all classes of fires. One disadvantage of the multipurpose chemical is that it is deliquescent (attracting moisture from the air) and, after discharge, forms a sticky residue which can necessitate a complete overhaul of any electric motors it contacts. The obvious safety aspects which this type extinguisher provides for personnel and the facility itself, however, make this a most desirable weapon for fighting fire and far outweigh its single disadvantage.

How to Order Fire Extinguishers

A. **Types Available**

1. *Two Standard Types*—Two types of portable fire extinguishers considered standard are:
   a. Type I–For Class A (wood, paper), Class B (oil), and Class C (electrical) Fire Areas.
   A 10-pound multipurpose dry chemical type (cartridge-operated) extinguisher.
   b. Type II–For Class B (oil) and Class C (electrical) Fire Areas.
   A 10-pound carbon dioxide type extinguisher. This extinguisher is to be used only in mechanical equipment rooms of large facilities.

2. *Nonstandard Types*—Extinguishers other than the two standard types are not recommended unless evaluated by a qualified safety or fire officer.

B. **How to Determine Your Needs**

Determine need and location of extinguishers as follows:

1. *For Class A (wood, paper) Fire Areas*—There are areas in which Class A fires are most likely to occur. There should be one extinguisher for each 5,000 square feet of floor space, or fraction thereof, located so as not to require a travel distance of more than 50 feet to the extinguisher when needed.

2. *For Class B (oil) and C (electrical) Fire Areas*—There are areas in which Class B or Class C fires may occur. There should be one extinguisher for each 2,500 square feet, or fraction thereof, located so as not to require a travel distance of more than 50 feet to the extinguisher when needed.
   Note: Lighting equipment, electrically operated office machines, and similar equipment are not to be considered as creating special hazards.

3. *Replacing Present Extinguishers*—Present extinguishers in good working condition are not to be replaced.

How to Install Fire Extinguishers for Best Use

A. Placing the Extinguishers

1. Fire extinguishers shall be mounted so that the top of the extinguisher is not more than 74 inches from the floor.
2. Extinguishers shall be placed where they will be accessible and visible from several different directions.
3. Access to the extinguisher shall be kept open at all times. The area below the equipment, as well as the area needed for approach to the extinguisher, must remain open and free. Outline, by a painted red line on the floor, the area to be kept open.
4. Water- or foam-type extinguishers shall not be located where they are subject to near-freezing temperatures (40°).
5. Nonconducting liquid (vaporizing) and carbon dioxide units should not be ordinarily exposed to 120° or over temperatures.

B. Making the Extinguishers Easy to Find

1. *Wall-Hung Extinguishers*—To make the locations of extinguishers hung on walls immediately apparent, paint a solid red rectangle extending 6 inches beyond each side of the extinguisher. Use a safety red paint. Where the location of the wall-hung extinguisher is partially hidden by columns or other obstructions, paint an additional red stripe, not less than 12 inches high and the same width as the extinguisher rectangle, at a height of not less than 12 feet from the floor, if such a spot is clearly visible and available. Otherwise, paint the additional red stripe at ceiling height.
2. *Extinguishers Hung on Columns*—Where the extinguisher is placed on a column, paint a solid red band circling the column. The band will extend 6 inches above and below the extinguisher. Where the location of the extinguisher is on a column, and is partially hidden by another column, the column involved must be further marked with a red band not less than 12 inches high at a height of not less than 12 feet from the floor, if such spot is clearly visible and available. Otherwise, the red band will be painted at the top of the column.

3. *Using Direction Signs*—Where conditions show their need, additional extinguisher identification in the form of directional signs should be used and appropriately located.

C. **Numbering and Tagging the Extinguishers**

1. Number serially all extinguishers at each installation to provide accurate servicing and inventory records. In multifloor structures the extinguisher number should be prefixed by the floor designation; e.g., 1-5 would mean extinguisher No. 5 on the first floor. Similarly, 2-1 would indicate extinguisher No. 1 on the second floor. The extinguisher number should be painted on the extinguisher (possibly on the bottom) and also on the bottom of the red-painted background where the extinguisher is mounted.
2. Wire an Inspection Tag, to every extinguisher. On the tag show the location number, 1-5, 2-6, the manufacturer's serial number, the last inspection date, and the *signature* of the inspector (Not his initials).

How to Keep Fire Extinguishers Ready for Use

A. **Responsibility for Maintaining Extinguishers**

The maintenance department, or safety director, is generally responsible for keeping the fire extinguishers in good working order for annual inspections, and recharging of fire extinguishers.

B. **Monthly Inspection of all Extinguishers**

Prepare a monthly checklist for inspection of all extinguishers to determine that:

1. They are properly located and readily accessible.
2. They have not been tampered with or damaged.
3. Nozzles are not clogged.
4. They are full.

C. **General Rules for Servicing Extinguishers**

1. Extinguishers must be inspected, taken apart, and recharged only by competent, experienced personnel.
2. All extinguishers must be inspected and recharged in accordance with instructions furnished by the manufacturer of each extinguisher. When there are no instructions available, the extinguishers shall be serviced as described in the following paragraphs.
3. See section "I" for hydrostatic shell pressure testing of extinguishers.

D. **Servicing Water-Type Extinguishers**

1. *Soda-Acid and Stored-Pressure Types*

a. Instructions Common to Both—KEEP EXTINGUISHERS FULL (TO FILLING MARK) AT ALL TIMES AND RECHARGE ANNUALLY AS WELL AS IMMEDIATELY AFTER USE. When recharging extinguishers, they should be emptied by discharging the unit. Discharge the unit, where possible, before an assembly of the fire brigade, as this will provide valuable training. If the extinguisher shell shows excessive mechanical damage, replace the unit with one in good working order. Wash the interior of the extinguisher shell thoroughly, removing any accumulated sediment from the bottom of the shell. Run water through the hose.

b. Soda-Acid Water-Type Extinguishers

(1) Mixture to be used–The amounts of ingredients required to re-charge a 2½ gallon soda and acid extinguisher are: 4 ounces of concentrated sulphuric acid, 1½ pounds of bicarbonate of soda, and 2½ gallons of plain water.

(2) Recharging the extinguishers

(a) Before recharging, inspect the gaskets, acid bottle, acid bottle rack, hose, hose connections, and nozzle. Replace any defective parts.

(b) Thoroughly dissolve in plain water the required amount of soda. Mix the material in a container *other* than the extingui-sher shell. Using the shell might damage its anticorrosion pro-tective lining.

(c) When the soda is dissolved completely, pour the mixture into the extinguisher shell.

(d) Fill the acid bottle to the acid mark on the bottle (4 ounces).

(e) Replace the lead stopper in the mouth of the bottle and place the bottle in its rack.

(f) Replace extinguisher cap, screwing it on until tight.

(g) Record charging date on record tag. Person responsible for su-pervision of recharging should sign tag.

(h) Refer to instructions on extinguisher shell if in doubt on any of the preceding steps.

(3) Harmful ingredients–Ingredients such as common salt, calcium chloride, and wetting agents must not be used in extinguishers of this type as they change the nature of the discharge, reduce its effectiveness, and corrode extinguishers, thereby making them dangerous to use.

(4) Supply of ingredients–Where these extinguishers are used, a quantity of chemical charges supplied by the extinguisher manu-facturer should be kept on hand so that the extinguishers may be promptly recharged after use.

(5) Semiannual inspections required–Schedule regular semiannual in-spections using the following procedures:

(a) Check record tags for recharge date.

(b) Remove extinguisher from bracket and remove cap.

(c) Note acid level in bottle.

(d) Note solution level in extinguisher.

(e) Check cap gasket.

(f) Check condition of hose.

(g) Check nozzle opening to see that it is free from obstruction.

(h) Correct defects at once.

c. Stored-Pressure Water-Type Extinguishers

(1) Recharging the extinguisher.

(a) Before recharging, inspect gaskets, hose, hose connections, nozzle springs, and pressure gage. Replace defective parts.

(b) Note pressure gage reading and, during discharge of water, note whether pressure gage reading responds to fall in contents.

(c) Completely discharge contents.

(d) Remove cap and note if extinguisher shell is clear of foreign matter.

(e) Check condition of hose, nozzle, gaskets, and gage.

(f) Correct defects at once.

(g) Fill with clear water (or approved antifreeze mixture if located where subject to freezing temperature).

(h) Restore compressed air charge to recommended reading.

(i) Place date and full name on attached inspection card.

(2) Using antifreeze solutions—Where an extinguisher is in an area exposed to freezing temperatures, an antifreeze solution must be used with the water. This solution must be approved by the manufacturers of the extinguisher.

2. *Pressure-Cartridge Water-Type Extinguishers*

a. Semiannual Inspections—Inspect pressure-cartridge water-type extinguishers semiannually. Do not discharge at inspection time unless it is part of a demonstration of fire extinguishing techniques directed by the safety officer.

Follow this procedure in semiannual inspection of this type unit:

(1) Disassemble all parts except hose and nozzle.

(2) Examine release mechanism, gaskets, seals, springs and pins, hose, and nozzle.

(3) Examine shell for damage. If it shows excessive mechanical damage, replace it.

(4) Weigh cartridge. Where the weight is ½ ounce less than that shown on the cartridge, it must be replaced. Return the used or defective cartridge to the manufacturer for a fully charged cartridge. Do not replace cartridge with any other make or type that may not fit exactly or may prevent the proper working of the extinguisher.

(5) Replace defective parts.

(6) Fill with water to proper level.

(7) Replace cartridge and reassemble.

(8) Place date and full name on attached inspection tag.

b. Supply of Replacement Cartridges–An adequate supply of replacement cartridges of proper make and type shall be kept on hand at larger installations.

c. Using Antifreeze Solutions–Where extinguisher is in an area exposed to freezing temperatures, an antifreeze solution must be used with the water. This solution must be one approved by the manufacturer of the extinguisher.

E. **Servicing Foam-Type Extinguishers**

1. *Annual Recharging Required*—Keep extinguishers full (to filling mark) at all times. Recharge annually as well as immediately after use. When foam-type extinguishers are recharged, they should be emptied by discharging the unit. The discharge of a unit, where possible, should be before an assembly of the fire brigade to provide valuable training.

2. *Recharging the Extinguisher*

a. Before recharging, inspect gaskets, inner chamber, stopper head, hose, hose connections, and nozzle. Replace defective parts. If the extinguisher shell shows excessive mechanical damage, replace it. Wash the interior of the extinguisher shell thoroughly; remove any accumulated sediment in the bottom of the shell. Run water through hose.

b. Use only approved recharge elements. (When in doubt, refer to instructions on the extinguisher and on recharge packages.)

c. Take package containing bicarbonate of soda and stabilizing agent (usually marked "B") and dissolve in *lukewarm* water. DO NOT USE HOT WATER. Use quantity of water specified on the package. Pour this solution into the outer chamber of the extinguisher.

d. Take package containing aluminum or ferric sulphate (usually marked "A") and dissolve in *hot* water. Use quantity of water specified on the package. Pour solution into the inner chamber of the extinguisher.

e. Replace the loose stopper head on inner chamber, and place the inner chamber in the outer chamber.

f. Replace extinguisher cap, screwing it on until tight.

g. If a pressure release device is provided, determine if it is serviceable.

h. Place date and full name on attached inspection tag.

3. *Semiannual Inspections*—Schedule a regular semiannual inspection, using the following procedures:

a. Check inspection tag for recharge date.

b. Remove extinguisher from bracket and remove cap.

c. Note solution level in inner and outer chamber.

d. Check cap gasket.

e. Check condition of hose.

f. Check nozzle opening to see it is not obstructed.

g. Correct defects at once.

4. *Harmful Ingredients*—Ingredients such as common salt, calcium chloride,

wetting agents, must not be used in extinguishers of this type, as they reduce the effectiveness of the discharge, change the nature of the discharge, and may corrode extinguishers making them dangerous to use.

5. *Supply of Ingredients*—Where extinguishers of this type are used, a quantity of the chemical charges supplied by the extinguisher manufacturer should be kept on hand so the extinguishers may be promptly recharged after use.

F. Servicing Dry Chemical Type Extinguishers (Including Multipurpose Type)

1. *Determining Need for Recharging*—Keep the extinguishers filled with the specified weight of dry chemicals at all times. In the case of cartridge-operated extinguishers, reweighing is the primary method of determining if the cartridge is fully charged. Some dry chemical extinguishers have a pressure gage to indicate the amount of pressure in the cartridge. Dry chemical extinguishers shall be recharged immediately after use even though they are only partly discharged.

   CAUTION: Dangerously high pressures can develop inside an extinguisher from the chemical reaction between even a trace of powder used in the conventional dry chemical extinguisher and the powder used in the multipurpose extinguisher. Use *only* the powder designed for each type extinguisher.

2. *Recharging the Extinguisher*

   a. Before recharging, complete the discharge of all material from the extinguisher. If the extinguisher shell shows excessive mechanical damage, replace it. The interior of the dry chemical extinguisher should never be washed as the moistening of the material would result in caking of the dry chemical compound and interfere with the usefulness of the unit as an extinguisher.

   b. Carefully examine gaskets, seals, hose, and nozzle. Replace defective parts.

   c. Place proper amount of dry chemical in the extinguisher.

   d. Insert a new cartridge of the right type and size manufactured especially for the extinguisher being recharged. *Note:* Weigh the cartridge to be sure it contains proper charge. The required weight appears on the cartridge.

   e. Carefully examine hose to make sure it is clear.

   f. Check all moving parts to be sure they are in good operating condition.

   g. Reassemble unit.

   h. Where extinguisher is equipped with a gage to indicate pressure existent in the cartridge, check reading on gage.

   i. The person responsible for inspection of this equipment must record charging date on inspection tag and sign his full name alongside the date.

3. *Semiannual Inspections*—Schedule regular semiannual inspections following this procedure:

    a. Weigh the cartridge or read the gage to determine if the cartridge is fully charged.

    b. Note level of compound in the cylinder and stir with a stick to be sure powder is loose and not caked.

    c. Check all moving parts to be sure they are in good working condition. Replace defective parts.

    d. Carefully examine condition of hose and nozzle. Be sure it is not blocked in any way.

    e. Check locking pin and sealing wire of cartridge.

    f. Whenever it is noted that the seal is broken or the pressure gage indicates incomplete charge, disassemble the unit and follow recharging procedure.

    g. Record inspection date, indicating whether extinguisher has been recharged and *sign* full name on inspection tag.

G. Servicing Nonconducting (Vaporizing) Liquid Type Extinguishers

1. *Keeping Filled with Liquid*—Keep extinguishers full at all times; after use they must be completely refilled. The exact name and source of the extinguishing fluid as specified by the manufacturer must be noted on the attached inspection tag. No other fluid may be used in the extinguisher. AVOID EXCESSIVE EXPOSURE TO THE TOXIC VAPOR OF THIS EXTINGUISHING LIQUID AS IT CAN BE INJURIOUS.

2. *Complete Recharging*

    a. When entire recharging is required, dismantle unit and thoroughly inspect gaskets, hose, nozzle, all moving parts, and shell. Replace any defective parts. If the extinguisher shell shows excessive mechanical damage, replace it.

    b. When recharging a vaporizing unit, keep in mind that this fluid does not deteriorate with age and any remaining liquid in the unit can be used. After examining, as set out in the preceding paragraph, refill the unit with approved liquid in accordance with instructions on the label affixed to the extinguisher. Place date and full name on the attached inspection tag.

3. *Required Inspections*

    a. Schedule inspections of vaporizing liquid extinguishers for every third month to see if they operate properly and are filled with approved liquid. Remove extinguisher from the bracket and pump a small amount of the fluid alternately upward and downward. Add amount of proper fluid necessary to fill completely.

    b. Record date and sign full name on attached inspection tag before restoring unit to bracket.

4. *Supply of Required Liquid*—Where extinguishers of this type are used, a

quantity of the special fire extinguishing liquid supplied by the manufacturer should be kept on hand so that extinguishers may be promptly recharged after use.

H.   Servicing Carbon-Dioxide Type Extinguishers

1. *Keeping Extinguisher Filled with Chemical*—Keep extinguishers full at all times. Reweighing is the only method of determining if the extinguisher is fully charged. These extinguishers shall be recharged immediately after use even though only partly discharged.

2. *Recharging the Extinguisher*—All recharging of carbon-dioxide extinguishers is generally performed by a local commercial recharging service, as special equipment is needed to recharge this type of unit.

3. *Scheduled Inspections*—Schedule at least two inspections each year using the following procedures:

    a. Weigh extinguishers to determine contents. The weight of the extinguisher is stamped on the valve. Follow the manufacturer's instructions for determining the correct weight as stamped on the cylinder band. If the carbon-dioxide charge is 10 per cent or more under rated capacity, it should be recharged at once.

    b. Replace defective parts and seal, if broken.

    c. Check condition of hose and nozzle.

    d. Check locking pin and sealing wire. If seal is broken, weigh extinguisher.

    e. Place date and sign full signature on attached inspection tag.

I.   Hydrostatic Shell Pressure Testing of Extinguishers

1. *Periodic Requirement for Testing*—The requirement for periodic hydrostatic shell pressure testing of the various types of fire extinguishers is as follows:

TYPE	INTERVAL (yrs.)
Soda-Acid	5
Stored Pressure Water	5
Pressure Cartridge Water	5
Foam	5
Dry Chemical (including multipurpose)	12
Nonconducting (vaporizing) Liquid	12
Carbon Dioxide	12

2. *Immediate Requirement for Testing*—Shell pressure test shall be conducted immediately upon indication of mechanical damage or corrosion to the extinguisher shell, without regard to the table above.

3. *Who May Conduct Shell Pressure Test*—Pressure test of fire extinguisher should be made by qualified maintenance personnel, safety director, or:

    a. The manufacturer of the extinguisher.

b. Qualified service agency.

c. Qualified testing laboratory.

(At offices having only one extinguisher, a temporary replacement should be rented.)

Fire Hose

A. *Characteristics*—Many buildings have fire hose racks. The fire hose is usually an unlined tube of linen fabric. Some leakage may occur when water is first turned on, but as the fabric becomes wet it swells and the leakage stops. Hose nozzles are without cutoff valves.

B. *How Used*—Fire hose should be used only for fighting fires and should be wetted at no other time. When a fire occurs in an area near the fire hose, the hose should be removed from the rack and laid out (without kinks) in the direction of the fire, ready for use only if the extinguisher fails to put out the fire. To start the water flow, open the hand valve connected to the standpipe.

C. *Maintenance and Inspection*

1. After Use–After the hose has become wet from use, it should be hung vertically in a well-ventilated area to assure complete drying and draining. The hose should not be returned to the rack until it is thoroughly dried.

2. Inspection

   a. At least once a year, the hose should be laid out and reracked in such a way that the folds are located in a different portion of the hose. Perform a "pull test," whereby two men pull the hose from opposite ends and then carefully examine the entire length of the hose for rot as indicated by threads having popped out. Check all fittings. If testing of standpipe valve is necessary, disconnect hose first since it is important to keep hose dry to avoid mildew and resultant deterioration.

   b. Wire an inspection tag to the hose rack and note the last reracking date and signature of the inspector. Frequent inspections of the hose and hose rack should be scheduled to make sure hose valves are completely shut off and the hose is absolutely dry.

3. Covering of Hoses–Unenclosed fire hoses should be covered with a clear polyethylene material.

Checklist

Following is a checklist of very important points:

A. All personnel responsible for maintaining fire extinguishing equipment must be thoroughly trained in their operation.

B. Follow implicitly the instructions of manufacturers of fire extinguishers in operating, charging, or maintaining extinguishers.

C. Use only dry chemical powder and cartridges furnished by the manufacturer of the unit in recharging extinguishers. Do not discard used or underweight cartridges. Return them to the manufacturer for credit on exchange or refilling.

D. Check all extinguishers monthly to determine that they are in their proper location, are readily accessible, and have not been tampered with. Examine further for deterioration, damages due to misuse, and to see that the openings of hose nozzles are not clogged. Flex the hose to check for excessive cracking or deterioration. Correct unsatisfactory conditions without delay.

E. Upon inspection, if an extinguisher or any part thereof shows evidence of deterioration or damage, and where the unit is obviously beyond repair, dispose of the damaged equipment. Where the damage seems repairable, return the equipment or parts to the manufacturer for examination and/or hydrostatic pressure tests.

F. After a fire hose has been used, carefully clean and dry it before rehanging on the hose rack.

G. If there is any question about inspection or the condition of fire extinguishing equipment, obtain assistance from local fire department personnel. Do not undertake repairs if you have any doubt as to parts or procedures involved.

H. Request local fire departments to examine outside standpipe and hose connections to insure that fire department equipment can be coupled to building system.

I. Hydrostatic pressure tests of fire extinguishers *must not* be conducted by incompetent personnel.

J. Fire inspection reports.

53 / FIRE PROTECTION

(*Sources:* "Safety Rules for Flammable Drum Storage Rooms," The Protectoseal Co., Chicago; *Buildings* magazine, *Third Annual Handbook of Building Operation*, pp. 104-105; *FIA Sentinel*; chart, "Use the Proper Extinguisher," reprinted with the permission of the Factory Insurance Association)

The object of fire protection is to safeguard life and property. The building should be of such construction that it will limit the progress and spread of any fire, with provisions made for the installations of alarms, sprinkler systems, and other fire fighting equipment.

Fire protection should be given proper consideration from the beginning of design to insure the best protection of its occupants, and possible substantial savings in insurance costs. Fire is likely to create panic. We cannot eliminate panic, however the conditions from which it is likely to arise, whether we occupy a new or old building or plant, can be avoided. Exits should be adequate, well-placed, marked properly, and built of material that will be fire resistant if fire strikes. Good fire prevention and protection facilities do not just happen. They have to be carefully planned, and this is accomplished most effectively when a new building is being planned or renovations made. Fire insurance companies generally maintain a staff to aid with fire protective problems and construction considerations.

The use of fireproof and fire resistant materials in a building prevents loss of life, property and inventories. It is instrumental in the preservation of employment. Because initial cost is sometimes prohibitive, all buildings are not made of fireproof or fire resistant materials. Insurance payments received as a result of fire losses are rarely adequate to cover the cost of new construction due to increased cost, for the loss of income incurred while a building is being reconstructed, or from loss of business resulting in diverting customers to new sources of supply in time of emergency. Capital is invested in buildings with the thought of securing a permanent and continuing income. For these reasons, fireproof and fire resistant materials must be considered in new construction, renovations, or when new equipment is purchased that requires the use of special building materials. A building may be considered fireproof or fire resistant, but its contents will generally burn.

Fires may occur in a fireproof building as a result of ignition of flammable materials which may be stored in the building. Therefore, it is imperative that these materials be stored properly, and a constant check made on their proper use. If we eliminate the *cause,* we will not be left with the *effects* of a fire.

It is important to know the various classes of fires, the type of extinguishing agent to use, and the capabilities of extinguishers. *Use the proper fire extinguisher.* (See Figure 53-1.)

The prime requisite for the protection of a building from the ravages of combustion, is a modern and efficient fire protection system. This system should incorporate fire detection devices, automatic and manual extinguishing systems, and manual fire alarm boxes.

The greatest bulwark of modern fire protection is the automatic sprinkler system. Properly installed, sprinkler systems will extinguish a fire quickly, or they will control the spread of a fire.

Types of Sprinkler Systems

Wet pipe system. The entire system is filled with water under pressure. Water is maintained to the sprinkler head, where a plug will melt at a low temperature and releases the water immediately. The wet pipe system should only be used in areas protected against freezing temperatures.

Dry pipe system. The pipe contains compressed air extending from the sprinkler head to a dry pipe valve. When the fuseable plug melts, the air pres-

USE THE PROPER EXTINGUISHER

CHART OF TYPICAL FIRE EXTINGUISHERS

TYPE OF EXTINGUISHER		Water pails (with or without cask or drum)	Pump Tank	Water Type	Soda and Acid	Anti-Freeze: Calcium Chloride	Anti-Freeze: Loaded Stream	Foam	Dry Chemical†	Liquefied Gas: Carbon Dioxide	Liquefied Gas: Freon 1301	Vaporizing Liquid††
SUITABLE FOR	CLASS A — Wood Paper Cloth	YES	YES	YES	YES	YES	YES	YES	Only with ABC powders.	No, except very small fires.	No, except very small fires.	No, except very small fires.
	CLASS B — Flammable Liquids	NO	NO	NO	NO	NO	NO	YES*	YES	YES	YES	YES
	CLASS C — Electrical Equipment	NO†	NO†	NO†	NO†	NO†	NO†	NO†	YES	YES	YES	YES
	CLASS D — Combustible Metals	NO	NO	NO	NO	NO	NO	NO	Special Powders	NO	NO	NO
Composition of extinguishing material		Water with or without calcium chloride.	Water with or without calcium chloride.	Water	Water with products of sodium bicarbonate solution and sulphuric acid reaction.	Water with dissolved calcium chloride and corrosion inhibitor.	Water with dissolved potassium carbonate and special salts.	Foam from reaction of aluminium sulphate and sodium bicarbonate solutions with foam stabilizer.	Usually powdered compounds of sodium or potassium bicarbonate, or ammonium phosphate base. Other formulations available for special hazards.	Carbon dioxide as gas and snow.	Monobromotrifluoromethane as gas.	Carbon Tetrachloride specially treated with corrosion inhibitor and freezing point depressant.
Principal extinguishing effect		Quenching	Quenching	Quenching	Quenching	Quenching	Quenching or Smothering	Smothering	Smothering	Smothering	Smothering	Smothering
Method of operating		Throw	Pump	Invert and bump on floor, or trigger type.	Invert	Invert and bump on floor, or trigger type.	Invert and bump on floor, or trigger type.	Invert	Invert and bump on floor, also plunger valve, or trigger type.	Valve, or trigger type.	Trigger type.	Pump, valve, or trigger type.
Means of expelling extinguishing agent		Throw	Hand Pump	Carbon dioxide cartridge, or stored pressure.	Chemical reaction to form carbon dioxide.	Carbon dioxide cartridge, chemical reaction, or stored pressure.	Carbon dioxide cartridge, or stored pressure.	Chemical reaction to form carbon dioxide.	Carbon dioxide cartridge or stored pressure.	Under pressure in extinguisher.	Under pressure in extinguisher plus stored pressure.	Hand pump or stored pressure.
Effective range		5-10 feet	30-40 feet	30-40 feet	30-40 feet	30-40 feet	30-40 feet	30-40 feet	5-20 feet	3-8 feet	4-6 feet	20-30 feet
Approximate duration of discharge at room temperature		Variable	5 gal.-2 minutes; 2½ gal.-1 minute	2½ gal.-1 minute	2½ gal.-1 minute	2½ gal.-1 minute	2½ gal.-1 minute	5 gal.-1 minute; 2½ gal.-1 minute	10-20 seconds	10-20 seconds	10 seconds	¾-1 minute
Approximate Weight fully charged		25-30 lbs.	2½ gal.-35 lbs.; 5 gal.-65 lbs.	2½ gal.-35 lbs.	2½ gal.-35 lbs.	2½ gal.-35 lbs.	2½ gal.-35 lbs.	2½ gal.-35 lbs.; 5 gal.-70 lbs.	12 lb.-40 lbs.; 20 lb.-55 lbs.	7½ lb.-35 lbs.; 15 lb.-50-60 lbs.; 20 lb.-58-75 lbs.	7 lbs.	1 qt.-7 lbs.; 1 gal.-25-35 lbs.; 3 gal.-75 lbs.
Requires protection from freezing		Yes, unless calcium chloride solution is used.	Yes, unless calcium chloride solution is used.	YES	YES	NO	NO	YES	NO	NO	NO	NO
Maintenance req.: Semi Annual / Annual		None Empty and refill	None Inspect, partly discharge, refill and tag.	Weigh cartridge and tag. Inspect pressure gauge.*	None Discharge, refill and tag.	Weigh cartridge, restore liquid and tag. Inspect pressure gauge.*	Weigh cartridge, restore liquid and tag. Inspect pressure gauge.*	None Discharge, refill and tag.	Weigh cartridge and tag. Inspect pressure gauge.*	Weigh and tag.	Weigh and tag.	None Inspect, partly discharge, refill and tag. Inspect pressure gauge.
Pressure Test: *5 years, **10 years, ***12 years			*	*	*	*	*	*	*	***	***	***

† Fires in electrical equipment can be safely attacked with extinguishers of this type ONLY AFTER electric current has been cut off.

* A special type foam charge should be used for "Polar Solvents" (alcohol, acetone, ketone, etc.).

‡ Effective for initial attack on fires in combustible fibers but should be followed up with fine water spray. Should not be used where open low voltage contacts are employed.

†† Extinguishers of this type should not be used in the vicinity of equipment which may be affected by the highly corrosive vapors resulting from their use.
Due to the toxic effect of gases or vapors generated by the use of this liquid, special precautions MUST BE TAKEN to avoid breathing these vapors particularly in confined or unventilated spaces.

Form No. N-90 Ed-63

CALCIUM CHLORIDE SOLUTIONS
(Calcium chloride furnished by extinguisher manufacturers contains a corrosion inhibitor)

Freezing Temperature	Water	Calcium Chloride	Degrees Baumé	Specific Gravity
10° F.	2 gals. 1 qt.	5 lbs.	17.7	1.139
Zero F.	2 gals. 1 pt.	6¼ lbs.	21.6	1.175
10° below zero F.	2 gals.	7 lbs. 6 oz.	24.7	1.205
20° below zero F.	2 gals.	8 lbs. 6 oz.	26.9	1.228
30° below zero F.	2 gals.	9 lbs. 6 oz.	28.6	1.246
40° below zero F.	2 gals.	10 lbs.	30.2	1.263

FACTORY INSURANCE ASSOCIATION
HARTFORD   CHICAGO   SAN FRANCISCO

Figure 53-1 / Typical chart of available fire extinguishers. (This chart, "Use the Proper Extinguisher," is reprinted with the permission of the Factory Insurance Association.)

sure will drop and water will flow through the sprinkler head. This system can be used in cold areas, however, the dry pipe valve itself must be located in a heated area in order to prevent freezing.

Preaction system. The pipe contains air, which may or may not be under pressure, and the water is held back by an automatic valve operated by a heat actuated device, which is generally more sensitive than the normal automatic sprinkler. These devices open the valve to allow water to flow into the pipes before the fuseable element of the sprinkler head melts. This sounds an alarm, which sometimes permits employees to extinguish the fire with portable units prior to the melting of the heads.

Deluge system. This system is used where unusual fire hazards exist and where large quantities of water are needed quickly over large areas. All heads are open, and water is held from the system by valves which open either manually or automatically by heat actuated devices.

Combined dry pipe and preaction sprinkler system. This system employs automatic sprinklers attached to a piping system containing air under pressure, with a supplemental heat responsive system of generally more sensitive characteristics than automatic sprinklers installed in the same area.

Carbon dioxide systems. Carbon dioxide, the colorless, odorless, heavier-than-air gas, has proved its value for extinguishing fires in electrical equipment, gasoline, oil, grease, and paints. Since carbon dioxide fire extinguishment leaves no residue, production can usually be resumed immediately. Fixed pipe systems, which vary in size from a few pounds of gas to many tons, provide means of rapid application of the gas to a fire.

Gas for *high pressure systems* is stored in steel cylinders of 50, 75, and 100 pounds, capacity, at approximately 850 psi at normal room temperature. Gas for low pressure systems is stored in refrigerated tanks at 0°F. and 300 psi. Tank sizes vary from one ton to one hundred and twenty-five tons.

Local application systems are designed to apply gas directly to a burning surface—the carbon dioxide extinguishing the fire by separating the flame from the surface supplying the combustible vapor, sometimes for only a few seconds. This action is commonly called "smothering the fire," and dilutes or displaces combustion air at the burning surface with the inert carbon dioxide. A *total flooding system* is designed to flood an enclosure with gas to reduce the oxygen percentage of the atmosphere below that needed to sustain combustion. *Extended discharge systems* are designed to continue gas discharge for the maintaining of a proper concentration until extinguishment is accomplished.

Carbon dioxide systems are usually automatic in operation with actuation by fixed-temperature thermostats, rate-of-rise (temperature) releases, or smoke detectors. It is always desirable to have separate reliable means of manual operation for use should the automatic system fail. Some merits of carbon dioxide systems are quick clean extinguishment on special hazard fires.

Automatic sprinklers are used to "back up" carbon dioxide systems because of the supply limitations of such systems, the more complicated release devices required, and the severity of hazards to be protected. As with any mechanical equipment a carbon dioxide system must receive periodic maintenance, preferably semiannually.

Automatic sprinkler alarms. Nearly all automatic sprinkler systems include an *alarm* which is operated by the flow of water, which operates a water alarm or sets off an electric alarm. Both systems can be connected to ring a bell, and the electric can be connected to a local fire alarm system. This type of system insures immediate notification of the local fire department.

Standpipe systems are another form of protection. This is generally a vertical pipe located in the stairtower or in some other central point in a building. Hose is connected on a swinging rack or reel to be used in case of fire. A control valve is located at each hose station and should be checked once a week. The water should be tested through the hose once every six months.

In large office buildings, plants, schools, or stores, this testing may be more convenient during the evening hours. In high buildings, the city water pressure is not always sufficient to supply water to upper stories and auxiliary pumps or tanks are often required to make the standpipe system operative for these stories. Pumps should be checked once every six months, and also placed on a periodic maintenance schedule.

Manual fire alarm boxes should be installed in prominent locations. They should have distinct instructions on them. An adequate number of fire extinguishers for given areas should be visible to the eye.

Sprinkler heads will accumulate dust. Generally, cleaning once each year will be sufficient. Extreme care must be taken not to trip one of the heads. A competent employee should be selected to accomplish this job. A small portable vacuum, or a small paint pail and a new paint brush may be used to remove the dust.

Auxiliary battery systems should be installed to set off fire alarms in the event of an electric power failure. These batteries should be tested once a month, and water added if required.

To insure that a fire protection system is completely operative, a weekly inspection of pressures, alarms, position of all valves, condition of equipment, and the accessibility of all valves and equipment should be made weekly.

All *hydrants* should be pressure tested once per year, along with the hoses. The drip valve is an important part of any hydrant. In most hydrants, the drip valve is designed to drain the entire barrel down to the level of the valve seat. This valve is generally a solid brass casting, leather faced, and works as a slide valve. This drip arrangement closes tight generally on the second turn in opening. It is not necessary to open the hydrant wide open to close the drip. In areas where a hydrant cannot drain automatically, the drip valve opening can be plugged in factory assembly at request. It is then necessary to pump the water from the hydrant barrel in freezing temperatures. Be sure to drain the hydrants every fall to prevent freezing and breaking. Antifreeze is sometimes added to the hydrant barrel during the fall, after draining or pumping, to insure protection.

Fire Prevention Procedure

Fire prevention is a year-round activity and a responsibility of the building superintendent, plant engineer, or safety chief. A fire committee should be formed and must have the support of top management. The committee should be headed by one or two fire marshals. Zone marshals should be appointed for each zone into which the building should be divided. A sufficient number of wardens should be appointed in each fire zone to assist the zone marshals.

A weekly fire inspection report should be made by the building superintendent or plant engineer when the inspection is made. Conditions should be noted and corrective action taken if required.

WEEKLY INSPECTION OF FIRE PROTECTION EQUIPMENT

Date ....

No.			Remarks
1	Two outside sprinkler valves separating direction of flow around building.	Open Shut	
2	Sprinkler valve outside security area. Pump pressure inside. Alarm setting.	Open Shut Satisfactory Checked	
3	Sprinkler valve outside warehouse. Pump pressure inside. Alarm setting.	Open Shut Satisfactory Checked	
4	Sprinkler valve outside treasurer's dept. Pump pressure inside. Alarm setting.	Open Shut Satisfactory Checked	
5	Sprinkler valve outside IBM room. Pump pressure inside. Alarm setting.	Open Shut Satisfactory Checked	
6	Sprinkler valve outside of boiler room. Pump pressure inside. Alarm setting.	Open Shut Satisfactory Checked	
7	Inspection of hose house and hydrant next to boiler room.	Checked	
8	Inspection of hose house outside of warehouse.	Checked	
9	Inspection of hose cart.	Checked	
10	Report on water pressures.	Good Fair Poor	
11	Condition of fire extinguishers.	Good Fair Poor	

General Remarks:

Signed .... (Inspector)

The training of fire teams should be the responsibility of the building superintendent or plant engineer if the members are from his department. If personnel from other departments are appointed, they should be instructed by a trained fire expert, with several periods of practical instruction. The fire committee with the building superintendent or plant engineer, should make an inspection once per month. Recommendations should be made by the committee, evaluated, and required action taken by the building superintendent or plant engineer. The basic purposes of the fire committee should be outlined and building and plant fire rules established. Rules for each building and plant will vary. Suggested basic building or plant rules are:

1. Safety dispose of trash, scraps, and debris.
2. Keep all equipment repaired.
3. Check all fire doors. Eliminate any obstructions.
4. Check all fire extinguishers. Recharge if necessary. Replace defective parts.
5. Replace all missing or defective hoses.
6. Check valves for easy operation and tight closure.
7. Clean and dispose of yard trash.
8. Test fire pumps.
9. Keep orderly hose houses.
10. Keep a check on all outside valves.
11. Check to see that welding and cutting tags are used.
12. Inspect to see that flammables are stored properly.
13. The circuit to a machine or motor should be shut off while repairs are being made.
14. Watch the fire teams in action and grade their performance.
15. Keep important records in fireproof cabinets. Inspect after normal working hours to see that records are not left on file cabinets, desks, tables, and other places.
16. Number file cabinets in office areas according to the value of their contents, to be removed in case of fire. The most important being numbered "1", the next "2", and so on.
17. Salvage covers should be purchased to place over material or equipment to minimize water damage.
18. The committee is responsible for Fire Prevention Week Activities. (See "Fire Prevention Week Activities.")

While the inspection is being made by the fire committee, all good practices in housekeeping should be kept in mind, and all discrepancies brought to the attention of the building superintendent, plant manager, or person responsible. If building or plant rules are adhered to, the chance of a fire will be minimized greatly. Effective *fire protection* in *any* building or plant should be kept on a day to day basis. Good housekeeping and accepted mechanical maintenance practices are the foundation of any fire protective system.

Fire security is dependent on capable leadership and good organization. The action taken immediately after a fire is discovered and before the arrival of the fire department greatly influences the net result of a fire. Good planning and organization is of the utmost importance for the protection of personnel as well as the building itself. Floor plans showing the flow of personnel to fire exits should be distributed. (See Figure 53-2.) A floor plan should also be made available showing the location and types of all fire extinguishers and alarms. (See Figure 53-3.) The building should be divided into zones with a zone marshal being responsible for these specific duties:

1. Evacuation of personnel (including the physically handicapped).
2. Preservation of company property; limiting the spread of fire.
3. Training of appointed wardens and their alternates in handling of fire procedures, such as fire fighting equipment, exits, and other responsibilities.
4. Fire prevention through reduction or elimination of hazardous materials or conditions.
5. Maintain order outside the building, moving employees from roadways and away from the buildings.
6. Persons should be designated to direct fire trucks to the fire.

A minimum of two fire drills each year should be held.

If a building or a plant is shut down during vacation, the following reminders should be kept in mind:

1. Plan by analyzing watchman service for adequate coverage.
2. Clean plant thoroughly, especially hazardous areas.
3. Repair fire doors and extinguishers. Be sure all fire doors are closed.
4. Protect outside, as thoroughly as interior. Clean all debris and control grass and weeds.
5. If design permits, shut off electrical power in each plant area. Maintain only circuits necessary for activities.
6. Check all fire pumps and automatic controls.
7. Shut off fuel supplies as close to source as possible.

Safety Factors

Welding and cutting. All welding and cutting should be restricted to safe areas. This equipment should not be used in the presence of flammable vapors, liquids, or tanks which have previously contained such materials. All combustibles should be at least thirty feet away. Men should be assigned to watch for sparks, and ample fire protection equipment should be on hand, with the area patrolled for at least one-half hour after completion.

Smoking. Enforce the NO SMOKING rule in restricted areas.
Training the watchman. The watchman should know:

1. How to quickly call the municipal fire department.
2. Location of alarm boxes.
3. Others to be called in times of emergencies, names and numbers.
4. How to start fire pump, and the location of all valves controlling sprinkler systems.
5. Location of valves controlling water, steam gases, fuel oil, or other flammables.
6. Location of switches that control power and lighting systems.

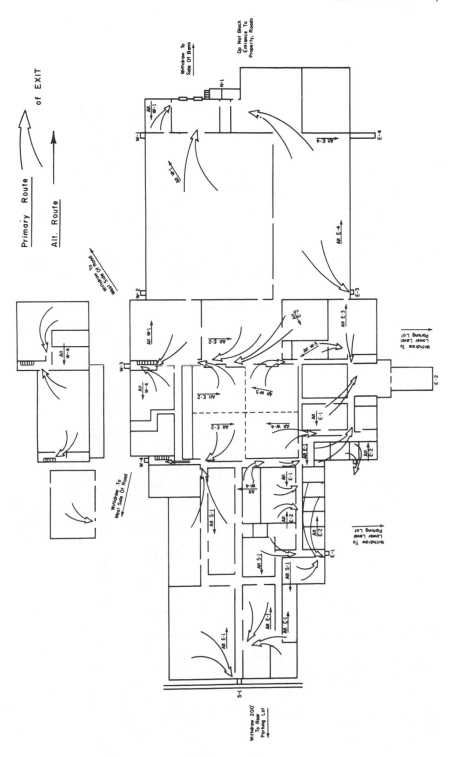

Figure 53-2

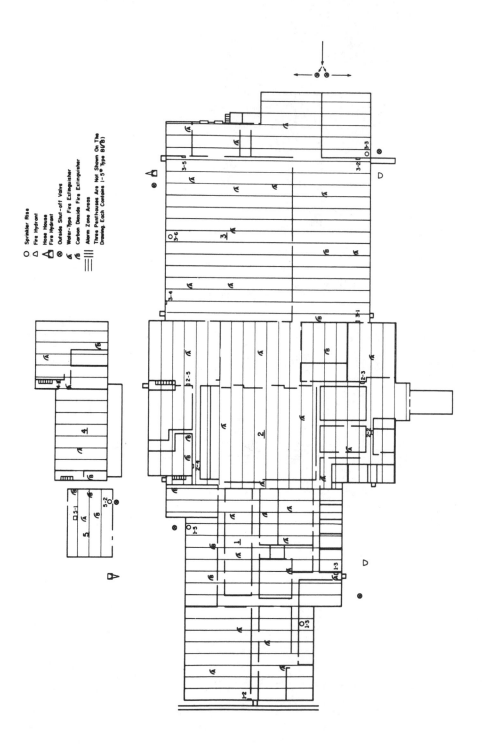

Figure 53-3

7. Proper fire fighting methods, and equipment and know how to use it.

Refresher courses should be held at least once per year with written exams conducted once every six months. This will instill the knowledge required in the watchman's mind and keep it refreshed.

Electrical equipment and wiring. The practice of employing competent and well-trained electricians cannot be overemphasized. The National Electric Code, which is revised and printed yearly, should be recognized as the guide for installation of electrical equipment. A copy should be secured each year with changes noted and referred to the electricians. A check should be made to insure their complete understanding of the changes. After electrical equipment and wiring have been installed, they must be kept in good condition by periodic inspections and by providing required maintenance.

Defective wiring causes many electrical fires. Wiring is affected by overloading, vibration, moisture, corrosion, heat, and cold. These conditions weaken the insulation and can cause a short circuit, arcing, and a fire. Common causes of fires at circuit breakers are pitted contacts, loose parts, and loose connections. In an air circuit breaker, arcing can produce molten pieces of metal which can escape if the box is not tight enough, or if the cover is not well secured. In an oil circuit breaker, dirty or low oil can cause arcing, ignition, or an expulsion of the oil. Burned out motors are caused by overloading, using types not suited to operating conditions, or by too frequent starting or jogging. Heavy dust and dirt clogging the ventilator space of a motor, or improper ventilation, can cause a motor to run hot. Fires can also start from electric lamps in contact with combustibles, or by ballasts on fluorescent fixtures, which can overheat and ignite combustible materials, liquids, or vapors in contact with them.

Static electricity. Where there is moving machinery and equipment, a volume of static electricity is built up which may discharge a spark, causing an explosion or fire. Static electricity is created by friction when two substances made of different materials are brought together then separated. Paper passing over rollers on a printing machine, and a belt moving on a pulley are two examples of how static electricity can be built up. In each instance, one substance comes in contact with the other, then moves away from it. Static buildups are greater in the winter months when the air is dry. The generation of static cannot be prevented. Spark discharging accumulations can be prevented by:

1. *Grounding Machines.* These are the simplest, most practical, and effective means. This is done by connecting a piece of wire from a metal part on the machine to the ground wire, a water pipe, or plate buried in the ground. This supplies a path for the static electricity to flow away from the machine.

2. *Humidification.* This can be secured by introducing a fine spray of steam into the air, or by installing humidifiers in large air handling units used for air conditioning. Small unit humidifiers for small installations may also be utilized.

3. *Ionization of Air.* This permits the static to flow out in all directions and can be accomplished by the use of gas flames.

Grounding of machinery is considered the best method. Periodic checks should be made to insure that all connections are intact, tight, and that a gap in the ground has not been made. Static discharges are extremely dangerous whenever flammable vapors or gases are present in the atmosphere.

Spontaneous combustion is one of the leading causes of plant and building fires. This is the result of improper storage of oily, solvent, or other chemically laden rags, clothes, or other materials. Ignition is produced by a chemical reaction, and starts with a slow oxidation which generates heat. This heat hastens the chemical action until it smolders and bursts into flame. The best preventative measure to stop fires by spontaneous combustion from rags or clothes containing oil, chemicals or solvents is to provide approved metal containers (Underwriters' Laboratory Approved). These cans must be emptied periodically and their contents disposed of so that they will not be a fire hazard.

Storage and handling of flammable liquids. Static electricity discharges are extremely dangerous wherever flammable vapors, gases, or combustible dust is present in the atmosphere. Conditions that cause static discharges have been discussed in the preceding paragraphs (Static Electricity). The most practical method of avoiding a fire or explosion, is by controlling the vapors and preventing them from accumulating to the amount that they become a fire hazard.

If flammable liquids, oil, or gases, are to be stored in a plant, special storage rooms should be constructed completely isolated from the main area. If an established plant has not the means of taking storage of flammables into consideration, it should be done *immediately*.

Walls, ceiling, and floors for flammable storage rooms should be of thick construction, to meet the requirements of the insurance company. They are generally made from concrete, brick, concrete block, or a combination of these materials. If iron and steel are used as supporting columns or beams, they should be covered with one or a combination of the following materials, concrete, or clay products.

The floor should be pitched toward the center of the room, where a drain should be located to carry away any spilled liquids or leaks from drums to an outside sump area. Windows should be of standard steel sash which open outward and contain wire glass. A good approved ventilating system should be installed to prevent any buildup of explosives or harmful vapors or gases. Light fixtures should be explosion-proof. The light switch and all fuse boxes or circuit breakers must be located outside the room. All entrances must be equipped with fireproof doors, and have a slightly raised sill to insure that any spills would not leave the room. Storage racks should be grounded, and provided with approved grounding accessories to accommodate all types of flammables stored. All drums with flammables should be grounded to the rack. If flammables are transferred from the drum to a safety can, the can should also be grounded. (See Figure 53-4.)

When drums are stored on racks, they should be placed on their sides with the side holes toward the top. The original cap should be replaced with a pressure or vacuum relief fitting. The pressure relief valve is used to automatically relieve excessive vapor pressures, and to close automatically to stop evaporation when the internal pressure drops below the hazard point. This fitting is generally equipped with a metal plate flame arrestor, which prevents exterior sources of ignition from igniting the vapor content within the drum. Excessive pressure can be built within the drum from the sun, or a fire, that can cause volatile liquids to expand and explode. A vacuum relief valve permits the rapid withdrawal of

GROUND DRUMS — Prevent Static Electric Spark

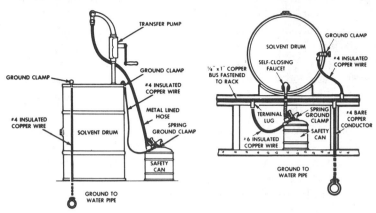

Figure 53-4 / **Method of grounding a drum.**

liquid from the drum, and maintains a balanced atmospheric pressure within the drum to permit a free flow of liquid from the drum. The greatest fire hazards occur when transferring flammable liquids from storage drums to various containers. When transferring liquids from a storage drum, a fully protected transfer pump or a self-closing faucet should be used. The pump should have fire baffles at the spout and inlet strainer inside the drum. The faucet should have a fire baffle and strainer in the spout. Both the pump and faucet should be made of nonferrous metals. The faucet should be a spring type that will dispense the liquid when depressed. Flexible hose can be attached to the faucet for containers that have small openings.

This room should be included on all watchman's rounds, to aid in detecting small leaks which may prevent a major disaster. Only trained electricians should be allowed to work on the equipment in this room, and periodic electrical inspections should be scheduled.

Storage of paints and oils. Paints and oils should be kept away from open flames and stored in metal lined, ventilated, locked closets. Carbon tetrachloride is not recommended for use in connection with electrical maintenance and should not be stocked. Trichloroethylene is a noninflammable solvent recommended for cleaning electrical parts.

Smoke Detectors

The earlier a fire can be detected, the earlier it can be controlled. Detection in the smoke stage, before appreciable heat buildup, increases the warning time and thus reduces the hazard to both life and property.

The detectors are generally designed to set off an alarm when either abnormal smoke accumulates to a point, or when the temperature reaches the alarm setting off the thermostat, which is usually available in the area of 135°F to 190°F. The settings should be discussed with the supplier and specified at the time of purchase so that normal products of combustion usually present in the building will not activate the smoke detection device.

Detectors are available for smoke, heat, optical flame, marine fire watch, and for aircraft fire detection. They can be engineered to control fire alarm systems, evacuation signals, shutoffs for ventilating fans, emergency lights, sprinkler systems, and other applications. Automatic self-contained smoke and flame fire detection and extinguishing systems for bank vaults, record centers, and other individual applications are also available.

A detector is now available which will sense the presence of fire by its flame, the heat it produces, or by the smoke arising from it, and extinguishes it through the use of gas. This is in contrast to extinguishers which smother or cool a flame. Once it senses a fire, the system releases the extinguishing gas from nozzles around the room (under the floor in computer rooms). The gas snuffs out flames by reacting with unstable chemical compounds in the flames which would other-wise build up and speed the fire. The considered use of this gas extinguishing system, or any other type of system, should be reviewed thoroughly with your insurance carrier for approval prior to purchase or installation.

A maintenance program should be developed to check the system, to include setting off the alarm on a scheduled basis. Codes which apply to the geographic location may dictate minimum frequency to cover the specified application.

Fire Prevention Week

Fire Prevention Week is observed each year during that week in which the anniversary date of the Great Chicago Fire (October 9, 1871) occurred, which destroyed an estimated $168,000,000 in property, and killed two hundred and fifty people. Despite the fact that fire prevention is a year round responsibility, this one week's observance does permit an exceptional opportunity to emphasize the extreme need for reducing the excessive destruction of property and loss of life by fire, and to re-emphasize those measures which can and should be taken.

Every building superintendent or plant engineer has a definite stake in this important project. Fire Prevention Week activities provide an opportunity for employee education through:

1. Demonstrating the fundamental causes of fire and proper method of extinguishment.
2. Revealing the cost of needless fire waste.
3. Showing what each employee can do to reduce the danger of fire in his work area.
4. Promoting a fire safe attitude.

The objectives are accomplished by:

a. Preparation of plant or building exhibits, which stress the importance of fire prevention, reporting and correcting fire hazards, turning in alarms promptly, and knowing the location and use of fire extinguishers.

b. Conducting a supervisors' training session during the week prior to Fire Prevention Week devoted to fire prevention as it applies to each supervisor's department.

c. Distribution of Fire Prevention Week folders and pamphlets to all employees.

d. Conduct demonstrations showing brigade drill, proper extinguishment methods and use of latest fire fighting equipment.

e. Presentation of fire films in the cafeteria during lunch periods.

f. Have a fire drill during the week.

The plant fire brigade should be instructed in the preparation of equipment for the winter months. Drills should be held in cooperation with the local municipal fire department. These concentrated activities make a lasting impression on employees so that the importance of fire prevention and protection during the remainder of the year is not soon forgotten.

54 / FLAG DISPLAY

(Source: U.S. Marine Corps)

How to Display the Flag

Laws have been written to govern the use of the flag and to insure proper respect for the Stars and Stripes. Custom has decreed certain other observances in regard to its use.

All the Services have precise regulations regarding the display of the National flag, which may vary somewhat from the general rules below.

Respect your flag and render it the courtesies to which it is entitled by observing the following rules:

The National flag should be raised and lowered by hand. Do not raise the flag while it is furled. Unfurl, then hoist quickly to the top of the staff. Lower it slowly and with dignity. Place no objects on or over the flag. A speaker's table is sometimes covered with the flag. This practice should be avoided.

When displayed in the chancel or on a platform in a church, the flag should be placed on a staff at the clergyman's right; other flags at his left. If displayed in the body of the church, the flag should be at the congregation's right as they face the clergyman.

1. When displayed over the middle of the street, the flag should be suspended vertically with the union to the north in an east and west street, or to the east in a north and south street.

2. When displayed with another flag from crossed staffs, the flag of the United States of America should be on the right (the flag's own right) and its staff should be in front of the staff of the other flag.

3. When it is to be flown at half-mast, the flag should be hoisted to the peak for an instant and then lowered to the half-mast position; but before lowering the flag for the day it should again be raised to the peak. Half-mast means hauling down the flag to one-half the distance between the top and the bottom of the staff. On Memorial Day display at half-mast until noon only; then hoist to the top of the staff.

4. When flags of states or cities or pennants of societies are flown on the same halyard with the flag of the United States of America, the latter should always be at the peak. When flown from adjacent staffs, the Stars and Stripes should be hoisted first and lowered last.

5. When the flag is suspended over a sidewalk from a rope extending from house to pole at the edge of the sidewalk, the flag should be hoisted out from the building, toward the pole, union first.

6. When the flag is displayed from a staff projecting horizontally or at any angle from the windowsill, balcony, or front of a building, the union of the flag should go to the peak of the staff (unless the flag is to be displayed at half-mast).

7. When the flag is used to cover a casket, it should be so placed that the union is at the head and over the left shoulder. The flag should not be lowered into the grave or allowed to touch the ground.

8. When the flag is displayed in a manner other than by being flown from a staff, it should be displayed flat, whether indoors or out. When displayed either horizontally or vertically against a wall, the union should be uppermost and to the flag's own right, that is, to the observer's left. When displayed in a window it should be displayed in the same way, that is, with the union or blue field to the left of the observer in the street. When festoons, rosettes, or drapings are desired, bunting of blue, white, and red should be used, but never the flag.

9. When carried in a procession with another flag or flags, the Stars and Stripes should be either on the marching right, or when there is a line of other flags, in front of the center of that line.

10. When a number of flags of states or cities or pennants of societies are grouped and displayed from staffs with our National flag, the latter should be at the center or at the highest point of the group.

11. When the flags of two or more nations are displayed, they should be flown from separate staffs of the same height and the flags should be of approximately equal size. International usage forbids the display of the flag of one nation above that of another nation in time of peace.

The flag should never be displayed with the union down except as a signal of dire distress.

Do not use the flag as a portion of a costume or athletic uniform. Do not embroider it upon cushions or handkerchiefs nor print it on paper napkins or boxes.

A federal law provides that a trademark cannot be registered which consists of, or comprises among other things, "the flag, coat of arms, or other insignia of the United States, or any simulation thereof."

When the flag is used in unveiling a statue or monument, it should not serve as a covering of the object to be unveiled. If it is displayed on such occasions, do not allow the flag to fall to the ground, but let it be carried aloft to form a feature of the ceremony.

Take every precaution to prevent the flag from becoming soiled. It should not be allowed to touch the ground or floor, nor to brush against objects.

The flag should not be dipped to any person or thing, with one exception: Navy vessels, upon receiving a salute of this type from a vessel registered by a nation formally recognized by the United States, must return the compliment.

When carried, the flag should always be aloft and free—never flat or horizontal.

Never use the flag as drapery of any sort whatsoever. Bunting of blue, white, and red—arranged with the blue above, the white in the middle, and the red below—should be used for such purposes of decoration as covering a speaker's desk or draping the front of a platform.

Do not use the flag as a receptacle for receiving, holding, carrying, or delivering anything. Never place upon the flag, or attach to it, any mark, insignia, letter, word, figure, design, picture, or drawing of any nature.

No other flag may be flown above the Stars and Stripes, except: (1) the United Nations flag at U.N. Headquarters; (2) the church pennant, a dark blue cross on a white background, during church services conducted by naval chaplains at sea.

Other Approved Customs

Highest honors are rendered to the National flag by all branches of the Armed Forces and the various patriotic societies throughout the country.

More than 50 years ago it was the custom to salute the National flag by uncovering; nowadays the hand salute is rendered by the entire personnel of the Armed Forces.

During the ceremony of hoisting or lowering the flag, or when the flag is passing in a parade or in a review, those present in uniform should render the right-hand salute. When not in uniform, men should remove their headdress with the right hand and hold it at the left shoulder, the hand being over the heart; women should place the right hand over the heart.

The flag should be displayed only from sunrise to sunset, or between such hours as may be designated by proper authority.

The flag can be displayed on all days when the weather permits, especially on New Year's Day, January 1; Inauguration Day, January 20; Lincoln's Birthday, February 12; Washington's Birthday, February 22; Easter Sunday (variable); Mother's Day, second Sunday in May; Armed Forces Day, third Saturday in May; Memorial Day (half-staff until noon), May 30; Flag Day, June 14; Independence Day, July 4; Labor Day, first Monday in September; Constitution and Citizenship Day, September 17; Columbus Day, October 12; Veterans Day, November 11; Thanksgiving Day, fourth Thursday in November; Christmas Day, December 25; such other days as may be proclaimed by the President of the United States; the birthdays of States (dates of admission); and on State holidays.

The custom of lowering the flag to half-mast or half-staff comes from the old

Figure 54-1

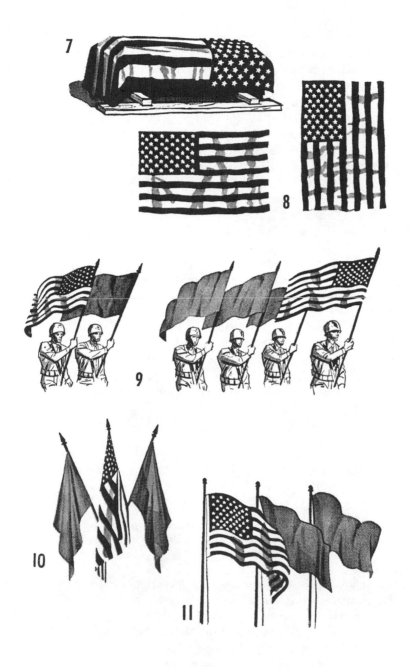

Figure 54-2

military practice of "Striking the Colors" in time of war as a sign of submission. It is known that as early as 1627 the flying of the flag at half-mast was a sign of mourning, and this has been continued to the present day.

If a serviceman or woman dies during a period of service, the flag is furnished by the Service. However, if he or she dies as an honorably discharged veteran, the flag is furnished by the Veterans Administration, Washington, D.C. The flag must be presented to the next of kin at the proper time during the burial service. If there is no relative, or one cannot be located, the flag must be returned to the Veterans Administration in the franked container for that purpose.

Many of the nation's dry cleaners, in cooperation with the American Legion, will dry-clean the National flag free of charge between June 1 and 12, provided the owner of the flag promises to fly it on Flag Day, June 14.

When the flag is in such a condition, through wear or damage, that it is no longer a fitting emblem for display, it should be destroyed in a dignified way, preferably by burning.

On suitable occasions repeat this pledge to the flag:

"I pledge allegiance to the flag of the United States of America, and to the Republic for which it stands, one Nation under God, indivisible, with liberty and justice for all."

This wording of the pledge varies slightly from the original, which was drawn up in 1892 in the office of *The Youth's Companion* in Boston. It was first used in the public schools in celebration of Columbus Day, October 12, 1892.

The pledge received official recognition by Congress in an Act of June 22, 1942. The phrase "under God" was added to the pledge by a Congressional Act of June 14, 1954. At that time, President Eisenhower said that "in this way we are reaffirming the transcendence of religious faith in America's heritage and future; in this way we shall constantly strengthen those spiritual weapons which forever will be our country's most powerful resource in peace and war."

55 / **GAS MASKS**

(*Source:* "The Place of the Gas Mask in Industry" by S. J. Pearce, Branch of Health Research, Bureau of Mines, U. S. Department of the Interior)

A person choosing respiratory protective equipment for a specific job or problem first should ask himself several questions:

Is there sufficient oxygen to support life?

What is the nature and concentration of the contaminent?

For how long will respiratory protection be needed?

What is the location of the contaminated area with respect to the fresh air?

How active is the wearer expected to be?

When one answers these questions, he can choose a suitable respiratory protective device—assuming he is familiar with the limitations and operating characteristics of the various types.

Gas masks are not designed to be used, nor can they be used safely, under all conditions. However, once the limitations of the gas mask are recognized, one finds that there is a range of conditions within which the gas mask can be used safely and relatively comfortably. In fact, it is the device of choice for use under such circumstances.

Limitations are not unique with gas masks; every respiratory protective device has some limitations.

1. Since a gas mask adds no oxygen to the air being breathed, it cannot be used in atmospheres containing less than sixteen per cent oxygen.

2. A gas mask is also limited to atmospheres that contain no more than two per cent, by volume, of a single harmful gas or a combination of gases (three per cent of ammonia).

3. Not all canisters will protect against all gases.

On the other hand, the gas mask has certain favorable features. The wearer is not hampered by trailing air lines; hence he can enter and leave a contaminated area through different openings. The gas mask is relatively light; its total weight, as worn, ranges from three and one-half pounds to seven pounds. It is mechanically simple and requires a minimum of maintenance. Training a person to wear a gas mask is much simpler than training him to wear the more complex type of respiratory protective devices.

There are many types of gas masks available. Some gas masks have canisters designed to protect against a single gas, such as ammonia; others protect against two classes of gases, such as acid gases and organic vapors; and the universal gas masks give protection against all industrial gases, vapors, and smokes.

Gas masks approved by the Bureau of Mines have passed rigid tests and are known to be reliable and capable of performing their intended function.

Gas masks are widely used by men in fumigating buildings and grain bins, in operating and maintaining chemical plants, in tank-gaging operations of the petroleum industry, in fire-fighting work, in plants where ammonia or other refrigerants are used, and in missile-fueling activities, to name only a few. Gas masks are generally kept in boiler rooms for general emergencies.

Gas masks should not be used as a substitute for good engineering practices, such as controlling toxic gases and vapors at the source or removing or dilution of them through proper ventilation. Hence, in industry, gas masks usually are not worn throughout the working day, but are used for short periods during repair or adjustment operations, or are kept ready for emergencies, such as fires or ruptured pipe lines.

If, after considering all the foregoing factors, you decide to use gas masks, the proper type and quantity should be purchased and the location for each mask should be chosen so the mask will be readily available when needed. It is advisable to store gas masks intended primarily for emergency use just outside the area where a gas emergency might occur. We know of instances where sudden outbursts of gas have occurred and all the available gas masks were in the gassed area and thus were inaccessible. What might have been a minor mishap became a serious matter, because the situation worsened while other gas masks were being obtained.

Training

Each person who might be called upon to use a gas mask should be properly trained in its use. The training should cover the following:

1. A discussion of the reasons for using the gas mask.
2. A description of its construction, operating principles, and limitations.
3. Instruction in the proper fitting of the gas mask to the wearer and in testing it for gas tightness.
4. Actual wearing of the gas mask in normal air while engaging in activities that would be expected in an emergency situation.
5. Wearing of the gas mask in an atmosphere containing an odorous material, such material being removable by the canister being used.
6. Instruction in the proper use and maintenance of the gas mask.

The instructor should be familiar with, and guided by, the material on the instruction card inside the lid of each gas mask case. He also should know the information given on the canister and gas mask labels.

Several important points should be observed in using gas masks:

The wearer of a gas mask should be certain that the atmosphere he is about to enter is not deficient in oxygen.

He should enter the contaminated area cautiously, and if the odor of the contaminant is noted, he should return to fresh air and correct the cause of the leakage in his gas mask.

He should always return to fresh air before replacing an exhausted gas mask canister with a fresh one.

He should be sure that he is in fresh air before removing the facepiece.

If a highly toxic atmosphere is to be entered, and there is any doubt as to the length of remaining life of a partially used canister, a fresh canister should be used. It is dangerous to try to squeeze the last few minutes of life out of a gas mask canister.

Wherever practicable, the "buddy" system should be used. One of the far-fetched ideas is that wearing of a gas mask will offset the harmful effects of gas inhaled prior to donning the mask. The *truth* is that a gas mask would prevent inhalation of additional gas, but it would have no curative properties to overcome the effects of the gas already inhaled. Further *false* notions follow:

False Notions

1. *If a person's face is covered he is protected against gas.* This is false. The idea may sound fantastic, yet we know or cases where men have put on gas mask facepieces *only* and have entered noxious atmospheres. They found out their mistake the hard way. A gas mask facepiece alone gives no protection against harmful gases.

2. *Any gas mask is better than none.* This false notion has prompted veterans of the armed forces to use surplus or souvenir service gas masks incorrectly for protection against ammonia or carbon monoxide or in oxygen deficient atmospheres.

3. *The efficiency of the canister is the only thing to consider in a gas mask.* This is false. An efficient and effective canister is essential, but it is equally essential that the proper facepiece be used, that it be in good repair, and that it be fitted to the wearer's face in a gas-tight manner as proven by the proper tests.

4. *Conventional-type prescription spectacles, with side pieces or bows, can be worn with gas mask facepieces without interfering with the gas-tightness of the face-fit, particularly if the bows are thin.* This is also false. Many tests made by men with varying facial shapes and sizes in a gas-filled chamber have proved this to be erroneous. Two gas mask facepieces now manufactured have provisions for supporting prescription eye glasses within the facepiece, but in neither facepiece is there anything between the facepiece and the wearer's face.

5. *If a gas mask will protect against the inhalation of a two per cent gas-air mixture for thirty minutes, it will protect against a four per cent, gas-air mixture for fifteen minutes or against a six per cent gas-air mixture for ten minutes under the same wearing conditions.* This is false. It is not correct to extrapolate to concentrations higher than two per cent (three per cent for ammonia), since canisters are designed to handle this concentration of gas, but might allow some of the gas to leak through the granular fill at higher concentrations, depending on the type and thickness of absorbent. Extrapolation to concentrations lower than two per cent is correct; for instance, if a gas mask will protect against the inhalation of a two per cent gas-air mixture for thirty minutes, it would be expected to protect against a one per cent gas-air mixture for sixty minutes or a 0.5 per cent gas-air mixture for one hundred and twenty minutes under the same wearing conditions.

6. *If the canister heats up, it is no good.* Here is another false idea. In general, the opposite is true, particularly if carbon monoxide is known to be present. When the catalyst in the universal gas mask canister performs its intended function of converting carbon monoxide to carbon dioxide, heat is liberated in proportion to the concentration of carbon monoxide. With concentrations of carbon monoxide above two per cent, the heat liberated can raise the temperature of the canister high enough to melt the solder on the canister. If excessive heating of the canister is noted, it is an indication that the wearer is in a high concentration of gas and that he should proceed with greater caution. The chemical or physical absorption of other gases and vapors, including water vapor, is accompanied by the liberation of heat. The amount of heat depends on the type and concentration of gas, and the type of absorbent. If a gas mask canister does not heat up in the presence of carbon monoxide, the wearer should return to fresh air immediately and replace the canister with a fresh one.

7. *A universal gas mask canister can be used safely for protection against the inhalation of carbon monoxide until its resistance increases significantly.* This is also false. The testing of nine universal gas mask canisters that had been used or had been unsealed for some time, showed that they gave no protection whatever against carbon monoxide. Yet the inhalation resistance of six of these canisters was

essentially the same as for new unsealed canisters, and that of two others was only slightly higher. However, if the inhalation resistance of a universal gas mask canister is found to be high, the canister probably is no good and should be discarded.

8. *Universal gas mask canisters will protect against the inhalation of all types of gases for two hours or until the window indicator changes color.* False again. The two-hour limitation, and the indication of the timer or window indicator apply only to protection against the inhalation of carbon monoxide. The only way to be sure if a gas mask canister is exhausted against other gases is by detecting the odor or irritating effect of those gases in the air entering the facepiece from the canister.

9. *The head harness of the gas mask facepiece should be adjusted only tight enough to be "comfortable."* Generally false. This may be true for persons with full, fleshy faces, but it is definitely false for persons with thin faces. The head harness should be tightened until the wearer obtains a gas-tight fit, preferably by the "positive pressure" test. This test for facepiece fit is made by closing off the exhalation valve of the facepiece and exhaling gently into it to build up a slight positive pressure within it. A gas-tight fit is indicated by the absence of any outward leakage of air around the periphery of the facepiece. If leakage of gas is detected by odor or irritating effect when a person enters a contaminated area while wearing a gas mask, he should return to fresh air to readjust the head harness. Such adjustment should not be made in the contaminated area because the facepiece could be pulled away from the face in so doing and thus allow a high concentration of the gas to enter the facepiece and possibly incapacitate the wearer.

The foregoing coverage of the gas mask may seem heavy on "do's" and "don'ts". We do not want to discourage those who are using or who may choose to use gas masks. Rather, we want to reveal the facts so that, being informed, such persons can use this most useful respiratory protective device more intelligently—and consequently more safely.

56 / **INSURANCE**

The operation of a building can depend upon the amount of settlement received from an insurance peril. For this reason, it is important that building owners and building superintendents review annually, the type and amount of protection they carry. The amount of coverage should be adequate to present day economic standards, as values change, replacement costs fluctuate, and various times there are changes in the types of coverage available. Where values frequently change, a six-month review is recommended. A similar review of liability insurance should be made annually and statistics secured from the carrier covering claims, accident rates, severity rate, cost per accident, and statistics from similar industries for comparison. Building owners and operators should be well schooled in the types of insurance available, their use, and application.

Review coverage to determine:
1. If the description of the building in the policy is up to date, and accurate.
2. If the character of the tenancy has changed.
3. If the neighborhood environment has changed.
4. If the economy has affected the value of the building.

Periodic inspections should be made to determine if fire hazards, or other areas that may result in a major breakdown, accident or injury are inspected thoroughly.

Fire Policy Provisions

The amount of collectible loss an owner will receive, in the event of damage to his building, is related to the adequacy of the amount of insurance carried. It is therefore important that the owner and the building superintendent understand the policy provisions.

Coinsurance clause. The coinsurance clause applies the principle that if the owner insures for less than a certain percentage of the actual value of the building, he assumes part of building coverage as a self or coinsurer with the company, the owner must assume the same proportionate share of the loss. With this principle, the company requires that the insured must carry insurance up to at least 80 per cent of the actual value of the property or he will not receive the full amount of the loss. The usual coinsurance requirement is 80 per cent at the time of loss. In some instances such as a sprinklered building, the amount required is 90 per cent.

Loss Adjustment Under Coinsurance

(80 per cent clause)

A. *Insurance exceeding 80 per cent of value is carried:*

Value of property	$10,000
Insurance required	8,000
Insurance carried	9,000

Losses up to $9,000—Will be paid in full.

B. *Insurance to 80 per cent of value is carried:*

Value of property	$10,000
Insurance required	8,000
Insurance carried	8,000

Losses exceeding $8,000—Face of policy paid ($8,000).
Losses under $8,000—Paid in full.

C. *Insurance to less than 80 per cent of value is carried:*

Value of property	$10,000
Insurance required	8,000
Insurance carried	5,000

Losses exceeding $8,000—Face of policy paid ($5,000).
Losses under $8,000—Paid in proportion of five-eighths of loss.

The following *insurable perils* should be thoroughly understood:

Fire

Smoke

Explosion

Boiler explosion

Robbery

Water damage

Civil authority

Public liability

Burglary

Fidelity

Windstorm and hail

Earthquake and volcanic eruption

Vandalism and malicious mischief

Sprinkler leakage

Riot, riot attending, strike, civil commotion

Aircraft and vehicles

Contractual liability

Employer's responsibility to employees

57 / MOVING

Plans for moving should take into consideration what is to be moved, the placement of the furniture and equipment in the new location, the selection of the mover, and the date for moving. The placement of furniture and equipment should be drawn into a colored, scaled floor plan. Floors or areas can be designated by special colors—1st floor blue, 2nd floor green, and so on. A piece of colored three inch paper tape, the same color as the one designated to a certain area, should then be applied to each piece of equipment in a specified place, such as lower left hand corner. If this is impossible on some pieces of equipment, use an obvious spot. The equipment should then be numbered in the order in which it will be moved into the new location. If a truck can handle twenty-eight items (to be computed by the square footage of each item), the twenty-eighth item should be loaded first, so that the number one item will come off the truck first. If further identification is required, the floor number could be added like 2, the bay A, item number 8, or 2-A-8. The floors of the new location should be marked with the same color tape, and corresponding numbers. *Example:* Desk 2-A-8 will be placed on the second floor, bay A, and the lower left hand corner of the desk will be placed over the number 8 in this area.

If several entrances are to be used, an entrance number can be added, or a color system adapted for one-story operations, having certain colors delivered to corresponding colored doors. The tape could be applied over or attached to the door. The paper tape will adhere to the desk, and is less likely to be removed or torn off. Many colors are available which aid in quick identification.

It is important to include the scheduling of delivery of new equipment in the moving plans. In some instances, it is wise to have medium-sized items delivered prior to the move, to the old location, so that employees may have an opportunity to acquaint themselves with the equipment. It will also present the opportunity of ironing out any problems with the equipment and to trade in or dispose of any excess equipment.

When selecting the moving contractor, it is good practice to obtain several bids. This will ascertain having a realistic appraisal of the job and the bidding.

A direct move without transfer should be specified, and a penalty clause inserted, if the date for completing the move is not kept. The contract should also cover the contractor's liability, and responsibility for correcting damage or loss that may occur in the course of the move. Insurance certificates should be submitted for amounts of damages specified.

Employee Participation

While planning the move, the employees should not be forgotten. They have an interest in the move and should be kept informed. Good communications will help them accept the change. This can be accomplished at management meetings, or employee magazines or memorandums.

To execute a successful move, you must have departmental participation, move sequence, and coordination. One person from each department should be assigned as a contact man. This man's responsibilities should be to attend the orientation meeting, and to tag all the equipment to be moved from his department, according to the approved layout which has been prenumbered, and to assist at the receiving of new equipment, and removal of excess equipment. In orientation meetings, the contact man can become familiar with the plans, so he can answer employee questions. He can record any missing equipment after the move and report it. He will be the clearing agent for all problems and questions, so that the number of persons who would have to contact the planners, would be kept to a minimum.

Although the contractor will do the actual work, a few employees should be on hand at the old location to help him identify the items to be loaded in proper sequence, which can be figured out in a special loading layout. Other employees should be located at the new location to spot the equipment in its proper location. The spotters should be provided with the approved new location layout.

If items show up that are not on the layout, they should be placed where they fit temporarily, placed to one side, or sent to a storage area. During such a move, it is generally impossible to make layout changes.

Other items to be considered are:

1. Installation of telephones prior to the move.
2. Notification of persons supplying services such as, office machines, deliveries, and other suppliers.
3. Is a change in insurance coverage required?
4. Change of address on stationery.
5. Was there a change in the telephone number?
6. Installation of electric outlets.
7. A system of requests for study of any problems which may arise should be typed and sent forward to the planners' group for study and recommendations.

All problems should receive quick follow-up. Moving problems should be solved promptly because of the lasting impression made on the personnel. Good service will help the employees to get settled more promptly. Some problems cannot be solved immediately. A new office means a change in working habits, and some employees require time to adjust to the changes.

A good moving job requires careful attention to the objective, careful planning of all the aspects of moving, communications with the employer prior to the move, providing the services required to operate at the new location, and the follow-up after the move has been completed.

58 / **RECORDS**

(*Sources:* Remington Rand Systems, New York; *Plant* magazine)

Machinery and equipment records are the basic requisites for establishing a sound maintenance program, as it is necessary to have a thorough knowledge of the type of equipment you have, and which portion of this equipment will require maintenance services. Properly designed record forms will supply historical data for justification of equipment replacement, repair costs, and initial cost. Other data supplied by the record card is as follows:

1. Lists equipment lubrication requirements.
2. Records a list of spare parts that should be carried in stock.
3. Supplies engineering specifications.
4. Gives the original purchase date, installation cost, purchase order number, and other upkeep costs.
5. Provides a basic record for identifying plant assets.

In order to start any maintenance program, a physical inventory of all machinery and equipment must be made.

A simple form should be used in taking an inventory of all plant or building equipment. A number system should be initiated and a number assigned to each piece of equipment. The number should be stamped on a metal tag and attached to the unit. A "Maintenance Descriptive Sheet" should be made on every piece of equipment, then on all new equipment as it is received.

See Figure 58-1, Figure 58-2, and Figure 58-3 for examples of Maintenance Record Cards which aid in keeping machines in better working conditions to produce better work, cost controls, and less machine downtime.

Work Orders provide:

The *production department* with a record of trouble spots, frequency of breakdowns, and production time lost.

The *maintenance department* with a record of trouble areas, that additional training may be required by the operator or maintenance personnel.

The *accounting department* with information of exactly what material and time were expended to maintain or repair a machine, and aid in inventory control.

Work orders also pinpoint responsibility, improve operations, and aid in assigning work in an orderly manner.

Accurate scheduling, adherence, and adjustment through experience, operating hours, and conditions, is the key to a successful maintenance program. When schedules are adhered to, defects are generally found in the early stages, when corrective measures normally can be made in a matter of minutes. If operation is

Figure 58-1

MAINTENANCE DESCRIPTIVE SHEET

Department ....
Inv. No. ....
Purchase Data ....

Item .... Manufacturer ....
Purchase Order No. .... From ....
Model .... Serial No. .... Type ....
H.P. .... R.P.M. .... Amps .... Volts ....
Phase .... Frame .... Cycles .... Cap. ....
Other: ....

....

Spare Parts on Hand: ....

....

....

....

....

Frequency of Periodic Inspections: ....

Remarks: ....

....

....

....

permitted until the failure stage, minor defects such as slight misalignment, vibration, over-lubrication, casting flaws, overheating, and many others, may result in many hours of downtime. It can be said that preventive maintenance is the balance between theoretical requirements and practical experience. This is where records can be used in measuring the results of a program. After a piece of equipment has been in operation for a period of time, the record may indicate that inspections should be made more frequently, or may be spaced further apart, as over-inspection can prove very costly and directly affects the purpose of your program. Frequency tables should be adjusted by experience to gain maximum dollar benefit from record keeping.

Figure 58-2

REQUISITION FOR MAINTENANCE SERVICES

Department:

Description of Work:

....

....

....

(For Maintenance Dept. Use) Job. No. ....
Labor: ....
Materials: ....
Other: ....
Total: ....

....

APPROVALS

....
Department Head   Maintenance Department

Data Processing

The use of data processing equipment in handling some of the scheduling, costing out jobs, inventory control, maintenance programs, and other duties of maintenance departments, keeps administrative cost to a minimum. The maintenance employee should be instructed in filling out a data processing card, using a code number for the type of labor, and the hours worked. Similar cards are filled out when materials are drawn from the stores department. These cards are sent daily to the data processing department. The progress of the job and a comparison of estimated time against actual time can be developed from this information. After recording the exact amount of material used, inventories can be adjusted. If various departments are charged for services, this can also be determined and charged immediately. Daily progress of jobs which require several days, is easily followed. If a job appears to be running excessively high, it will be spotted immediately. The system of charging other departments for services will make operating departments more maintenance conscious, as the cost of repairs can be quickly shown.

The use of these records will aid in training others who must estimate jobs,

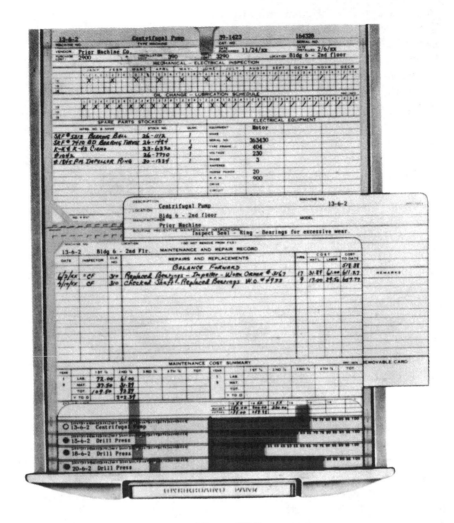

Figure 58-3

as estimated figures against actual figures will be readily accessible. Scheduling and preventive maintenance programs can be put on data processing equipment, and exact daily, weekly, and monthly records and reports will be easily available.

Considerations for Retaining Records

Statutory Limitations. Certain federal statutes, such as the Social Security Act, Wage and Hour Law and state statutes, require that certain records be retained for specified periods.

Business Reference Value. What is their direct value in conducting the business?

Record Copy (the original or a duplicate copy).
Temporary Value.
Legal Value.
Supporting Record.
Research Value.
Is the Record Vital?
Determination of What a Record Should Do.
Historical Value. Facts and events relating to the growth and progress of a company should be retained.

Protection of Vital Records

Vulnerability to Disaster. Flood, fire, explosion, attack, or nuclear explosion.
After the Disaster. The possibilities with which you may be confronted are:
1. Resisting unjust claims.
2. Determining accounts payable.
3. Collection of accounts payable.
4. A plan to resume vital operations as soon as possible.
5. Being able to prove the entity of the business.
6. Collection of insurance indemnity.

Recommendations. Select a minimum number of records with which to accomplish the six situations outlined above. Store these records in a separate location, keep them up to date and establish a retention schedule.

Record Storage Considerations

Location. Should be as close to the office as possible.
Size. Suitable to handle all inactive records with 25 per cent expansion allowance.
Layout. Large enough to handle storage racks and laid out to aid in removing records through retention schedule.
Ventilation. A humidity range of 55 per cent to 75 per cent is desirable, and circulation adequate to prevent a heat buildup which may cause the records to dry out.
Aisles. Should be a 30 to 36-inch width. One service aisle of five feet should be included.
Fire Protection. Maintain a rigid fire prevention rule. Sprinkler systems are a matter of choice due to the possibility of accidental discharge of a head. Fire extinguishers should be installed and clearly marked. The "No Smoking" rule should be strictly enforced, with signs posted to that effect.
Equipment. Standard steel shelving purchased to handle the particular weight of the items to be stored need not be a deluxe type. A simple style will suffice. Sturdy cardboard boxes are sufficient to hold the records. These are generally sized 10 by 12 by 15 inches.
Lighting. The installation of lighting fixtures should conform to the aisle arrangement. 20–25 foot-candles are sufficient.

Figure 58-4

DAILY WORK REPORT

AREA	Broom Swept	Dry Mopped	Damp Mopped	Machine Scrubbed	Waxed	Machine Buffed	Dusted	Vacuumed	Walls	Windows	Woodwork

Indicate jobs done by "x"

Employees Assigned: ....

....

Date: ....

Hours: ....

Total ....

....

Foreman

Floor Load. Should be calculated according to the weight of the items to be stored, plus the weight of the storage racks, generally on a per foot basis.

Security. The area should offer safe storage from dirt, dust, fire, flooding, and unauthorized entry. If a watchman's tour is made, consideration should be given to placing a station in the area designated for security. Depending upon the nature of the records, other facilities may be installed, such as burglar alarms and outside protection service.

Reference. A reference room or area should be provided with a table, chairs, and other necessary items for personnel who come to the area to examine or search for records.

Structure. Should be fireproof or be divided from the other part of the building by means of a fire wall.

Materials. In most cases should be fireproof in the doors, roof, and floors.

Special Considerations. In planning a security area, considerations should be given to the storage of microfilm, magnetic tapes, and punch cards. These items generally require special temperature and humidity requirements. With the constant increase of these types of systems in all types of office buildings and plants, their inclusion is most desirable.

Figure 58-5

CLEANING SCHEDULE

Month ....

AREA

....

Job	Frequency
Broom sweep	
Dry mop	
Damp mop	
Scrub (machine)	
Wax	
Buff (machine)	

Figure 58-5 *(Continued)*

Dust	
Vacuum	
Walls	
Windows	
Woodwork	

Foreman

_____

59 / SAFETY

(*Source:* "Ten Commandments of Safety for Supervisors," reproduced by special permission of the American Management Association. Further reproduction prohibited.)

Accidents do not happen, they are caused. More accidents are caused by people than by machines. The most effective means by which the accident rate may be reduced, is by preventing human error and carelessness. Accidents can be reduced by incorporating safety into every detail of a building or plant operation. Success in safety depends on knowing what an accident is, how, and why it occurred, prevention methods, and how they can be achieved. A safety program must have the approval and support of "top management". The responsibility for a safety program rests with supervisory personnel. Safety work may be carried on by line organizations, a safety director, or a committee system. Regardless of the system employed, an organization must be formed, and the following points established:

1. Scope of activity and extent of authority.
2. Time, place, and frequency of meetings.
3. Planned order of program, and mandatory attendance.

Organization Goals

To create and maintain a continuing interest in safety by *all* employees, both management and hourly.

To conduct regular inspections in order to prevent accidents, and to ascertain that all safe practices are being followed. This applies to the actual building, machinery, equipment, and occupants.

To investigate and find all facts concerning all accidents.

To make a complete analysis of the causes, and to list them.

To select the best method of correction, and to see that it is applied and continued in the interest of safe practices.

To plan for the purchase of safe equipment, covering proper guards, and to see that safety is included in the design of equipment and building layout.

To establish disciplinary procedures, and to study accident reports.

Employee Participation

The *greater* the *employee participation* in the safety program, the more effective it will be. Suggested activities for the employee are:

1. Safety suggestion system.
2. Safety campaigns and contests.
3. First aid training.
4. Safety meetings.
5. Plant fire bridges.
6. Participation on safety committees, and in organizations.

Inspection

Most accidents are the responsibility of the building superintendent or plant manager. Accidents may be prevented by discovering and eliminating the cause. He should not be the "cause" by his failure to accept the responsibility of safety and accident prevention in his building or plant. Well planned inspections, systematically applied, are an effective means of discovering hazardous conditions.

Inspection of the following items may be considered the beginning of an effective safety program:

1. Ladders and scaffolds.
2. Dust, fumes, gases, and vapors.
3. Housekeeping.
4. Stair towers.
5. Need for machine safety guards.
6. Eye protection equipment.
7. Protective clothing and gloves.

8. Methods of handling material.
9. Hand tools.
10. Cranes and hoists.
11. Elevators.
12. Grounding of electric equipment.
13. Short circuits and overloads.
14. Acids in open containers.
15. Proper storage of paints, gasoline, and other liquids.
16. Oil on floors.
17. Pressure vessels.
18. Oiling methods.
19. Slippery floors.

Many other items can be added to this list. A list of the hazards in the building should be made and checked through regularly scheduled inspections. This listing will vary for each type of building or plant, according to function each is performing.

Types of Accidents

The following list shows the various types of accidents for which a building superintendent, plant manager, or other supervisory personnel are responsible.

Type	Cause
OBJECT	Machines, electrical equipment, chemicals, and motor vehicles.
ACCIDENT	Falls, slips, drowning, to be wedged or pinned.
UNSAFE ACTS	Operation without authority. Operating or working at unsafe speeds. Making safety devices inoperative. Failure to use safety equipment. Unsafe loading practices.
PERSONAL CAUSES	Disregard of instructions. Improper attitude. Lack of knowledge. Mental deficiencies. Fooling or horseplay.
ENVIRONMENTAL	Improper guarding. Improper illumination. Defective parts. Improper ventilation. Improper dress. Poor housekeeping.

All the accidents listed and many others can be prevented through *inspection, training, proper maintenance, suitable layout, proper ventilation* and *illumina-*

tion, *supervision*, and *discipline*. All of *these* corrections are the responsibility of management, with the building superintendent or plant manager as their representative, being responsible for the majority of the accident causes listed.

Handling of Materials

The handling of materials is the principal single source of accidents and injuries in American industry. Twenty-five per cent of all injuries are caused from some form of materials handling. Accident analysis shows that the substitution of mechanical facilities for handling materials over manual methods, reduces accidents greatly.

Mechanical Methods:

1. Hoisting apparatus.
2. Overhead travelling cranes.
3. Conveyors.
4. Mechanical shovels.
5. Elevators.
6. Chutes.
7. Electric trucks, tractors, and hand trucks.
8. Sliding materials on skids or rollers.
9. Wheelbarrows.

Some form of hand labor is generally the cause for the largest number of permanent injuries. These injuries are generally caused by the handling of objects such as sheet metal, machinery, lumber, freight, sheet iron, and steel. Special safety *practices, training* and regular *inspections* should be devoted to these types of materials.

With mechanized handling, strain from lifting excessive loads is almost obsolete. When lifting, it should be remembered that the back is to be kept as *erect* as possible, so that the body is *over* the load. This will utilize the heavy leg muscles and places minimum tension on body parts subject to injury from lifting.

Costs of Accidents

Accidents are costly not only to employees, but to employers, the injured man, his family, and to society as a whole.

Indirect Costs:

1. Cost of time lost by injured employee.
2. Cost of time lost by other employees who stop work out of curiosity, sympathy, or to offer assistance.
3. Cost of time lost by foreman's investigations, reports, hearings, and training new employee.
4. Cost of damage to product, machinery, and building.
5. Cost of in-plant medical aid not covered by insurance.

Direct Cost of Accidents:

1. Hospitalization.
2. Doctor bills.
3. Compensation.

Safety Performance

Appraisal of safety performance can be attained only by the knowledge of at least three basic items.

1. The accident frequency rate of the firm:

$$\text{Accident Frequency Rate} = \frac{\text{Number of Disabling Injuries} \times 1,000,000}{\text{Total Number of Man-Hours Worked}}$$

2. The accident severity of the firm:

$$\text{Accident Severity Rate} = \frac{\text{Number of Days Lost} \times 1,000}{\text{Total Number of Man-Hours Worked}}$$

3. The comparison of these rates, with other rates of the same or similar industries employing good practice.

Example: What are the accident frequency and severity rates of a firm with 90 workers averaging 40 hours a week each, if in six months, four workers were injured, and if jointly they lost 100 days from work?

$$\text{Accident Frequency Rate} = \frac{4 \text{ Injuries} \times 1,000,000 \text{ Man-Hours}}{90 \text{ Workers} \times 40 \text{ Hours per Week} \times 26 \text{ Weeks}} = 4.2+$$

$$\text{Accident Severity Rate} = \frac{100 \text{ Days Lost} \times 1,000}{90 \text{ Workers} \times 40 \text{ Hours per Week} \times 26 \text{ Weeks}} = 1.19+$$

A disabling injury is usually defined in practice, as one causing loss of working time beyond the normal shift, during which the injury was incurred.

Uses of Accident Rates:

1. To measure the accident experiences of departments, divisions, or firms.
2. To determine from month to month, or for any specific period, if accidents are increasing or decreasing.
3. For comparative purposes, and for accident prevention contests.

Safety Training

Unless employees know how to use it, and where it is located, safety equipment is of no use. Employees should be trained in the proper use of equipment, through instruction periods giving complete explanation and demonstration of the equipment. Each employee should be given the opportunity to use it, as the purposes of practice are proficiency and retention. The employee will learn by doing. This equipment should be identified by special signs, using specific colors, or markings placed near the equipment. The color will identify the specific type of equipment. If equipment, such as fire extinguishers, are mounted on a pole, the color code should encircle the pole, in order to make it visible to everyone in the immediate area.

When an employee is hired, it should be made certain that he understands that *all* accidents should be reported. Accident reports and records are a necessity. These records should be permanent, and based on accurate information. This form must be designed carefully, so that the safety engineer or person responsible for the safety program can interpret the record, to determine the cause of the accident. It should contain the required information for insurance companies carrying compensation risk, government organizations, industrial commissions, or the insurance carrier. The responsibility for the reporting of accidents should rest on the foreman or department head, as he is responsible for the actions of his employees. The cooperation of the individual making the report can be secured by explaining to him, that reporting absolute facts of the accident will be beneficial to his operation, and may prevent someone else from being injured.

Safety is the responsibility of every employee and employer for twenty-four hours a day, and for every day in the week. Its importance cannot be over-emphasized. An everlasting interest must be maintained by all employees. A safety organization is a *must*. It pays many dividends to the employer.

Have you checked your *accident rates lately, indirect* and *direct* costs of accidents, and *compensation payments?*

Ten Commandments of Safety for Supervisors

(*Source:* "Ten Commandments of Safety for Supervisors," reproduced by special permission of the American Management Association. Further reproduction prohibited.)

1. You are a supervisor and thus, in a sense, have two families. Care for your people at work as you would care for your people at home. Be sure each of your men understands and accepts his personal responsibility for safety.

2. Know the rules of safety that apply to the work you supervise. Never let it be said that one of your men was injured because you were not aware of the precautions required on his job.

3. Anticipate the risks that may arise from changes in equipment or methods. Make use of the expert safety advice that is available to help you guard against such new hazards.

4. Encourage your men to discuss with you the hazards of their work. No job should proceed where a question of safety remains unanswered. When you are receptive to the ideas of your workers, you tap a source of firsthand knowledge that will help you prevent needless loss and suffering.

5. Instruct your men to work safely, as you would guide and counsel your family at home—with persistence and patience.

6. Follow up your instructions consistently. See to it that workers make use of the safeguards provided them. If necessary, enforce safety rules by disciplinary action.

Do not fail the company, which has sanctioned these rules—or your workers, who need them.

7. Set a good example. Demonstrate safety in your own work habits and personal conduct. Do not appear as a hypocrite in the eyes of your men.

8. Investigate and analyze every accident—however slight—that befalls any of your men. Where minor injuries go unheeded, crippling accidents may strike later.

9. Cooperate fully with those in the organization who are actively concerned with employee safety. Their dedicated purpose is to keep your men fully able and on the job and to cut down the heavy personal toll of accidents.

10. *Remember:* Not only does accident prevention reduce human suffering and loss; from the practical viewpoint, it is no more than good business. Safety, therefore, is one of your prime obligations—to your company, your fellow managers, and your fellow men.

By leading your men into "thinking safety" as well as working safely day by day, you will win their loyal support and cooperation. More than that, you will gain in personal stature. Good men do good work for a good leader.

—Charles P. Boyle

60 / SCHEDULING

One of the greatest cost reductions that can be made in a plant or building is achieved by detail planning which leads to proper scheduling. In order to prevent interruptions in production schedules due to breakdown in machines or building facilities, maintenance work of a preventive nature to eliminate failure of machines or equipment through lack of repair or attention must be performed. There is sometimes a conflict of ideas between production and maintenance departments that creates special problems in the scheduling of preventive maintenance. Production management desires to obtain maximum machine and facility utilization at all times, therefore they are reluctant to give up machine and facility production time to permit maintenance work. Maintenance management wishes to gain maximum utilization of maintenance personnel on a planned and scheduled basis, avoiding unscheduled emergency work. The decisions have to be made:

1. To run machines and use facilities at maximum utilization levels with no scheduled maintenance, other than emergency repairs?

or

2. Have a preventive maintenance program to reduce emergency repairs, which will cut down output on a scheduled basis, rather than risk long delays with breakdowns?

The answer to these questions can be supplied by an analysis of historical data, or starting a system to provide such data, to be gathered over a six month to a one year period, then evaluated to determine your break even point.

Objectives of a Planning
and Scheduling Program

1. To set up a maintenance program on a scheduled basis which will integrate production schedules with the best utilization of maintenance and production personnel.

2. To obtain optimum maintenance at least cost. This would mean the lowest maintenance cost per finished product.

3. To reduce downtime and breakdown time. When a plant or a department is completely shut down for a certain period, consideration should be given to the expected life of machine and facility components in critical machines or equipment.

4. To maintain accurate records of equipment and facilities covering man hours, parts replacements, and costs. Good records will reveal if nonessential jobs are being done, or if your scheduling is not correct.

5. To provide required inspections for essential equipment that may cause a shutdown of production or endanger life or safety practices.

6. To eliminate conflicts or craft restrictions, and to allow for *proper* inventories of parts, tools, and other articles, being careful not to overstock.

Most manufacturers furnish recommended maintenance schedules with their equipment, along with maintenance manuals and parts catalogs. Their recommendation should be followed for the first few cycles, then *judgment should be made* if *more* or *less* maintenance is required for your particular operation. The average hours of running time will have to be known for the cycle to estimate the correct maintenance scheduling. Delivery dates on replacement parts should be determined to aid in proper scheduling. Maintenance manuals are of utmost importance and provide a good basis for training of personnel and in recommending proper techniques.

Schedules should be reviewed once each year to determine their accuracy. Changes may be required through purchases of new and better equipment, new tools, relocations, and better methods.

Recommended Maintenance Schedules
for Certain Building and Plant Facilities

Weekly

Air Compressor. Drain water.
Boilers: Clean panel box.
       Turn filters.
       Clean cones.
       Clean scanner tube.

Blow down.
Clean outside (front and back).
Clean motor belts.
Check boiler water treatment (also cooling tower).
Check all drains.
Check sprinkler valves and pump pressures.

Biweekly

Check heater relays in boiler.
Check boiler stacks.
Check grease and water in electric trucks.
Check water in fire alarm batteries.
Check lawn mower (change oil, clean, and adjust).
Check packing on engine room pumps.
Check tractor (oil and grease).

Monthly

Boilers: Check fire bricks.
        Check electrodes.
        Clean programmer.
        Test safety valves.
Climate changers, oil and grease.
Check conveyor.
Clean motors, drain and clean cooling tower.
Clean, oil, adjust pumps in engine room.
Run purge compressor (October through April).

Bimonthly

Check filters.
Check strainers in all faucets.
Grease, oil, check hydraulic level in lift dock.
Rotate transfer pumps.

Every Three Months

Change oil in air compressor.
Check battery charger.
Oil overhead outside doors.
Grease and oil inside fans in lavatories.
Oil filters.
Clean and inspect floor machines and vacuum cleaners.
Check roof fans.
Purge compressor (oil in crankcase).
Check warehouse hand trucks.
Oil, grease, and clean transfer pumps.
Clean drains in all water fountains.
Clean and spray drains in lavatory floors.

Every Four Months

Grease and oil fans in lavatories (outside).
Oil and grease warehouse roof fans.
Check unit heaters.

Every Six Months

Rotate air compressors.
Clean out boiler heaters.
Tighten all electric and telephone boxes.
Clean outside of electric trucks and paint if necessary.
Weight carbon dioxide fire extinguishers.
Check caulking on flag poles.
Check guard fence.
Check oil manhole.
Clean oil strainers on boilers and transfer pumps.
Clean sprinkler heads.
Check unit heaters and thermostats in penthouse.
Clean soap dispensers.
Clean inside water fountains.
Blow down water heaters.
Check inside pump couplings.

Yearly

Change to auxiliary engine room pump.
Wash ceilings and window sills in all lavatories.
Check boiler valves.
Vacuum all ceiling panes.
Wash all desks.
Clean condenser tubes.
Remove condenser head.
Dismantle and clean motors on electric trucks.
Wash all file cabinets.
Check all fire extinguishers.
Shut water off and drain hose bibs.
Check all light tubes.
Dismantle and clean roof fan motors.
Check toilet seats and fasten loose rubbers.
Wash wastepaper baskets.
Wash glass partitioning.
Inspect roof.

Every Two Years

Replace all light tubes.
Vacuum drapes.
Vacuum panels around anemostats.
Clean purge compressor float valve.

(*Source:* "The Watchmen's Handbook"—This material used by permission of American District Telegraph Co., copyright owner.)

The Watchman

Promptness. It is important to always report for duty on time. A watchman who is ill or unable to report for any other reason, should be sure to communicate with his employer in time to enable him to get a substitute.

Reporting for duty. Upon reporting for duty a watchman should make his presence known to his superior and ask if there are any special orders. If so, written notes should be made so that they will not be forgotten.

Smoking. There should be no smoking during tour of duty.

Alcohol. Never drink anything intoxicating before or during your period of duty.

Check your equipment. Before starting on the first round, all equipment should be checked. See that you have your watchman's clock, or tour key, flashlight, extra batteries and bulbs are available, and other allied equipment. Check your watch to make sure it is correct. If a revolver is provided, it should be inspected carefully.

Never leave the building. During your period of duty (unless your tour or patrol requires it) never leave the building.

Visitors. Never allow unauthorized persons to enter the premises. Never allow personal acquaintances to visit you during your period of duty.

Report broken rules. If you see anyone breaking the *no smoking* rules or any other safety rule; or if you see any evidence that such rules have been broken, such as cigarette butts on the floor, report such violations of safety rules to your superior.

Packages. Never allow anyone to leave the premises after closing hours with a bundle or package unless you have been notified by someone in authority to do so, in which case, for your own protection, you should ask for a written order permitting the removal of the package from the premises.

Telephone. Before starting your first round, check the telephone (or telephones, if there are several connections) to make sure which are connected and whether they are in working order.

Know your premises. Your first duty is to know your premises thoroughly. Check carefully through the following list and if there is anything on it that you do not know, ask your superior to tell you. You should know the following:

1. Where all stairways and doors lead to.
2. Locations of all fire and other emergency exits.
3. What stocks or materials are stored on the different floors, and in the various rooms and closets, throughout the premises.
4. Locations of fire alarm boxes.

5. Location of telephones.

6. Location of all emergency fire fighting equipment.

7. Locations of light switches, so that if emergency demands you can turn on the lights in any part of the premises without delay.

8. Locations of fire standpipe lines, fire hydrants and hose reels, so that in case of a fire you can show the fireman to this equipment without delay.

9. Location of control switches of elevator motors.

10. If the premises are sprinklered, you should know the locations of all valves controlling the supply of water to the sprinkler system and where extra sprinkler heads are kept.

11. Locations of fuse boxes, power control switches, steam valves, hydrants and any other control devices relating to the machinery and operations within the premises. In case of fire or other emergency, it often is important for the fireman or other emergency forces to operate or close such devices and you should be able to give them any needed information without delay.

12. You should know how to operate all of the above equipment and devices that you might need to use in case of emergency. If you do not know this, ask your superior to arrange for the engineer or other competent person to give you any instructions you need.

13. You should know the locations of control room and shut off switches for air-conditioning or ventilating systems.

14. If your duties include attending to the heating system, make sure that you thoroughly understand its operation, so that if anything goes wrong, you will be able to detect it and take proper action. If unable to cope with the situation yourself, notify your superior.

15. To insure that you know your plant thoroughly, you should make a complete tour of the premises during the daytime, asking your superior or other employees for information about equipment, machinery and operations that you may need to assist you in your work.

16. The following information, names, addresses and telephone numbers should be kept posted at your headquarters:

a. Location of nearest street fire alarm box if your plant is not directly connected to your local fire bureau.

b. Telephone numbers of the police department and fire department.

c. The name, address, and telephone numbers of your superior or other employees that special conditions in your plant might require you to call in emergencies, such as the plant superintendent, engineer, foreman, and others. Consult with your superior to get any of this information that you need.

Making your rounds. You should always make your rounds on time, and you should follow the exact route that has been laid out for you. Of all your patrols, the first round is the most important. If anything is wrong, it is important to discover it as soon as possible. The following is a checklist of some main points to watch for on your first round.

1. Has any motor or machine been left running that should have been turned off at the end of the normal shift? If so, attend to the condition yourself if you can. If you are in doubt what to do, notify your superior immediately.

2. Have any water faucets or other water outlets been left running? If so, shut them off.

3. Have any unnecessary lights been left burning? If so, turn them out.

4. Test the telephone to make sure that it is in working order.

5. Check all doors, windows, and skylights, and see that they are closed, to prevent possible damage from rain or snow, as well as to protect the premises against intruders.

6. If any windows are broken, fix them temporarily, and notify your immediate superior.

7. Check to see that all passageways leading to exits, fire doors, and equipment, are unobstructed. If they are try to move the item, if not possible report it to your superior.

8. Check to see that all fire doors that should be closed are closed.

9. Investigate any unusual odors, especially odors of smoke or gas.

10. Check to see that boxes, rubbish or other dangerous materials that might spread a fire have not been left near boilers or other heating units. Oily rags that might cause spontaneous combustion should be placed in a metal container or removed from the building.

Taking care of yourself. Always direct the beam of your flashlight ahead of you. Even if you know your plant well enough to go through a dark room without a light, someone may have rearranged the furniture.

Be on guard against open elevator doors, shafts and other floor openings. You should know where the first aid equipment is located and have access to it, so you can help yourself in case of slight injury. If you suffer a more serious injury or become ill, summon aid immediately.

Watchmen's Patrol Systems

Most employers equip their premises with some type of system which enables the watchman to record his patrols of the premises. These systems are of two main types: (1) Time recording systems, usually known as "Watchclock" systems and, (2) Central Station Watchman's Reporting Service.

Watchclock systems. This type of system usually consists of a portable clock which the watchman carries with him on his tour and a series of stations or keys distributed about the premises. Inserting the key into the clock records upon a paper roll, chart or dial inside the clock, the time when each station is visited. The following day the watchman's superior examines the tape in the clock to see if the rounds were made on time and in the order prescribed, during the preceding night.

If you are supplied with such a watchclock system, be sure that you visit your stations on time, and in the order prescribed. Do not take "shortcuts", omit visiting any station or change the order in which you are scheduled to visit them. If an emergency arises that compels you to delay or otherwise depart from the routine that has been set for you, be sure to inform your superior of this fact and the reason, when you turn in the clock upon completing your period of duty.

Central Station Watchman's Reporting System. With this type of system, the watchman is in contact with outside personnel, which are ready to come to his aid in time of emergency. This system consists of one or more preliminary and transmitting stations distributed throughout the premises, and a register key which the watchman carries and inserts and turns in the stations during his rounds. With this type of system the stations must be visited in regular order, the key is so constructed that it will fit in the station only if the preceding ones have been visited in their correct order. Signals from the transmitting stations must be on time or the *Central Station* (outside contact) will investigate.

What to Do
Upon Discovering a Fire

Always remember "the first five minutes at a fire are worth the next five hours." If the firemen are summoned *right away* and start fighting the fire in its earliest stages, they can nearly always put it out before it causes much damage.

If you discover a fire that is so small that you are *sure* you can put out with sand, water or a fire extinguisher, you should extinguish the fire. Having done so, examine carefully the area of the fire, to make sure that it is completely out, with no danger of breaking out again. If there is the slightest doubt, call the fire department. It is far better to be safe than sorry.

How to summon the fire department. The best way to summon the fire department in your case will depend on what type of protection has been provided your building.

1. If the premises are protected by a *manual fire alarm system* connected to the local fire department, all you have to do is pull the box.

2. Some buildings are equipped with *Local Interior Fire Alarm Systems.* Operating a box of this type system does not summon the fire department, but merely sounds an alarm on the premises, to warn the occupants. *Make sure* that you know what type fire alarm system your premises are equipped with and that you know what it will do, how to operate it and the locations of all the alarm boxes.

3. Telephone or pull a street fire alarm box. If you phone make certain that in the excitement of reporting the fire you give the name of your premises and address distinctly. The telephone number of the fire department should be memorized by all the watchmen. The location of the nearest fire alarm box should also be memorized by all watchmen.

Security Survey

Outdoor. A survey of your outdoor areas should be made to determine if security can be improved. Windows that could be used to gain entrance should be screened or barred. In some cases it may be desirable to wire the windows to set off an alarm, should they be broken or forced, doors may also be wired. By designating certain doors for entrance and exit pilferage can be controlled.

Parking lots may be fenced in for authorized cars only, with entrance and exit made through a guard gate. An employee identification system can be installed, and authorization required by all visitors. Trucks should have authorized passes, and be routed through a guard gate. The perimeter of the plant should be lit up by floodlight at night.

Internal. A survey of your interior area should be made to determine if security can be improved. Windows that could be used to pass material out should be screened. Employee entrances and exits can be designated to insure that all employees pass through guarded doors only. It is recommended that all personal packages be checked with the guard when entering the plant, and picked up when leaving. Shipping areas should be screened and locked so that drivers would not have access to any merchandise when authorized personnel are not present.

62 / MAINTENANCE STANDARDS

(*Source:* "Techniques of Plant Maintenance and Engineering," Clapp & Poliak, Incorporated, New York)

It is generally considered impossible to secure 100 per cent productivity at all times from the maintenance department. However, it is possible to decrease the unproductive times considerably through good supervision, better planning and scheduling, and through work simplification and sampling methods, along with work measurement techniques.

Sound planning and scheduling procedures should be put into operation. This is the direct responsibility of *every* maintenance supervisor. This will insure that the job can be done when scheduled, and that the correct *number* of men have been assigned to the job. Tool requirements should be established, and crafts dispatched to the job in proper sequence along with the knowledge of their next job and its location, in order to avoid wasting time waiting for further assignments.

Methods of Determining Standards

Estimates. The most common way to determine maintenance standards is by estimating. It is fairly easy to estimate the cost of materials, and sometimes rather difficult to estimate the time required to perform the job. The labor estimate can only be defined as the probable time required to perform a job, based on the best judgement of the person making the estimate. Therefore the personal experience, knowledge, and ability of the estimator will determine the quality of the estimate. Employees will not respond to poor estimates. If the estimates are too tight, they lose interest in trying to meet the goal; if they are too loose, they lose confidence in the program.

Historical records. This is a refined method of estimating using the average time values based on past experience. Historical work orders are assembled showing all maintenance work that has been performed with average times arrived at through the grouping of similar jobs and obtaining standard times. The disadvantage of this method is that its accuracy is low and reflects only what has been done in the past. If you induce a new method of performance, secure better tools, or new equipment, the standard would then become loose, and you may have difficulty adjusting the standard. In this way past inadequacies become built into the system.

Measurement by standard time data. This is the most accurate method of determining job times. The standard data should be gathered through the use of actual time studies, or through synthetic time such as measurement time method or work factor. The use of this method makes it possible to compute the job time by dividing the standard time by the actual time.

Work sampling. The primary objective of a work measurement program is to determine the things over which the workman has no control and that are reducing his effectiveness, and therefore his efficiency. The analysis of work sampling survey reports will provide many kinds of information that may be useful to management. You will be able to determine the average number of man-hours required to do specific jobs. This information can be used to set up performance standards, and also to estimate costs on future jobs. Work sampling is a method of obtaining facts about machine or human activities. Two methods of work sampling are the random sampling and the continuous observation. The latter method requires greater man power assignment and cost, but provides more representative data more quickly.

Introducing the Program

Prepare a manual explaining the program. State how and by whom it would be executed. Outline the role each person is to play—engineer, maintenance foreman, supervisor, and other personnel. A representative of each of these groups should receive a copy of the manual before publication and should be asked for comments prior to publication.

Inform all personnel and union representatives of your plan and solicit their cooperation and assistance. Distribute copies of the manual to persons taking part in the execution of the program mentioned above. Several days after the manuals have been distributed, a meeting should be called to go over the survey procedures in detail and to answer any questions they may have.

Select the person or group to be surveyed and arrange for the observers who are generally supervisors or engineers from either the maintenance or operations department. It is recommended that a minimum of 50 per cent of a department or group be surveyed at one time to prevent the work from being loaded on to the observed persons.

Observers are required to be at the plant gate a half hour before the shift starts. When they arrive each is told which workman he will follow and what the man's classification is. The observer introduces himself to the employee at the gate, advises him he will be with him all day, and stays with him for the entire shift. The employees' supervisor does not know that his group is being surveyed until his workmen show up in the shop or on the job accompanied by the observer.

In his report, the observer accounts for every minute of the employee's day from the time he enters the plant until he leaves at the end of the shift, except for the official lunch period, noting the starting time to the nearest minute for every change in operation or activity in the left-hand column of his form. He also writes a brief description of the activity in the right-hand column.

The observer stays out of the way of the workman and asks him as few questions as possible. As the shift progresses, the "general remarks" section of the survey report is filled in. The remarks may consist of the observer's personal opinions, and deal with such things as shoddy work, unsafe practices, lack of instructions, and similar information, particularly the matters that may have affected the workman's effectiveness. The observer may also report unsafe practices, poor workmanship, loafing, and so on, on the part of other workmen in the general area although they may have no connection with the work he is observing.

The next day the observer goes over his survey report, coding each entry and computing the elapsed time for each activity. To assist in the coding he uses the "survey time breakdown code". Then the "time analysis summary" is prepared to summarize total time by each code, from the data shown on the observer's report, the "time study check list" and the table shown in part. The observer also goes over his "general remarks", and clarifies them wherever necessary.

Using the Findings

As soon as possible after completion of a survey, the survey group supervisor goes over every report in detail with the supervisor of the shop or group surveyed and his section supervisor. The purpose of this review is to direct the supervisor's attention to the things that can be done immediately to improve the effectiveness of the workmen. Situations that require top management consideration and decision are reviewed with the supervisors for informational purposes, but nothing is done about them until management directs that action be taken.

When the survey results show that representative information has been obtained, then conclusions may be drawn. The number of surveys that must be run will vary. We consider that we have representative information when our "time analysis summaries" show practically no change in statistics from previous surveys.

When representative information is obtained a written report is prepared, listing situations that need top management's attention and suggesting solutions. This report is reviewed by the supervisor of the survey group and the assistant project director. The latter then discusses the findings with plant supervisors, tells them what corrective action he expects them to take, and sets a date for accomplishment.

In conclusion, the importance of the "time breakdown code" is stressed. It is essential to allow adequate time for drawing up this survey, since it is the yardstick for measuring the work.

Survey Time Breakdown Code

Code   *Productive Time*

A   1. *Actual Time:*

        a. To include only that time actually spent working directly upon the equipment being repaired or job assigned.

b. Any work with tools necessary and incidental to the job. *Note:* Some job assignments do not involve direct work on equipment or making repairs. In such cases use the proper productive code. The craft and job classification plus a brief description of job assignment in the survey notes will assist in coding each survey report entry. *For example:* A tool crib operator's productive duties consist of material procurement. So code as M. A crane operator's main productive activity is transportation of equipment or rigging. So code as T.

C 2. *Communication Time*—to include the following:

    a. Supervisor to worker instruction.
    b. Reports by workers to supervisors.
    c. Discussions between craft worker and other craft workers regarding the equipment.
    d. Discussions between craft worker and unit operators in regard to equipment.
    e. Telephone calls for material.
    f. Writing up time sheets.
    g. Filling out work orders.
    h. Writing orders for material.
    i. Tagging out of equipment to be worked on.
    j. Studying drawings and specifications, requisitions, bills of materials, or any written or printed information pertinent to assigned job.
    k. Records or check lists required for PM program.

M 3. *Material Procurement Time*—to include the following:

    a. Gathering material or tools.
    b. Hunting for material or tools.
    c. Putting away material or tools.
    d. HP survey to determine that material or tool is not contaminated and may be used or returned to storage.

T 4. *Transportation and Handling Time*—to include the following:

    a. Walking (includes walking to get HP permit, materials, tools, and instructions).
    b. Riding.
    c. Loading or unloading tools, material, or equipment.
    d. Rigging.
    e. Climbing.
    f. Use of crane to lift materials or equipment into place.
    g. Driving to and from work to haul materials or tools or equipment.

CL 5. *Cleaning Time*—to include the following:

    a. Clean-up of equipment before, during, and after repair, including cleaning and decontamination of parts.

b. Clean-up of tools after use, including decontamination.

c. Gathering and disposing of used materials and other refuse in a working area.

d. Time to scrub body or hands to remove chemicals or contamination.

e. Washing hands that have become so soiled that they must be cleaned in order to go on with job. *Note:* Removal of machine gaskets is considered cleaning time.

SHP   6. Safe Work Permit and HP Permit (preparing and obtaining, issuance of dosimeter). Does not include waiting on safety engineer of HP. Code WOC for waiting time.

Non-Productive Time

1. *Delay Due to Others:*

WO   a. Waiting for operator. *For example:* Waiting for operator to get equipment ready for repair. (Actual time to turn equipment over to craftsmen. Does not include waiting time before operator starts turnover procedure. This time is coded WOC.)

WT   b. Waiting for transportation.

WM   c. Waiting for material delivery.

WS   d. Waiting for supervisor. *For example:* Waiting for supervisor to make decision as to method of repair.

WOC   e. Waiting for other crafts. *For example:* (1) Waiting for other crafts to complete their portion of a job so that the observed worker can proceed; (2) Radio-active contamination and clean-up; and (3) Waiting on HP safety engineer, infirmary nurse, late bus at start of shift, and materials or toolroom clerks.

WWA   f. Waiting for work assignment. *For example:* Worker runs out of scheduled work and has notified supervisor but has not been given a new assignment.

WCW   g. Waiting for co-worker. Lost time due to assignment of two people to a one-man job. (Necessary in many cases.)

PC   h. Protective clothing, (dress and undress) personal or anti-C clothing.

U   2. *Unavoidable Delays:*

a. Attempting to use telephone but it was busy, cannot get person called.

b. Waiting to make the several necessary packing adjustments to a pump just put back in service after a repacking job.

c. Repair or malfunctioning equipment or tools.

3. *Other Delays:*

L   a. Lunch—includes preparation, coffee or coke break.

S   b. Smoking—includes travel to and from processing area to clean area.

MIS   c. Miscellaneous—includes the following:

(1) Washing hands for lunch and at end of shift.

(2) Personal needs.

(3) Idle time when not waiting for others to perform work or for work assignments. *Note:* Record and code all miscellaneous time. A description of miscellaneous activity in the survey notes is unnecessary. Any excessive use of miscellaneous time can be noted under General Remarks.

SM   d. Safety meetings.

UB   e. Union business (workman's committee, gripe sessions, phone calls, posters, talking with shop steward).

Figure 62-1

Survey No. .... Observer: ....

1. Worker used open end wrench to tighten tubing connections. This is correct tool.

2. Worker had to work on floor when bending tubing. A bench would have made it easier.

3. 8:45—Co-worker studied drawings, apparently trying to figure out what to do.

4. 8:55—Made mistake. Scrapped tubing, about nine or ten inches.

5. Metal and burnable waste observed in waste containers. Metal scrap mostly copper tubing.

6. 9:15—Maintenance engineer visited job.

7. Bottom of D/P cabinet dirty. Covers for other instruments laying in bottom of cabinet.

8. Worker disconnected piece of tubing from another instrument from which water drained. He was checking to see how tubing fittings were made up.

9. 10:35—Up to now co-worker has done little helping. Helper or co-worker is actually a 2/c mechanic fitter (fitter).

10. 10:45—Co-worker measured for a tubing run.

11. Foreman found tubing fitting; worker went to sponsor's warehouse to get cutter. He found it in sponsor's warehouse.

12. 11:16—Two electricians delivered materials to general work area. Parked material in front of instrument panel.

13. 11:22—Operations operator took lunch box and left for lunch.

14. 12:30—Another workman used desk top as work bench on which to saw piece of instrument.

15. 12:30—Co-worker trying to make up a piece of copper tubing.

16. There were not enough welding shields to protect eyes of surrounding workers from arc flashes.

17. 12:45—Two electricians showed up to work on installing square duct in or on instrument console. Area already congested before they arrived. They were in way of men working on tubing.

18. 1:02—Had to rework a piece of tubing on which he had made a mistake.

Figure 62-2

Time Analysis Summary

Survey No. .... Observer: ....

Craft .... Classification .... Date ....

SUMMARY (Analysis of Time Used)	ACTUAL SURVEY TIME (Recorded)		ACTUAL EFFECTIVE TIME (Estimated)		POTENTIAL EFFECTIVE TIME (Estimated)	
	Time Man-Min.	Per cent of 8-hr. Shift	Time Man-Min.	Per cent of 8-hr. Shift	Time Man-Min.	Per cent of 8-hr. Shift
Productive Time						
1. Actual time.	250	52.1				
2. Communication time.	82	17.1				
3. Material procurement time.	25	5.2				
4. Transportation time.	46	9.6				
5. Safe work and/or HP permit.						
6. Cleaning time.	4	0.8				
7. Estimate available due increase effective (−15 per cent).						
Total Productive Time	407	84.8				
Non-Productive Time						
1. *Delays Due to Other Groups:*						
a. Waiting for operator.						
b. Waiting for transportation.						
c. Waiting for material.						
d. Waiting for supervisor.						
e. Waiting for other craft.	9	1.9				
f. Waiting for work assignment.						
g. Waiting for co-worker.	1	0.2				
2. *Unavoidable Delays:*						
3. *Time For:*						
a. Protective clothing.	21	4.4				
b. Union business.						
c. Procedures.						
d. Defective equipment or tools.						
e. Safety rules or meeting.						

SUMMARY (*Analysis of Time Used*)	ACTUAL SURVEY TIME (*Recorded*)		ACTUAL EFFECTIVE TIME (*Estimated*)		POTENTIAL EFFECTIVE TIME (*Estimated*)	
	Time Man-Min.	*Per cent of 8-hr. Shift*	*Time Man-Min.*	*Per cent of 8-hr. Shift*	*Time Man-Min.*	*Per cent of 8-hr. Shift*
4. *Other Delays:*						
a. Lunch.	24	5.0				
b. Smoking.						
c. Miscellaneous.	18	3.7				
5. *Ineffective Time.*						
6. *15 per cent of estimate available increase.*						
Total Non-Productive Time	73	15.2				

Figure 62-3

STANDARD JOB TIMES FOR FLOOR CLEANING OPERATIONS
(*Time in Minutes per 1,000 sq. ft.*)

Sweeping		Machine Scrub	
Unobstructed	9	Unobstructed	25
Slightly obstructed	10	Slightly obstructed	35
Obstructed	12	Obstructed	40
Heavily obstructed	16	Heavily obstructed	45

Dust Mopping		Machine Polish	
Unobstructed	7	Unobstructed	15
Slightly obstructed	9	Slightly obstructed	25
Obstructed	12	Obstructed	30
Heavily obstructed	16	Heavily obstructed	35

Damp Mopping		Vacuum (wet pick-up)	
Unobstructed	16	Unobstructed	20
Slightly obstructed	23	Slightly obstructed	27
Obstructed	27	Obstructed	31
Heavily obstructed	32	Heavily obstructed	35

Wet Mop and Rinse		Vacuum (dry pick-up)	
Unobstructed	35	Unobstructed	14
Slightly obstructed	45	Slightly obstructed	17
Obstructed	50	Obstructed	19
Heavily obstructed	55	Heavily obstructed	23

Hand Scrub		Strip and Rewax	
Unobstructed	240	Unobstructed	100
Slightly obstructed	300	Slightly obstructed	120
Obstructed	330	Obstructed	140
Heavily obstructed	360	Heavily obstructed	180

Hand Scrub (long brush)	
Unobstructed	75
Slightly obstructed	105
Obstructed	120
Heavily obstructed	135

TIME STUDY CHECK LIST

Survey No. MTR-ETR #119 Date 4-28-60

Craft Mechanic Fitter Observer P. C. Leahy

PRODUCTIVE TIME CODES									NON-PRODUCTIVE TIME CODES																
									Waiting For								Time For					Other			
Actual Time			Communi-cation	Material Procurement	Transport-ation		Safe Work HP Permit	Cleaning	Operator	Trans-portation	Material	Supv'r.	Other Crafts	Work As-signment	Co-Worker	Unavoid-able Delay	Protective Clothing	Union Business	Procedure	Defective Tools or Equipment	Safety Rules or Meeting	Lunch & Coffee	Smoking	Miscel-laneous	
A			C	M	T		SHP	CL	WO	WT	WM	WS	WOC	WWA	WCW	U	PC	UB	P	DET	SM	L	S	MIS	
4	2	3	2	4	5	1		1					3		1		5					15		2	
5	2	3	1	2	1	1		3					2				2					4		5	
4	3	15	1	4	1	5							4				8					5		1	
3	2	2	3	2	2												6							1	
2	3	6	1	1	1																			4	
6	1	10	10	2	2																			5	
3	2	3	1	1	1																				
5	4	1	1	1	1																				
7	1	1	2	8	1																				
4	2	5	9		1																				
3	8	2	1		1																				
3	1	1	1		1																				
1	7	7	1		2																				
7	5	2	9		1																				
1	3	5	1		2																				
3	1	14	2		2																				
2	5		1		1																				
1	2		1		1																				
2	19		1		3																				
2	1		5		1																				
5	5		2		2																				
3	3		4		1																				
1	1		1		1																				
1	1		2		2																				
1	2		1		1																				
1	4		2		1																				
			4																						
			1																						
			1																						
			5																						
			1																						
250			82	25	46			4					9		1		21					24		18	
Column Totals																									
Grand Total																									

Figure 62-4

Figure 62-5

STANDARD JOB TIME LIST
(*Time in Seconds*)

Dusting

Ash Tray	15
Book Cases	
13x35x12 inches	22
36x30x8 inches	33
12x40x12 inches	26
42x24x11 inches	49
Cabinets	
36x77x18 inches	106
30x66x18 inches	42
Calculators	
Small	7
Large	9
Chairs	
Large	63
Medium	35
Stenographer's	22
Cigarette Stand	25
Clock, Desk	8
Clock, Wall	20
Desks	
Large	48
Medium	43
Small	38
Desk Items (miscellaneous	3
Doors	
Without glass	25
With glass	40
Elevator Cabs (inside)	196
Files	
4 drawer	22
5 drawer	27
Fire Extinguishers	16
In and Out Trays	8
Lamps and Lights	
Wall Fluorescent	8
Desk Fluorescent	18
Table lamp, shade	35
Floor lamp, shade	35
Partitions, Glass-50 sq. ft. per unit	60
Rack, Coat, and Hat (6)	90
Radiators and Window Ledge (124x15 inches)	45

Radiator (flush with wall) (40"x30"x6')	21
Sand Urns	60
Spittoons	180
Tables	
Large	60
Medium	35
Small	22
Telephone	9
Typewriter (covered)	7
Vending Machine	60
Venetian Blinds (standard)	210
Wastebaskets	15

Lavatory Items

Cleaning Commode (with partition)	180
Door (spot wash)	50
Door Latch	10
Mirrors	
25x49 inches	20
60x21 inches	20
88x31 inches	40
Napkin Dispenser	13
Napkin Disposal	10
Paper Towel Dispenser	7
Paper Towel Disposal	10
Shelving	
20 inches long	8
126x6 inches	60
Urinals (complete)	120
Wainscoting	
75 to 100 feet long	25
Wash Basin, Soap Dispenser	120

Washing

Glass Partitions	
Clear—8 sq. ft. per unit.	60
Opaque—20 sq. ft. per unit.	60

Miscellaneous

Door (washing)	150
Drinking Fountain	90
Vacuuming	
Large Divan	190

Figure 62-6

TABLE SHOWING MINUTES AS PERCENTAGE OF
A STANDARD 8-HOUR (480 MINUTES) SHIFT

	0+	100+	200+	300+	400+
1	.208	21.042	41.875	62.708	83.542
2	.417	21.250	42.083	62.917	83.750
3	.625	21.458	42.292	63.125	83.958
4	.833	21.667	42.500	63.333	84.167
5	1.042	21.875	42.708	63.542	84.375
6	1.250	22.083	42.917	63.750	84.583
7	1.458	22.292	43.125	63.958	84.792
8	1.667	22.500	43.333	64.167	85.000
9	1.875	22.708	43.542	64.375	85.208
10	2.083	22.917	43.750	64.583	85.417
11	2.292	23.125	43.958	64.792	85.625
12	2.500	23.333	44.167	65.000	85.833
13	2.708	23.542	44.375	65.208	85.042
14	2.917	23.750	44.583	65.417	86.250
15	3.125	23.958	44.792	65.625	86.458
16	3.333	24.167	45.000	65.833	86.667
17	3.542	24.375	45.208	66.042	86.875
18	3.750	24.583	45.417	66.250	87.083
19	3.958	24.792	45.625	66.458	87.292
20	4.167	25.000	45.833	66.667	87.500
21	4.375	25.208	46.042	66.875	87.708
22	4.583	25.417	46.250	67.083	87.917
23	4.792	25.625	46.458	67.292	88.125
24	5.000	25.833	46.667	67.500	88.333
25	5.208	26.042	46.875	67.708	88.542
26	5.417	26.250	47.083	67.917	88.750

63 / TRAINING

If you were asked to define your position as a sanitarian, building manager, building superintendent, or by whatever title your position is designated, what would your answer be? It may be described as follows:

"You are charged to protect the capital investment of your employer and his building, and to provide a clean and healthful atmosphere."

Part of your responsibility, I am sure, is the training of your personnel. Why is training required? Training is required because of increased labor costs; because of the constant introduction of new products and machines into the market which require special techniques in order to use them, and above all, to insure that your employer is getting maximum utilization of every dollar spent for labor and materials. These points are true, whether applied to a sanitation program or to a building maintenance program covering equipment.

Where does our training program begin? The training program begins with you, your supervisor and foreman, and with your employees. First let us consider "you". Are you satisfied that you are now operating as a modern supervisor? Have you kept up with new products, materials, and techniques? You are responsible for planning the work of your unit. It is upon these plans which the performance of your employees depend. It is also the way in which you organize the work, the way you divide and assign the work load, that makes it possible for your employees to handle the entire program efficiently. You are responsible for motivating your employees towards performing a willing and capable job. I ask, "Is this you?" If not, it may be time for you to reconsider how you *plan, organize, control, motivate* or *coordinate*. Does your subordinate management personnel understand these functions fully, and know exactly how you want them performed? If they do not, training should begin immediately.

Your employer feels that you are responsible to find new methods covering the following:

1. Increase productivity.
2. Control quality of work.
3. Train your employees.
4. Plan the work.
5. Promote safety and improve housekeeping.
6. Control costs and reduce waste.
7. Maintain machines and equipment; set up periodic inspections on floor machines and other mechanical tools used, including physical building equipment and mechanical building equipment.
8. Keep good discipline.
9. Handle grievances.
10. Write reports.

Many other items for which you are responsible, may be added to this list. If you feel that you are not handling any one of the items mentioned, in a proper manner it is probably time for you to consider further training for yourself. This can be accomplished through reading, demonstrations by manufacturers, work seminars, correspondence education, and college training.

The next consideration is the performance of management personnel and supervisors under you. Their training depends upon what you want to accomplish. Training may be given to them through the above mentioned recommendations for your own training, or by direct instruction from you. (Instructional consideration should be given to these people in the field of "work sampling" and "work simplification".)

The training of your employees should begin the moment they walk through the front door and are interviewed by the personnel department. From this very moment, habits, practices, and working rules should be put into effect. Training in building sanitation and maintenance is not effected only through instruction periods or courses. Situations change constantly in the sanitation and maintenance field, which makes perpetual training a requirement. You will be constantly making on-the-spot decisions, and these should always conform with good practices. If a sanitation employee is required to make decisions for himself, and if in your opinion they are not in line with good practices, on-the-spot correction must be made immediately, and a precedent and pattern set forth on the particular situation upon which you made judgement.

Let us consider the point of plant safety. Many dollars are spent by firms for the promotion of safety programs. If we continually train our employees in safety on the job, the need for special safety programs will be reduced. This is one example where perpetual training is required. The need for safe practices along with training is always present, and is your direct responsibility. Emergency situations, new practices, and new tools are changed more frequently in your department than in any other department within the same company.

It must be remembered that your employees should be selected properly. Training is easier, and many training problems are eliminated if the right person is in the right job. If your employee is properly trained, he will have developed correct habits and skills. This will result in a good possibility of your getting things accomplished. If your employee is not trained properly, he will develop undesirable habits and will prove to be ineffective, and a hindrance to your operation. It is important to consider mentioning standards in the initial interview with an employee. This will impress upon him that he must carry his work load.

We are now ready for "Training Program Considerations". We should establish, what training is to be given to the new employee, and to what depth we are going to train him.

Training Program Considerations

It is important to have every employee in the building maintenance force trained to do every job, if possible. After this has been accomplished, it can be determined which of the employees can do certain jobs in the most efficient and thorough manner. After every employee has been thoroughly trained in all positions, the jobs should be assigned according to performance. Advantages derived from this method are:

1. Provision for required depth in the department in order to operate effectively and efficiently.
2. Better understanding by each employee, of the job done by his fellow worker.
3. Training of the employee to serve as a replacement during vacation, in time of sickness, death in the family, or other reasons.

Every employee should understand the job of his fellow employee in order to have a better understanding of the work load of the other employees. The work load is determined by management and for psychological purposes of the employee, the positions should be kept equal. Special consideration should be given to physically handicapped employees.

It is recommended that the training be done by the building superintendent, as he is responsible for the employees' performance. (In some organizations this training may be delegated to the service of a training department or a training director.) Training his employees gives the building superintendent a thorough understanding of all jobs and functions which will result in his being able to communicate more intelligently with higher management concerning his needs and department operations. He will also develop a better understanding of his organization for contract negotiations and budget purposes.

Another important factor in the training being conducted by the building superintendent is that in the testing of new products, he will be able to instruct his personnel in the application of these products and can better determine their net result and usefulness. This leads the building superintendent into maintaining a constant evaluation of new products, which in turn if accepted, may require changing or rescheduling jobs if the product proves to be more effective than the one being utilized.

When a new employee is hired, a thorough orientation between the employee and the immediate supervisor should take place. The following reminders are suggested:

A. *Job Factors:*

1. *Job Title (and Grade).* If job is under "Job Evaluation Program," use job evaluation book to show specific job title and grade. Otherwise, give job title and indicate how this job relates to other jobs in the department.

2. *Duties.* Explain briefly the functions of the department, so that employee can see the broad picture; give general idea of individual's work.

3. *Pay Rate and Pay Basis.* See that employee knows hourly rate; when pay days occur; when his *first* pay day occurs; and time and method of pay check distribution.

4. *Hours of Work.* Give starting and stopping time and lunch period times as well as total normal hours per day and per week.

5. *Time Card Use.* See that the employee knows the location and use of time card rack where his time card will be located. Demonstrate use of time clock. Explain rules and regulations concerning use of time card and time clock; and relation of time card to pay check. Explain what to do when errors in time card use occur.

6. *Probationary Period.* Tell employee length of formal probation in the department; explain why such an arrangement exists. Tell employee that the company is also "on probation" as his employer; urge employee to discuss matters frankly with you.

7. *Absence and Tardiness.* Explain company and departmental rules and procedures concerning notification in advance of absences; explain tardiness policy. Point out that company loses when an employee is not on the job since company overhead and fixed costs must be met even though employee does not get paid.

8. *Supervision.* Be sure employee knows the name of and meets all management persons in the department with whom he will be in contact.

9. *Smoking.* Explain smoking rules and reasons therefor; name or show nearest "smoking" area and any regulations concerning its use.

10. *Rest Period, Location of Rest Areas and Lavatories.* If definite rest periods are established, give times and procedures; if no definite rest periods, explain company policy in this respect. Indicate nearest lavatory location.

11. *Sanitary Regulations.* Explain to the employee we are vitally interested in

sanitary conditions. Acquaint him with the rules and regulations governing sanitary conditions in such places as his work area, rest area and lavatories.

12. *Leaving Department.* The supervisor should know where each employee is at all times—*be sure* that this is understood.

13. *Telephone Use and Personal Letters.* Explain company policy concerning telephone calls, both incoming and outgoing; explain that personal letters are not to be written, sent, or received on company time or through company channels or using company materials.

14. *Visits to First Aid Room.* Acquaint employee with location of first aid room. Explain the importance of reporting an accident or illness to his supervisor in order that he may be referred to the first aid room immediately.

15. *Fire Drills.* Explain to the employee that fire drills are held from time to time for the purpose of protecting the well-being of all employees. Tell him what he should do when a fire drill occurs.

16. *Accident Prevention.* Point out existing types of accident hazards and emphasize the need for care. Explain that all accidents should be reported to the supervisor immediately. Explain importance of this in relation to workmen's compensation and other employee's benefits.

17. *Educational Opportunities.*

18. *Union Status.* Advise employee if he or she will be eligible to join a union; if so, name union, tell whether joining is obligation or not (union or open shop); advise length of waiting or probationary period, or other facts that he or she should know about membership.

19. *Work Policy.* Explain to the employee that each person within a department is expected to carry his own work load. Failure to do this can place unnecessary burdens on others.

B. A complete explanation of the job should be given to the employee, including a schedule with the time allotted for each specific job and the areas to be covered. Explain what tools are to be used, how they are to be cared for and in what condition they are to be stored when the employee is through using them.

C. The employee is then assigned to work with the supervisor or one of the better workers for a period of one week, with an assigned area equal to that of the other employees. This will have the new employee working with another employee and teaches him to carry his own work load. After one week the employee should be able to handle his assigned duties by himself.

D. A demonstration should be arranged showing the tools with which an employee will work. He should be provided with a complete schedule of the frequency of operations, explaining whether the job is to be done daily, weekly, or monthly.

E. After one month on a job, the employee is assigned to a new area or another job. He is trained with another employee as described previously. If there are more than five different jobs per shift, the training on five jobs should be sufficient to give the department enough depth in personnel qualifications.

F. A check on the progress of the employee should be made weekly during the first month, and semimonthly while training on other jobs. The immediate supervisor should be consulted on the employee's progress and of any deficiencies in his work. His strong points should be noted and these should be discussed.

If he is acceptable, he should be offered encouragement. The employee should also be encouraged to discuss any deficiencies that he may see in the cleaning program and any new ideas that he may have relating to his work.

G. Training is recommended by conferences, lectures, demonstrations, training films, film strips and slides. There are many sources of supply for this type training. After each period of instruction, a discussion should be held, followed by a class conducted on practical applications, under close supervision, with on-the-spot corrections.

H. Various manufacturers, if their products are purchased, will demonstrate the correct way in which their products are to be used.

Caution: These demonstrations should be observed by the building superintendent before the manufacturer's representative is permitted to describe his product and method to the working force, in order to determine whether or not the method is acceptable to him.

I. A re-evaluation of jobs and job descriptions should be made at least once each year. The maintenance program may change as the building ages and many physical changes may be made each year. Several jobs may have to be revised and rescheduled. If a complete revision or rescheduling is not necessary, with only small changes made in any of the jobs, it is important that these changes be noted on the employee's schedule. If an employee should suddenly become ill and is unable to report for work, this would facilitate providing the employee who will be temporarily handling the job with a complete up-to-date schedule.

J. Many times, if a building superintendent has a building under good control, he is lax in sampling his employees' work. It is recommended that employees' work be sampled once per year. Job procedures should also be simplified once per year. In reviewing procedures, method and order of sequence should be included for each job. This will eliminate the wandering of employees to all sections of the building and will result in greater efficiency. The employee should be required to perform his job in proper sequence and his whereabouts known in times of emergency. A check should also be made to determine that the employee has the proper tools available to him. Make certain that he is trained to carry the necessary tools with him to eliminate his retracing his steps. Employees should be checked for unavoidable delays which may be eliminated through rescheduling or with a conference held with the originator of the delay.

The practice of making on-the-spot corrections cannot be overemphasized. Do not wait for someone else to do it or tell you about it. This is *your department, your building, your responsibility*—then you can be sure of *your job.*

K. Introduce the employee to a sufficient number of people employed in his immediate work area in order to eliminate any unnecessary strangeness in his starting to work.

Methods of Instruction

Every building maintenance manager must constantly keep himself informed of new methods, techniques, systems, and products. This information must be transmitted to all operating personnel. It is essential that he be able to present effective instruction, to insure that his personnel understands exactly in what manner he expects his building and its equipment maintained.

A good instructor is a person who plans his work carefully, in order to be effective. Teaching is an art, not a science. In the acquisition of this art, it is possible to establish flexible principles or fundamentals, which may be used as guides toward the goals of proficiency and excellence. It is upon these established fundamentals of learning, that a good instructor must build his instruction, in order to qualify himself as a teacher.

An instructor must know his students. He must understand the persons he will teach and the way in which they learn. He must see the course of instruction from the viewpoint of the student. He must realize that his students are interested in the "Why", "Who", or "What" of what they are asked to do. He should keep in mind that good instruction and instructors remain in the mind of the student. Students are quick to detect the incompetent person. Most persons are capable of mastering the essentials of building maintenance training if they are taught well. *An instructor should have a thorough knowledge of his subject.*

Fundamentals of Instructional Situations

The following facts are presented to the instructor for thoroughly training a student for maintenance work.

The five stages of the teaching process are as follows:

1. Understanding.
   a. Principles of learning.
   b. Stages of instruction.
   c. Preparation.
2. Presentation.
3. Application.
4. Examination.
5. Review and/or critique.

Understanding

Principles of Learning

The only justification for instruction is learning. What causes a man to learn? We know from experience, that to learn, we must have a reason or motive for learning. We must understand what we are learning and be able to practice it.

Three fundamentals of learning on which instructional methods are based are *motivation, understanding,* and *practice.*

Motivation. When training a person, we cannot depend on his inclinations. To expose a person to knowledge is not enough to guarantee that he will open his mind to new ideas. Something must be done to cause the employee to apply himself mentally and physically to the instruction received. In choosing a method of instruction, a type should be used that will instill motivation, create his desire to learn, and make him realize that the instruction fills his personal need. He will be influenced by both his own and the instructor's attitude, when factors favorable to learning predominate.

Clarify the objective. A person's attitude at the beginning of an instruction period is favorable to receive the instruction because he is curious to find out what is going to happen. From this point, what he sees and hears, determines his response.

Understanding. Instruction must be planned so that understanding in your personnel's mind is accomplished with the shortest expenditure of time. The best way to help a person to understand a subject is to progress from the known to the unknown. All learning must start at a point of knowledge already possessed by the instructor. Aiding people in understanding can be effective through speech, however this can be supplemented, and greater achievements realized, when the five senses are used to full capacity when teaching. Understanding can also be accomplished by the functional approach. This is generally accomplished through associating the subject with practical examples and application.

Practice. The purposes of practice are proficiency and retention. The purpose of practice is to learn by doing. Supervision is of prime importance, as you must not permit incorrect habits to be formed. Length and number of practice periods should be sufficient to insure the understanding of what is to be accomplished by every employee.

Stages of Instruction

The presentation accomplishes motivation and understanding by clarifying, observing, and explaining ideas through lectures, discussions, and demonstrations. Application will perfect the understanding of skills and techniques through practice. Examination verifies the understanding and efficiencies, and provides motivation and guides for learning.

Preparation

Every lesson which is taught should have a basic *outline or lesson plan.* In the preparation of the instructor's manuscript, the student represents the framework around which the instructor will base his lesson.

The steps in lesson preparation are based on a logical thought process. Good instruction depends directly on the care and thought the instructor puts into its preparation. The key to an effective lesson preparation is knowing the six following steps:

1. Consider the lesson objective.
2. Gather necessary information and materials.
3. Study and analyze the material.
4. Determine and write the lesson plan.
5. Conduct rehearsals.
6. Check final arrangements.

It should always be remembered that the value an employee derives from his instructor, is care and thought which an instructor puts into his preparation.

Presentation

The presentation stage of instruction is important in the teaching process because it represents the teaching and showing ways of instruction. The actual delivery of the prepared lesson to the class during this stage of instruction puts the instructor's preparation to test.

Methods of conducting the presentation. There are three methods of conveying instruction to a class. They are the *lecture*, the *conference*, and the *demonstration*. Every presentation has an introduction. The introduction should gain the employees' interest, and prepare his mind for receiving the instruction. The introduction is your first chance for motivation. It should arouse interest, clarify the objective, and relate to a previous lesson or experience.

The *lecture* or the talk is an oral discourse by the instructor. Its advantage lies in the fact that many ideas can be presented in a short time, and it is adaptable to instructing large groups, or for a general orientation. One of the disadvantages of the lecture type of instruction is that it has complete lack of student participation. If possible, applicable training aids, such as charts, slides, and other available presentations should be used. A talk should not exceed fifty minutes in length. Short talk periods of fifteen to twenty minutes, combined with demonstrations and practical work periods are preferred to the lecture type method of instruction.

The *conference* is an oral discussion by the instructor, augmented by active class participation. It has the distinct advantage over the lecture method, in that it enables the instructor to determine by class participation, whether or not the instruction is making the correct impression. It allows the student to clear up any point he does not understand by asking questions. This type of instruction stimulates thinking and expression of thought. There are two general types of conferences; the information type and the development or direct discussion type.

Information Type. During this type of conference the instructor states facts which he wishes the employees to learn, then answers questions asked by them, covering points which are not understood. The information conference is generally used when the instructor is sure that the students will understand his presentation.

Development or Direct Discussion Type. The development or direct discussion conference requires the student to develop the ideas which the instructor desires to teach. This is accomplished by the instructor's putting leading questions to the students, which he must answer by using his own reasoning and judgement powers. This type of conference generally requires the student to have sufficient basic knowledge of the subject, so as to have reasonable answers to leading questions.

Demonstration is an accurate portrayal of a procedure, technique, or operation. It is the showing phase of instruction. Demonstrations are used to illustrate, either step by step, or in its entirety. Because it exemplifies the demonstrator's skill, the demonstration must be well prepared in detail, and must be rehearsed to secure proficiency. *Note:* A demonstration may constitute a complete lesson in itself.

Combination of Methods. The three basic methods are not mutually exclusive of each other. Any one of the three may be used to present a complete lesson, however, they are used to best advantage in combination with each other.

Application

The application stage of instruction is that part of the teaching process in which the student learns by doing the practical work. Learning by doing is one of the easiest and most natural ways of learning. It is readily adaptable to teaching almost every maintenance subject. By performance of an action, the student completes his understanding, and by repetition of it, he becomes proficient. The application stage may comprise a part of a lesson or a complete period or periods of instruction. If it is practical in your training, it is recommended that 15 per cent of the time be spent in telling, 25 per cent in showing, and 60 per cent in doing.

In preparing a period of practical work, the instructor first determines what kind of a training problem confronts him. If the problem is to train individuals, he may use one or a combination of three basic methods:

1. *Controlled (Group Performance) Method.* The instructor explains and demonstrates step by step. Class "imitates" works with instructor.
2. *Independent Method.* Instructor explains and demonstrates. Employee then performs at his own rate of speed. Corrections are made by instructor or assistants.
3. *Coach and Pupil Method.* Instructor explains and demonstrates. Employees are paired off as a coach and pupil. The pupil performs and the coach supervises. The coach becomes the pupil, and the pupil, the coach. Instructor and assistants supervise.

The three methods mentioned above can be used in combination with each other. Two or more used in the same period will add variety and inject interest into the instruction.

Examination

Knowing that he will be judged by the result of his examination or test, the employee's pride spurs him on to make more of an effort to learn.

Purposes of testing:

1. To bring instructional process to a logical conclusion.
2. To determine whether the student understands the subject and has developed the ability to apply it.

Types of tests:

1. Oral.
2. Written.
3. Performance.

Good questions will have a specific purpose and will be understood by the entire class. One point should be emphasized in each question, and each question should require a definite answer. This should stimulate thinking and discourage guessing the answers.

The *oral* test is based on the questions and answers generally used at the end of a class. Good questioning technique enables the instructor to impress the essentials on the employee's mind, and develops confidence in his ability to learn.

A good questioning technique is to ask the question in a clear voice, pause, look at the entire class, then call on an individual by name.

The *written* test is one in which the student writes his answers or solutions to the questions. The written test may be one of two types—essay or objective.

The *performance* test is conducted by having the student perform the job. A test of this type will definitely answer the question, "Can he do it?" There are three main parts to this test:

1. Good directions to the individual.
2. Directions to the checker.
3. Check list and score sheet for recording results.

In preparing an oral, written, or performance test, the instructor must first decide what he is trying to measure. He then decides the best method of measuring, and selects the best testing method, or a combination of tests which he will use.

Characteristics of a good test:

1. Valid.
2. Reliable.
3. Comprehensive.
4. Easy to take.
5. Easy to score.
6. Fair.

Review and/or Critique

The final stage of instruction which always follows the application or examination is the review and/or critique. The term "critique" is usually referred to the review given after the examination or applicatory exercise. The instructor reviews the knowledge, skills, and techniques acquired during the instruction period. This stage completes the picture by clarifying any phases of instruction which were not completely understood and any main points which should be re-emphasized.

NOTE: All tables shown that do not denote permission have been supplied and are reprinted with permission of International Correspondence Schools, Division of Intext, Scranton, Pennsylvania.

64 / DECIMALS, EQUIVALENTS, FRACTIONS, PER CENTS

EQUIVALENTS OF MILLIMETERS
IN INCHES

mm.	In.	mm.	In.	mm.	In.	mm.	In.
1	.039	39	1.535	76	2.992	114	4.488
2	.079	40	1.575	77	3.032	115	4.528
3	.118	41	1.614	78	3.071	116	4.567
4	.157	42	1.654	79	3.110	117	4.606
5	.197	43	1.693	80	3.150	118	4.646
6	.236	44	1.732	81	3.189	119	4.685
7	.276	45	1.772	82	3.228	120	4.724
8	.315	46	1.811	83	3.268	121	4.764
9	.354	47	1.850	84	3.307	122	4.803
10	.394	48	1.890	85	3.346	123	4.843
11	.433	49	1.929	86	3.386	124	4.882
12	.472	50	1.969	87	3.425	125	4.921
13	.512	51	2.008	88	3.465	126	4.961
14	.551	52	2.047	89	3.504	127	5.000
15	.591	53	2.087	90	3.543	128	5.039
16	.630	54	2.126	91	3.583	129	5.079
17	.669	55	2.165	92	3.622	130	5.118
18	.709	56	2.205	93	3.661	131	5.158
19	.748	57	2.244	94	3.701	132	5.197
20	.787	58	2.283	95	3.740	133	5.236
21	.827	59	2.323	96	3.780	134	5.276
22	.866	60	2.362	97	3.819	135	5.315
23	.906	61	2.402	98	3.858	136	5.354
24	.945	62	2.441	99	3.898	137	5.394
25	.984	63	2.480	100	3.937	138	5.433
26	1.024	64	2.520	101	3.976	139	5.472
27	1.063	65	2.559	102	4.016	140	5.512
28	1.102	66	2.598	103	4.055	141	5.551
29	1.142	67	2.638	104	4.095	142	5.591
30	1.181	68	2.677	105	4.134	143	5.630
31	1.220	69	2.717	106	4.173	144	5.669
32	1.260	70	2.756	107	4.213	145	5.709
33	1.299	71	2.795	108	4.252	146	5.748
34	1.339	72	2.835	109	4.291	147	5.787
35	1.378	73	2.874	110	4.331	148	5.827
36	1.417	74	2.913	111	4.370	149	5.866
37	1.457	75	2.953	112	4.409	150	5.906
38	1.496			113	4.449		

FRACTIONS AND DECIMALS OF AN INCH
AND MILLIMETERS

Fractions of an Inch	Decimals of an Inch	Millimeters	Fractions of an Inch	Decimals of an Inch	Millimeters
1/64	.0156	.397	33/64	.5156	13.097
1/32	.0312	.794	17/32	.5312	13.494
3/64	.0468	1.190	35/64	.5468	13.890
1/16	.0625	1.587	9/16	.5625	14.287
5/64	.0781	1.984	37/64	.5781	14.684
3/32	.0937	2.381	19/32	.5937	15.081
7/64	.1093	2.778	39/64	.6093	15.478
1/8	.125	3.175	5/8	.625	15.875
9/64	.1406	3.572	41/64	.6406	16.272
5/32	.1562	3.969	21/32	.6562	16.669
11/64	.1718	4.366	43/64	.6718	17.065
3/16	.1875	4.762	11/16	.6875	17.462
13/64	.2031	5.159	45/64	.7031	17.859
7/32	.2187	5.556	23/32	.7187	18.256
15/64	.2343	5.953	47/64	.7343	18.653
1/4	.25	6.350	3/4	.75	19.050
17/64	.2656	6.747	49/64	.7656	19.447
9/32	.2812	7.144	25/32	.7812	19.844
19/32	.2968	7.541	51/64	.7968	20.240
5/16	.3125	7.937	13/16	.8125	20.637
21/64	.3281	8.334	53/64	.8281	21.034
11/32	.3437	8.371	27/32	.8437	21.431
23/64	.3593	9.128	55/64	.8593	21.828
3/8	.375	9.525	7/8	.875	22.225
25/64	.3906	9.922	57/64	.8906	22.622
13/32	.4062	10.319	29/32	.9062	23.019
27/64	.4218	10.716	59/64	.9218	23.415
7/16	.4375	11.112	15/16	.9375	23.812
29/64	.4531	11.509	61/64	.9531	24.209
15/32	.4687	11.906	31/32	.9687	24.606
31/64	.4843	12.303	63/64	.9843	25.003
1/2	.5	12.700	1	1.0	25.400

TEMPERATURE CONVERSIONS

Cent.	Fahr.	Cent.	Fahr.	Cent.	Fahr.
0	32	170	338	480	896
5	41	175	347	490	914
10	50	180	356	500	932
15	59	185	365	510	950
20	68	190	374	520	968
25	77	195	383	530	986
30	86	200	392	540	1004
35	95	210	410	550	1022

40	104	220	428	560	1040
45	113	230	446	570	1058
50	122	240	464	580	1076
55	131	250	482	590	1094
60	140	260	500	600	1112
65	149	270	518	610	1130
70	158	280	536	620	1148
75	167	290	554	630	1166
80	176	300	572	640	1184
85	185	310	590	650	1202
90	194	320	608	660	1220
95	203	330	626	670	1238
100	212	340	644	680	1256
105	221	350	662	690	1274
110	230	360	680	700	1292
115	239	370	698	710	1310
120	248	380	716	720	1328
125	257	390	734	730	1346
130	266	400	752	740	1364
135	275	410	770	750	1382
140	284	420	788	760	1400
145	293	430	806	770	1418
150	302	440	824	780	1436
155	311	450	842	790	1454
160	320	460	860	800	1472
165	329	470	878	820	1508

Degrees Cent. x 9/5 plus 32 = Fahr.

Degrees Fahr. minus 32 x 5/9 = Cent.

(Reprinted with permission of International Correspondence Schools, Division of Intext, Scranton, Pennsylvania)

PERCENTAGE

Table of Per Cents and Values in Decimals and Fractions

Per Cent	Decimal	Fraction	Per Cent	Decimal	Fraction
1	.01	1/100	150	1.50	150/100 or 1½
2	.02	2/100 or 1/50	500	5.00	500/100 or 5
5	.05	5/100 or 1/20	¼	.0025	¼/100 or 1/400
10	.10	10/100 or 1/10	½	.005	½/100 or 1/200
25	.25	25/100 or ¼	1½	.015	1½/100 or 3/200
50	.50	50/100 or ½	8-1/3	.08-1/3	8-1/3/100 or 1/12
75	.75	75/100 or ¾	12½	.125	12½/100 or 1/8
100	1.00	100/100 or 1	16-2/3	.16-2/3	16-2/3/100 or 1/6
125	1.25	125/100 or 1¼	62½	.625	62½/100 or 5/8

When a per cent, written %, is used in a calculation, it is always changed to a fraction or to a decimal. The above table gives a number of different values in per cent changed to equivalent values in decimals and fractions.

(Reprinted with permission of International Correspondence Schools, Division of Intext, Scranton, Pennsylvania)

Decimal Equivalents of 64ths

The decimal fractions printed in large type give the exact value of the corresponding fraction to the fourth decimal place. A given decimal fraction is rarely exactly equal to any of these values, and the numbers in small type show which common fraction is nearest to the given decimal. Thus, lay off the fraction .1330 in 64ths. The nearest decimal fractions are .1250 and .1406. The value of any fraction in small type is the mean of the two adjacent fractions. In this instance the mean fraction is .1328, and as .1330 is greater than this, .1406 or 9/64 will be chosen. In the same manner, the nearest 64ths corresponding to the decimal fractions .3670 and .8979 are found to be 23/64 and 57/64, respectively.

Fraction	Decimal	Fraction	Decimal	Fraction	Decimal	Fraction	Decimal
	.0078		.2578		.5078		.7578
1/64	.0156	17/64	.2656	33/64	.5156	49/64	.7656
	.0235		.2735		.5235		.7735
1/32	.0313	9/32	.2813	17/32	.5513	25/32	.7813
	.0391		.2891		.5391		.7891
3/64	.0469	19/64	.2969	35/64	.5469	51/64	.7969
	.0547		.3047		.5547		.8047
1/16	.0625	5/16	.3125	9/16	.5625	13/16	.8125
	.0703		.3203		.5703		.8203
5/64	.0781	21/64	.3281	37/64	.5781	53/64	.8281
	.0860		.3360		.5860		.8360
3/32	.0938	11/32	.3438	19/32	.5938	27/32	.8438
	.1016		.3516		.6016		.8516
7/64	.1094	23/64	.3594	39/64	.6094	55/64	.8594
	.1172		.3672		.6172		.8672
1/8	.1250	3/8	.3750	5/8	.6250	7/8	.8750
	.1328		.3828		.6328		.8828
9/64	.1406	25/64	.3906	41/64	.6406	57/64	.8906
	.1485		.3985		.6485		.8985
5/32	.1563	13/32	.4063	21/32	.6563	29/32	.9063
	.1641		.4141		.6641		.9141
11/64	.1719	27/64	.4219	43/64	.6719	59/64	.9219
	.1797		.4297		.6797		.9297
3/16	.1875	7/16	.4375	11/16	.6875	15/16	.9375
	.1953		.4453		.6953		.9453
13/64	.2031	29/64	.4531	45/64	.7031	61/64	.9531
	.2110		.4610		.7110		.9610
7/32	.2188	15/32	.4688	23/32	.7188	31/32	.9688
	.2266		.4766		.7266		.9766
15/64	.2344	31/64	.4844	47/64	.7344	63/64	.9844
	.2422		.4922		.7422		.9922
1/4	.2500	1/2	.5000	3/4	.7500	1	1.0000
	.2578		.5078		.7578		1.0078

(Reprinted with permission of International Correspondence Schools,
Division of Intext, Scranton, Pennsylvania)

Decimal Equivalents

Decimals of a Foot for Each 1-32 of an Inch

Inch	0"	1"	2"	3"	4"	5"	6"	7"	8"	9"	10"	11"
0	0	.0833	.1667	.2500	.3333	.4167	.5000	.5833	.6667	.7500	.8333	.9167
1/32	.0026	.0859	.1693	.2526	.3359	.4193	.5026	.5859	.6693	.7526	.8359	.9193
1/16	.0052	.0885	.1719	.2552	.3385	.4219	.5052	.5885	.6719	.7552	.8385	.9219
3/32	.0078	.0911	.1745	.2578	.3411	.4245	.5078	.5911	.6745	.7578	.8411	.9245
1/8	.0104	.0937	.1771	.2604	.3437	.4271	.5104	.5937	.6771	.7604	.8437	.9271
5/32	.0130	.0964	.1797	.2630	.3464	.4297	.5130	.5964	.6797	.7630	.8464	.9297
3/16	.0156	.0990	.1823	.2656	.3490	.4323	.5156	.5990	.6823	.7656	.8490	.9323
7/32	.0182	.1016	.1849	.2682	.3516	.4349	.5182	.6016	.6849	.7682	.8516	.9349
1/4	.0208	.1042	.1875	.2708	.3542	.4375	.5208	.6042	.6875	.7708	.8542	.9375
9/32	.0234	.1068	.1901	.2734	.3568	.4401	.5234	.6068	.6901	.7734	.8568	.9401
5/16	.0260	.1094	.1927	.2760	.3594	.4427	.5260	.6094	.6927	.7760	.8594	.9427
11/32	.0286	.1120	.1953	.2786	.3620	.4453	.5286	.6120	.6953	.7786	.8620	.9453
3/8	.0312	.1146	.1979	.2812	.3646	.4479	.5312	.6146	.6979	.7812	.8646	.9479
13/32	.0339	.1172	.2005	.2839	.3672	.4505	.5339	.6172	.7005	.7839	.8672	.9505
7/16	.0365	.1198	.2031	.2865	.3698	.4531	.5365	.6198	.7031	.7865	.8698	.9531
15/32	.0391	.1224	.2057	.2891	.3724	.4557	.5391	.6224	.7057	.7891	.8724	.9557
1/2	.0417	.1250	.2083	.2917	.3750	.4583	.5417	.6250	.7083	.7917	.8750	.9583
17/32	.0443	.1276	.2109	.2943	.3776	.4609	.5443	.6276	.7109	.7943	.8776	.9609
9/16	.0469	.1302	.2135	.2969	.3802	.4635	.5469	.6302	.7135	.7969	.8802	.9635
19/32	.0495	.1328	.2161	.2995	.3828	.4661	.5495	.6328	.7161	.7995	.8828	.9661
5/8	.0521	.1354	.2188	.3021	.3854	.4688	.5521	.6354	.7188	.8021	.8854	.9688
21/32	.0547	.1380	.2214	.3047	.3880	.4714	.5547	.6380	.7214	.8047	.8880	.9714
11/16	.0573	.1406	.2240	.3073	.3906	.4740	.5573	.6406	.7240	.8073	.8906	.9740
23/32	.0599	.1432	.2266	.3099	.3932	.4766	.5599	.6432	.7266	.8099	.8932	.9766
3/4	.0625	.1458	.2292	.3125	.3958	.4792	.5625	.6458	.7292	.8125	.8958	.9792
25/32	.0651	.1484	.2318	.3151	.3984	.4818	.5651	.6484	.7318	.8151	.8984	.9818
13/16	.0677	.1510	.2344	.3177	.4010	.4844	.5677	.6510	.7344	.8177	.9010	.9844
27/32	.0703	.1536	.2370	.3203	.4036	.4870	.5703	.6536	.7370	.8203	.9036	.9870
7/8	.0729	.1562	.2396	.3229	.4062	.4896	.5729	.6562	.7396	.8229	.9062	.9896
29/32	.0755	.1589	.2422	.3255	.4089	.4922	.5755	.6589	.7422	.8255	.9089	.9922
15/16	.0781	.1615	.2448	.3281	.4115	.4948	.5781	.6615	.7448	.8281	.9115	.9948
31/32	.0807	.1641	.2474	.3307	.4141	.4974	.5807	.6641	.7474	.8307	.9141	.9974

(Reprinted with permission of International Correspondence Schools, Division of Intext, Scranton, Pennsylvania)

Linear Measure

12 inches (in.)	= 1 foot . . . . ft.
3 feet	= 1 yard . . . . yd.
5½ yards	= 1 rod . . . . rd.
40 rods	= 1 furlong . . . : . . . . fur.
8 furlongs	= 1 mile . . . . mi.

in.		ft.		yd.		rd.		fur.		mi.
36	=	3	=	1						
198	=	16½	=	5½	=	1				
7,920	=	660	=	220	=	40	=	1		
63,360	=	5,280	=	1,760	=	320	=	8	=	1

Surveyor's Measure

7.92 inches	= 1 link . . . . li.
25 links	= 1 rod . . . . rd.
4 rods = 100 links = 66 feet	= 1 chain . . . . ch.
80 chains	= 1 mile . . . . mi.

mi.		ch.		rd.		li.		in.
1	=	80	=	320	=	8,000	=	63,360

Square Measure

144 square inches (sq. in.)	= 1 square foot . . . . sq. ft.
9 square feet	= 1 square yard . . . . sq. yd.
30¼ square yards	= 1 square rod . . . . sq. rd.
160 square rods	= 1 acre . . . . A.
640 acres	= 1 square mile . . . . sq. mi.

sq. mi.		A.		sq. rd.		sq. yd.		sq. ft.		sq. in.
1	=	640	=	102,400	=	3,097,600	=	27,878,400	=	4,014,489,600

Surveyor's Square Measure

625 square links (sq. li.)	= 1 square rod . . . . sq. rd.
16 square rods	= 1 square chain . . . . sq. ch.
10 square chains	= 1 acre . . . . A.
40 acres	= 1 square mile . . . . sq. mi.
36 square miles (6 mi. square)	= 1 township . . . . Tp.

sq. mi.		A.		sq. ch.		sq. rd.		sq. li.
1	=	640	=	6,400	=	102,400	=	64,000,000

The acre contains 4,840 sq. yd., or 43,560 sq. ft., and in form of a square is 208.71 ft. on a side.

Cubic Measure

1,728 cubic inches (cu. in.)	= 1 cubic foot . . . . cu. ft.
27 cubic feet	= 1 cubic yard . . . . cu. yd.
128 cubic feet	= 1 cord . . . . cd.
24¾ cubic feet	= 1 perch . . . . P.

cu. yd.		cu. ft.		cu. in.
1	=	27	=	46,656

Measures of Angles or Arcs

60 seconds (")	= 1 minute . . . . '
60 minutes	= 1 degree . . . . o
90 degrees	= 1 rt. angle or quadrant . . . . □
360 degrees	= 1 circle . . . . cir.

1 cir. = 360o = 21,600' = 1,296,000"

Avoirdupois Weight

437½ grains (gr.) . . . . = 1 ounce . . . . oz.
16 ounces . . . . = 1 pound . . . . lb.
100 pounds . . . . = 1 hundredweight . . . . cwt.
20 hundredweight or 2,000 lb. . . . . = 1 ton . . . . T.

T.		cwt.		lb.		oz.		gr.
1	=	20	=	2,000	=	32,000	=	14,000,000

The avoirdupois pound contains 7,000 grains.

Long Ton Table

16 ounces . . . . = 1 pound . . . . lb.
112 pounds . . . . = 1 hundredweight . . . . cwt.
20 hundredweight, or 2,240 lb. . . . . = 1 ton . . . . T.

Troy Weight

24 grains (gr.) . . . . = 1 pennyweight . . . . pwt.
20 pennyweights . . . . = 1 ounce . . . . oz.
12 ounces . . . . = 1 pound . . . . lb.

lb.		oz.		pwt.		gr.
1	=	12	=	240	=	5,760

Dry Measure

2 pints (pt.) . . . . = 1 quart . . . . qt.
8 quarts . . . . = 1 peck . . . . pk.
4 pecks . . . . = 1 bushel . . . . bu.

bu.		pk.		qt.		pt.
1	=	4	=	32	=	64

The "U.S. struck bushel" contains 2,150.42 cu. in. = 1.2444 cu. ft. Its dimensions are, by law, 18½ in. in diameter and 8 in. deep.

The "heaped bushel" is equal to 1¼ struck bushels, the cone being 6 in. high.

For approximations, the bushel may be taken at 1¼ cu. ft.; or 1 cu. ft. may be considered 4/5 bu.

The "British bushel" contains 2,218.19 cu. in. = 1.2837 cu. ft. = 1.032 U. S. bushels.

The "dry gallon" contains 268.8 cu. in., being 1/8 bu.

Liquid Measure

4 gills (gi.) . . . . = 1 pint . . . . pt.
2 pints . . . . = 1 quart . . . . qt.
4 quarts . . . . = 1 gallon . . . . gal.
31½ gallons . . . . = 1 barrel . . . . bbl.
2 barrels . . . . or 63 gallons . . . . . = 1 hogshead . . . . hhd.

hhd.		bbl.		gal.		qt.		pt.		gi.
1	=	2	=	63	=	252	=	504	=	2,016

The "U.S. gallon" contains 231 cu. in. = .134 cu. ft., nearly; or 1 cu. ft. contains 7.48 gal.

The following cylinders contain the given measures very closely:

	Diam. Inches	Height Inches		Diam. Inches	Height Inches
Gill	1¾	3	Gallon	7	6
Pint	3½	3	8 Gallons	14	12
Quart	3½	6	10 Gallons	14	15

With water at its maximum density, 1 cu. ft. weighs 62.425 lb. and 1 gal. of pure water weighs 8.345 lb.

For approximations, 1 cu. ft. of water is considered equal to 7½ gal., and 1 gal. as weighing 8-1/3 lb.

The "British imperial gallon," both liquid and dry, contains 277.274 cu. in. = .16046 cu. ft., and is equivalent to the volume of 10 lb. of pure water at 62°F.

To reduce U. S. to British liquid gallons, divide by 1.2. Conversely, to convert British into U. S. liquid gallons, multiply by 1.2; or, increase the number of gallons 1/5.

Miscellaneous Table

12 articles	=	1 dozen		20 quires	=	1 ream
12 dozen	=	1 gross		1 league	=	3 mi.

12 gross	=	1 great gross	1 fathom = =	6 ft.
2 articles	=	1 pair	1 hand =	4 in.
20 articles	=	1 score	1 palm =	3 in.
24 sheets	=	1 quire	1 span =	9 in.

1 knot (U.S.) = 6,080 ft. = 1.16 mi. (roughly).
1 meter = 3 ft. 3-3/8 in. (nearly).

(Reprinted with permission of International Correspondence Schools, Division of Intext, Scranton, Pennsylvania)

Measures of Length

12 inches (in.)	= 1 foot	ft.
3 feet	= 1 yard	yd.
5½ yards	= 1 rod	rd.
5,280 feet	= 1 statute mile	mi.
6,080 feet	= 1 nautical mile	mi.
6 feet	= 1 fathom	f.
7.92 inches	= 1 link	li.
100 links	= 1 chain	ch.
80 chains	= 1 statute mile	mi.
1 mil	= .001 inch	
1 microinch	= .000001 inch	

Measures of Area

144 square inches (sq. in.)	= 1 square foot	sq. ft.
9 square feet	= 1 square yard	sq. yd.
30¼ square yards	= 1 square rod	sq. rd.
272¼ square feet	= 1 square rod	
160 square rods	= 1 acre	A.
640 acres	= 1 square mile	sq. mi.
1 circular inch	= .7854 square inch	
1 square inch	= 1.2732 circular inches	
1 circular mil	= area of a circle .001 inch in diameter	
1 square inch	= 1,273,239 circular mils	

A square 208.71 ft. on a side contains 1 A., or 43,560 sq. ft.
A square mile, or 640 A., is known as a "section."

Avoirdupois Weight

437.5 grains (gr.)	= 1 ounce	oz.
16 ounces (7,000 grains)	= 1 pound	lb.
2,000 pounds	= 1 short ton	T.
2,240 pounds	= 1 long ton	T.

Apothecaries' (Druggists') Weight

20 grains (gr.)	= 1 scruple	sc.
3 scruples	= 1 dram	dr.
8 drams	= 1 ounce	oz.
12 ounces (5,760 grains)	= 1 pound	lb.

The grain has the same value in all three of these systems of weight.

Cubic Measure

1,728 cubic inches (cu. in.)	= 1 cubic foot	cu. ft.
27 cubic feet	= 1 cubic yard	cu. yd.
128 cubic feet	= 1 cord (of wood)	cd.
16½ to 25 cubic feet	= 1 perch (of masonry)	P.
100 cubic feet	= 1 register ton	T.
1 U.S. shipping ton	= 40 cubic feet	
	= 31.16 imperial bushels	
	= 32.143 U.S. bushels	

1 British shipping ton . . . . = 42 cubic feet
                     = 32.719 imperial bushels
                     = 33.75 U.S. bushels

United States Liquid Measure

4 gills (gi.) . . . . = 1 pint . . . . pt.
2 pints . . . . = 1 quart . . . . qt.
4 quarts . . . . = 1 gallon . . . . gal.
31½ gallons . . . . = 1 barrel . . . . bbl.

    The U. S. gallon contains 231 cu. in.; hence, there are 7.481 gal. in a cu. ft.

British Imperial Measure

4 gills . . . . = 1 pint . . . . =   34.683 cu. in.
2 pints . . . . = 1 quart . . . . =   69.366 cu. in.
4 quarts . . . . = 1 gallon . . . . =   277.463 cu. in.
2 gallons . . . . = 1 peck . . . . =   554.926 cu. in.
4 pecks . . . . = 1 bushel . . . . = 2,219.704 cu. in.

    One imperial gallon contains approximately 1.2 U.S. gal.; and one imperial bushel contains approximately 1.03 U.S. bu.

Circular Measure

60 seconds (") . . . . = 1 minute . . . . '
60 minutes . . . . = 1 degree . . . .°
90 degrees . . . . = 1 quadrant . . . .□
360 degrees . . . . = 1 circle .O

Measures of Time

60 seconds (sec.) . . . . = 1 minute . . . . min.
60 minutes . . . . = 1 hour . . . . hr.
24 hours . . . . = 1 day . . . . da.
7 days . . . . = 1 week . . . . wk.
365 days . . . . = 1 year . . . . yr.

Miscellaneous Measures

12 articles . . . . = 1 dozen   1 league . . . . = 3 miles
12 dozen . . . . = 1 gross   1 fathom . . . . = 6 feet
12 gross . . . . = 1 great gross   1 hand . . . . = 4 inches
20 articles . . . . = 1 score   1 palm . . . . = 3 inches
24 sheets . . . . = 1 quire   1 span . . . . = 9 inches
20 quires . . . . = 1 ream

    One cu. ft. of pure water at the temperature at which it is densest, weighs 62.425 lb. For ordinary calculations, the weight of a cu. ft. of water is taken as 62½ lb. A gallon of pure water at the point of maximum density weighs 8.345 lb. For ordinary calculations, the weight of a gallon of water is taken as 8-1/3 lb.

    Under normal conditions at sea level, the pressure of the atmosphere is 14.7 lb. per sq. in.

(Reprinted with permission of International Correspondence Schools, Division of Intext, Scranton, Pennsylvania)

Areas and Volumes

Circles, Ellipses, Spheres, Cylinders,
Prisms, Cones, and Pyramids

Circle: Diameter $= \begin{cases} 2 & \text{X Radius} \\ 1.128 & \text{X Side of Equivalent Square} \end{cases}$

Circumference	=	$\begin{cases} 3.1416 \\ \dfrac{4 \times Area}{Diameter} \\ 6.283185 \\ 3.545 \\ .866 \end{cases}$	X Diameter X Radius X Side of Equivalent Square X Perimeter of Equivalent Square
Area	=	$\begin{cases} 3.1416 \\ .7854 \\ .07958 \\ \tfrac{1}{2} \text{ Diameter} \\ Circumference \end{cases}$	X Square of Radius X Square of Diameter X Square of Circumference X ½ Circumference X ¼ Diameter
Side of Equivalent Square	=	$\begin{cases} .8861 \\ \\ .282 \end{cases}$	X Diameter X Circumference
Perimeter of Equivalent Square	=	1.128	X Circumference
Side of Inscribed Square	=	$\begin{cases} .7071 \\ \\ .225 \end{cases}$	X Diameter X Circumference
Area of Circumscribed Square	=	$\begin{cases} \text{Square of Diameter} \\ \\ 1.2732 \end{cases}$	 X Area
Ellipse: Area	=	.7854	X Sum of Diameters
Sphere: Area	=	$\begin{cases} 3.1416 \\ .3183 \\ Diameter \\ 1/3 \text{ Diameter} \end{cases}$	X Square of Diameter X Square of Circumference X Circumference X Area
Volume	=	$\begin{cases} .5236 \\ 4.1888 \\ .016887 \end{cases}$	X Cube of Diameter X Cube of Radius X Cube of Circumference
Side of Inscribed Cube	=	1.1547	X Radius
Cylinder or Prism: Volume	=	Altitude	X Area of One End
Cone or Pyramid: Volume	=	½ Altitude	X Area of Base

66 / METRIC SYSTEM, WEIGHTS AND MEASURES

(Reprinted with permission of International Correspondence Schools, Division of Intext, Scranton, Pennsylvania)

METRIC SYSTEM

The metric system is based on the meter, which, according to the U.S. Coast and Geodetic Survey Report of 1884, is equal to 39.370432 inches. The value commonly used is 39.37 inches, and is authorized by the U.S. government. The meter is defined as one ten-millionth of the distance from the pole to the equator, measured on a meridian passing near Paris.

There are three principal units—the meter, the liter (pronounced lee-ter), and the gram—which are the units of length, capacity, and weight, respectively. Multiples of these units are obtained by prefixing to the names of the principal units the Greek words deca (10), hecto (100), and kilo (1,000); the submultiples, or divisions, are obtained by prefixing the Latin words deci (1/10), centi (1/100), milli (1/1,000). These prefixes form the key to the entire system. In the following tables, the abbreviations of the principal units of these submultiples begin with a small letter, while those of the multiples begin with a capital letter; they should always be written as here printed.

Measures of Length

1,000 microns	= 1 millimeter	mm.
10 millimeters	= 1 centimeter	cm.
10 centimeters	= 1 decimeter	dm.

```
10 decimeters . . . . . . . . . . . . . . . . = 1 meter . . . . . . . . . . . . . . . . . . . . m.
10 meters. . . . . . . . . . . . . . . . . . . = 1 decameter . . . . . . . . . . . . . . . . . Dm.
10 decameters . . . . . . . . . . . . . . . . = 1 hectometer . . . . . . . . . . . . . . Hm.
10 hectometers . . . . . . . . . . . . . . . = 1 kilometer . . . . . . . . . . . . . . . . Km.
```

Measures of Surface (Not Land)

```
100 square millimeters (mm.²) . . . . . . . = 1 square centimeter . . . . . . . . . . . . . cm.²
100 square centimeters . . . . . . . . . . . = 1 square decimeter . . . . . . . . . . . . . dm.²
100 square decimeters . . . . . . . . . . . = 1 square meter . . . . . . . . . . . . . . m.²
```

Measures of Capacity

```
10 milliliters (ml.) . . . . . . . . . . . . . = 1 centiliter . . . . . . . . . . . . . . . . . cl.
10 centiliters . . . . . . . . . . . . . . . = 1 deciliter . . . . . . . . . . . . . . . . . dl.
10 deciliters . . . . . . . . . . . . . . . . . = 1 liter . . . . . . . . . . . . . . . . . . . . l.
10 liters . . . . . . . . . . . . . . . . . . . = 1 decaliter . . . . . . . . . . . . . . . . . Dl.
10 decaliters . . . . . . . . . . . . . . . . = 1 hectoliter . . . . . . . . . . . . . . . . Hl.
10 hectoliters . . . . . . . . . . . . . . . . = 1 kiloliter . . . . . . . . . . . . . . . . . Kl.
```

The liter is equal to the volume that is occupied by 1 cubic decimeter

Measures of Volume

```
1,000 cubic millimeters (mm³.) . . . . . . . = 1 cubic centimeter . . . . . . . . . . . . . cm³.
1,000 cubic centimeters . . . . . . . . . . = 1 cubic decimeter . . . . . . . . . . . . . dm³.
1,000 cubic decimeters . . . . . . . . . . = 1 cubic meter . . . . . . . . . . . . . . . m³.
```

Measures of Weight

```
10 milligrams (mg.) . . . . . . . . . . . . . = 1 centigram . . . . . . . . . . . . . . . . cg.
10 centigrams . . . . . . . . . . . . . . . . = 1 decigram . . . . . . . . . . . . . . . . . dg.
10 decigrams . . . . . . . . . . . . . . . . = 1 gram . . . . . . . . . . . . . . . . . . . g.
10 grams . . . . . . . . . . . . . . . . . . . = 1 decagram . . . . . . . . . . . . . . . . . Dg.
10 decagrams . . . . . . . . . . . . . . . . = 1 hectogram . . . . . . . . . . . . . . . . Hg.
10 hectograms . . . . . . . . . . . . . . . . = 1 kilogram . . . . . . . . . . . . . . . . . Kg.
1,000 kilograms . . . . . . . . . . . . . . . = 1 ton (metric) . . . . . . . . . . . . . . . M.T.
```

The gram is the weight of 1 cubic centimeter of pure distilled water at a temperature of 39.2°F.; the kilogram is the weight of 1 liter of water; the ton is the weight of 1 cubic meter of water.

METRIC CONVERSION FACTORS

Measures of Length

```
1 millimeter . . . . . . . . . . . . . . . . . . . . = .03937 in.
1 centimeter . . . . . . . . . . . . . . . . . . . . = .3937 in.

                                                    ⎧ = 39.37 in.
1 meter . . . . . . . . . . . . . . . . . . . . . . . ⎨ = 3.28083 ft.
                                                    ⎩ = 1.0936 yd.

                                                    ⎧ = 3,280.83 ft.
1 kilometer . . . . . . . . . . . . . . . . . . ⎨ = 1,093.61 yd.
                                                    ⎩ = .62137 mi.

                                                    ⎧ = 25.4 mm.
1 inch . . . . . . . . . . . . . . . . . . . . . . . ⎨ = 2.54 cm.
                                                    ⎩ = .0254 m.

                                                    ⎧ = 304.8 mm.
1 foot . . . . . . . . . . . . . . . . . . . . . . . ⎨ = .3048 m.

1 yard . . . . . . . . . . . . . . . . . . . . . . . = .9144 m
1 mile . . . . . . . . . . . . . . . . . . . . . . . = 1.609 km.
```

Measures of Area

1 square millimeter	=	.00155 sq. in.
1 square centimeter	=	.155 sq. in.
1 square meter	=	10.764 sq. ft.
	=	1.196 sq. yd.
1 are	=	.0247 acre
	=	1,076.4 sq. ft.
1 hectare	=	2.471 acres
	=	107,640 sq. ft.
1 square kilometer	=	.3861 sq. mi.
	=	247.1 acres
1 square inch	=	645.2 sq. mm.
	=	6.452 sq. cm.
1 square foot	=	929 sq. cm.
	=	.0929 sq. m.
1 square yard	=	.836 sq. m.
1 acre	=	40.47 ares
	=	.4047 hec.
1 square mile	=	2.5899 sq. km.

Measures of Volume and Capacity

1 liter	=	1 cu. dm.
	=	61.023 cu. in.
	=	.0353 cu. ft.
	=	2.202 lb. water at 62°F.
	=	1.0567 U.S. qt.
	=	.2642 U.S. gal.
1 cubic centimeter	=	.061 cu. in.
1 cubic meter	=	264.2 U.S. gal.
	=	1.308 cu. yd.
	=	35.314 cu. ft.
1 cubic inch	=	16.387 cu. cm.
1 cubic foot	=	28.317 cu. dm.
	=	28.317 l.
	=	.02832 cu. m.
1 cubic yard	=	.7645 cu. m.
1 U.S. gallon	=	3.785 l.
1 British gallon	=	4.543 l.
1 U.S. quart	=	.946 l.

Measures of Weight

1 gram	=	15.432 gr.
1 kilogram	=	2.2046 lb. (av.)
	=	35.274 oz. (av.)
1 metric ton	=	2,204.6 lb. (av.)
	=	.9842 long ton
	=	1.1023 short ton
1 grain	=	.0648 g.
1 ounce (av.)	=	28.35 g.
1 pound (av.)	=	.4536 kg.
1 short ton	=	.907 M. T.
1 long ton	=	1.016 M. T.
	=	1,016 kg.

Miscellaneous Conversion Factors

1 horsepower	=	33,000 ft.-lb. per min.
	=	550 ft.-lb. per sec.
	=	2,546 B.t.u. per hr.
	=	42.4 B.t.u. per min.
	=	.71 B.t.u. per sec.
	=	746 watts

```
1 kilowatt . . . . . . . . . . . . . . . . . . . = 1,000 watts
                                               = 1.34 hp.
                                               = 44,250 ft.-lb. per min.
                                               = 57 B.t.u. per min.
1 watt . . . . . . . . . . . . . . . . . . . . = Unit of electrical power
                                               = .00134 hp.
                                               = 44.25 ft.-lb. per min.
                                               = .74 ft.-lb. per sec.
                                               = 3.42 B.t.u. per hr.
1 degree Fahrenheit . . . . . . . . . . . . . . = .555ºC.
1 degree Centigrade . . . . . . . . . . . . . . = 1.8ºF.
1 B.t.u. . . . . . . . . . . . . . . . . . . . = 777.5 ft.-lb.
1 calorie . . . . . . . . . . . . . . . . . . . = 3.968 B.t.u.
1 pound per square inch . . . . . . . . . . . . = .0703 kg. per sq. cm.
1 gram per square millimeter . . . . . . . . . = 1.422 lb. per sq. in.
1 pound per square foot . . . . . . . . . . . . = 4.882 kg. per sq. m.
1 inch mercury . . . . . . . . . . . . . . . . = 1.133 ft. water
                                               = .4912 lb. per sq. in.
```

Measures of Length

Name		Meters		U.S. In.		U.S. Ft.
Millimeter (mm.)	=	.001	=	.039370	=	.003281
Centimeter (cm.)	=	.010	=	.393704	=	.032809
Decimeter (dm.)	=	.100	=	3.937043	=	.328087
Meter (m.)	=	1.000	=	39.370432	=	3.280869
Decameter (Dm.)	=	10.000	=		=	32.808690
Hectometer (Hm.)	=	100.000	=		=	328.086900
Kilometer (Km.)	=	1,000.000	=	.621 mi.	=	3,280.869000
Myriameter (Mm.)	=	10,000.000	=	6.214 mi.	=	32,808.690000

The centimeter, meter, and kilometer are the units in practical use, and may be said to occupy the same position in the metric system as do inches, yards, and miles in the U.S. and English system of measurement.

Measures of Area

Name		Sq. M.		Sq. In.		Sq. Ft.		A.
Sq. millimeter (sq. mm.)	=	.0000010	=	.001550				
Sq. centimeter (sq. cm.)	=	.0001000	=	.155003	=	.00107641		
Sq. decimeter (sq. dm.)	=	.0100000	=	15.5003	=	.10764100		
Sq. meter, or centare (sq. m., or ca.)	=	1.000000	=	1,550.03	=	10.764100	=	.000247
Sq. decameter, or are (sq. dm., or a.)	=	100.0000	=	155,003	=	1,076.4101	=	.024710
Hectare	=	10,000.00	=		=	107,641.01	=	2.47110
Sq. kilometer	=	.3861099 sq. mi.			=	10,764,101	=	247.110
Sq. myriameter	=	38.61090 sq. mi.					=	24,711.0

Measures of Volume

Name		Cu. M.		Cu. In.		Cu. Ft.	Cu. Yd.
Cu. centimeter) = (c.c. or cu. cm.)		.000001	=	.061025			
Cu. decimeter) = (cu. dm.))		.001000	=	61.0254			
Centistere	=	.010000	=	610.2540	=	.35316	

Decistere	=	.100000	=	3.53156	
Stere (cu. m.)	=	1.000000	=	35.3156	= 1.308
Decastere	=	10.000000	=	353.156	= 13.080

Measures of Capacity

Name	Liters	Liq. Meas.	Dry Meas.
Milliliter (ml. =) (c.c.)	= .00100	= .008454 gi.	= .001816 pt.
Centiliter (cl.)	= .01000	= .084537 gi.	= .018162 pt.
Deciliter (dl.)	= .10000	= .845370 gi.	= .18162 pt.
Liter (1) (= cu. dm.)	= 1.0000	= 1.05671 qt. .264179 gal.	= .11351 pk.
Decaliter (Dl.) (= centistere)	= 10.000	= 2.64179 gal.	= 1.1351 pk.
Hectoliter (Hl.) (= decistere)	= 100.00	= 26.4179 gal.	= 2.83783 bu.
Kiloliter (Kl.) (= cu. m., or stere)	= 1,000.0	= 264.179 gal.	= 28.3783 bu.
Myrialiter (Ml.) (= decastere)	= 10,000	= 2641.79 gal.	= 283.783 bu.

The milliliter (or cubic centimeter) and the liter are the units most commonly used. A liter of pure water at 4°C., or 39.2°F., weighs 1 kg.

Metric Weights

The gram is the basis of metric weights, and is the weight of a cubic centimeter of distilled water at its maximum density, at sea level, Paris, barometer 29.922 inches.

Name	Grams	Grains	Av. Oz.	Av. Lb.
Milligram (mg.)	.001	.01543		
Centigram (cg.)	.010	.15432		
Decigram (dg.)	.100	1.54323		
Gram (g.)	1.000	15.43235	.03527	.0022046
Decagram (Dg.)	10.000		.35274	.0220462
Hectogram (Hg.)	100.000		3.52739	.2204622
Kilogram (Kg.)	1,000.000		35.27395	2.2046223
Myriagram (Mg.)	10,000.000			22.0462234
Quintal (Q.)	100,000.000			220.4622341
Tonneau (T.)	1,000,000.000			2,204.6223410

The gram and the kilogram (called "kilo") are the units in common use.

Factors for Conversion

For approximations, it may be useful to remember the following:

1 cm.	=	.4 in. (nearly)
1 m.	=	40 in. (roughly)
1 km.	=	5/8 mi. (nearly)
1 l.	=	1 liq. qt. (nearly)
1 l.	=	1/9 pk. (nearly)
1 gr.	=	15.4 gr.
1 kg.	=	2-1/5 lb.

British Weights and Measures

Avoirdupois Weight

16 drams	=	1 ounce	=	437½ grains
16 ounces	=	1 pound (lb.)	=	7,000 grains
28 pounds	=	1 quarter (qr.)		
4 quarters	=	1 hundredweight (cwt.) (112 lb.)		
20 cwt.	=	1 ton		

Troy Weight

24 grains . . . . = 1 pennyweight (dwt.) (= 1.555 grammes)
20 dwt. . . . . = 1 ounce (= 31.1035 grammes)

For gold and silver the ounce, divided decimally and not into grains, is the sole unit of weight. The Troy ounce is the same as the Apothecaries' ounce = 480 Avoirdupois grains (31.1035 grammes) in weight.

Equivalents of Electrical Units

1 Kilowatt . . . . = 1,000 Watts
1 Kilowatt . . . . = 1.34 H.P.
1 Kilowatt . . . . = 44,260 Foot-Pounds per minute
1 Kilowatt . . . . = 56.89 B.Th.U. per minute
1 H.P. . . . . = 746 Watts
1 H.P. . . . . = 33,000 Foot-Pounds per minute
1 H.P. . . . . = 42.41 B.Th.U. per minute
1 B.Th.U. . . . . = 778 Foot-Pounds
1 B.Th.U. . . . . = 0.2930 Watt-Hour
1 Joule . . . . = 1 Watt-Second

Square, Surface, or Land Measure

The Square Foot contains 144 square inches

Yard . . . . =	9 feet . . . . =		1,296 inches
Rod, Pole, or Perch . . . . =	30¼ yards . . . . =		272¼ feet
Chain . . . . =	16 rods . . . . . = 484 yards . . . . . . =		4,356 feet
Rood . . . . =	40 rods . . . . = 1,210 yards . . . . . =		10,890 feet
Acre . . . . =	4 roods . . . . . = 160 rods . . . . =		4,840 yards
Yard of Land . . . . =	30 acres . . . . =		120 roods
Hide . . . . =	100 acres . . . . =		400 roods
Mile . . . . =	640 acres = 2,560 roods = 6,400 chains = 102,400 rods, poles, or perches, or 3,097,600 square yards.		

An acre roughly stated has four equal sides of 69½ yards.

Metric Weights and Measures

Measures of Length

1 Centimetre . . . . = 0.394 inch
1 Metre . . . . = 3.28 feet . . . . = 1.094 yards
1 Kilometre . . . . = 1,093.6 yards . . . . = 0.62 mile
1 Inch . . . . = 2.54 centimetres
1 Yard . . . . = 0.914 metre
1 Mile . . . . = 1.609 kilometres

Approximate Metric Equivalents

1 Decimetre . . . . = 4 inches
1 Metre . . . . = 1.1 yards
1 Kilometre . . . . = 5/8 of a mile
1 Hectare . . . . = 2½ acres
1 Stere, or Cub. Metre . . . . . = ¼ of a cord
1 Litre . . . . = 1.06 qt. liquid . . . . = 0.9 qt. dry
1 Hectolitre . . . . = 2¾ bush.
1 Kilogramme . . . . = 2-1/5 lb.
1 Metric Ton . . . . = 2,200 lb.

Measures of Weigh

1 Gramme . . . . = 15.43 grains
1 Kilogramme . . . . = 2.205 lb.
1 Tonne . . . . = 0.98 ton
1 Ounce Troy . . . . = 31.1 grammes

```
1 Ounce Avoirdupois . . . . . . = 28.352 grammes
1 Lb. . . . . . . . . . . . . . . = 0.4536 kilogramme
1 Ton . . . . . . . . . . . . . . = 1.016 tonnes
```

Measures of Capacity

```
1 Litre. . . . . . . . . . . . . = 1.760 pints . . . . . . . . . . . . . . . . = 0.22 gallon
1 Hectolitre . . . . . . . . . . = 2.749 bushels
1 Pint . . . . . . . . . . . . . = 0.568 litre
1 Gallon . . . . . . . . . . . . = 4.546 litres
1 Cubic Cm. (water) . . . . . . = 1 gramme
```

Measures of Land

```
100 Sq. Metres . . . . . . . . . = 119.6 sq. yards
1 Hectare . . . . . . . . . . . = 2.47 acres
1 Sq. Kilometre . . . . . . . = 0.386 sq. mile
1 Acre . . . . . . . . . . . . . = 0.404 hectare
1 Sq. Mile . . . . . . . . . . . = 2.59 sq. kilometres
```

67 / PROPERTIES OF MATERIALS

(Reprinted with permission of Publishing Division, Intext, Scranton, Pennsylvania)

Capacities of Cylindrical Tanks

Inside Diameter of Tank, Feet	Capacity per Foot of Shell Length, Gallons	Deduction or Addition for Various Heads in Gallons			
		Type A Add	Type B Add	Type C Deduct	Type D Add
1	5.84	0.80	none	0.80	none
1½	13.15	2.65	,,	2.65	,,
2	23.43	6.73	,,	6.73	,,
2½	36.70	18.18	,,	18.18	,,
3	52.75	23.26	,,	23.26	,,
3½	72.00	34.00	,,	34.00	,,
4	93.80	52.00	,,	52.00	,,
4½	118.75	75.00	,,	75.00	,,
5	146.80	105.00	,,	105.00	,,
6	211.10	175.00	,,	175.00	,,
7	287.00	285.00	,,	285.00	,,
8	376.00	415.00	,,	415.00	,,
9	475.00	600.00	,,	600.00	,,
10	587.00	815.00	,,	815.00	,,
12	845.00	1,400.00	,,	1,400.00	,,

NOTE. Type A, two bumped heads; Type B, one bumped head and one dished head; Type C, two dished heads; Type D, two flat heads.

(Reprinted with permission of Publishing Division, Intext, Scranton, Pennsylvania)

Capacities of Round Tanks in Gallons
per Foot of Water Depth

Diameter of Tank, Feet	Contents per Foot of Water Depth, Gallons
1	5.84
1½	13.15
2	23.43
2½	36.70
3	52.75
3½	72.00
4	93.80
4½	118.75
5	146.80
6	211.00
7	287.00
8	376.00
9	475.00
10	587.00
12	845.00
14	1,150.00
16	1,502.00
18	1,905.00
20	2,343.00

(Reprinted with permission of Publishing Division, Intext, Scranton, Pennsylvania)

Capacities of Square and Rectangular Tanks

Size of Tank Width x Length, Feet	DEPTH OF WATER IN TANK—FEET									
	1	2	3	4	5	6	7	8	9	10
2 x 2	30	60	90	120	150	180	209	239	269	299
2 x 3	45	90	135	180	224	269	314	359	404	449
2 x 4	60	120	180	239	299	359	419	479	539	598
2 x 5	75	150	224	299	374	449	524	598	673	748
2 x 6	90	180	269	359	449	539	628	718	808	898
3 x 3	67	135	202	269	337	404	471	539	606	673
3 x 4	90	180	269	359	449	539	628	718	808	898
3 x 5	112	224	337	449	561	673	785	898	1010	1122
3 x 6	135	269	404	539	673	808	942	1077	1212	1346
3 x 7	157	314	471	628	785	942	1100	1257	1414	1571
3 x 8	180	359	539	718	897	1057	1255	1436	1616	1795
3 x 9	202	404	606	808	1010	1212	1414	1615	1818	2020
4 x 4	120	239	358	479	598	718	838	957	1077	1197
4 x 5	150	299	449	598	748	898	1047	1197	1346	1497
4 x 6	180	359	539	718	897	1057	1255	1436	1616	1795
4 x 7	209	419	628	838	1047	1256	1466	1676	1885	2094
4 x 8	240	478	716	958	1196	1436	1776	1914	2154	2394
4 x 9	269	539	808	1077	1346	1616	1885	2154	2424	2693
4 x 10	300	598	898	1196	1496	1796	2094	2394	2692	2994

(Reprinted with permission of Publishing Division, Intext, Scranton, Pennsylvania)

*Quantities of Water Contained in Pneumatic Tanks at
Various Pressures in Per Cent of Total Tank Capacity*

(Without the Use of Initial Air)

Pressure Psi	Per Cent of Tank Capacity	Pressure Psi	Per Cent of Tank Capacity	Pressure Psi	Per Cent of Tank Capacity
0	0	35	70	70	82
5	25	40	73	75	83
10	40	45	75	80	84
15	50	50	77	85	85
20	57	55	79	90	86
25	63	60	80	95	86
30	67	65	81	100	87

(Reprinted with permission of Publishing Division, Intext, Scranton, Pennsylvania)

Friction Loss per 100 Feet of Pipe, in Feet

Gallons per Minute	Pipe Sizes in Inches							
	½	¾	1	1¼	1½	2	2½	3
1	2.1							
2	7.4	1.9						
3	15.8	4.1	1.3					
4	27.0	7.0	2.1					
5	41.0	10.5	3.3					
6	57.0	14.7	4.6	1.20				
7	76.0	19.5	6.0	1.6				
8	98.0	25.0	7.8	2.0				
9		31.2	9.4	2.5	1.2			
10		38.0	11.7	3.1	1.4			
15		80.0	25.0	6.4	3.0	1.1		
20			42.0	11.1	5.2	1.8		
25			64.0	16.6	7.8	2.7		
30			89.0	23.5	11.0	3.8	1.3	
35				31.2	14.7	5.1	1.7	
40				40	18.8	6.6	2.2	
50				60	28.4	9.9	3.3	1.4
60				85	39.6	13.9	4.7	1.9
70					53.0	18.4	6.2	2.6
100						35.8	12.0	5.0
120						50.0	16.8	7.0
140						67.0	22.3	9.2
160						86.0	29.0	11.8
180							35.7	14.8
200							43.1	17.8
220							52.0	21.3
240							61.0	25.1
260							70.0	29.1
280							81.0	33.4
300							92.0	38.0

Friction Loss in 90° Elbows, in Equivalent Number of Feet of Straight Pipe

Size of Elbow, in Inches

½	¾	1	1¼	1½	2	2½	3
5	6	6	8	8	8	11	15

(Reprinted with permission of Publishing Division, Intext, Scranton, Pennsylvania)

Quantities of Water Contained in Pneumatic Tanks at Various Pressures in Per Cent of Total Tank Capacity

(With the Use of Initial Air)

Initial Air Gage Pressure Psi	Gage Pressure in Pounds per Square Inch														
	5	10	15	20	25	30	35	40	45	50	55	60	65	70	75
5	0	20	33	42	50	55	60	64	67	69	71	73	75	77	78
10		0	17	28	38	45	50	54	58	62	65	67	69	71	72
15			0	14	25	33	40	46	50	54	57	60	62	65	67
20				0	13	23	30	36	42	46	50	53	57	59	61
25					0	11	20	27	34	39	43	47	50	63	66
30						0	10	18	25	31	36	40	44	47	50
35							0	9	17	23	28	33	37	41	44
40								0	9	15	21	27	31	35	39
45									0	8	14	20	25	29	33
50										0	7	13	19	24	28
55											0	7	13	19	24
60												0	6	12	17
65													0	6	11
70														0	5
75															0

(Reprinted with permission of Publishing Division, Intext, Scranton, Pennsylvania)

Dimensions of Standard Cylindrical Steel Pressure Tanks

Type of Tank	Capacity in Gals.	Diameter x Length Inches	65 lb. Water Pressure		100 lb. Water Pressure	
			Thickness of Convex Heads-In.	Thickness of Shell Inch	Thickness of Convex Heads-In.	Thickness of Shell Inch
	66	20 x 48	¼	3/16	¼	3/16
	85	20 x 60	¼	3/16	¼	3/16
	100	24 x 48	¼	3/16	¼	3/16
	120	24 x 60	¼	3/16	¼	3/16
WELDED	140	24 x 72	¼	3/16	¼	3/16

	150	30 x 48	¼	3/16	5/16	¼
	180	30 x 60	¼	3/16	5/16	¼
	220	30 x 72	¼	3/16	5/16	¼
	250	30 x 84	¼	3/16	5/16	¼
	295	30 x 96	¼	3/16	5/16	¼
	315	36 x 72	¼	¼	3/8	¼
	365	36 x 84	¼	¼	3/8	¼
	420	36 x 96	¼	¼	3/8	¼
	750	78 x 96			7/16	¼
RIVETED	940	48 x 120			7/16	¼
	1,420	54 x 144			7/16	5/16
	1,890	54 x 192			7/16	5/16
	2,920	60 x 240			7/16	5/16
	4,240	72 x 240			½	5/16
	5,760	84 x 240			½	5/16
	9,000	96 x 288			5/8	3/8

(Reprinted with permission of International Correspondence Schools, Division of Intext, Scranton, Pennsylvania)

Contents of Pipes and Cylindrical Tanks per Foot of Length

Diam. in Inches	Cubic Feet	U.S. Gallons	Diam. in Feet	Cubic Feet	U.S. Gallons
¼	0.0003	0.0025	1.00	0.7854	5.875
3/8	0.0008	0.0057	1.25	1.227	9.180
½	0.0014	0.0102	1.50	1.767	13.22
5/8	0.0021	0.0159	1.75	2.405	17.99
¾	0.0031	0.0230	2.00	3.142	23.50
7/8	0.0042	0.0312	2.25	3.976	29.74
1	0.0055	0.0408	2.50	4.909	36.72
1¼	0.0085	0.0638	2.75	5.940	44.43
1½	0.0123	0.0918	3.00	7.069	52.88
1¾	0.0167	0.1249	3.25	8.296	62.06
2	0.0218	0.1632	3.50	9.621	71.97
2¼	0.0276	0.2066	3.75	11.045	82.62
2½	0.0341	0.2550	4.00	12.566	94.00
2¾	0.0412	0.3085	4.25	14.19	105.20
3	0.0491	0.3672	4.50	16.10	120.30
3½	0.0668	0.4998	4.75	17.22	128.80
4	0.0873	0.6528	5.00	19.60	146.90
4½	0.1104	0.8263	6.00	28.22	211.50
5	0.1364	1.020	7.00	38.42	287.90
5½	0.1650	1.234	8.00	50.30	376.00
6	0.1963	1.469	9.00	63.60	475.40
6½	0.2304	1.724	10.00	78.50	587.50
7	0.2673	1.999	11.00	95.00	710.90
7½	0.3068	2.295	12.00	113.00	846.00
8	0.3491	2.611	13.00	132.80	992.90
8½	0.3941	2.948	14.00	154.00	1151.50
9	0.4418	3.305	15.00	176.70	1321.90
9½	0.4922	3.682	16.00	201.00	1504.10
10	0.5454	4.080	18.00	254.20	1903.60
10½	0.6013	4.498	20.00	314.20	2350.10
11	0.6600	4.937	22.00	380.00	2843.60
11½	0.7213	5.396	24.00	452.00	3384.10
12	0.7854	5.875	25.00	477.00	3672.00

(Reprinted with permission of Publishing Division, Intext, Scranton, Pennsylvania)

Horsepower Transmitted by Manila Ropes

Diameter of Rope (In.)	Velocity, FPM (Feet per Minute)									
	1000	1500	2000	2500	3000	3500	4000	4500	5000	5500
	Horsepower									
¾	2.3	3.3	4.3	5.2	6.0	6.6	7.2	7.3	7.4	7.3
7/8	3.0	4.5	5.9	7.0	8.2	9.0	9.6	9.8	10.0	9.6
1	4.0	5.9	7.7	9.2	10.6	11.8	12.7	12.9	13.0	12.7
1-1/8	5.0	7.5	9.7	11.6	13.5	14.9	16.0	16.3	16.7	16.5
1¼	6.3	9.1	12.0	14.3	16.7	18.5	20.0	20.2	20.7	20.1
1-3/8	7.5	10.8	14.4	17.4	20.0	22.1	23.7	24.5	24.6	24.0
1½	9.0	13.5	17.4	20.7	23.0	26.3	28.7	29.0	29.5	28.6
1-5/8	10.5	15.5	20.1	24.3	27.9	30.8	32.9	34.1	34.3	33.3
1¾	12.3	18.0	23.6	28.2	32.7	36.4	38.5	39.4	40.5	38.7
2	16.0	23.2	30.6	36.8	42.5	46.7	50.0	51.7	52.8	50.6
2¼	20.0	29.6	38.6	46.6	53.6	59.2	63.6	65.8	66.3	64.4
2½	25.0	36.6	47.7	57.5	66.0	71.2	78.0	80.0	81.0	79.0

(Reprinted with permission of Publishing Division, Intext, Scranton, Pennsylvania)

Breaking Strengths of Wire Ropes

Diameter of Rope (In.)	Haulage Rope 6 Strands, 7 Wires Each				Hoisting Rope 6 Strands, 19 Wires Each			
	Iron	Cast Steel	Extra-Strong Cast Steel	Plow Steel	Iron	Cast Steel	Extra-Strong Cast Steel	Plow Steel
	Ultimate Breaking Strength, Tons							
¼	–	2.00	2.15	2.35	0.97	2.10	2.3	2.5
9/32	1.17	2.52	2.72	2.95	–	–	–	–
5/16	1.43	3.10	3.35	3.65	1.43	3.20	3.50	3.90
3/8	2.05	4.30	4.70	5.15	2.05	4.50	5.00	5.50
7/16	2.76	5.80	6.30	6.90	2.76	6.00	6.60	7.30
½	3.57	7.50	8.20	9.00	3.57	7.70	8.50	9.40
9/16	4.49	9.40	10.30	11.30	4.49	9.60	10.60	11.70
5/8	5.52	11.50	12.60	13.80	5.52	11.80	13.10	14.40
¾	7.86	16.50	18.10	19.80	7.86	16.80	18.70	20.60
7/8	10.50	22.40	24.60	16.80	10.60	22.80	25.40	28.00
1	13.70	29.00	31.90	34.80	13.70	29.50	33.00	36.50
1-1/8	17.20	36.40	40.00	43.60	17.20	37.00	41.50	46.00
1¼	21.00	44.50	48.70	53.00	21.00	46.00	51.00	56.50
1-3/8	25.20	53.00	58.20	63.50	25.20	55.00	61.50	68.00
1½	29.70	62.50	68.70	75.00	29.70	65.00	72.50	80.50
1-5/8	–	–	–	–	34.80	76.00	85.00	94.00
1¾	–	–	–	–	40.10	88.00	98.00	108.00
2	–	–	–	–	51.80	114.00	127.00	140.00
2¼	–	–	–	–	64.80	144.00	160.00	176.00
2½	–	–	–	–	79.10	176.00	195.00	214.00

(Reprinted with permission of Publishing Division, Intext, Scranton, Pennsylvania)

Reduction in Strength in Wire Ropes Due to Bending

Diameter of Rope (In.)	Type of Rope	Diameter of Sheave or Drum, Ft.							
		1	1½	2	2½	3	4	5	6
		Reduction in Strength, Lb.							
½	6 x 7	—	3,680	2,800	2,240	1,840	1,400	1,120	920
	6 x 19	2,720	1,860	1,360	1,120	920	680	560	460
	6 x 37	1,840	1,240	920	760	620	460	—	—
	8 x 19	1,720	1,120	860	680	560	920	340	280
¾	6 x 7	—	—	9,440	7,520	6,240	4,720	3,760	3,120
	6 x 19	9,440	6,400	4,720	3,760	3,200	2,360	1,880	1,600
	6 x 37	6,240	4,160	3,120	2,520	2,080	1,560	1,260	1,040
	8 x 19	5,760	3,840	2,880	2,320	1,920	1,440	1,160	960
1	6 x 7	—	—	—	—	14,960	11,200	8,960	7,480
	6 x 19	—	14,880	11,200	8,960	7,440	5,600	4,480	3,720
	6 x 37	15,040	10,080	7,520	6,000	5,040	3,760	3,000	2,520
	8 x 19	—	9,120	6,840	5,480	4,560	3,420	2,740	2,280

(Reprinted with permission of Publishing Division, Intext, Scranton, Pennsylvania)

Constant Properties of Metals

Metal	Symbol	Melting Point Degrees F	Specific Heat at 68° F	Density Grams per Cubic Centimeter	Hardness (Brinell)	Coefficient of Linear Expansion per Degree F at 68° F	Electrical Conductivity Copper = 100
Aluminum	Al	1218	0.214	2.70	16	0.0000136	65
Antimony	Sb	1166	.0505	6.62	42	.0000064	4
Cadmium	Cd	610	.055	8.65	35	.0000166	24
Chromium	Cr	2940	.11	7.14	70	.0000046	66
Cobalt	Co	2700	.1	8.90	75	.0000068	17
Copper	Cu	1981	.092	8.94	30	.0000094	100
Gold	Au	1945	.031	19.30	25	.0000082	71
Iron	Fe	2795	.107	7.87	45	.0000065	17
Lead	Pb	621	.0306	11.34	2.9	.0000163	8
Magnesium	Mg	1204	.246	1.74	—	.0000150	38
Manganese	Mn	2300	.121	7.44	—	.0000130	35
Mercury	Hg	−38	.03325	13.55	—	.0001011	2
Molybdenum	Mo	4750	.065	10.20	160	.0000020	36
Nickel	Ni	2646	.105	8.90	70	.0000071	25
Platinum	Pt	3225	.0324	21.45	42	.0000050	16
Silicon	Si	2590	.181	2.40	—	.0000039	0.004
Silver	Ag	1761	.0558	10.50	27	.0000107	105
Sodium	Na	208	.295	0.97	0.07	.0000390	38
Tin	Sn	449	.054	7.30	14	.0000124	15
Titanium	Ti	3175	.13	4.50	160	.0000039	26
Tungsten	W	6100	.034	19.3	230	0.0000020	31
Vanadium	V	3130	.1153	5.68	250	—	—
Zinc	Zn	787	0.0925	7.14	35	0.0000172	30

(Reprinted with permission of Publishing Division, Intext, Scranton, Pennsylvania)

Safe Working Loads for Hoisting Chains (ASTM Specification A-56-30)

Wrought-Iron Crane Chain

Diameter of Link Bar (In.)	Weight per 100 Ft. (Lb.)	Length (In.)	Width (In.)	Breaking Load (Psi)	Safe Working Load (Psi)
¼	78	1-27/64	1	3,535	1,060
5/16	115	1-11/16	1-3/16	5,520	1,655
3/8	166	1-29/32	1-7/16	7,950	2,385
7/16	220	2-5/32	1-5/8	10,830	3,250
½	275	2-13/32	1-13/16	14,145	4,240
9/16	350	2¾	1-15/16	17,895	5,370
5/8	430	3	2-3/16	22,095	6,630
¾	615	3-7/16	2-9/16	31,800	9,540
7/8	820	4-1/16	3-1/16	43,245	12,960
1	1,045	4-5/8	3-7/16	56,550	16,950
1-1/8	1,310	5-3/16	3-15/16	66,800	20,040
1¼	1,600	5-5/8	4-5/16	82,500	24,750
1-3/8	1,930	6-7/16	4-11/16	99,800	29,910
1½	2,335	6-15/16	5-1/16	118,700	35,600
1-5/8	2,740	7-9/16	5-7/16	139,500	41,800
1¾	3,180	8-5/16	5-13/16	161,600	48,450
1-7/8	3,650	9-1/16	6-5/16	185,500	55,300
2	4,100	9-13/16	6-11/16	211,100	63,300

Steel Proof-Coil Chain

Diameter of Link Bar (In.)	Weight per 100 Ft. (Lb.)	Length (In.)	Width (In.)	Breaking Load (Psi)	Safe Working Load (Psi)
¼	70	1-9/16	1-1/16	3,400	850
5/16	105	1-51/64	1-3/16	5,300	1,325
3/8	158	2-3/64	1-7/16	7,700	1,925
7/16	210	2-5/16	1-11/16	10,500	2,625
½	265	2-9/16	1-7/8	13,700	3,425
9/16	335	2-15/16	2-1/16	17,300	4,325
5/8	410	3-3/16	2-5/16	21,400	5,350
¾	580	3-11/16	2-11/16	30,700	7,675
7/8	780	4-5/16	3-3/16	41,800	10,450
1	1,000	4-13/16	3-9/16	54,700	13,675

(Reprinted with permission of International Correspondence Schools, Division of Intext, Scranton, Pennsylvania)

Wire and Sheet Metal Gages

Gage Number	American Wire Gage	Birmingham Wire Gage	United States Steel Wire Gage	United States Standard Sheet Gage	American Wood Screw Standard	American Machine Screw Standard
4/0	0.460	0.454	0.3938	0.406		
3/0	.410	.425	.3625	.375		
2/0	.365	.380	.3310	.344		
0	.325	.340	.3065	.312	0.060	0.060
1	.289	.300	.2830	.281	.073	.073
2	.258	.284	.2625	.266	.086	.086
3	.229	.259	.2437	.250	.099	.099
4	.204	.238	.2253	.234	.112	.112
5	.182	.220	.2070	.219	.125	.125
6	.162	.203	.1920	.203	.138	.138
7	.144	.180	.1770	.188	.151	
8	.128	.165	.1620	.172	.164	.164
9	.114	.148	.1483	.156	.177	
10	.102	.134	.1350	.141	.190	.190
11	.091	.120	.1205	.125	.203	
12	.081	.109	.1055	.109	.216	.216
13	.072	.095	.0915	.094		
14	.064	.083	.0800	.0821	.242	
15	.057	.072	.0720	.070	.	
16	.051	.065	.0625	.062	.268	
17	.045	.058	.0540	.056		
18	.040	.049	.0475	.050	.294	
19	.036	.042	.0410	.0437		
20	.032	.035	.0348	.0375	.320	
21	.0285	.032	.0317	.0344		
22	.0253	.028	.0286	.0312		
23	.0226	.025	.0258	.0281		
24	.0210	.022	.0230	.0250	.372	
25	.0179	.020	.0204	.0219		
26	.0159	.018	.0181	.0188		
27	.0142	.016	.0173	.0172		
28	.0126	.014	.0162	.0156		
29	.0113	.013	.0150	.0141		
30	.0100	.012	.0140	.0125		
31	.0089	.010	.0132	.0109		
32	.0080	.009	.0128	.0101		
33	.0071	.008	.0118	.0094		
34	.0063	.007	.0104	.0086		
35	.0056	.005	.0095	.0078		
36	.0050	.004	.0090	.0070		
37	.0045		.0085	.0066		

(Reprinted with permission of Publishing Division, Intext, Scranton, Pennsylvania)

Factors of Safety

Material	Steady or Gradually Applied Loads	Suddenly Applied Loads	Shocks
Brick and stone	15	25	30
Cast iron	6	15	20
Steel castings	6	8	20
Structural aluminum	3	6	10
Structural steel	3	6	10
Timber	4	5	7
Wrought iron	4	6	10

(Reprinted with permission of Publishing Division, Intext, Scranton, Pennsylvania)

Strength and Properties of Iron, Steel, and Aluminum

Material	Ultimate Strength (Psi)			Modulus of Elasticity (Psi)		Elastic Limit (Psi)	Modulus of Rupture (Psi)
	Tension	Compression	Shear	Tension or Compression	Shear		
Cast iron	20,000	80,000	18,000	15,000,000	6,000,000	–	30,000
Steel castings	70,000	70,000	60,000	30,000,000	11,000,000	32,000	65,000
Structural aluminum	60,000	–	36,000	10,300,000	3,800,000	37,000	60,000
Structural steel	64,000	64,000	50,000	30,000,000	11,000,000	35,000	64,000
Wrought iron	50,000	50,000	40,000	27,000,000	10,000,000	30,000	45,000

(Reprinted with permission of International Correspondence Schools, Division of Intext, Scranton, Pennsylvania)

Coefficients of Friction for Various Materials

Description of Surfaces in Contact	Type of Lubrication	Coefficient of Friction
Wrought iron on cast iron	Dry	.192
Wrought iron on bronze	Grease	.160
Cast iron on cast iron	Dry	.152
Cast iron on bronze	Grease	.132
Bronze on bronze	Dry	.199
Bronze on cast iron	Dry	.213
Bronze on cast iron	Grease	.098
Steel on bronze,		
Bearing velocity, 50 r.p.m.	Bath	.0011
Bearing velocity, 500 r.p.m.	Bath	.0023
Bearing velocity, 2,000 r.p.m.	Bath	.0044
Bearing velocity, 4,000 r.p.m.	Bath	.0062
Steel on cast iron	Grease	.108
Steel on gun metal	Bath	.0011

(Reprinted with permission of Publishing Division, Intext, Scranton, Pennsylvania)

Table of Cube Roots

Number	Cube Root	Number in Thousands	Cube Root	Number in Millions	Cube Root
1	1.0000	10	21.5	190	575
2	1.2599	50	36.8	200	585
3	1.4422	100	46.4	210	594
5	1.7100	200	58.5	220	604
8	2.0000	300	66.9	230	613
11	2.2240	400	73.7	240	621
16	2.5198	500	79.4	250	630
21	2.7589	600	84.3	260	638
27	3.0000	700	88.8	270	646
34	3.2396	800	92.8	280	654
43	3.5034	900	96.6	290	662
53	3.7563	1,000	100	300	669
64	4.0000	5,000	171	310	677
77	4.2543	10,000	215	320	684
91	4.4979	15,000	247	330	691
107	4.7475	20,000	271	340	698
125	5.0000	25,000	292	350	705
166	5.4959	30,000	311	360	711
216	6.0000	35,000	327	370	718
275	6.5030	40,000	342	380	724
343	7.0000	45,000	356	390	731
422	7.5007	50,000	368	400	737
512	8.0000	55,000	380	410	743
614	8.4994	60,000	391	420	749
625	8.5499	65,000	402	430	755
729	9.0000	70,000	412	440	761
857	9.4986	75,000	422	450	766
900	9.6549	80,000	431	460	772
1,000	10.0000	85,000	440	470	778
1,331	11.0000	90,000	448	480	783
1,728	12.0000	95,000	456	490	788
2,197	13.0000	100,000	464	500	794
2,744	14.0000	110,000	479		
3,375	15.0000	120,000	493		
4,096	16.0000	130,000	507		
4,913	17.0000	140,000	519		
5,832	18.0000	150,000	531		
6,859	19.0000	160,000	543		
8,000	20.0000	170,000	554		
9,000	20.8000	180,000	565		

(Reprinted with permission of International Correspondence Schools, Division of Intext, Scranton, Pennsylvania)

Specific Heats of Solids, Liquids, and Gases

Solids	Specific Heat	Liquids	Specific Heat	Gases	Specific Heat Constant Pressure	Constant Volume
Coal	.314	Water	1.0000	Air	.23751	.16902
Copper	.0951	Alcohol	.6200	Oxygen	.21751	.15507
Gold	.0324	Wood Spirit	.6009	Nitrogen	.24380	.17273
Aluminum	.2143	Proof Spirit	.9730	Hydrogen	3.40900	2.41226
Wrought Iron	.1138	Mercury	.0333	Superheated Steam	.48050	.34600
Steel (soft)	.1165	Benzine	.4500	Carbon Monoxide	.24790	.17580
Steel (hard)	.1175	Lead (melted)	.0402	Carbon Dioxide	.21700	.15350
Zinc	.0956	Sulfur (melted)	.2340			
Brass	.0939	Tin (melted)	.0637			
Glass	.1937	Sulfuric Acid	.3350			
Cast Iron	.1298	Oil of Turpentine	.4260			
Lead	.0314	Glycerine	.5550			
Nickel	.1089	Sea Water	.9400			
Platinum	.0324	Petroleum	.5110			
Silver	.0570					
Tin	.0562					
Ice	.5040					
Sulphur	.2026					
Charcoal	.2410					
Graphite	.201					
Granite	.192					
Asbestos	.195					
Limestone	.216					
Sand	.191					
Rubber	.481					
Wood	.327					

(Reprinted with permission of International Correspondence Schools, Division of Intext, Scranton, Pennsylvania)

Average Weight and Specific Gravity of Materials

Woods

Name of Substance	Average Specific Gravity	Weight per Cubic Foot, Pounds	Weight per Cubic Inch, Pounds
Ash	.56	35	.020
Beech	.60	37	.021
Birch	.65	40	.023
Cedar	.35	22	.013
Cherry	.48	30	.017
Chestnut	.42	26	.015
Cork	.24	15	.009
Cypress	.44	27	.016
Ebony	1.22	71	.041
Elm	.48	30	.017
Fir, Balsam	.36	22	.013
Hemlock	.49	31	.018

Hickory	.65	41	.024
Lignum Vitae	1.25	78	.045
Mahogany	.75	47	.027
Maple	.85	41	.024
Oak, Live	.87	54	.031
Oak, Red	.60	37	.021
Oak, White	.63	39	.023
Pine, White	.41	26	.015
Pine, Yellow	.54	34	.020
Spruce	.40	25	.015
Walnut	.55	34	.019

(Reprinted with permission of International Correspondence Schools, Division of Intext, Scranton, Pennsylvania)

Specific Gravities
Weight per Cubic Foot in Pounds

Water, pure at 32°F	62.417
Water, pure at 39.1°F	62.425
Water, pure at 62°F	62.355
Water, pure at 212°F	59.700
Water, sea	64.080
Ice	57.400
Snow, fresh	5 to 12
Snow, wet	15 to 50

Weights and Specific Gravities of Metals

Name of Metal	Weight per Cubic Inch, Pounds	Weight per Cubic Foot, Pounds	Specific Gravity
Aluminum	.096	166	2.66
Antimony	.242	418	6.70
Bismuth	.355	613	9.82
Brass, cast	.292	504	8.07
Brass, rolled	.302	523	8.38
Copper, cast	.319	550	8.81
Copper, rolled	.321	555	8.89
Gold, 24-carat	.697	1,204	19.29
Iron, cast	.260	450	7.21
Iron, wrought	.277	480	7.69
Lead, commercial	.412	712	11.41
Mercury, 60°F	.490	846	13.55
Silver	.378	655	10.50
Steel	.283	490	7.85
Tin, cast	.265	458	7.34
Zinc	.253	437	7.00

Weights and Specific Gravities of Masonry

Masonry	Weight per Cubic Foot, Pounds	Specific Gravity
Common brickwork, cement mortar	130	2.08
Common brickwork, lime mortar	120	1.92
Granite or limestone rubble, dry	138	2.21
Granite or limestone rubble	150	2.40
Granite or limestone, dressed	165	2.64
Pressed brickwork	140	2.24
Sandstone rubble	145	2.32
Terra-cotta masonry	112	1.79

Specific Gravities

Weights of Dry Woods

Name of Wood	Weight per Cubic Foot, Pounds	Name of Wood	Weight per Cubic Foot, Pounds
Evergreens		*Hardwoods*	
Cedar, canoe	23	Birch, red	35
Cedar, red	30	Chestnut	28
Cypress	29	Cherry	36
Fir, balsam	23	Elm	34
Fir, red of California	29	Gum, sweet	37
Fir, white	22	Hickory	51
Hemlock	26	Locust.	42
Pine, Oregon	32	Maple, hard	43
Pine, long-leaf	38	Oak, red	45
Pine, short-leaf	32	Oak, white	50
Pine, sugar	22	Poplar, yellow	26
Pine, white	24	Sycamore	35
Redwood, California	26	Walnut, black	38
Spruce, black	28		
Spruce, Norway	29	*Rare Woods*	
Spruce, Sitka	26	Boxwood	50
Spruce, white	25	Ebony	76
		Mahogany	32-60
Hardwoods		Rosewood	68
Ash, black	39	Teak	50
Ash, white	39	Walnut, Circassian	35
Basswood	28	Walnut, Persian	36
Beech	42		

Weights and Specific Gravities of Building Materials, Etc.

Name of Material	Weight per Cubic Foot, Pounds	Specific Gravity
Bluestone	160	2.56
Brick, pressed	150	2.40
Brick, common	125	2.00
Cement, Portland (packed)	100−120	1.60−1.92
Cement, Portland (loose)	70−90	1.12−1.44

Cement, Slag (packed)	80–100	1.28–1.60
Cement, Slag (loose)	55–75	.88–1.20
Chalk	156	2.50
Charcoal	15–34	.24–.54
Cinder concrete	110	1.76
Clay, ordinary	120–150	1.92–2.40
Coal, hard, solid	93.5	1.50
Coal, hard, broken	54	.867
Coal, soft, solid	84	1.35
Coal, soft, broken	54	.87
Coke, loose	23–32	.37–.51
Concrete, cement, stone, or gravel	140–155	2.24–2.48
Earth, rammed	90–100	1.44–1.60
Granite	165–170	2.64–2.72
Gravel	117–125	1.87–2.00
Lime, quick (ground loose)	53	.85
Limestone	170	2.72
Marble	164	2.62
Plaster of Paris (cast)	80	1.28
Sand	90–106	1.44–1.70
Sandstone	151	2.42
Shale	162	2.60
Slate	160–180	2.56–2.88
Terra-cotta	110	1.76
Trap rock	170	2.72

(Reprinted with permission of International Correspondence Schools, Division of Intext, Scranton, Pennsylvania)

Liquids

Name of Substance	Specific Gravity	Weight per Cubic Foot, Pounds	Weight per Cubic Inch, Pounds
Acid, Muriatic	1.2	74.7	.0434
Acid, Nitric	1.54	96.	.0556
Acid, Sulfuric	1.841	114.8	.0665
Alcohol, Commercial	.833	52.	.0301
Alcohol, Pure	.792	49.3	.0286
Benzine	.71	44.	.0256
Chloroform	1.513	95.	.055
Ether	.73	45.4	.0263
Gasoline, (motor)	.729	45.5	.0263
Kerosene	.80	50.	.0288
Oil, Lard	.92	57.3	.0332
Oil, Mineral	.915	57.	.033
Turpentine	.87	54.2	.0312
Water, 39.2°F	1.00	62.428	.0361
Water, 212°F	.9584	59.830	.0346
Water (ice)	.90	56	.0324
Water (snow)	.125	8	.0046
Water, Sea	1.025	64	.037

Gases and Vapors

Name of Substance	Specific Gravity	Cubic Feet per Pound
Air	1.000	12.388
Acetylene	.898	13.792
Ammonia	.59	21.036
Carbon Monoxide	.967	12.810
Carbon Dioxide	1.529	8.152
Ethylene	.967	12.806
Helium	.137	90.430
Hydrogen	.069	178.83
Marsh Gas	.554	22.429
Nitrogen	.97	12.77
Oxygen	1.105	11.209
Sulfur Dioxide	2.213	5.598
Water Vapor	.623	19.922

Metals

Name of Substance	Specific Gravity	Weight per Cubic Foot, Pounds	Weight per Cubic Inch, Pounds	Approximate Melting Point, Degrees F
Aluminum	2.70	168.56	.0976	1,217.7
Antimony	6.62	413.2	.255	1,166
Bismuth	9.79	610.6	.354	519.8
Brass	8.5	532	.307	1,650 (about)
Bronze	8.84	533	.32	1,650 (about)
Cadmium	8.65	540	.3125	609.6
Copper	8.89	555	.322	1,981
Gold	19.3	1,204.9	.196	1,945
Iridium	22.43	1,399.7	.81	4,262
Iron, Cast	7.08	441.8	.256	2,050
Iron, Wrought	7.85	489.8	.284	2,800
Lead	11.3	707.1	.410	621
Manganese	7.2	449.5	.260	2,246
Mercury	13.5	845.3	.491	37.98
Nickel	8.85	552.5	.320	2,645
Platinum	21.38	1,334.1	.775	3,191
Silver	10.5	655.5	.380	1,760.6
Steel	7.84	488	.283	2,370 to 2,550
Tin	7.29	455	.264	449.4
Zinc	7.14	445.8	.258	786.9

(Reprinted with permission of Publishing Division, Intext, Scranton, Pennsylvania)

Fluids
Recommended Thickness of Insulation

(Temperature Difference, Covered Surface to Surroundings, Deg F)

Nominal Pipe Size	100	200	300	400	500	600 HT	600 Mag	700 HT	700 Mag	800 HT	800 Mag	900 HT	900 Mag	1000 HT	1000 Mag
			85% Magnesia												
1 in.	S	S	1½	2	2	2	..	2	..	2	..	2½	..	2½	..
2 in.	S	S	1½	2	2	1½	1½	1½	2	2	2	2½	2	2½	2½
3 in.	S	S	1½	2	2	1½	1½	1½	2	2	2	2½	2	2½	2½

4 in.	S	S	1½	2	2	1½	1½	1½	2	2	2	2½	2	2½	2½
5 in.	S	S	2	DS	DS	1½	2	1½	2	2	2	2½	2	2½	2½
6 in.	S	S	2	DS	DS	1½	2	1½	2	2	2	2½	2	2½	2½
8 in.	S	S	2	DS	DS	1½	2	2	2	2	2½	2½	2½	3	2
10 in.	S	S	2	DS	DS	1½	2	2	2	2	2½	2½	2½	3	2
12 in.	S	S	2	DS	DS	1½	2	2	2	2	2½	2½	2½	3	2
14 in.	S	S	2	DS	DS	1½	2	2	2	2	2½	2½	2½	3	2
16 in.	S	S	2	DS	DS	1½	2	2	2	2	2½	2½	2½	3	2
18 in.	S	S	2	DS	DS	1½	2	2	2	2	2½	2½	2½	3	2
20 in.	S	S	2	DS	DS	1½	2	2	2	2	2½	2½	2½	3	2
Flat	1½	1½	2	2½	3	1½	2	2	2	2	2½	2½	2½	3	2

Note: HT = high-temperature covering; S = standard thickness; DS = double standard

(Reprinted with permission from *Betz Handbook of Industrial Water Conditioning*. Copyright 1962, by Betz Laboratories, Inc.)

COMMON IMPURITIES FOUND IN WATER

Constituent	Chemical Formula	Difficulties Caused	Means of Treatment
Turbidity	None—expressed in analysis as units.	Imparts unsightly appearance to water. Deposits in water lines, process equipment, boilers, etc. Interferes with most process uses.	Coagulation, settling and filtration.
Color	None—expressed in analysis as units.	May cause foaming in boilers. Hinders precipitation methods such as iron removal, hot phosphate softening. Can stain product in process use.	Coagulation and filtration. Chlorination. Adsorption by activated carbon.
Hardness	Calcium and magnesium salts expressed as $CaCO_3$.	Chief source of scale in heat exchange equipment, boilers, pipelines, etc. Forms curds with soap, interferes with dyeing, etc.	Softening. Distillation. Internal boiler water treatment. Surface active agents.
Alkalinity	Bicarbonate (HCO_3), carbonate (CO_3), and hydrate (OH), expressed as $CaCO_3$.	Foaming and carry-over of solids with steam. Embrittlement of boiler steel. Bicarbonate and carbonate produce CO_2 in steam, a source of corrosion.	Lime and lime-soda softening. Acid treatment. Hydrogen zeolite softening. Demineralization. Dealkalization by anion exchange. Distillation.
Free Mineral Acid	H_2SO_4, HCl, etc. expressed as $CaCO_3$.	Corrosion.	Neutralization with alkalies.
Carbon Dioxide	CO_2	Corrosion in water lines and particularly steam and condensate lines.	Aeration. Deaeration. Neutralization with alkalies. Filming and neutralizing amines.
pH	Hydrogen ion concentration defined as: $pH = \log \dfrac{1}{(H+)}.$	pH varies according to acidic or alkaline solids in water. Most natural waters have a pH of 6.0-8.0.	pH can be increased by alkalies and decreased by acids.

Constituent	Chemical Formula	Difficulties Caused	Means of Treatment
Sulfate	$(SO_4)^{--}$	Adds to solids content of water, but, in itself, is not usually significant. Combines with calcium to form calcium sulfate scale.	Demineralization. Distillation.
Chloride	Cl^-	Adds to solids content and increases corrosive character of water.	Demineralization. Distillation.
Nitrate	$(NO_3)^-$	Adds to solids content, but is not usually significant industrially. High concentrations cause methemoglobinemia in infants. Useful for control of boiler metal embrittlement.	Demineralization. Distillation.
Fluoride	F^-	Cause of mottled enamel in teeth. Also used for control of dental decay. Not usually significant industrially.	Adsorption with magnesium hydroxide, calcium phosphate, or bone black. Alum coagulation.
Silica	SiO_2	Scale in boilers and cooling water systems. Insoluble turbine blade deposits due to silica vaporization.	Hot process removal with magnesium salts. Adsorption by highly basic anion exchange resins, in conjunction with demineralization. Distillation.
Iron	Fe_{++} (ferrous) Fe_{+++} (ferric)	Discolors water on precipitation. Source of deposits in water lines, boilers, etc. Interferes with dyeing, tanning, paper mfr., etc.	Aeration. Coagulation and filtration. Lime softening. Cation exchange. Contact filtration. Surface active agents for iron retention.
Manganese	Mn_{++}	Same as iron.	Same as iron.
Oil	Expressed as oil or chloroform extractable matter.	Scale, sludge, and foaming in boilers. Impedes heat exchange. Undesirable in most processes.	Baffle separators. Strainers. Coagulation and filtration. Diatomaceous earth filtration.
Oxygen	O_2	Corrosion of water lines, heat exchange equipment, boilers, return lines, etc.	Deaeration. Sodium sulfite. Corrosion inhibitors.
Hydrogen sulfide	H_2S	Cause of "rotten egg" odor. Corrosion.	Aeration. Chlorination. Highly basic anion exchange.
Ammonia	NH_3	Corrosion of copper and zinc alloys by formation of complex soluble ion.	Cation exchange with hydrogen zeolite. Chlorination. Deaeration.
Conductivity	Expressed as micromhos, specific conductance.	Conductivity is the result of ionizable solids in solution. High conductivity can increase the corrosive characteristics of a water.	Any process which decreases dissolved solids content will decrease conductivity. Examples are demineralization, lime softening.
Dissolved Solids	None.	"Dissolved solids" is a measure of total amount of dissolved matter, determined by evaporation. High concentrations of dissolved solids are objectionable because of process	Various softening processes, such as lime softening and cation exchange by hydrogen zeolite, will reduce dissolved solids. Demineralization. Distillation.

Constituent	Chemical Formula	Difficulties Caused	Means of Treatment
		interference and as a cause of foaming in boilers.	
Suspended Solids	None.	"Suspended Solids" is the measure of undissolved matter, determined gravimetrically. Suspended solids plug lines, cause deposits in heat exchange equipment, boilers, etc.	Subsidence. Filtration, usually preceded by coagulation and settling.
Total Solids	None.	"Total Solids" is the sum of dissolved and suspended solids, determined gravimetrically.	See "Dissolved Solids" and "Suspended Solids."

(Reprinted with permission of Publishing Division, Intext, Scranton, Pennsylvania)

Dimensions of Commercial Steel Pipe

Pipe Size, in Inches Nominal Diameter	Outside Diameter	10	20	30	40	60	80	100	120	140	160
							Wall Thickness, in Inches				
½	0.840	...	...	...	0.109	...	0.147	...	...	...	0.187
¾	1.05	...	...	...	0.113	...	0.154	...	...	...	0.218
1	1.315	...	...	...	0.133	...	0.179	...	...	...	0.250
1¼	1.660	...	...	...	0.140	...	0.191	...	...	...	0.250
1½	1.900	...	...	...	0.145	...	0.200	...	...	...	0.281
2	2.375	...	...	...	0.154	...	0.218	...	...	...	0.343
2½	2.875	...	...	...	0.203	...	0.276	...	...	...	0.375
3	3.500	...	...	...	0.216	...	0.300	...	...	...	0.437
3½	4.000	...	...	...	0.226	...	0.318	...	...	...	...
4	4.500	...	...	...	0.237	...	0.337	...	0.437	...	0.531
5	5.563	...	...	...	0.258	...	0.375	...	0.500	...	0.625
6	6.625	...	...	...	0.280	...	0.432	...	0.562	...	0.718
8	8.625	...	0.250	0.277	0.322	0.406	0.500	0.593	0.718	0.812	0.906
10	10.75	...	0.250	0.307	0.365	0.500	0.593	0.718	0.843	1.000	1.125
12	12.75	...	0.250	0.330	0.406	0.562	0.687	0.843	1.000	1.125	1.312
14	14.00	0.250	0.312	0.375	0.437	0.593	0.750	0.937	1.062	1.250	1.406
16	16.00	0.250	0.312	0.375	0.500	0.656	0.843	1.031	1.218	1.437	1.562
18	18.00	0.250	0.312	0.437	0.562	0.718	0.937	1.156	1.343	1.562	1.750
20	20.00	0.250	0.375	0.500	0.593	0.812	1.031	1.250	1.500	1.750	1.937
24	24.00	0.250	0.375	0.562	0.687	0.937	1.218	1.500	1.750	2.062	2.312

(Reprinted with permission of International Correspondence Schools,
Division of Intext, Scranton, Pennsylvania)

Strength of Materials

Tension

Material	Modulus of Elasticity E_t Lbs. per Sq. In.	Yield-Point Stress L_t Lbs. per Sq. In.	Ultimate Tensile Strength S_t Lbs. per Sq. In.	Elongation s_t per °Cent.
Steel				
Soft steel	29,000,000	30,000	55,000	30
Steel tubing	30,000,000	59,500	65,000	30
Boiler shell plates	30,000,000	32,500	65,000	30
Ship plates	30,000,000	29,000	58,000	26
Structural steel	30,000,000	27,500	55,000	22
Rivet bars	30,000,000	25,000	50,000	30
Untreated forgings	30,000,000	37,500	75,000	18
Annealed forgings	30,000,000	40,000	80,000	20
Cold-rolled bars for bolts, studs	30,000,000	35,000	70,000	10
Cold-rolled axles	30,000,000	50,000	85,000	20
Tires for locomotives	30,000,000		105,000	12
Tool steel	30,000,000	132,000	184,000	20
Steel castings (soft)	30,000,000	42,500	65,000	22
Steel castings (hard)	30,000,000	45,000	80,000	20
Rail steel	30,000,000	57,000	115,000	13
Wrought Iron				
Plates	25,000,000	26,000	49,000	16
Welded pipes	25,000,000	24,000	40,000	12
Staybolt iron	25,000,000	30,000	50,000	30
Malleable Iron				
Castings	25,000,000	28,500	50,000	15
Cast Iron				
Gray cast iron	13,000,000		30,000	
White cast iron	17,000,000		45,000	
Copper				
Sheets and plates	16,500,000	21,000	35,000	25
Rods and bars	15,000,000	24,000	40,000	15
Hard-drawn wire	17,000,000	36,000	60,000	
Castings	15,000,000		25,000	20
Brass				
Castings	13,000,000	6,500	22,000	18
Sheets (thin)	14,000,000	40,000	65,000	
Seamless tubing	13,500,000		40,000	25
Bronze				
Gun metal	15,000,000	12,000	30,000	10
Aluminum bronze	16,800,000	33,000	85,000	29
Bearing metal	13,000,000		20,000	10
Car journal bearings	13,000,000		41,800	34
Spring wire	15,000,000		130,000	
Silicon bronze	13,500,000		65,000	
Tobin bronze	13,500,000	42,000	75,000	
Wood				
Oak	1,500,000		8,000	
Pitch pine	1,900,000		7,500	
Leather	25,000	2,100	4,200	

Strength of Materials

Shear

Material	Modulus of Elasticity E_s Pounds per Square Inch	Ultimate Shearing Strength S_3 Pounds per Square Inch
Steel rivet bars and firebox plates	12,000,000	49,000
Steel plates (boiler, shell, ship, and structural)	12,000,000	52,000
Steel forgings, or rolled bars (mild)	12,000,000	55,000
Steel forgings, or rolled bars (hard)	12,000,000	65,000
Wrought-iron rolled bars	10,000,000	40,000
Wrought-iron plates	10,000,000	40,000
Cast iron	6,000,000	20,000 to 40,000
Bronze castings	5,800,000	34,000
Copper castings		17,000
Copper plates		22,000
Oak, across the grain		4,400
Pitch pine, across the grain		4,500
Oak, with the grain		440
Pitch pine, with the grain		450

Strength of Materials

Compression

Material	Modulus of Elasticity E_c Pounds per Square Inch	Yield-Point Stress L_c Pounds per Square Inch	Ultimate Compressive Strength S_c Pounds per Square Inch*
Steel			
Soft steel	29,000,000	30,000	30,000
Steel tubing	30,000,000	59,500	59,500
Boiler shell plates	30,000,000	32,500	32,500
Ship plates	30,000,000	29,000	29,000
Structural plates	30,000,000	27,500	27,500
Rivet bars	30,000,000	25,000	25,000
Forgings	30,000,000	38,000	38,000
Rolled bars	30,000,000	35,000	35,000
Tool steel	30,000,000	132,000	132,000
Steel castings (soft)	30,000,000	42,500	42,500
Steel castings (hard)	30,000,000	45,000	45,000
Wrought Iron			
Plates	25,000,000	26,000	26,000
Welded pipes	25,000,000	24,000	24,000
Staybolt iron	25,000,000	30,000	30,000
Cast Iron			
Ordinary castings	15,000,000		90,000
Hard castings	17,000,000		150,000
Brass			
Castings	13,000,000	28,000	
Sheet brass	14,000,000	28,000	
Brass tubing	13,500,000	28,000	

Bronze		
Hard cast bronze	13,000,000	70,000
Wood		
Oak (with the grain)	1,500,000	8,000
Pitch pine (with the grain)	1,900,000	6,000
Stone		
Slate slabs		15,000
Sandstone		8,000
Brick		2,500

*For steel and wrought-iron columns subjected to combined compressive and bending stresses, the yield point is used as the value for the ultimate compressive strength.

Strength of Materials

Bending Strengths of Materials

Material	Ultimate Bending Strength, Pounds per Square Inch
Cast iron	30,000
Wrought iron	45,000
Steel	65,000
Stone	1,200
Concrete	700
Ash	8,000
Hemlock	3,500
Oak, white	6,000
Pine, white	4,000
Pine, yellow	7,000
Spruce	3,000
Chestnut	4,500

(Reprinted with permission of Publishing Division, Intext, Scranton, Pennsylvania)

STORAGE BATTERIES

Freezing Points of Electrolyte

Sp. Gr.	Temp. Deg. F	Sp. Gr.	Temp. Deg. F	Sp. Gr.	Temp. Deg. F
1.060	25	1.160	2	1.260	−75
1.080	22	1.180	−6	1.280	−90
1.100	18	1.200	−16	1.300	−95
1.120	14	1.220	−31	1.320	−80
1.140	8	1.240	−51		

(Reprinted with permission of Publishing Division, Intext, Scranton, Pennsylvania)

The Meaning of pH Numbers

Hydrogen- Ion Concentration, in Grams per Liter (1)	Column 1 Expressed as a Fraction (2)	Reciprocal of Column 2 (3)	Logarithm of Column 3 (4)	pH (5)	Acidity or Alkalinity of Solution Relative to a Neutral Solution (6)	
1.0	1	1	0	0	10,000,000	
0.1	$1/10^1$	10^1	1	1	1,000,000	Acidity
0.01	$1/10^2$	10^2	2	2	100,000	
0.001	$1/10^3$	10^3	3	3	10,000	
0.0001	$1/10^4$	10^4	4	4	1,000	
0.00001	$1/10^5$	10^5	5	5	100	
0.000001	$1/10^6$	10^6	6	6	10	
0.0000001	$1/10^7$	10^7	7	7	1 ← Neutral (pure water)	
0.00000001	$1/10^8$	10^8	8	8	10	
0.000000001	$1/10^9$	10^9	9	9	100	
0.0000000001	$1/10^{10}$	10^{10}	10	10	1,000	Alkalinity
0.00000000001	$1/10^{11}$	10^{11}	11	11	10,000	
0.000000000001	$1/10^{12}$	10^{12}	12	12	100,000	
0.0000000000001	$1/10^{13}$	10^{13}	13	13	1,000,000	
0.00000000000001	$1/10^{14}$	10^{14}	14	14	10,000,000	

68 / REFRIGERATION AND AIR CONDITIONING

(Reprinted with permission of Publishing Division, Intext, Scranton, Pennsylvania)

Refrigeration and Air Conditioning
Pipe Sizes in Ammonia Systems

Heat Transfer, B. t. u. per Hour	Internal Diameter of Discharge Pipe, Inches	Internal Diameter of Suction Pipe, Inches	Weight of Ammonia to Fill 100 Feet of Discharge Pipe, Pounds
36,000	¼	½	2
72,000	¼	¾	2
132,000	3/8	1	3
204,000	3/8	1¼	3
300,000	½	1½	5
528,000	¾	2	12
756,000	1	2½	21

1,200,000	1	3	21
1,560,000	1¼	3½	32
2,040,000	1¼	4	32
3,024,000	1½	5	46
4,800,000	2	6	82

(Reprinted with permission of Publishing Division, Intext, Scranton, Pennsylvania)

Refrigeration and Air Conditioning
Pipe Sizes in "Freon" System

Heat Transfer B. t. u. per Hour	Internal Diameter of Discharge Pipe, Inches	Internal Diameter of Suction Pipe, Inches	Weight of "Freon" to Fill 100 Feet of Discharge Pipe, Pounds
9,600	¼	½	3
19,200	¼	¾	3
36,000	3/8	1	6
55,200	½	1¼	11
79,200	¾	1½	24
138,000	1	2	43
198,000	1	2½	43
312,000	1¼	3	69
420,000	1½	3½	97
540,000	2	4	172
792,000	2	5	172
1,248,000	2½	6	256

(Reprinted with permission of Publishing Division, Intext, Scranton, Pennsylvania)

Refrigeration and Air Conditioning
Sizes of Pressure Relief Valves

Refrigerating Capacity, Tons	Diameter of Valve, Inches
15 or less	3/8
16 to 30	½
31 to 60	¾
61 to 100	1
101 to 175	1¼
176 to 250	1½
251 to 450	2
451 to 900	2 (two valves)

(Reprinted with permission of Publishing Division, Intext, Scranton, Pennsylvania)

Comparison of Systems of Units

Quantity	MKS	CGS	Modified
Length	Meter	Centimeter	Centimeter
Time	Second	Second	Second
Current	Ampere	Abampere	Ampere
Potential	Volt	Abvolt	Volt
Magnetic flux	Weber	Maxwell or Lines	Lines
Magnetic flux density	Weber per square meter	Gauss = Maxwell per square centimeter	Gauss = Lines per square centimeter
Magnetomotive force	Pragilbert	Gilbert	Ampere-turn
Magnetic intensity	Praoersted	Oersted = Gilbert per centimeter	Ampere-turn per centimeter
Permeability, for non-magnetic materials	1.257×10^{-6}	1	$4\pi/10 = 1.257$

Summary of Basic Information

	Electric	Magnetic	Dielectric
Motion	I = amperes I = coulombs/second	ϕ = magnetic flux ϕ = lines	Q = dielectric flux Q = coulombs
Force	E = volts	F = ampere-turns	E = volts
Impedance	R = resistance R = ohms	R = reluctance	S = elastance
Ohm's Law	E = IR	F = ϕR	E = QS
Admittance	g = 1/R = conductance	P = 1/R = permeance	C = 1/S = capacitance
Impedance Law	L = length $R = p\dfrac{L}{A}$ p = constant	A = area $R = K\dfrac{L}{A}$ K = constant	$S = K\dfrac{L}{A}$ K = constant
Ohm's Law	$E = Ip\dfrac{L}{A}$	$F = \phi K\dfrac{L}{A}$	$E = QK\dfrac{L}{A}$
Gradients	G = E/L	H = F/L	G = E/L
Densities	D = I/A	β = ϕ/A	D = Q/A
Unit Ohm's Law	G = pD	H = Kβ	G = KD
Energy	EIt	$\frac{1}{2}LI_m^2$	$\frac{1}{2}CE_m^2$

Induction-Motor Synchronous Speeds

Poles	Synchronous Speed	
	25 Cycles	60 Cycles
2	1,500	3,600
4	750	1,800
6	500	1,200
8	375	900
10	300	720
12	250	600

(Reprinted with permission of Publishing Division, Intext, Scranton, Pennsylvania)

Data for Annealed Copper Wire

Size (B&S)	Diam. (Mils.)	Area (Cir. Mils.)	Lbs. per 1,000 Ft.	Ohms per 1,000 Ft. 20C	75C
0000	460.	211,600	641	0.0490	0.0596
000	410.	167,800	508	0.0618	0.0752
00	365.	133,100	403	0.0779	0.0948
0	325.	105,500	320	0.0983	0.1195
1	289.	83,700	253	0.1239	0.1507
2	258.	66,400	201	0.1563	0.1900
3	229.	52,600	159	0.1970	0.2396
4	204.	41,700	126	0.2485	0.3022
5	182.	33,100	100	0.3130	0.3810
6	162.	26,250	79.5	0.395	0.4805
7	144.3	20,820	63.0	0.498	0.606
8	128.5	16,510	50.0	0.628	0.764
9	114.4	13,090	39.6	0.792	0.963
10	102.0	10,380	31.4	1.000	1.215
11	90.7	8,230	24.9	1.260	1.532
12	80.8	6,530	19.8	1.588	1.931
13	72.0	5,180	15.7	2.003	2.436
14	64.1	4,110	12.4	2.525	3.070
15	57.1	3,260	9.86	3.184	3.870
16	50.8	2,580	7.82	4.02	4.88
17	45.3	2,050	6.20	5.06	6.16
18	40.3	1,624	4.92	6.38	7.76
19	35.9	1,288	3.90	8.05	9.79
20	32.0	1,022	3.09	10.15	12.35
21	28.5	810	2.45	12.80	15.57
22	25.3	642	1.95	16.14	19.63
23	22.6	510	1.54	20.36	24.76
24	20.1	404	1.22	25.67	31.20
25	17.9	320	0.97	32.40	39.40
26	15.9	254	0.769	40.8	49.6
27	14.2	202	0.610	51.5	62.6
28	12.6	160	0.484	64.9	78.9
29	11.3	127	0.384	81.8	99.5
30	10.0	100	0.304	103.2	125.5
31	8.9	79.7	0.241	130	158.2
32	8.0	63.2	0.191	164	199.5
33	7.1	50.1	0.152	207	251.6
34	6.3	39.8	0.120	261	317.3
35	5.6	31.5	0.0954	329	400.0
36	5.0	25.0	0.0757	415	505
37	4.5	19.8	0.0600	523	636
38	4.0	15.7	0.0476	660	802
39	3.5	12.5	0.0377	832	1,012
40	3.1	9.9	0.0299	1,015	1,276

Resistivity and Temperature Coefficients of Resistance

Material	a_o	Ohms per mil-foot at $0°$ C
Aluminum	0.00407	15.72
Carbon (incandescent lamp)	−0.00030	24,000
Copper (annealed)	0.00427	9.55
Eureka	0.00005	282.60
German silver (Cu 60, Zn 25, Ni 15)	0.00036	199.00
Gold	0.00364	13.67
IaIa (Cu 60, Ni 40) soft	0.000005	283.00
Iron (electrolytic)	0.00680	52.80
Manganin (Cu 84, Mn 12, Ni 4)	0.000006	290.00
Nichrome (Ni-Fe-Cr alloy)	0.00186	646.00
Nickel (electrolytic)	0.00618	41.65
Platinum	0.00370	65.80
Silver	0.00407	8.84
Steel (soft)	0.00680	83.60
Steel (4% silicon)	0.00318	282.00
Tin	0.00458	782.00
Tungsten (annealed)	0.00510	26.25